Foundations of Higher Mathematics

Exploration and Proof

Foundations of Higher Mathematics

Exploration and Proof

Daniel Fendel
San Francisco State University

with
Diane Resek
San Francisco State University

Addison-Wesley Publishing Company
Reading, Massachusetts   Menlo Park, California   New York
Don Mills, Ontario   Wokingham, England   Amsterdam   Bonn
Sydney   Singapore   Tokyo   Madrid   San Juan

Sponsoring Editor: Charles B. Glaser
Production Coordinator: Christopher Keane
Production Services: Ann Kilbride/*Cambridge Studio*
Text Designer: Joyce Weston
Copy Editor: Joseph Fineman
Illustrator: ST Associates
Art Consultant: Loretta Bailey
Manufacturing Supervisor: Roy Logan
Cover Design: Marshall Henrichs

Library of Congress Cataloging–in–Publication Data

Fendel, Daniel M.
Foundations of higher mathematics : exploration and proof / by Daniel Fendel with Diane Resek.
p. cm.
Includes index.
ISBN 0-201-12587-0
1. Mathematics. I. Resek, Diane. II. Title.
QA37.2.F46 1990
510—dc20 89-36078
CIP

ISBN 0-201-12587-0

3 4 5 6 7 8 9 10-CRW-99989796

In memory of my father,
and in appreciation of my mother.

D. F.

In memory of my parents.

D. R.

Contents

Preface

This text is designed for sophomore and junior level students who need a transitional course to bridge the gap between the standard calculus sequence and a variety of upper-division mathematics courses, where the ability to think like a mathematician is critical. This book proceeds at an intuitive pace and provides the reader with many opportunities to stop and think about the material presented. We have provided a blend of actual mathematics and discussion about mathematics, in an effort to teach the reader specific mathematical concepts, as well as how mathematicians learn and do mathematics. To achieve this blend, the book contains formal, symbolic definitions, as well as pictures, metaphors, and intuitive descriptions. It provides both technical details and broad outlines of proofs. It shows how to investigate a problem and develop ideas, as well as how to test those ideas and, if appropriate, prove that they are correct.

In short, our goal is to teach our readers how to think the way mathematicians think.

The Theme of Exploration

Throughout this text, we emphasize mathematical exploration. This emphasis is our solution to a problem mathematics educators have long observed: namely, the inability of many students to think about a problem. All too often, students want to be told "how to do it," which simply will not work when it comes to working with complex mathematical concepts. The ability to explore a mathematical idea is critical to students' success in upper division mathematics, and involves a desire and ability to experiment, investigate, work with examples, ask questions, make guesses, and test conjectures.

We hope that our exploration theme will help students develop good habits when working with theorems and formal proofs. We ask students to look at a theorem intuitively before attempting to write a proof, to first convince themselves that a statement is really true, and to develop a healthy skepticism before assuming that it is true. In addition, we ask students to participate in the building of formal definitions, rather than simply presenting them with definitions in their evolved mathematical form, which sometimes seems to students like pulling a rabbit out of a hat. Our experience has been that this exploratory approach to theorems and definitions enables students to own the mathematics they are learning, providing both stronger motivation and deeper understanding.

The Role of Logic—"The Working Tools of the Mathematician"

Our experience has shown that students also have problems with logic. They tend to be confused about what can be assumed in the process of writing a proof, how a proof should be organized, and how they can tell when a proof is finished. The fundamental idea of an "if . . . , then . . .", or "conditional", statement is often perplexing to students. They find the formal truth table definition of implication artificial, and it does not make much sense to them when they are told that the statement "if $2 + 2 = 7$, then $5^2 = 9$" is considered true.

We have realized that what students really need are the working tools of logic, the aspects of logic that a mathematician uses regularly, rather than a study of logic for its own sake. They need an intuitive understanding of predicate logic, rather than the formalism of propositional logic. Students benefit from a thorough discussion of the meaning of the conditional statement, and can understand and appreciate the concept of a counterexample. Since predicate logic is what the working mathematician uses, we have found it most effective to put aside propositional logic and formal truth tables.

Section 1.3 of Chapter 1, which contains our introduction to the use of the conditional statement, emphasizes the principle that such a statement is considered true unless there is a counterexample. (For those instructors who wish to introduce their students to propositional logic, we have included a brief introduction to this topic in Appendix A.)

In addition, some students have difficulty understanding logic that involves the use of complex logical ideas within definitions, especially the use of quantifiers. Students tend not to appreciate the fact that many important definitions involve additional, dummy variables, as well as those variables to which the defined terminology applies directly. (For example, defining the statement "m is a divisor of n" involves the existence of a third object, t, such that $m \cdot t = n$.) We have introduced the concept of a *hidden variable* to describe this phenomenon. Section 3.1 of Chapter 3 contains a thorough introduction to the use of quantifiers, which is then reinforced in subsequent discussions.

Students are also confused by the fact that the "if . . . , then . . ." form can be used to express definitions or properties of objects as well to state theorems. (For example, the statement "A is a subset of B" can be defined by the conditional sentence "if $x \in A$, then $x \in B$".) We have found that identifying such universally quantified definitions, and comparing them with existentially quantified definitions (such as that of "divisor") is helpful to the student. Theorems involving universally quantified definitions in their hypotheses generally require proofs using the method called *specialization,* while theorems involving existentially quantified definitions in their hypotheses generally require proofs using a method we call "name it, then use it." Section 4.1 discusses these two types of proofs, and the role of quantifiers in setting up proofs.

The Purpose of the EXPLORATIONS, and How to Use Them

To accomplish our goal of generating active involvement by the reader, we have developed several pedagogical features. These features are discussed briefly in the introduction to Chapter 1. One important feature is the Exploration.

The primary goal of the Explorations is not to teach the student some specific mathematics, but to teach the student a way of thinking. The learning that takes place is intended to be primarily process-oriented rather than content-based. We want the reader to learn how to cope with open-ended, unstructured questions; how to formulate conjectures; how to evaluate the reasonableness of a statement; how to make plausibility arguments; how to make constructive use of examples, and so on.

But time is often severely limited. Therefore, the text is also set up so that any Exploration is optional. Any single Exploration can be omitted, without creating a significant problem later on. (In the cases where the material from an Exploration is referred to later on, a brief introduction will suffice.)

But we encourage students to do some of the Explorations, especially at the beginning of the course. The process of working on Explorations gives students a head start in both developing their intuition and learning how to approach the proof of a theorem.

Explorations usually benefit from a general introduction of the topic via class discussion. After this introduction, we suggest whole class discussion, small group investigation, or assignment as homework. For more specific suggestions relative to the use of EXPLORATIONS, please see the Instructor's Manual that accompanies this text.

Creating a Syllabus

There is far more material in this text than can be included in a one-semester course. We have designed the later chapters to be generally independent of each other, so that instructors have considerable flexibility in developing a syllabus that suits their needs. Our discussion here is designed to aid instructors in developing a syllabus that best meets the needs of their school and students.

Essentials of Part I

The first four chapters constitute PART I—"Thinking Mathematically." PART I mixes ideas about mathematics with actual mathematics. The following Sections are essential, either because of specific content or because of the philosophy of the book.

CHAPTER 1
- Introduction—all
- Section 1.1—at least one of the Explorations
- Section 1.2—all
- Section 1.3—all, except "Proof for Certainty and Proof for Understanding"

CHAPTER 2—all

CHAPTER 3
- Section 3.1—all
- Section 3.2—Topic 2

CHAPTER 4

Section 4.1—at least one example of "Specialization" and at least one example of "name it—then use it." (If you skip Topic 1 of Section 3.2, you will need to avoid examples that use this material, or simply define "upper bound" and "bounded above.")

Section 4.2—"Proof by Contradiction," "Proof by Contrapositive," and "If and Only If."

Part II

The next three chapters, Chapters 5, 6, and 7, constitute our PART II—"Core Mathematics." The ideas in these chapters are *universal mathematics*—that is, concepts and language commonly used in almost every branch of mathematics. Although almost all of this material is essential to the mathematician, the following items represent our priorities.

Section 5.1—all
Section 5.2—all, except "Relationships among the principles"
Section 6.1—all

The remainder of these chapters—Sections 5.3, 6.2, 7.1, and 7.2—are essentially independent of one another, except that the discussion of $\mathbf{Z}_n$ in 7.1 requires the introductory material from 6.2. We would give first priority to Section 6.2, although classes that emphasize the Linear Diophantine Equation problem in Sections 2.3 and 2.4 may wish to do at least the first part of Section 5.3. An alternative would be to select the material you want to cover from the final three chapters, then teach the necessary prerequisites from PART II.

PART III

The final three chapters constitute our PART III—"Topics in Mathematics." These chapters, which are completely independent from one another, give students a taste of upper division mathematics. These chapters emphasize the careful development of sophisticated definitions such as set equivalence, limit, and isomorphism. We want students to realize that the concepts they are studying were not created arbitrarily out of thin air. They were developed with a purpose, and each detail is there for a specific reason.

Note: Section 9.3—"Limits of Functions and Continuity"—covers material that students perhaps should learn in Calculus I, and the material in Section 9.1 and 9.2 is also part of the usual calculus sequence. Students with a strong calculus background might skip this material, and go straight to Section 9.4. But the depth of the discussion in Sections 9.1–9.3 is considerably beyond that usually found in a calculus class, and uses the quantifier language developed in earlier chapters. Students will need this level of understanding to succeed in the usual course in real analysis. Section 9.4 introduces the reader to real analysis proper, with a discussion of two topics that are not part of the calculus sequence.

The following list shows what you need from PART II to discuss properly each of the chapters from PART III.

Chapter	*Prerequisite material from PART II*
CHAPTER 8 Cardinality	Section 6.1—through "More about Functions" Section 6.2—except "Congruence mod *n*" Section 7.2—through "Partial Ordering" Also, brief use is made of uniqueness of prime factorization from Section 5.3
CHAPTER 9 Limits and Continuity	none of PART II is required, but Topic 1 from Section 3.2 is needed
CHAPTER 10 Groups	Section 6.1—"More about Functions" Section 6.2—"Congruence mod *n*" Section 7.1—all

Acknowledgements

We wish to thank the following people who have contributed to this book.

Our publishing team—editors Tom Taylor, whose enthusiasm got Addison-Wesley involved, and Chuck B. Glaser, who led the project through to fruition; production coordinator, Chris Keane; and the reviewers who read either some or all of the manuscript at various stages:

Professor David Barnette
University of California—Davis

Professor Nicholas Branca
San Diego State University

Professor Carl C. Cowen Jr.
Purdue University

Professor Robert G. Dean
Stephen F. Austin State University

Professor Christopher Ennis
Carleton College

Professor Michael Evans
North Carolina State University

Professor William Finzer
Software Developer

Professor Joel K. Haack
Oklahoma State University

Professor Umberto Neri
University of Maryland

Professor K. C. Ng
California State University at Sacramento

Professor Holly B. Puterbaugh
University of Vermont

Professor Roger Waggoner
University of Southwestern Louisiana

Our colleagues, especially those who used portions of this book during development: Elaine Kasimatis, at Sacramento State University; and Sergei Ovchinnikov, at San Francisco State University.

Our students, who helped find mistakes, pointed out places where we were unclear, and gave us encouragement by learning from our manuscript.

Our mentors — people from whom we learned about mathematics and about teaching, especially Robert B. Davis, Ira Ewen, Leon Henkin, and William F. Johntz.

Our friends and families, who gave us personal encouragement, occasional advice, and lots of love over the course of the writing.

PART I

Thinking Mathematically

1

About Exploration and Proof

Introduction

Many of us have been taught that mathematics consists of finding numerical answers to an endless series of individual exercises. From arithmetic through algebra, and even in calculus, the focus of much of mathematics teaching is on "finding the right answer". Even in traditional geometry classes, where students supposedly learn about proofs, there is often a "right answer" approach. Students are expected to "learn *the* proof".

In this book, we have a different perspective. One of the main themes of this text is *exploration*. In brief, exploration involves examining a situation, with or without a particular question in mind, and discovering whatever you can about it. It involves "messing around" with mathematical ideas—trying one thing and then another, looking at examples, making guesses, asking questions.

This is the way mathematicians really work. We are trying to teach you how to approach mathematics like a mathematician. You are probably planning to study more mathematics after you are finished with this book. In order to understand such subjects as modern algebra and real analysis, you cannot just read a book and listen to lectures—you will need to *think* about the mathematical ideas.

Mathematics is not a spectator sport. Even when you are just reading mathematics, you need to be an active participant, or you won't understand. Advanced mathematics texts expect you to explore, though they don't explicitly tell you to do this. For example, when a mathematics book gives you a new definition, you must "play with it". That's the best way you can come to understand a new idea. Advanced textbooks will expect you to have an understanding of the new idea as they proceed. Therefore, you should have pencil and paper at your side every time you pick up this book, and you should be prepared to use them. That is, *you must do some work!*

(In this connection, we mention a story told about Euclid and the Egyptian monarch Ptolemy I. The king inquired whether there was a simpler way to learn geometry than to go through the details of Euclid's *Elements*. Euclid is reported to have replied: "There is no royal road to geometry". His remark applies equally well to all of mathematics.)

We also encourage you to share your ideas with classmates and friends. Mathematicians don't work in isolation. They may go off on their own for a while to puzzle something out, but then they bring back what they've discovered, or think they've discovered, and ask a colleague to give them feedback. Often they work in pairs or larger groups, bouncing ideas off each other, until something fruitful emerges.

To help you develop the habit of exploration, we have interspersed the text with several devices to encourage you to become personally involved. For instance, we will sometimes pose questions that look like this:

What do you suppose this fancy type means?

When you see this large type, you are being asked to stop reading and to respond to some question that has just been posed. Perhaps we want you to reflect on something that has just been said. At other times, we will ask you to come up with an example, or to construct the next step of a proof. Sometimes the question is fairly specific; at other times, it is more general.

Often we will ask you to "Get Your Hands Dirty". This means that you must stop being an armchair mathematician and get to work—test out the ideas just discussed, do some examples, use the technique just introduced, experiment a little. Often the problems suggested in a "GYHD" will be fairly routine, of the sort that would normally appear in the exercise set at the end of a section (and they may reappear there as well). But we don't want you to wait until the end of a section. You need to *do something* in order to really absorb the ideas being developed.

At other times, we will pose a more substantial **Question.** Mathematics is not a straight and narrow path — it includes many choices. We would like you to feel like a participant in those decisions, to develop a feeling for how the world of mathematical ideas came to be the way it is.

Finally, there will be some times when we set up an **Exploration.** This is a chance for you to let your imagination soar, and to examine the full range of your mathematical experience. The questions posed in an Exploration will generally be very open-ended. They will usually have many possible "right answers", often more than any one reader (or author) could possibly come up with.

All of these pedagogical devices are a means toward an end — for you to become an active reader and an active "doer" of mathematics.

Caution: In many cases, a question we pose in one section is discussed further in a later section. Please resist the temptation to look ahead to find "the answer". The point is for you to become familiar with the ideas involved in the question. Knowing the answers is not nearly as important as developing a familiarity with the ideas. Such a familiarity will only come by applying your mind to the problem. We repeat, *you must do some work.*

The second major theme of this book is *proof.* Proof is the process of attaining and demonstrating certainty about what you have discovered, and so it is a form of communication and explanation. But it is a specialized form of communication, with particular attention to precision of thought. Mathematical proof has some special language and symbolism, which have been adopted by mathematicians to prevent ambiguity. You will have to learn these conventions.

But we want to emphasize that, despite this necessary formalism, the process of working with proofs should be seen as growing out of the process of exploration. At all times, when you develop, write, or read a proof, you should be thinking beyond the formalism, and asking "What's really going on here? What is this really saying?" and "Is there some way I can explain this more clearly or give someone a better insight into the situation?"

In Section 1.1 of this chapter, we will explain in more detail what the idea of exploration is all about and illustrate this idea with three examples. Section 1.2 discusses the transition from exploration to proof, and why we prove things. Section 1.3 is an intuitive introduction to the actual writing of mathematical proofs. Beginning with a simple example, it discusses the special language in which mathematical statements are made, especially the meaning of "if . . . , then . . ." statements.

Note: Appendix B contains basic terminology and notation about sets, which will be used throughout this book. It also includes a list of fundamental facts about arithmetic that we will be assuming. You may wish to review this material before proceeding. We will refer you to Appendix B as each idea is introduced into the main text.

1.1 Exploration

The best way to explain what exploration is about is by means of some examples. This section features explorations of three separate situations: two numerical games, and one more traditional mathematical topic. In each case, we will present the general situation, and ask you to explore it. In the first exploration ("SubDivvy"), we will give you extensive guidance as to how to begin and how to proceed with the process. In the second example ("Sets closed under addition or subtraction"), we will give you some hints, but then put you on your own. In the third exploration ("Make It More"), we will just explain the game and ask you to explore.

The Joy and Frustration of Exploration

Before we begin, it may be helpful to say a few words about the *psychology* of exploration—the attitudes and emotions that go with this approach to doing mathematics.

In these explorations, as with others to follow, our main goal will be to point you on the way to understanding the situation. This book is not going to solve most of these problems for you (although some exploration topics will be followed up elsewhere in the book because of their mathematical importance). We would rather leave them as puzzles for you to play with in your spare time, so that you can get the pleasure of discovering the solutions for yourself. Your main interest should be in the *process* of thinking about mathematical problems, rather than on the final answers.

You may find it upsetting or frustrating to leave an exploration without a complete answer. This is the reality of doing mathematics. Try to see this in a positive light: an unsolved problem can be a pleasant thing to have around when you're sitting on the bus, waiting in line, or otherwise killing time. Treasure your unsolved problems. Think of the satisfaction when you finally solve them (and you will eventually solve some of them).

You also may experience some frustration as you're working on a problem. One constructive response is to take a break. Let your mind focus on something else, and come back to the topic later. You may discover that you learned something unconsciously, or that you can find a fresh approach.

Another way to deal with frustration is to share it. Talk to someone else about the mathematical situation you are exploring. In the process of explaining it, you will be forced to clarify your thoughts, and in that process, things may come together. Your friend may also have some helpful questions or ideas.

A third approach is to focus on the fact that you are "just exploring". Say to yourself, "I'm not 'solving the problem'. I'm just trying to increase my understanding of a particular situation." This adjustment in attitude may take some of the pressure off, and allow you to think more freely and enjoy it more. A bunch of guesses or some scratch paper covered with scribblings can be ample evidence that you really have "done the assignment".

Perhaps the best thing you can do about the feeling of frustration that comes with exploration is recognize that it is a normal part of almost any worthwhile challenge.

And now, on to the explorations!

SubDivvy: A Game

SubDivvy is a numerical game for two players, played according to the following rules:

1. A whole number greater than 1—let's call it N—is chosen at random. (You may wish to set an upper limit, say 100, for N, and use a spinner or similar device for choosing the number.)

2. Player 1 chooses a positive integer different from N that is a divisor of N (i.e., that divides "evenly" into N, so that the remainder is zero), and subtracts that divisor from N. The result of that subtraction is given to player 2.

3. Player 2 works with the result of the subtraction just as player 1 worked with N: by choosing a positive divisor of that number different from the number itself, and subtracting the chosen divisor from it. The result of the subtraction is given to player 1.

4. Play continues, with players taking turns choosing divisors and subtracting, until the result reaches the number 1. The player who produces the result 1 is the winner.

It is helpful to refer to the number a player has just been given as that player's *position*. The number selected at random at the start of the game will be called the *starting number*, so the starting number is player 1's first position.

Example 1. **A Sample Game of SubDivvy** The rules by themselves may be hard to follow, so here's an example of what the game looks like, with comments describing what is happening at each step. (This example is not necessarily played well by either player.)

68	The number 68 is randomly selected as the starting number N (so 68 is now player 1's position).
-4	Player 1 chooses 4 (which is a divisor of 68), and subtracts.
64	The result of the subtraction is 64 (which is given to player 2, so player 2's position is now 64).
-16	Player 2 chooses 16 (which is a divisor of 64), and subtracts.
48	The result of the subtraction is 48 (which is given to player 1, so player 1's position is now 48).
-24	Player 1 chooses 24 (which is a divisor of 48), and subtracts.
24	The result of the subtraction is 24.
-8	Player 2 chooses 8 (which is a divisor of 24), and subtracts.
16	The result of the subtraction is 16.
-8	Player 1 chooses 8 (which is a divisor of 16), and subtracts.
8	The result of the subtraction is 8.
-4	Player 2 chooses 4 (which is a divisor of 8), and subtracts.
4	The result of the subtraction is 4.
-2	Player 1 chooses 2 (which is a divisor of 4), and subtracts.
2	The result of the subtraction is 2.
-1	Player 2 chooses 1 (which is a divisor of 2), and subtracts.
1	The result of the subtraction is 1, so the game ends, and player 2 is the winner. □

Now that you have seen a sample game of SubDivvy, we want you to explore the game in general. This discussion will begin to explain what that process means.

First of all, you should be sure you understand how the game works—that is, be sure you can follow the rules. Probably the best way to do this is to find a partner and play it a few times. Playing is an important part of exploration.

Next, you should begin to formulate some questions. What would you like to know about SubDivvy? Questions may occur to you before you begin to play, as you play, or afterward. Get into the habit of writing down your questions.

And then you should try to answer the questions.

This is not easy. It is all so vague. But that's typical of exploration, and, indeed, typical of mathematics, once you really get into it.

Suggestion: After you have played with a friend, find two more people, and play in teams of two, so that you and your partner can discuss what moves to make. This may inspire you with some questions or insights.

Exploration 1: SubDivvy

Before you read any further, you should do what you can to carry out this exploration, as outlined above. Don't be discouraged if you don't get very far in this first exploration.

What can you say about the game of SubDivvy?

•EXPLORE•

Reminder: Whenever you see this "explore" symbol, or similar instructions in large type, it means you should stop reading and do some thinking about the question, situation, or problem under discussion. Usually it will mean that you will use pencil and paper. In this case it also means find a friend or friends to play SubDivvy with. Learn as much as you can about the game. Allow yourself enough time to really examine the Exploration topic. When you've had enough, come back to the book.

Some Hints on SubDivvy

Because this is the first of our explorations, you may appreciate some further guidance. But you should at least have actually played several games of SubDivvy and made up one or two questions before reading on.

Let's look at some possible questions.

Perhaps the question that comes to mind first is:

> Which player will win?

This question is very broad, and doesn't have a simple answer. Who wins will depend on what the starting number is and what numbers each player chooses to subtract at each turn. We can reformulate the question this way:

> What is the best strategy for each player to use in order to be most likely to win?

By *strategy*, we mean a comprehensive plan of play, which spells out in complete detail exactly what the player will do in every possible situation. One example of a strategy is to always subtract the largest permissible number. Another strategy is to always subtract the smallest prime divisor of the given position. A third strategy would be to combine these, doing the first if the position is odd, and the second if it is even.

The crucial thing to keep in mind is that a strategy must be complete as described above—it must tell you what to do in every possible situation. A strategy is called a *winning strategy* if it leads to a victory no matter what choices the opponent makes.

We can make the strategy question more specific by restating it as follows:

> Are there certain starting numbers for which player 1 can create a winning strategy? Are there others for which player 2 can create a winning strategy?

In some cases, neither player has any choices, and so the question of strategy becomes irrelevant. For example, what happens if the starting number is 3?

Work it out. Play the game.

Player 1 must subtract 1, leaving 2, and then player 2 wins by subtracting 1. No "strategy" is required.

Notice that it doesn't matter at what stage in the game the position 3 occurs. The player whose turn it is at that point will lose. Therefore it makes sense to call the number 3 a "losing position". Similarly, 2 is a "winning position".

A slightly more interesting situation occurs if the starting number is 4. If player 1 subtracts 2, then player 2 has the winning position of 2, and can win by subtracting 1. On the other hand, if player 1 begins by subtracting 1, then player 2 will have the losing position of 3. Thus a player with the position of 4 can win by playing intelligently. There is a strategy for winning if your position is 4.

We can summarize the idea of the last paragraph by giving a precise meaning to the terms "winning position" and "losing position":

A number is called a *winning position* if there is a winning strategy for a player who has that position.

A number is called a *losing position* if a player getting that position must give the opponent a winning position. (In other words, no matter what the player with the given position does, the opponent will then be able to win by making the right choices.)

As we showed above, 2 is a winning position: all you have to do is subtract 1 and you win. On the other hand, 3 is a losing position: the only thing you can do is subtract 1, and that gives your opponent the winning position of 2, so you lose. The number 4 is also a winning position: if you make the correct choice of subtracting 1, you guarantee yourself a victory.

Warning: If you have a winning position, that does not mean that you will necessarily win. It only means that you *can* win by making the right moves. Similarly, a losing position doesn't mean you necessarily will lose. But it means that, no matter what you do, your opponent has a winning strategy available. You can only win if your opponent makes a blunder. Your opponent can get a victory by making the right moves. (Verify this warning by reviewing the case $N = 4$.)

Example 2. **A Sample Strategy** Consider the case $N = 15$. The following discussion shows that this is a losing position. It describes a winning strategy for the second player:

Player 1 must choose 1, 3, or 5. We look at each possibility:

Case 1. Player 1 chooses 1 (giving player 2 the position of 14):
Player 2 should choose 7, giving player 1 the position of 7.
*Player 1 must then choose 1, giving player 2 the position of 6.
**Player 2 should then choose 3, giving player 1 the position of 3.
Player 1 must then choose 1, giving player 2 the position of 2.
Player 2 should then choose 1, winning the game.

Case 2. Player 1 chooses 3 (giving player 2 the position of 12):
Player 2 should choose 3, giving player 1 the position of 9.
Player 1 then can choose either 1 or 3, and we must consider both possibilities:

Case 2a. Player 1 chooses 1, giving player 2 the position of 8.
Player 2 should then choose 1, giving player 1 the position of 7.
At this point, player 1 is in the situation indicated at * in case 1: player 1 must choose 1, and play proceeds as described above, with player 2 winning.

Case 2b. Player 1 chooses 3, giving player 2 the position of 6.
At this point, player 2 is in the situation indicated at ** in case 1: player 2 can then proceed as in case 1, from the position of 6, with player 2 winning.

Case 3. Player 1 chooses 5 (giving player 2 the position of 10):
Player 2 should choose 5, giving player 1 the position of 5.
Player 1 must then choose 1, giving player 2 the position of 4.
Player 2 should then choose 1, giving player 1 the position of 3.
As in case 1, player 1 must choose 1, giving player 2 the position of 2, and then player 2 chooses 1, winning the game.

This strategy for player 2 considers every possible move that player 1 can make, and provides a move for player 2 which leads to a winning game.

There may be other moves that player 2 could make that would lead to a winning game, but a strategy only needs to provide one winning move for each possible situation. □

Get Your Hands Dirty 1

Is 6 a winning position or a losing position? (You should examine every possible sequence of moves, and determine if a player whose position is 6 has a winning strategy.) □

You should do this "Get Your Hands Dirty" task now (and do others as they appear). It will help make the concepts in the discussion so far more real for you, and prepare you for the discussion to come. As we stressed in the Introduction, you can't learn mathematics without doing some work. You may have to go back and reread the discussion above, but that's natural.

Here are two more specific suggestions for examining the problem of strategy:

1. Save the arithmetic sequences from your games, and examine them, looking for patterns. Ask: *What mathematical concepts might be relevant to an understanding of these patterns?* For example, the divisors of some numbers might have special properties that are relevant to SubDivvy. Anything you know about numbers and their divisors might be useful here.
2. Work backwards: when you examine your games, figure out at what point in the game you were sure you (or your opponent) would win. Ask: *Which numbers represent winning or losing positions for each player?*

There are other interesting questions about SubDivvy besides the issue of strategy. Here is one other line of inquiry:

How many turns will it take for the game to end?

Of course, that will depend on the starting number, and on the choices the players make. So we might try a more specific question:

Given a particular starting number, what is the smallest (or largest) number of moves possible until the game ends?

We ask you now to return to Exploration 1, with the specific questions we have raised in mind, and keeping open to the possibility of others as they arise.

• E X P L O R E •

General Techniques of Exploration

This example illustrates several general principles or methods that can be applied to most explorations:

a) Exploration often involves *guesswork*. For example, we suggested examining types of divisors for various numbers. One possibility that might come to mind is prime numbers. Maybe if the starting number is a prime, then player 1 will win. Then again, maybe not. A mathematical guess is called a *conjecture*.

b) Exploration often requires looking at lots of *examples*. One important part of becoming a good "explorer" is learning how to choose examples well. This sometimes means examining especially simple examples, but other times means avoiding cases that may be overly simple and therefore misleading. Examples are particularly important in testing conjectures. An example that shows that a particular conjecture is wrong is called a *counterexample*.

In the course of learning more about mathematics, you will gradually build up a "catalog" of examples—representatives of different kinds of mathematical objects that you can use to test your conjectures. At this stage, many of the items in your catalog will be specific real numbers or sets of real numbers. Even within that particular framework, you can achieve considerable variety: negative numbers as well as positive, irrational numbers as well as rational, prime numbers as well as composite, finite sets as well as infinite, and so on. As we move through this book, we will have other kinds of examples to work with: functions, relations, operations, and so on. You may wish to keep a "catalog" in which you keep track of interesting examples and comment on the variety of possibilities within various categories of objects.

c) Exploration often depends on *prior knowledge*. This exploration requires you to have an understanding of basic arithmetic. It wouldn't make sense to show this game to children in early elementary school—they wouldn't know enough about division even to be able to play. As you move on in mathematics, the prior knowledge required may increase, although some very advanced problems can be stated that require little mathematical background. We will look at other explorations later on that assume more mathematics than this one.

d) Exploration is often given clearer direction by *introducing appropriate terminology*. In this example, the concepts of "winning position" and "losing position" provide a focus and frame of reference for the discussion, and make communication about the game much easier. (This terminology, or something like it, is actually fairly commonly used in mathematics.) Essentially what we are doing here is introducing a shorthand way of saying something complex.

We could have defined these terms some other way, if we had wanted to. We are always free to introduce new terminology that will simplify or clarify a discussion, and can define that terminology in whatever way will be convenient. The only requirement is that we use our terminology according to the meaning we have given it. This is easier to do if the formal definition is similar to the everyday usage of the words. When you come across definitions in a mathematics text, you will often have to do some playing around to figure out what's going on, and what the new terminology really means. Examining a definition is an important type of exploration.

e) Exploration is often more successful if it is done in cooperation with others. In this example, you had a partner to play the game with, so that you could discuss strategy. In other situations, you may simply want someone else who is thinking about the same problem, so you can compare notes with each other, and share ideas. In either case, you have a chance to articulate your ideas. Talking about something out loud, or explaining it to someone else, often leads to increased understanding. Making mathematics a social as well as an intellectual activity can make it both more productive and more enjoyable.

f) Exploration is usually assisted by good record keeping. If you keep track of your games of SubDivvy, you will later have a collection of data to use in checking your conjectures. In some explorations, you may want to try to organize your data in a way that makes patterns more visible.

Sets Closed under Addition or Subtraction

We turn now to another topic for exploration. This exploration may appear to be more mathematical than the SubDivvy example. It concerns the set $\mathbf{Z}$ of *integers*—i.e., $0, \pm 1, \pm 2, \pm 3, \pm 4, \ldots$—and involves examining *subsets* of the set of integers. (See Appendix B, if necessary.)

Here are some subsets of the integers which we will refer to in our discussion:

the positive integers: $1, 2, 3, 4, \ldots$;
the multiples of 3: $0, \pm 3, \pm 6, \pm 9, \ldots$;
the integers ending in 7: $\pm 7, \pm 17, \pm 27, \pm 37, \ldots$.

Each of the first two of our examples of subsets of the integers has the following interesting property:

No matter what two numbers in the set you choose (including the case where the two numbers chosen are the same), their sum will also be in the set.

• V E R I F Y •

A set with the property described just above is called *closed under addition*. Notice that the third set is not closed under addition: for example, the sum of 17 and -37 is -20, which is not in the set.

Similarly we can define a set to be *closed under subtraction* if the *difference* of two of its elements, for every choice of the numbers, is in the set. The second set above — the set of multiples of 3 — is closed under subtraction. (Verify.) However, the other two sets above are not: the difference of the positive integers 9 and 16 (in that order) is -7, which is not a positive integer. And the difference of 7 and -27 is 34, which does not end in 7.

Notice that, for a set to be closed under addition or subtraction, a certain condition has to be true for *every* pair of numbers from the set. However, just one pair is needed to show that a set is not closed. For example, the pair of numbers 17 and -37, as discussed above, shows that the third set is not closed under addition.

Exploration 2: Sets Closed under Addition or Subtraction

We've defined two concepts — "closed under addition" and "closed under subtraction" — that mathematicians consider interesting.

• E X P L O R E •

Comment: You might consider it unfair to be told to explore when there isn't a question yet. But remember that part of exploration is asking questions. So perhaps your first objective should be to formulate a question.

Some Hints

The main question that we want to focus on is simply this: which subsets of the integers are closed under addition and which are closed under subtraction? If you formulated that question in your own mind, and began to consider it, then read on. However, if that question didn't occur to you, or you didn't give it any serious thought yet, don't feel bad. Just resume your exploration now, focusing specifically on that question before reading further.

Here are some hints for continuing your exploration of the question:

1. Look at lots of examples. Try as many different sets as you can think of. Keep track of which ones are closed under each of the operations, and which are not.
2. Look for patterns among your "winning" examples. Another key aspect of exploration is *generalization*. In this context, it might mean looking at *some sets* that are closed under a particular operation, finding a property that they have in common, and then making the conjecture that *all sets* with that property are closed under the given operation.
3. Start with a small, finite set — for example, two integers — and examine what else you would need to put into that set to make it closed under either addition or subtraction. Continue the process until you get a closed set. Think about what has happened. Repeat the process for a different initial set.

4. *Warning:* The "addition" part of this question is actually much more complex than the "subtraction" part. There's no way you could be expected to predict that, especially since addition is usually simpler than subtraction.

•EXPLORE•

COMMENTS

1. One of the key elements in this exploration topic is the initial formulation of the concept of a set being "closed under" some operation. It is not a complicated idea, but introducing *new terminology*—simply giving the idea a name—helps give it clarity. This naming process also immediately suggests possibilities for generalization, such as changing the operation. We will see other situations where introducing a *special notation* for an idea plays a similar role of clarifying and focusing, so that we can better understand a situation.

2. The concept of a set being closed under an operation is important in abstract algebra. The particular example of subsets of **Z** that are closed under subtraction will prove helpful in understanding the notion of greatest common divisor and in solving simple equations involving the integers. Exploring the concept intuitively and informally here is important groundwork for being able to work with it more abstractly in later chapters.

Make It More: A Game

Here's another two-person game to explore. The game is called "Make It More". Each of two players has a three-digit number to fill in. A playing sheet might look like this:

Player 1: ____ ____ ____
Player 2: ____ ____ ____

Player 1 begins by choosing any digit except 0 or 5, and writing it in his ones column.

Player 2 doubles the digit player 1 just entered, and looks at the ones digit of the doubled value. (If the doubled value is less than 10, the doubled value is the same thing as its ones digit.) Player 2 then picks any digit (except 0 or 5) which is less than or equal to that ones digit, and writes it in her ones column.

Now player 1 doubles the digit player 2 just entered, and, as in the previous step, looks at the ones digit of the doubled value. Player 1 then picks any digit (except 0 or 5) which is less than or equal to that ones digit, and writes it in his tens column.

This process continues, with player 2 picking a tens digit, then player 1 picking a hundreds digit, and finally player 2 picking a hundreds digit.

The winner is the player whose final 3-digit number is larger.

Example 3. **A Sample Game of Make It More** A sample game of Make It More might look like this:

Move 1:

Player 1:	___	___	4	(Player 1 chooses 4 for the ones column.)
Player 2:	___	___	___	

Move 2:

Player 1:	___	___	4	
Player 2:	___	___	7	(Twice 4 is 8; player 2 chooses a number, 7, which is less than or equal to 8, for the ones column.)

Move 3:

Player 1:	___	3	4	(Twice 7 is 14, and the ones digit of this doubled value is 4; player 1 picks a number, 3, which is less than or equal to 4, for the tens column.)
Player 2:	___	___	7	

Move 4:

Player 1:	___	3	4	
Player 2:	___	6	7	(Twice 3 is 6; player 2 chooses a number, 6, which is less than or equal to 6, for the tens column.)

Move 5:

Player 1:	1	3	4	(Twice 6 is 12, and the ones digit of this doubled value is 2; player 1 picks a number, 1, which is less than or equal to 2, for the hundreds column.)
Player 2:	___	6	7	

Move 6:

Player 1:	1	3	4	
Player 2:	2	6	7	(Twice 1 is 2; player 2 chooses a number, 2, which is less than or equal to 2, for the hundreds column.)

Therefore player 2 wins this game. □

Exploration 3: Make It More

What can you say about the game of Make It More?

•EXPLORE•

EXERCISES

1. (Exploration 1) The discussion of SubDivvy implicitly assumed that every game would have a winner. Is this true? Write a statement explaining your answer. Identify any basic facts about the whole numbers that you need.
2. (Exploration 1) What are the possibilities for the last move of SubDivvy? Why are there no other possibilities? Explain carefully, identifying any basic facts about the whole numbers that you need.
3. (Exploration 1) Give a complete winning strategy for SubDivvy for the appropriate player, using each of the numbers from 2 through 10 as the starting number.
4. (Exploration 1) In the hints for Exploration 1, we mentioned the question of how many turns a game of SubDivvy might take. Here are some specific avenues to explore relating to that idea:
   a) In terms of the starting number, what is the largest number of turns the game can take? (This question is pretty easy.)
   b) In terms of the starting number, what is the smallest number of turns the game can take? (This question is much harder. Don't expect to get a complete answer. Try to answer the question for some special cases.)
   c) In terms of a specified number of turns, what is the smallest number you can start with? (This is similar to part a, but the roles of the starting number and the number of turns have been reversed. It's just as easy as part a.)
   d) In terms of a specified number of turns, what is the largest number you can start with? (This seems to resemble part b in the same way that part c resembles part a, but it is actually considerably easier than part b. It is, however, still somewhat harder than part c.)
5. (Exploration 2) Give as many examples as you can find of sets that are closed under either multiplication or division. Describe any general families of such sets that you discover.
6. (Exploration 3) In Make It More, the digits 0 and 5 are not allowed. Why do you suppose that is? What might happen if those digits were permitted? What would be a winning strategy if 0 were allowed (but not 5)? What would be a winning strategy if both 0 and 5 were allowed? What if 5 were allowed but not 0?
7. (Exploration 3) Which player has a winning strategy for Make It More under the original rules given in the exploration? What is that strategy?
8. Consider the following game for two players, called Multiple Elimination: A number N is chosen, and the integers from 2 through N are written out. Player 1 picks one of the numbers, and then crosses out all the multiples of that number (including the number itself). (For instance, if player 1 picks 3, then $3, 6, 9, 12, \ldots$, all get crossed out.) Then player 2 picks a number that is not crossed out, and crosses out all of its multiples (again, including the number itself). Play continues, with players alternating, picking a number that remains and crossing out that number's multiples (always including the number itself). The player who crosses out the last number is the winner.
   a) For each of the cases $N = 2$ through $N = 9$, decide which player has a winning strategy, and describe that strategy. (*Note:* it will probably be harder to *find* the winning strategy than it is to *describe* it.)
   b) Repeat part a for the case $N = 10$.

9. Consider the following game for two players, sometimes called "Linear Nim". At the beginning, there are a certain number N of marks on the paper, say $N = 15$:

|||||||||||||||

Each player, in turn, crosses out either 1, 2, or 3 marks. For example, player 1 could cross out 3, and player 2 might then cross out 1, leaving this situation:

~~|||~~~~|~~|||||||||||

Play continues until all marks are crossed out. The player crossing out the last mark is the winner.

- **a)** What is a winning strategy for this game? Which player should win? How would the strategy change if N were 10? 25? 12?
- **b)** What if the number of marks crossed out were allowed to go up to 4?

10. In your own words, summarize what you think are the main ideas of this section. What have you learned about exploration?

1.2 From Exploration to Proof

The main focus of exploration is on examining what is going on in a particular situation. It generally proceeds intuitively, sometimes haphazardly. It is filled with guesses and examples. Your main interest is in acquiring information and gaining insight. After you have explored for a while, you will very likely come up with some ideas about what's going on. You're ready to solidify some of your thinking. Eventually, you will want to try to prove your conjectures.

But before we talk about how to do proofs, and the reasons for doing proofs, it may help to identify another aspect of the exploration process. After the open-ended stage of exploration, you will need to make some decisions about how to proceed. Perhaps you have come up with various ideas about what is going on. Sometimes, however, "conjecture" is too optimistic a word for what you have in mind—your ideas may simply be *possibilities* that occurred to you.

Making Decisions

Before you invest time and energy in trying to prove that a particular statement is true, you might like to be more confident that it is correct. You need to decide if it seems likely or not, and whether you should pursue it further. You will have to decide whether to try to prove some conjecture true, or look for a counterexample. The next exploration focuses on this decision-making aspect of the exploring process.

Exploration 1: "Do You Think It's True?"

Here are some mathematical statements for you to look at. Some are fairly difficult, others easier. Experiment with each of them—explore them—until you feel ready to make an informed guess as to whether each is true or false.

1. There are integers x and y, which are solutions to the equation $23x + 37y = 52$.
2. Given a line L (remember that a line is infinite) and a point P not on that line, there is a unique line through P, which does not intersect L.
3. If you pick N distinct points on the circumference of a circle, and draw line segments connecting them all with each other, then the interior of the circle will be divided into 2^{N-1} portions (see Figure 1.1).
4. The formula $(a + b)^2 = a^2 + b^2$ is correct for all real numbers a and b.
5. If ABC is a right triangle, with a right angle at vertex B, and D is the midpoint of the hypotenuse, then the line segment connecting vertex B to D is half the length of the hypotenuse (see Figure 1.2).

Remember that you should approach these statements as if you were asked for an informed opinion, rather than an authoritative ruling. As you decide on your answers, notice with each one how sure you are of your answer. As always, try to keep track of the thinking you did along the way.

• EXPLORE •

Between True and False

You may have found yourself with more than two categories of answers to Exploration 1; that is, the statements may not have fallen neatly into a true-or-false classification. Here are some other possibilities:

1. "I don't know." This is perfectly legitimate. It means that you really have no sense one way or the other.
2. "The statement is ambiguous." For example, statement 2 doesn't specify that the line through P is supposed to be in a plane with L. Whether or not the statement is true might depend on this extra condition.
3. "It's true in some cases, false in others." For example, this might apply to statement 3, where the truth of the statement might depend on the value of N. We will discuss this kind of answer briefly in Section 1.3 of this chapter, when we talk about "truth sets".
4. "It seems to be true (or false), but I don't understand why and I'm really not certain." This is a situation ripe for further exploration and perhaps for the discussion of proof.

Exploration 1 represents an intermediate stage in doing mathematics: in the first stage you explore for a while, and reach some tentative ideas. In the middle stage, as in Exploration 1, you look at those ideas further, and become convinced in your own mind of what is going on. You're ready to call your ideas "conjectures". But you may not be sat-

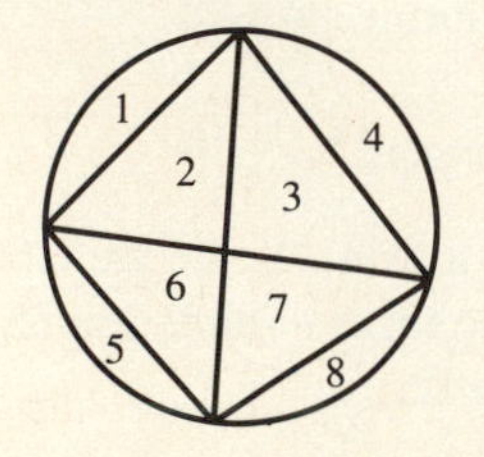

Figure 1.1 Four points on the circumference are all connected to each other, creating $8 = 2^{4-1}$ portions of the circle.

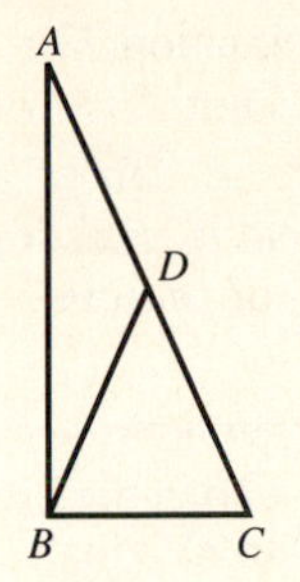

Figure 1.2 D is the midpoint of hypotenuse AC. Is segment BD half the length of AC?

isfied with saying "I think that such and such is the case". You may want to be able to say "I know that such and such is the case".

That's where the idea of *proof* comes in. All stages of doing mathematics are concerned with acquiring understanding, and the separations between the stages are not always sharply defined. But the hallmark of the proof stage is that it is primarily concerned with acquiring *certainty*.

A proof can vary in style from very informal and intuitive to highly technical and sophisticated. The difference between an "explanation" and a "proof" is often a matter of degree of detail and formality. In many contexts a good intuitive explanation is just as valuable as a more formal argument. We will be doing a great deal of both in the course of this book.

Why Prove Things?

We have already noted one of the main reasons for proving mathematical statements: *certainty*. It is often important for us to know if a statement is simply plausible, or is something we can rely on.

But there are several other reasons why mathematicians go to the trouble of proving statements.

CORRECTING MISTAKES It often happens that an exploration will lead to an incorrect conclusion. For example, an examination of several cases of statement 3 in Exploration 1 (concerning subdividing a circle) might suggest that the statement is true. In fact, even if you continue up to as many as 5 points on the circumference, the statement holds true: 5 points will produce 2^4 portions of the interior of the circle. But 6 points can produce at most 31 portions, not the 32 ($= 2^5$) required.

A mathematician attempting to prove that the statement is true would run into trouble. Difficulty in creating a proof is a clue that the statement you're trying to prove may be false. So the wary mathematician might go back to the problem and explore some more, discovering the counterexample when $N = 6$.

GAINING INSIGHT Creating a proof forces you to examine the situation more carefully than you did before. For example, consider statement 5 of Exploration 1, concerning a segment to the midpoint of the hypotenuse of a right triangle. Experimentation and measurement of several triangles would probably convince you that the statement is true. But the result might seem like a mystery to you.

The traditional proof suggests viewing the right triangle as half of a rectangle, with the hypotenuse as a diagonal (see Figure 1.3). The segment in question is then seen to actually be half of the other diagonal, so statement 5 follows from the intuitive fact that the diagonals of a rectangle are of equal length.

If you picture the triangle in this context, there is no longer a mystery about the statement. The process of proving the statement has helped to understand it.

GENERALIZATION Often if you examine a proof, you will notice that it will apply to many situations other than the one you originally were considering. For example, if you try to prove that the set of multiples of 3 is closed under addition, you may notice that the number 3 itself did not play any special role: any integer in that situation would have worked just as well.

Thus, in attempting to prove the initial assertion, we come up with the following much more general conclusion: For any integer N, the set of multiples of N is closed under addition.

COMPUTATIONAL GUIDANCE Sometimes a mathematical theorem is a statement that something *exists*—for example, you learned in elementary algebra that a quadratic equation always has a solution (either real or complex). One way to prove this statement is to use the quadratic formula, which not only shows that there is a solution, but demonstrates how to find it. There are many similar instances even in advanced mathematics where the proof of a general statement actually constitutes a computational guide to solving some specific problem.

Formulation of a proof will generally begin by building on a previous exploration. As you begin to work on a proof or explanation, you may find yourself needing to explore some more. As you explore, you may find yourself making further conjectures and wanting to explain them more fully. The dividing line between exploration and proof is not a sharp one, and the two processes should be thought of as mutually supportive—that is, work on each will be helpful with the other.

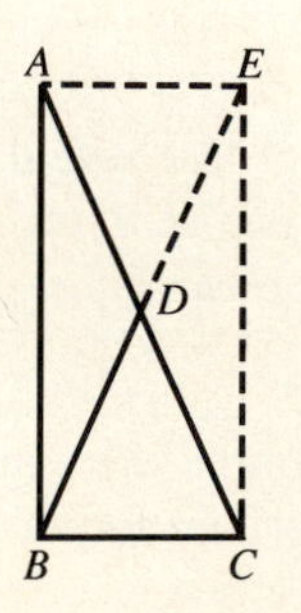

Figure 1.3 AC and BE are the diagonals of rectangle $ABCE$. The two diagonals AC and BE bisect each other at D, so the length of BD is half the length of AC.

EXERCISES

In these exercises, pay attention to making decisions, acquiring information, developing a feeling for the situation. You should also be asking yourself "why is it so?" What does your intuition tell you? But don't worry about a formal proof of your responses to these questions—that will come later.

1. Statement 1 of Exploration 1 asks whether a particular linear equation in two variables has an integer solution.

a) For that specific equation, the answer is yes. Find an integer solution for that equation.
b) Find a linear equation in two variables, using integer coefficients and constant term, which does *not* have an integer solution. (There is a difference between the statements "there is no solution" and "I can't find a solution". You should have a reason to be fairly certain that there is no solution to the equation you give. Don't just give an equation for which you can't find a solution.)
c) What can you say about which equations of this type have integer solutions and which don't?

2. Examine the sequence of squares of positive integers: $1, 4, 9, 16, 25, \ldots$. What do you notice about differences between consecutive terms of this sequence? What about differences between terms that are two steps apart in the sequence (such as 4 and 16)? What about terms that are three steps apart? Do you think your observations describe patterns that continue throughout the sequence of squares, or are they just coincidences?

3. Examine the ones digits in successive powers of 2 $(2, 4, 8, 16, \ldots)$. Do you see any patterns? Do the same thing for powers of 3. What is similar about the patterns and what is different? What do you think would happen with powers of other one-digit numbers? Do you find any patterns if you examine the tens digits of successive powers of 2? of 3? other one-digit numbers?

4. For each of the numbers from 2 through 20, count how many divisors the given number has. (Both 1 and the number itself are considered divisors.) Do you see any patterns? Are there any rules for determining the number of divisors in special cases? Can you develop an informed guess about how to determine the number of divisors without actually listing them all?

5. (Based on Problem 1.2 in *Mathematics: Problem Solving Through Recreational Mathematics*, by B. Averbach and O. Chein, Freeman, 1980) Ms. X, Ms. Y, and Ms. Z—an American woman, an Englishwoman, and a Frenchwoman, but not necessarily in that order—were seated around a circular table, playing a game of Hearts. Each passed three cards to the person on her right. Ms. Y passed three hearts to the American. Ms. X passed the queen of spades and two diamonds to the person who passed her cards to the Frenchwoman. Who was the American? The Englishwoman? The Frenchwoman? (*Note:* We will ask you to *prove* your answer for this problem in Section 1.3. Your focus here should be on the process of analysis.)

6. The statement "if $a < b$ and $b < c$, then $a < c$" is correct for all real numbers a, b, and c. State as many other general principles about inequalities as you can. Also, state some *false* principles—statements that seem reasonable, and perhaps are true for certain special cases or categories of real numbers, but are not true for all real numbers. (Learning to distinguish between the two kinds of statements in this exercise is very important.)

1.3 Proof: An Introduction

We have already pointed out that a proof is a special kind of explanation of some mathematical idea or observation—an explanation that usually is precise and formal, rather than intuitive and informal.

The standards for judging whether something is a complete proof vary with the context. A mathematician writing for other mathematicians may skip many steps, which the reader is expected to be able to fill in easily. A student writing for an introductory course may be expected to provide greater detail. You should use the many proofs given in this text as a guideline for what level of completeness is appropriate. With experience you will learn to adjust your style to suit the intended audience.

We need to look carefully at exactly what it *means* for a mathematical statement to be considered true, as well as at what constitutes an acceptable *argument* that it is true. We will begin with an example of a simple proof, and discuss some principles that it illustrates.

A Simple Proof

Before you read Example 1, take a minute to think about this question:

Why is the sum of even integers an even integer?

? **?** **?**

Reminder: This string of question marks means "stop reading and think about the question."

Now read the proof.

Example 1 **A Simple Proof** In the following proof, we have numbered the steps for the purpose of our later discussion.

Theorem 1.1. If x and y are even integers, then $x + y$ is an even integer.

Proof:

1. We begin by assuming that x and y are even integers.
2. What exactly does being an even integer mean? An even integer is a number that is a multiple of 2, or, equivalently, is twice some integer. In other words, x can be written as $2m$, for some integer m, and y can be written as $2n$, for some integer n.
3. What does that say about $x + y$? It means that $x + y$ is equal to $2m + 2n$, which equals $2(m + n)$. Since $m + n$ is itself an integer, this tells us that $x + y$ is an even integer. ∎

This example introduces several important aspects of formal mathematical proofs. We will discuss each of these elements in detail, but first, we want to give an overview of the proof.

THE STATEMENT OF THE THEOREM A *theorem* is an assertion that can be proved, based on specific assumptions. Theorem 1.1 is formulated in typical "if . . . , then . . ." form. The "if . . . , then . . ." type of statement is also known as a *conditional sentence*.

STEP 1 The first step illustrates the basic method of *assuming the hypothesis and deriving the conclusion,* which is at the heart of most proofs of "if . . . , then . . ." statements. We will discuss this at length, because the meaning of "if . . . , then . . ." statements in mathematics often causes students difficulty.

Another important aspect of the first step is the use of the letters x and y, known as *variables*. We will look at such questions as: what exactly do these letters mean? where do they come from? what are the rules for working with them?

STEP 2 The next stage of the proof in Example 1, after stating the assumption, is to *interpret* the assumption, to make use of it in some constructive way. The essence of step 2 is to translate the hypothesis from words into an algebraic form. This involves using the definition of the term "even integer". Definitions are often the natural starting place when you are trying to prove some basic theorem.

Note that we have chosen to interpret the assumption "x is even" by translating it into the equation "$x = 2m$". Your first inclination may have been to think of the condition "x is even" as saying something like "x is evenly divisible by 2". Though correct, that alternative interpretation turns out to be less useful in a formal proof. It says that, if you divide x by 2, the result is some integer. What our interpretation has done is to give that "some integer" a name. This naming of the object being referred to is a very helpful technique, since it allows us to manipulate the object in equations. We will look at the formal definition of "even integer" and "divisibility" in Section 2.3.

Note: A standard formal textbook proof would probably express step 2 more concisely, saying something like: "therefore, x and y can be written as $2m$ and $2n$, for integers m and n respectively".

STEP 3 The proof reaches fruition in step 3. As suggested by the conclusion of the theorem, we look at $x + y$. We can use the notation introduced in step 2 to get an expression for $x + y$ in terms of the new objects m and n. The simple algebraic step that $2m + 2n = 2(m + n)$ allows us to draw the conclusion we seek: that $x + y$ is itself an even integer.

As we have already suggested, this step is primarily a formalization of our intuitive understanding of why the sum of even integers is even. When you thought about the question preceding Example 1, you probably had some ideas similar to what is expressed here.

This step, brief as it is, is the *work* of the proof. The first two steps set us up to work with the objects involved, and now we actually do something with them. The distributive property, which tells us that $2m + 2n$ is equal to $2(m + n)$, is the key concept behind the formal proof of this theorem. (This property is one of our basic assumptions about arithmetic, listed at the end of Appendix B.)

In most proofs, the work will be much more complex than this. It is in this step that your intuition will be needed, and for which the process of exploration will play an important preparatory role. Though there are some proofs that almost "write themselves", because they follow the definitions so closely, those will become rarer and rarer as you move along in mathematics. Before you can write a good proof, you usually have to understand intuitively what's going on. The essence of writing a proof is to formalize that intuitive understanding.

The following task is intended to reinforce the ideas we've just discussed:

Get Your Hands Dirty 1

Prove the following theorem using Example 1 as a model:

If x is an even integer, and y is a multiple of x, then y is even. □

With these preliminary remarks, we now turn to a fuller discussion of the individual components of the proof in Example 1.

Conditional Sentences

The theorem in Example 1, like many you will come across in mathematics, is stated in "if . . . , then . . ." form. Here are some other examples of statements of this type:

Example 2 "If . . . , then . . ." Statements

1. If x and y are odd integers, then xy is odd.
2. If two of the sides of a triangle have equal length, then two of the angles have equal measure.
3. If a and b are real numbers, then $|a + b| \leq |a| + |b|$.
4. If a function is differentiable, then it is continuous. □

A statement of this type is called a *conditional sentence*. The first part of the statement—the part that goes with "if"—is called the *hypothesis*, and the second part—the part that goes with "then"—is called the *conclusion*. By its nature, the hypothesis is describing something that might be true—depending on the particular choice of objects under discussion. That's why we say "if". A conditional sentence is saying that, *in those cases where the hypothesis turns out to be true*, the conclusion will also be true.

We can think of a conditional sentence as a relationship between two ideas or possibilities. In brief, a conditional sentence tells us that, under certain conditions or circumstances (expressed in the hypothesis), there will be certain consequences that will result (as expressed in the conclusion). We sometimes represent the hypothesis and conclusion of the sentence symbolically by letters such as P and Q, and the conditional sentence is read "if P, then Q".

When we are trying to prove that a statement of the form "if P, then Q" is true, we are only interested in the situation in which P is true (in our case, when x and y are both even). Of course there are examples in which x or y (or both) will not be even. But those situations have no relevance to this theorem. A case where the hypothesis is false cannot affect the truth or falsity of the conditional sentence. It may turn out in such a case that $x + y$ is even (for example, if $x = 3$ and $y = 7$), or it may turn out that $x + y$ is odd (for example, if $x = 8$ and $y = 13$).

But we don't care what happens in those cases. If statement P is not true in a particular case, then that case does not affect the conditional sentence at all, regardless of whether Q is true or false.

Another way to say this is that proving "if P, then Q" is the same as assuming P and then using that assumption to derive Q.

Students often get confused when a proof starts out by saying "assume such and such" because they feel that an unjustified step is being taken. But assuming the hypothesis of a conditional sentence you're trying to prove, as in step 1 of Example 1, is completely correct. We are not saying "P is true"; we are asking "what if P were true?"

This key idea is worth summarizing:

Proving a conditional sentence is the same as assuming its hypothesis and then deriving its conclusion. Therefore you can begin the proof of a conditional sentence by assuming that its hypothesis is true. The aim of the rest of the proof is to show, using this assumption, that the conclusion is true.

Variation in the Form of Conditional Sentences

A given conditional sentence can be rephrased in different ways.

Example 3 **Alternative Phrasings** Here are some alternative ways in which the conditional sentence "if x and y are even integers, then $x + y$ is even" might be stated:

$x + y$ is even whenever x and y are even integers.
Evenness of $x + y$ follows from evenness of x and y.
x and y are both even integers only if $x + y$ is even.
x and y being even integers implies that $x + y$ will be even.

We will look at these and other alternatives again in Section 4.2. □

Sometimes the "conditional" nature of a statement is hidden by the fact that its hypothesis is not explicit. For example, recall statements 3 and 4 from Example 2:

3. If a and b are real numbers, then $|a + b| \leq |a| + |b|$.
4. If a function is differentiable, then it is continuous.

The ideas expressed in these conditional sentences could have been stated as follows:

3′. $|a + b| \leq |a| + |b|$.
4′. A differentiable function is continuous.

In the first case, the hypothesis has been completely omitted, and just been taken for granted. In the second case, the hypothesis and the conclusion have been blended together into a simple declarative sentence. Identifying the hypothesis and conclusion of a conditional sentence can sometimes be an important step in writing a proof.

Get Your Hands Dirty 2

Write each of these statements in "if . . . , then . . ." form, introducing any necessary variables, and identify the hypothesis and conclusion:

a) The sum of two real numbers is always real.
b) A nonempty set has at least two subsets.
c) Every nonvertical line has a y-intercept. □

Variables

In Theorem 1.1, the hypothesis says "x and y are even integers". Which ones? We don't know. The "x" and "y" here represent "generic" objects, and all we know about them is that they are even integers. These two objects named x and y have somehow entered the picture, and the fact that they are even integers is the total information we have available about them. Nothing more.

Here's an example of how *not* to prove Theorem 1:

> Assume that x and y are even integers, say 8 and 34. Then the sum $x + y$ is $8 + 34$, or 42. Since 42 is even, the theorem is true.

The point is that x and y might be 8 and 34, but they might not. If we did assume that they were 8 and 34, we would end up proving something about 8 and 34, which is not our goal. We want to prove something about *all* even integers. Therefore we cannot use any information about x and y that isn't a consequence of their simply being even integers.

It is precisely this principle that makes this conditional sentence a universal statement. By not assuming anything else about x and y beyond their evenness, we gain the right to say that whatever we are able to deduce applies to all even integers.

The generic elements used in sentences like these are called *variables*. Typically, the conclusion of a conditional sentence will refer to the same variables introduced in the hypothesis, as is illustrated by the statements in Example 2. It is these shared objects that create the connection between the two parts of the sentence.

You may find it helpful to use the following image for thinking about the role of such variables:

Metaphor: Introducing Variables

Some objects have floated in the window. All you know about these mysterious objects is that they make the hypothesis true. You can use that fact in any way you like, and you can give the objects names such as x and y, but you cannot assume any other information about the objects. Your task is to show that these objects make the conclusion true.

This image of the objects represented by the variables as coming from "outside", or as being given to us by "someone else", is used commonly in the writing of proofs, in such phrases as "suppose we are given two numbers x and y such that . . . ." This type of language will be used freely in examples of proofs given in this book, and you should learn to use it in writing your own proofs as well.

Open Sentences

In order to better understand how a conditional sentence works, it's helpful to look at the hypothesis and conclusion as entities in themselves. We saw just above that both the hypothesis and the conclusion of a conditional sentence may involve generic objects called variables. Until those variables are replaced by specific objects, these two parts of the conditional sentence are individually neither true nor false. Incomplete statements like these are sometimes called *open sentences*.

Example 4 **Some Open Sentences** Here are some examples of open sentences concerning sets and numbers:

1. $S \subseteq \mathbf{Z}$ (S is a variable).
2. $a > b$ (a and b are variables).
3. $A \cap B = \varnothing$ (A and B are variables).
4. $y^2 - 5y + 6 = 0$ (y is a variable).
5. $z \in C \cup D$ (z, C, and D are variables).

[The set notations "$\subseteq$" (subset), "$\cap$" (intersection), "$\varnothing$" (empty set), "$\in$" (element of), and "$\cup$" (union) are introduced and defined in Appendix B, along with other standard notation and terminology. Following the appendices, there is a complete list of the notation used in this book.] □

Substitution and Truth Sets

Earlier in this section, we referred to a conditional, or "if . . . , then . . .", sentence as "a relationship between two ideas or possibilities". The two ideas or possibilities are expressed by its hypothesis and its conclusion, each of which is generally an open sentence, usually involving the same variables.

One useful way to express the relationship between hypothesis and conclusion is to look at each open sentence as describing a set of objects—the objects that, when used in place of the variables, produce true statements.

The process of replacing a variable in an open sentence by an appropriate specific object is called *substitution* or *replacement*. For example, consider the open sentence "x is an even integer". If we replace x by 4, we get the statement "4 is an even integer", which is true. On the other hand, if we replace x by 11, we get the statement "11 is an even integer", which is false.

If a variable is used more than once in a given open sentence, then each occurrence of that variable must be replaced by the same value. For example, in the open sentence "$2x$

is a divisor of $5x - 4$", if we replace x by 4, we get the statement "8 is a divisor of 16" (which is true), while if we replace x by 7, we get "14 is a divisor of 31" (which is false). The open sentence "$2x$ is a divisor of $5x - 4$" is considered to have only one variable, since only one number can be used to make a replacement.

The characterization of a statement as either true or false is called its *truth value*. Thus, substituting specific values for the variables of an open sentence results in a statement with a particular truth value. The collection of objects that make a particular open sentence true is called its *truth set*. This idea is basically the same as that used in set-builder notation (see Appendix B). In set-builder notation, the set being defined is often simply the truth set of the open sentence that appears at the end of the notation. For example, $\{x : x \text{ is an even integer}\}$ is another way of designating the truth set for the open sentence "x is an even integer", namely, the set of even integers. We will look at this set and related sets and examine set-builder notation more carefully in Chapter 2.

It is important to have a *universe*, or *universe of discourse*, from which the replacement values are to be selected. For example, in the open sentence "$x < 4$", we need to know if we are talking about real numbers, integers, rational numbers, or some other category, in order to determine the truth set. We often specify this universe at the beginning of set-builder notation: for example, $\{x \in \mathbf{R} : x < 4\}$ is the interval $(-\infty, 4)$, while $\{x \in \mathbf{N} : x < 4\}$ is the set $\{1, 2, 3\}$. ($\mathbf{R}$ is the set of real numbers; $\mathbf{N}$ is the set of natural numbers—i.e., positive integers; see Appendix B.)

When we want to talk about a generic universe, we will use the symbol $\mathbf{U}$.

Example 5 **Truth Sets** Give the truth set for each of these open sentences, using the universe indicated, and listing the elements explicitly.

a) p is a prime less than 20 (universe $= \mathbf{N}$).
b) $x^2 - 4x - 5 = 0$ (universe $= \mathbf{R}$).
c) y is the square of an integer (universe $= \{1, 2, \ldots, 50\}$).
d) $u^2 = -9$ (universe $= \mathbf{R}$).

Write down your answers before reading our solutions.

SOLUTIONS

a) $\{2, 3, 5, 7, 11, 13, 17, 19\}$.
b) $\{-1, 5\}$.
c) $\{1, 4, 9, 16, 25, 36, 49\}$.
d) $\varnothing$. □

Get Your Hands Dirty 3

List the elements in the truth set for each of these open sentences, using the universes indicated:

a) $0 < x < 10$ (universe $= \mathbf{Z}$).
b) $(t - 2)(t^2 - 9) = 0$ (universe $= \mathbf{R}$).

c) B is a subset of $\{2, 4, 6, 8, 10\}$ with exactly two distinct elements (universe = the set of subsets of $\{2, 4, 6, 8, 10\}$). □

Notation such as "$p(x)$" (read "p of x") is often used to represent some unspecified, "generic" open sentence involving the variable "x", or as a shorthand for some specific open sentence. So $p(x)$ might be the statement "x is an even integer", or the statement "x is a subset of $\mathbf{Z}$", or the statement "there is a real number whose square is x".

Open sentences can be combined using words like "and", "or", and "not" to create more complex open sentences. Generally, this involves open sentences with the same variable(s). For example, if "$p(x)$" is the open sentence "x is odd" and "$q(x)$" is the open sentence "$x < 10$" (both with the universe $\mathbf{N}$), then "$p(x)$ and $q(x)$" is the open sentence "x is odd and $x < 10$", whose truth set is $\{1, 3, 5, 7, 9\}$. We will examine this process more fully in Chapter 3.

Example 6 **Representing Open Sentences Symbolically** We can represent the open sentences from Example 4 as follows:

$q(S)$: the open sentence "$S \subseteq \mathbf{Z}$".
$r(a, b)$: the open sentence "$a > b$".
$s(A, B)$: the open sentence "$A \cap B = \varnothing$".
$t(y)$: the open sentence "$y^2 - 5y + 6 = 0$".
$u(z, C, D)$: the open sentence "$z \in C \cup D$". □

The notation $p(x)$ is intended to resemble the standard notation $f(x)$ for working with functions. Just as we might "plug in" a number into a function, and get out a number, so here we "plug in" an object into an open sentence, and get out a statement with a specific truth value.

Example 7 **Substituting into Symbolic Open Sentences** Here are some examples of what happens to the symbolic open sentences of Example 6 when their variables are replaced by specific objects:

If we let $S = \{1, 2, 3\}$, then $q(S)$ becomes $q(\{1, 2, 3\})$, which is the true statement "$\{1, 2, 3\} \subseteq \mathbf{Z}$".
If we let $a = 5$ and $b = 8$, then $r(a, b)$ becomes $r(5, 8)$, which is the false statement "$5 > 8$".
If we let $A = \mathbf{N}$ and $B = \mathbf{Q}$, then $s(A, B)$ becomes $s(\mathbf{N}, \mathbf{Q})$, which is the false statement "$\mathbf{N} \cap \mathbf{Q} = \varnothing$". ($\mathbf{Q}$ is the set of rational numbers.)
If we let $y = 2$, then $t(y)$ becomes $t(2)$, which is the true statement "$2^2 - 5 \times 2 + 6 = 0$".
If we let $z = 3$, $C = \{1, 2\}$, and $D = \{2, 3, 4\}$, then $u(z, C, D)$ becomes $u(3, \{1, 2\}, \{2, 3, 4\})$, which is the true statement "$3 \in \{1, 2\} \cup \{2, 3, 4\}$".

Thus, for example, the truth value of $q(\{1, 2, 3\})$ is "true" and the truth value of $r(5, 8)$ is "false". □

Conditional Sentences and Truth Sets

With the concept of truth set in mind, we return to our description of a conditional sentence as a relationship between hypothesis and conclusion.

Consider the following:

QUESTION 1 If a conditional sentence as a whole is true, what does that tell us about the relationship between the truth set for the hypothesis and the truth set for the conclusion?

In thinking about this question, try the following conditional sentence as an example: "If x is a prime number greater than 2, then x is an odd integer". What is the truth set of its hypothesis? of its conclusion? How are these sets related?

Then ask yourself how the example can be generalized.

? ? ?

The answer to this question can be expressed very simply, and offers us a helpful way of thinking about conditional sentences (if you didn't really work on the question, go back and do so):

> *For a conditional sentence to be true means that the truth set for its hypothesis is a* ***subset*** *of the truth set for its conclusion; in other words, that any object which makes the hypothesis true will also make the conclusion true.*

That, intuitively, is precisely what an "if . . . , then . . ." statement ought to mean: that *if* x represents an object that makes the hypothesis true, *then* x must make the conclusion true as well. Otherwise, the conditional sentence is false.

Notice that we have actually replaced one "if . . . , then . . ." statement by another, since the subset relationship is itself formally defined in terms of a conditional statement: "$A \subseteq B$" means "if $x \in A$, then $x \in B$" (see Definition 2.2 in Section 2.1). Thus, the subset relationship can be thought of as a "generic conditional sentence", that is, an "if . . . , then . . ." statement can be rephrased as an assertion that one set is a subset of some other. For example, statement 4 of Example 2 ("if a function is differentiable, then it is continuous") can be viewed as saying that the set of differentiable functions is a subset of the set of continuous functions.

Counterexamples

The conditional sentences of our earlier examples have all been true. Another important perspective to consider about conditional sentences is this: under what circumstance would you consider a conditional sentence false? Try to answer that question, using the following statement, which is clearly false, as a guide: "if x is a positive integer, then x is even".

Precisely what is it that makes the preceding sentence false?

We know that the given statement is false because there is a *counterexample*. By this we mean that there is a number for x (actually there are lots of them), which is a positive integer but is not even—for example, $x = 7$. We found a value to replace the variable, which made the hypothesis true and the conclusion false.

This is an idea we will use over and over again. So we repeat it for emphasis:

A conditional sentence is considered false if values can be found for the variables that make the hypothesis true and the conclusion false.

Of course, when a variable appears in both the hypothesis and the conclusion, we must use the same value in both places in order to have a counterexample. The fact that there are numbers which are positive and that there are numbers which are not even is not enough to make the statement false. But the fact that there is at least one number that is both positive and not even does make it false.

Notice that, with true conditional statements (as well as with false ones), we can often find objects to substitute that will make the hypothesis or the conclusion individually false. For example, consider the following statement (from Example 1):

If x and y are even integers, then $x + y$ is an even integer.

The substitution $x = 5$, $y = 7$ makes the hypothesis false and the conclusion true; the substitution $x = 8$, $y = 3$ makes the hypothesis false and the conclusion false; and the substitution $x = 10$; $y = 42$ makes the hypothesis true and the conclusion true. The only combination we cannot create for this sentence is "hypothesis true, conclusion false". That's because the conditional sentence as a whole is true.

This special way in which the truth or falsity of a statement depends on the truth or falsity of its hypothesis and conclusion is very important. We will present this idea in more formal terms in Chapter 3, but for now, we will repeat the basic principle one more time, slightly rephrased:

A conditional sentence is considered true if replacing its variables by specific objects can only give a combination from the following list:

- *hypothesis true, conclusion true;*
- *hypothesis false, conclusion true;*
- *hypothesis false, conclusion false.*

The conditional sentence is considered false only if there is a way to replace its variables by specific objects that make the hypothesis true and the conclusion false.

Notice that particular examples in which the hypothesis is false or in which the conclusion is true cannot assure us of the truth of a conditional sentence. However, a single example in which the hypothesis is true and the conclusion is false makes the conditional sentence false, and it is the absence—nonexistence—of any such example that will make the sentence true.

Thus the statement "if x is an even integer, then x^2 is an even integer" is considered true because there is no counterexample, i.e., because there is no even integer whose square is not even. Similarly, statement 1 of Example 2 is true because there is no pair of odd integers whose product is not odd. But the conditional sentence "if x is a positive integer, then x is even" is false, because there is at least one positive integer that is not even (for example, 7).

Working with Definitions

We noted in our discussion of Example 1 that *definitions* play an important role. It is impossible for us to prove anything in mathematics unless we know what the words mean.

When we define a term, we are describing the precise circumstances in which it applies. For example, we might say:

DEFINITION: An *even integer* is an integer that has 2 as a factor.

Because we have labeled this as a definition, it has a "two-directional" meaning: it not only means that all even integers have 2 as a factor; it also means that every integer with 2 as a factor is even.

In other words, definitions give perfect equivalents. The expressions "an even integer" and "an integer that has 2 as a factor" are intended to mean exactly the same thing.

Compare the above example with the following statement:

An integer that is divisible by 4 is an even integer.

This is not a definition. While it is true that every integer divisible by 4 is even, it is not the case that every even integer is divisible by 4. The two expressions "an integer that is divisible by 4" and "an even integer" are not intended to be synonymous.

In Section 4.2, we will discuss the use of the phrase "if and only if" to describe the "synonymous" nature of definitions.

Proof for Certainty and Proof for Understanding

We conclude this section by looking at two problems involving proof. The first involves the game SubDivvy 1 (Exploration 1 in Section 1.1). Like the theorem in Example 1 of this section, the SubDivvy problem that we are about to do involves a result which is probably intuitively clear. In thinking about how to develop a proof, we will see how the examination of specific cases can be used as a guideline for developing a general argument. The main purpose of the proof is to establish certainty. Though you may gain some additional insights into the situation by working out the proof, the main goal is to guarantee that we haven't overlooked some exception to the general situation.

The approach in the second problem is different. The goal in this problem is two-fold: to develop a "counting formula" for a specific physical situation, and to gain some insight into the reasons why the formula works. We will describe how you might obtain the formula through "experimental evidence", and then discuss the role of proof in helping to understand why the formula is correct.

Proof for Certainty: The Last Move of SubDivvy

Exercise 2 of Section 1.1 asked what the last move in a game of SubDivvy could be. If you have not yet thought about that problem, do so now.

? ? ?

Some experimentation should make you fairly confident that the last move must be from 2 to 1. We will prove this now. In order to do so, we need to first state this result formally. One helpful way to do so is to set it up as an "if . . . , then . . ." statement with a variable, as follows:

THEOREM 1.2. If the last move in a game of SubDivvy is from n to 1, then $n = 2$.

Rather than just writing out the proof, we will describe some of the exploratory thinking that might go into it. First, however, you should make some attempts on your own to develop a proof.

Try it.

(In what follows, the actual proof is marked by a vertical line on the left margin. The text that has no margin line is exploratory discussion.)

We might begin by trying a number for n which is bigger than 2, e.g., $n = 5$. Why can't we move from 5 to 1? In SubDivvy, each move involves a subtraction. In this case, in order to end up at 1, we would have to subtract 4 from 5. That suggests the following as a first step of the proof:

Suppose that the final move is from n to 1. Then this move must consist of subtracting $n - 1$ from n.

Now why can't we subtract 4 from 5? The reason is that 4 isn't a divisor of 5, and the rules require that the number being subtracted be a divisor of the number being subtracted from. We can state this idea briefly as follows:

In order to subtract $n - 1$ from n in SubDivvy, $n - 1$ must be a divisor of n. In other words, $n/(n - 1)$ must be an integer.

We now need to figure out why, in general, $n/(n - 1)$ can't be an integer unless $n = 2$. Let's look at some examples:

$$n = 6: \quad \frac{n}{n-1} = \frac{6}{5} = 1\frac{1}{5};$$

$$n = 10: \quad \frac{n}{n-1} = \frac{10}{9} = 1\frac{1}{9};$$

$$n = 3: \quad \frac{n}{n-1} = \frac{3}{2} = 1\frac{1}{2}.$$

We are getting fractions between 1 and 2. In fact, we get $1 + 1/(n - 1)$ each time. For what values of n is this an integer? Only for $n = 2$ (see comment 1 below).

But

$$\frac{n}{n-1} = 1 + \frac{1}{n-1}$$

(by ordinary algebra or long division), and this can be an integer only if $1/(n - 1)$ is an integer. This happens only if $n - 1 = 1$, so we must have $n = 2$. ∎

To summarize, here are the theorem and its proof, without the commentary:

THEOREM 1.2. If the last move in a game of SubDivvy is from n to 1, then $n = 2$.

Proof: Suppose that the final move is from n to 1. Then this move must consist of subtracting $n - 1$ from n.

In order to subtract $n - 1$ from n in SubDivvy, $n - 1$ must be a divisor of n. In other words, $n/(n - 1)$ is an integer.

But

$$\frac{n}{n-1} = 1 + \frac{1}{n-1}$$

(by ordinary algebra or long division), and this can only be an integer if $1/(n - 1)$ is an integer. This only happens if $n - 1 = 1$, so we must have $n = 2$. ∎

COMMENTS

1. We took for granted the fact that the only positive integer values of x for which the fraction $1/x$ is an integer is the value $x = 1$. This can be proved from our "Basic Assumptions about Arithmetic" in Appendix B (see Exercise 6).
2. When we wrote "suppose that the final move is from n to 1", we were assuming that the hypothesis is true. Or, more precisely, we were assuming that n is an integer that makes the hypothesis true, i.e., an element of the truth set of the hypothesis.
3. The proof ended when we showed that n makes the conclusion true, i.e., that n is in the truth set of the conclusion.
4. Working with examples made the arithmetic patterns clear enough that they could be expressed generally—that is, algebraically.
5. You may have found other ways to state the result, other ways to discover the proof of the given result, or other ways to write the actual proof. The fact that your way is different is perfectly O.K.

Proof for Understanding: String Cutting

Imagine a single piece of string, which can be bent back and forth. It might look like Figure 1.4 (here the string is bent so that it has three "layers").

Imagine now that we take scissors and cut across the bent string. The result will be several pieces, and might look like Figure 1.5 (in this case, there are four pieces). We

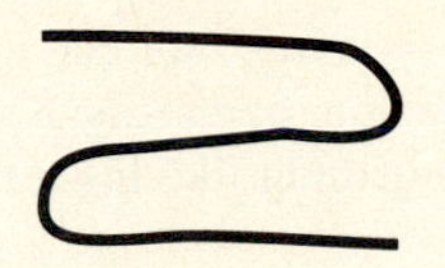

Figure 1.4. A bent piece of string.

could have made more than one cut across the bent string, giving more pieces: see Figure 1.6, where there are two cuts. We can count and see that there are seven pieces as a result of this process. Figure 1.7 shows a case of four layers and three cuts, and has 13 pieces.

As we vary the number of layers and the number of cuts, what will happen to the total number of pieces?

Find a formula.

COMMENT Analyzing this situation may not be easy. For example, you may have noticed that in Figure 1.7, the portion to the left of the first cut has three pieces, the two middle portions four each, and the right end only two. A similar analysis of Figure 1.6 gives the numbers 2, 3, and 2. But it isn't clear at all how this information can be generalized. Not everything you notice is necessarily helpful.

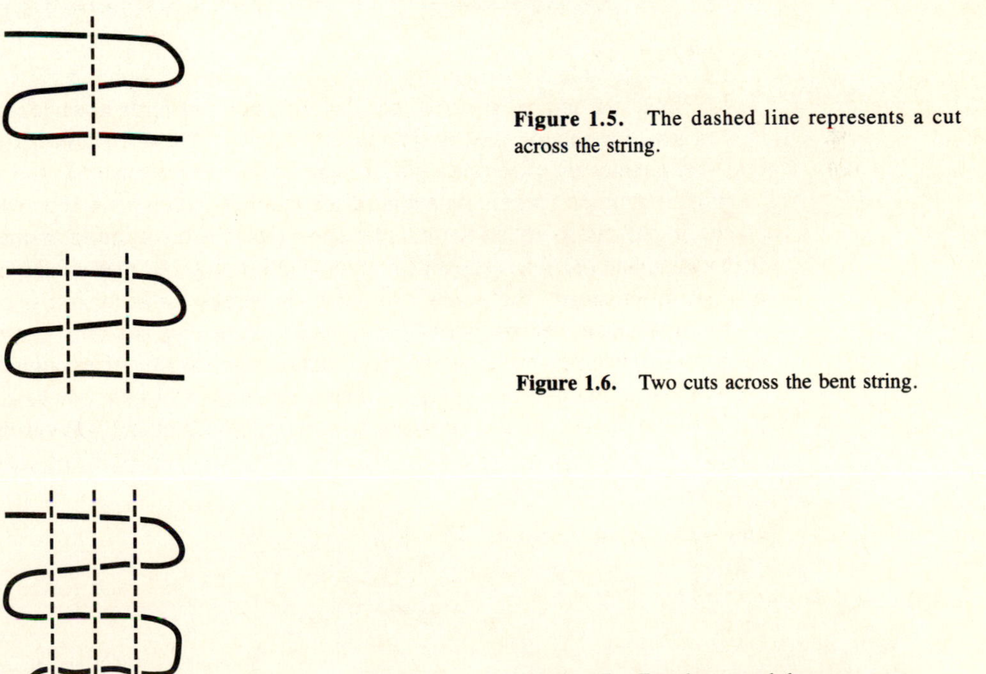

Figure 1.5. The dashed line represents a cut across the string.

Figure 1.6. Two cuts across the bent string.

Figure 1.7. Four layers and three cuts.

One approach to this difficulty is *data gathering* — that is, we can try numerous examples, collect the information in some organized fashion, and analyze the results, looking for numerical patterns. We can create a chart something like Figure 1.8 (the entries are based on Figures 1.4–1.7).

What is the relationship between the last column and the first two?

Find the relationship.

If necessary, a few more entries in the chart should suggest the following formula:

$$\text{No. of pieces} = (\text{No. of layers}) \cdot (\text{No. of cuts}) + 1.$$

This completes one stage of the process of working with this problem. We have developed the desired formula.

But that still leaves us in the dark as to why the formula works — we don't really understand what is going on in this problem. Equally important, we are not really certain that the formula is correct. It fits our evidence, but that may just be a coincidence.

We can use our informed guess at the formula as a clue to establishing certainty and gaining understanding. Knowing what the correct formula apparently is, we can now look for a proof or explanation, and perhaps find the insight that was missing at the previous stage.

Why does the formula work?

? ? ?

If you were able to answer the last question, you probably have what amounts to a proof. Since the terms "layer", "cut", and "piece" haven't been precisely defined, the idea of proof in this context is somewhat informal. All we're looking for is an explanation that enables us to see at a glance why every example will fit the formula.

The "+1" of the formula is a hint: if the number of cuts is 0, we just have the original piece. So we can focus on the number of *new* pieces resulting from the cuts. Picture the string stretched out in a straight line: you can see that, in that situation, each time there is a cut in the string, the number of pieces increases by 1.

In our problem, each cut across the *bent* string may actually cut the string several times. Since each cut goes across every layer, the number of new pieces created by the cut is equal to the number of layers. The total number of new pieces is therefore (no. of layers) · (no. of cuts). Adding this to the original piece of string gives the formula above.

No. of layers	No. of cuts	No. of pieces
3	0	1
3	1	4
3	2	7
4	3	13

Figure 1.8 This table shows the number of pieces of string for certain combinations of layers and cuts.

COMMENTS

1. We might have been satisfied with finding the formula, and not worried about the proof. In that case, we might not have gained the understanding of the physical problem which resulted from the further examination. An important part of the motivation for developing a proof is to help us to come to grips with the issues of insight and understanding.

2. When a result such as this formula is presented, someone may ask "how do you know?" The question has more than one interpretation. It may mean "how did you find out?" In the context of the string-cutting problem, we know the answer because of our experimentation. Careful examination of the information resulting from several examples suggested the particular numerical relationship. The same "how do you know" question is also used to mean: "what makes you certain?" or "why should I rely on your assertion?" From that point of view, it is the explanation which is most important. We know that the formula above is correct because we understand why it works. Both perspectives are important. The first kind of knowledge is often the result of exploration; the second is more closely associated with proof.

EXERCISES

1. (GYHD 1) Prove the following:

   If x is an even integer, and y is a multiple of x, then y is even.

2. (GYHD 3) List explicitly the elements of the truth set for each of these open sentences, using the universes indicated:
   - **a)** $0 < x < 10$ (universe = $\mathbf{Z}$).
   - **b)** $(t - 2)(t^2 - 9) = 0$ (universe = $\mathbf{R}$).
   - **c)** B is a subset of $\{2, 4, 6, 8, 10\}$ with exactly two distinct elements (universe = the set of subsets of $\{2, 4, 6, 8, 10\}$).

3. For each of the following conditional statements, state whether each of the given substitutions represents a counterexample.
   - **a)** If n is a prime, then n is odd.
     - **i)** $n = 2$.
     - **ii)** $n = 6$.
     - **iii)** $n = 9$.
   - **b)** If x is an even integer, then x is a multiple of 4.
     - **i)** $x = 5$.
     - **ii)** $x = 2$.
     - **iii)** $x = 12$.
   - **c)** If $A \subseteq B$, then A and B have an element in common.
     - **i)** $A = \{1, 2, 3\}$, $B = \{5, 6, 7\}$.
     - **ii)** $A = \varnothing$, $B = \varnothing$.
     - **iii)** $A = \{2\}$, $B = \mathbf{Z}$.
   - **d)** If $t^2 = 9$, then $t > 2$.
     - **i)** $t = -3$.

ii) $t = 3$.
iii) $t = 5$.
e) If $a > b$ and $c > d$, then $ac > bd$.
i) $a = 3, b = 1, c = 8, d = 5$.
ii) $a = 2, b = 4, c = 10, d = -2$.
iii) $a = -2, b = -4, c = 5, d = -3$.
f) If u is an odd integer, then u is even.
i) $u = 0$.
ii) $u = 3$.

4. For each of the following statements, write an "if . . . , then . . ." statement, introducing appropriate variables, which makes the same assertion (all are intended to be interpreted so that they are true statements):
a) The square of a real number is not negative.
b) Any point that is in the first quadrant has positive coordinates.
c) A proper subset of a set with three elements can't have more than two distinct elements.
d) All pairs of lines in the plane that aren't parallel are identical or meet in just one point.

5. For each of the following statements, first decide if it is true or false. Then, if it is true, give an informal explanation. If it is false, give a counterexample.
a) Any divisor of an odd integer is odd.
b) Any divisor of an even integer is even.
c) The quotient of two even integers is an even integer.
d) If a, b, and c are real numbers with $a < b$ and $a < c$, then $a < b + c$.
e) If a and b are real numbers with $a > b$, then $a > b - 1$.
f) If a, b, and c are real numbers with $a < bc$, then either $a < b$ or $a < c$.

6. Prove the following, using the "Basic Assumptions about Arithmetic" from Appendix B:
If n is a positive integer such that $1/n$ is an integer, then $n = 1$.

7. In our discussion of the string-cutting problem, we commented on the number of pieces in each portion of Figures 1.6 and 1.7, and noted that this information might be hard to use. Develop the pattern described there, and discuss how that pattern could be used to construct the formula that solves this problem.

8. Read Exercise 5 of Section 1.2. In that exercise, you are asked to decide which woman has which nationality. If you haven't already done so, do that exercise. Then prove your answer. In other words, show that your solution is consistent with the information given, and also show that no other solution is possible. (Though the problem is phrased to suggest that there is a unique solution, there is no guarantee of this. Much of mathematics is concerned with proving "uniqueness" of solutions to various problems.)

9. Read Exercise 9 of Section 1.1. In that exercise, you are asked to develop a winning strategy for the game Linear Nim. If you haven't already done so, examine that exercise for the case $N = 15$. Then:
a) describe a winning strategy, and
b) prove your answer.

In other words, if you think player 1 should win, state what the first move should be, how player 1 should respond to each possible move by player 2 at each stage, and prove that this strategy will result in victory for player 1. On the other hand, if you think player 2 should win, describe how player 2 should respond to each possible move by player 1 at each stage, and prove that this strategy will result in victory for player 2.

2

Exploration and Proofs about Sets and the Integers

Introduction

In this chapter, we will look at two central topics in mathematics: *sets* and *the integers*. Both topics are building blocks and tools for more advanced mathematical concepts.

The four sections of this chapter come in pairs: exploration and proofs about sets, and then exploration and proofs about the integers. In the "explore" sections, we will introduce key concepts and give you a chance to investigate some questions on your own. The "prove" sections use the new concepts to illustrate how to work with the formalism of mathematical proofs.

Section 2.1 introduces the ideas of *power set* and *Cartesian product,* discusses problems related to the *empty set,* and presents a useful investigatory tool—*Venn diagrams*. This material provides the content for the principles discussed in Section 2.2, including how to apply a known theorem, how to use symmetry within the hypothesis of a theorem to avoid repetition, and how to break up a proof into cases.

Section 2.3 presents formal definitions of the concepts of *multiple* and *divisor*. These familiar notions are used in several explorations, including a look at *linear Diophantine equations*. They are then used in Section 2.4 to illustrate two important proof techniques: the use of *existence* statements and the method of proof by *contradiction*. We also introduce the concept of *common divisor,* and use that to begin resolving one of the explorations of the previous section.

2.1 Exploring Sets

We begin this section by introducing two important methods, besides those of Appendix B, for "creating" new sets out of old ones.

Reminder: You may wish to review Appendix B before beginning this section.

Power Set

In our examples of Chapter 1, we had occasion to ask about various subsets of the integers or subsets of other sets. It is often helpful to be able to talk about *all* the subsets of a given set. We have the following definition:

DEFINITION 2.1: For a set A, the *power set* of A, written "$\mathcal{P}(A)$", is the set consisting of all the subsets of A.

For example, if $A = \{1, 4, 6\}$, then the set $\{1, 4\}$ is an *element* of $\mathcal{P}(A)$. Since any set is a subset of itself, we also have $\{1, 4, 6\} \in \mathcal{P}(\{1, 4, 6\})$.

If we want to know all of the elements of $\mathcal{P}(\{1, 4, 6\})$, we are faced with a subtle question:

QUESTION 1 Is the empty set a subset of $\{1, 4, 6\}$?

What exactly does "subset" mean?

Intuitively, one set A is considered a subset of another set B if A is "contained in" or "included in" B. That description may leave some room for ambiguity, so we introduce a formal definition to try to resolve this question:

DEFINITION 2.2: For sets A and B, "A is a *subset* of B" means: if $x \in A$, then $x \in B$.

But we still have the potential for confusion: if A is the empty set, then the hypothesis, "$x \in A$," can't be true. What do we do with this situation?

What exactly does "if..., then..." mean?

As we discussed in Section 1.3, a conditional sentence is only considered false if there is a counterexample, that is, a choice of variables that makes its hypothesis true and its conclusion false. But there is no way to make this happen in our example. In a sentence of the form "if x is in $\varnothing$, then x is in B", there is no element which will make the hypothesis true. Thus, since there can be no counterexample, the conditional statement is considered true. So our definition of the word "subset" combines with our definition of "if..., then..." to tell us that the answer to Question 1 must be "yes". Indeed, the empty set is a subset of every set. We state this formally:

THEOREM 2.1. For any set A, $\varnothing \subseteq A$.

COMMENT If this doesn't fit your intuitive picture for the word "subset", then you must recognize that mathematicians are using that word differently from the way you would. It doesn't mean that your picture is "wrong"—it just means that your picture is not the one that mathematicians use. In order to communicate with the mathematical world, you will have to use the word "subset" their way.

Because this false-hypothesis situation can be somewhat troublesome, we repeat the key idea here for emphasis.

> According to our definition of "truth" for conditional sentences, *any conditional sentence whose hypothesis cannot be true is a true statement.*

Such a conditional sentence is sometimes called *vacuously true*—the conditional sentence is true, but the truth set of its hypothesis is the empty set. Since there will never be a situation to which this conditional sentence can be applied, it may seem like a rather useless result. But the fact that we consider it true saves us from having to make exceptions to all the theorems that begin "if $x \in A$" where A might be the empty set.

The following example gives other instances of vacuously true statements.

Example 1 **Vacuously True Statements** The following conditional sentences are vacuously true, because their hypotheses have empty truth sets:

1. If x is a real number with $x^2 < 0$, then $x = 17$.
2. If n is a prime number divisible by 10, then n is also divisible by 3.
3. If ABC is a triangle in the plane containing two right angles, then ABC is equilateral.
4. All the teenagers who were elected President of the US prior to 1980 were at least six feet tall. (*Note:* This is not in the form of a conditional sentence, but it can be interpreted as one whose hypothesis has an empty truth set.) □

Returning to the subject of power sets, we can list all of the subsets of our sample set, $A = \{1, 4, 6\}$:

$$\mathcal{P}(A) = \{\emptyset, \{1\}, \{4\}, \{6\}, \{1,4\}, \{1,6\}, \{4,6\}, \{1,4,6\}\}.$$

Thus, the cardinality of $\mathcal{P}(A)$ is 8; that is, it contains exactly eight elements. Our notation for this is "$|\mathcal{P}(A)| = 8$" (see Appendix B).

Get Your Hands Dirty 1

List the elements of $\mathcal{P}(X)$ and find the cardinality $|\mathcal{P}(X)|$ for each of the following sets:

a) $X = \{1\}$.
b) $X = \{1, 2\}$.
c) $X = \{1, 2, 3, 4\}$. □

We have seen that the empty set is considered a subset of every set. What, then, is the power set of the empty set, i.e., what is $\mathcal{P}(\emptyset)$?

? ? ?

We want to know what sets are subsets of $\emptyset$. The empty set itself is one possibility: "$\emptyset \subseteq \emptyset$" is a true statement. And nothing else will work: if A is a nonempty set, then

"$A \subseteq \varnothing$" will be false, because there will be some object which is an element of A but is not an element of $\varnothing$.

Thus the only subset of $\varnothing$ is $\varnothing$ itself. In other words, $\mathcal{P}(\varnothing) = \{\varnothing\}$.

Note carefully $\mathcal{P}(\varnothing)$ is $\{\varnothing\}$, not $\varnothing$. It is important to distinguish between an object and the set containing that object. Just as the number 3 is different from the set $\{3\}$, so also $\varnothing$ is different from $\{\varnothing\}$. The cardinality of $\mathcal{P}(\varnothing)$ is 1, not 0.

Metaphor: Sets as Containers It may be helpful to visualize a set as a "container" of some kind. The empty set—$\varnothing$—is a box with nothing in it. But $\{\varnothing\}$ is a box which has something in it—namely, an empty box. Boxes can be "nested" to various degrees: we can have a box within a box within a box, etc.

Note that a single box might contain some numbers, and also contain a box of numbers, and even contain a box with several other boxes in it. In that way we get sets that have "mixed type," as illustrated in the following example:

Example 2 **A Set of "Mixed Type"** Consider the following set:

$$Y = \{1, 2, \{3, 4\}, \{5, 6, 7\}, \{\{2, 3, 4\}, 6\}\}.$$

The cardinality of Y is 5: the number 1 is one of its members; the set $\{3, 4\}$ is another of its members; and the set $Z = \{\{2, 3, 4\}, 6\}$ is yet another member of Y. It is true that $1 \in Y$, that $\{3, 4\} \in Y$, and that $\{1, 2\} \subseteq Y$, but it is not true that $\{3, 4\} \subseteq Y$ or that $\{1, 2\} \in Y$. □

This sort of set is somewhat unusual in ordinary mathematics. In most situations, within a given set, the objects are of the same "type". In exploring a problem or planning a proof, it's often very helpful to think about the types of the objects involved: are they numbers, sets, functions, ordered pairs, etc.? Such care in thinking about and writing a proof can help you avoid going off in a wrong direction.

Here is another interesting example:

$$S = \{\{1, 2\}, \{1, 3\}\}.$$
$$T = \{\{2, 3\}, \{1\}\}.$$

Although each *element* of S has a member in common with each *element* of T, nevertheless the *sets* S and T are disjoint from each other—there is no object which is a member of both S and T.

Get Your Hands Dirty 2

Is it possible for something to be both an element and a subset of the same set? In other words, can you create an example of two "things"—V and W—such that "$V \in W$" and "$V \subseteq W$" are *both* true statements? Think about the "type" of thing that V and W might each be. □

Cartesian Product

In ordinary algebra, we discuss the graphs of equations like "$3x + 7y = 4$". The graph is a set of "points", and each point is represented by an "ordered pair"—a pair of numbers such as $(-1, 1)$. We saw a similar idea in Section 1.3, where we had open sentences involving more than one variable. For example:

$r(a, b)$: the open sentence "$a > b$".
$s(A, B)$: the open sentence "$A \cap B = \varnothing$".
$u(z, C, D)$: the open sentence "$z \in C \cup D$".

QUESTION 2 How do we express the truth sets for open sentences like these? What do we use for the universe?

? ? ?

We need a set whose elements are pairs, triples, etc., with their components, or "coordinates", drawn from specific sets. This is one of the motivations behind the following definition:

> **DEFINITION 2.3:** For sets A and B, the *Cartesian product of A and B* (written "$A \times B$") is the set consisting of all ordered pairs (x, y) in which x is an element of A and y is an element of B. In brief:
>
> $$A \times B = \{(x, y) : x \in A \text{ and } y \in B\}.$$
>
> "A" and "B" are called *factors* of $A \times B$, and "x" and "y" are called *coordinates* of the pair (x, y).

For example, if $A = \{1, 2, 4\}$ and $B = \{2, 5\}$, then we have

$$A \times B = \{(1, 2), (1, 5), (2, 2), (2, 5), (4, 2), (4, 5)\}.$$

The Cartesian product of sets is also known as the *cross product* or simply the *product*.

COMMENTS

1. The order of the factors affects the product. In our example, we have $(1, 2) \in A \times B$, but $(1, 2) \notin B \times A$.
2. There is no reason why the coordinates of an ordered pair can't be the same. In our example, since $2 \in A \cap B$, we have the element $(2, 2)$ in $A \times B$.
3. Definition 2.3 can be generalized to apply to more than two sets; e.g., $A \times B \times C = \{(x, y, z) : x \in A,\ y \in B, \text{ and } z \in C\}$.

Get Your Hands Dirty 3

List the elements and find the cardinality of each of the following Cartesian products:

a) $\{1, 2, 3\} \times \{2, 4, 6\}$.
b) $\{1, 2\} \times \{1, 2, 3\} \times \{2, 3\}$. □

Historical note: The term "Cartesian product" comes from the name of the French philosopher and mathematician, René Descartes (1596–1650). Descartes is often considered the inventor of analytic geometry, in which the standard coordinate system for graphing equations is used to combine algebraic methods with geometrical insight. He has also been called "the father of modern philosophy".

Even if the elements of the sets are not real numbers, we can use the idea of a "coordinate system" to picture the elements of the Cartesian product as "points". For example, if $S = \{a, b, c\}$ and $T = \{d, e\}$, we can picture "axes" for each set, as shown in Figure 2.1, giving six "points" in $S \times T$.

Returning to Question 2, we can express the truth sets for open sentences with more than one variable as subsets of an appropriate Cartesian product. For example, if x and y are each supposed to be real numbers, then the universe for the open sentence "$x + y > 0$" will be $\mathbf{R} \times \mathbf{R}$, and its truth set is a set of ordered pairs which we can describe using set-builder notation as $\{(x, y) \in \mathbf{R} \times \mathbf{R} : x + y > 0\}$. This set can be represented as a graph, as in Figure 2.2.

For the open sentence $u(z, C, D)$—"$z \in C \cup D$"—we might use something like $\mathbf{N} \times \mathcal{P}(\mathbf{N}) \times \mathcal{P}(\mathbf{N})$ as the universe. One element of the truth set would be the triple $(3, \{1, 2\}, \{2, 3, 4\})$.

Get Your Hands Dirty 4

Give five elements of the truth set for each of the following open sentences:

a) $3x - 7y = 12$ (universe $= \mathbf{R} \times \mathbf{R}$: make "$x$" the first coordinate).
b) $r + 2s + 3t = 17$ (universe $= \mathbf{N} \times \mathbf{N} \times \mathbf{N}$: make "$r$" the first coordinate and "s" the second coordinate).
c) $a \neq b$ (universe $= \{1, 2, 3, 4\} \times \{2, 3, 4, 5\}$: make "$a$" the first coordinate). □

The case of the empty set makes for an interesting question for the Cartesian product, just as it did for power sets:

Get Your Hands Dirty 5

What is the result if one of the factors in a Cartesian product is the empty set? In other words, if A is any set (possibly empty), what are $A \times \varnothing$ and $\varnothing \times A$? □

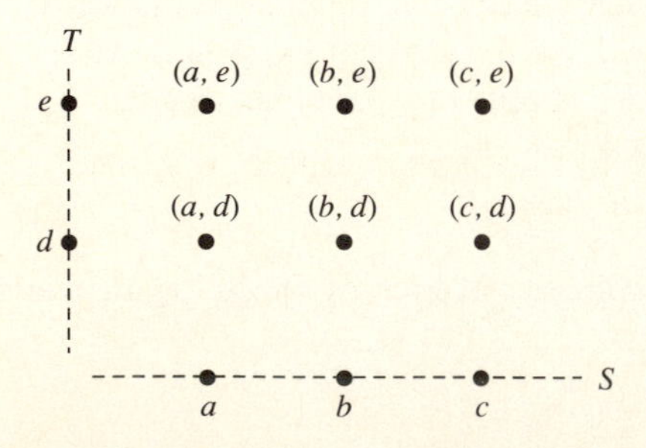

Figure 2.1. The "S-axis" has three elements, and the "T-axis" has two elements, so $S \times T$ has six "points".

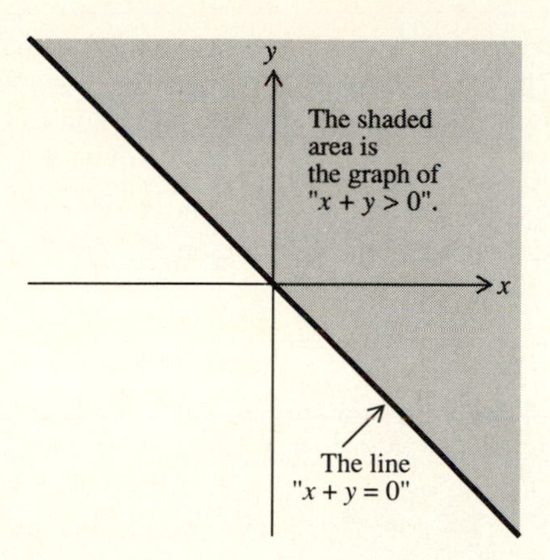

Figure 2.2. The graph of the open sentence "$x + y > 0$".

Set-Builder Notation

We have used set-builder notation, and the related concept of the truth set of an open sentence, as a way of specifying sets that interest us. The following metaphor may be a useful way to think about this method:

Metaphor: "Getting into the Club"

We can think of a set as a club, which has certain membership requirements. In order to gain entry into the club, you need to demonstrate that you satisfy those requirements. On the other hand, if you know that some object has been admitted to the club, you can be sure that the object does fulfill the requirements.

For example, consider $U = \{x : x \in \mathbf{N} \text{ and } x < 10\}$). Picture the number 7 walking up to the door of the U club. The guard at the door says to 7: "Are you a natural number? Are you less than 10?" Since 7 can truthfully answer "yes" to both questions, the guard lets 7 into the club.

If you are trying to *prove* that some object is in a set defined by means of set-builder notation, then you must "convince the guard" that your object meets the requirements, whatever they happen to be.

On the other hand, if you *know* that some object is in a particular set, then you can picture that object as having a "certificate of membership" for that "club". This certificate states the requirements that the object had to satisfy in order to get into the club. You can use that membership certificate as part of a proof if the information on the certificate is relevant to the given problem.

The use of the set-builder method, however, is more complex than our examples so far have indicated. Consider the set $S = \{2x : x \in \mathbf{Z}\}$. What are its elements?

? ? ?

S is not the truth set for the open sentence "$x \in \mathbf{Z}$", but it is based on that open sentence. We begin by selecting an object for x which is in that truth set, but this object x is not necessarily itself an element of S. Instead we *use* x to create an object, namely $2x$, that is in S.

Thus, for example, 7 is in the truth set for "$x \in \mathbf{Z}$", and so 2×7, or 14, is an element of S. However, 7 itself is not an element of S, because there is no object x in the truth set of "$x \in \mathbf{Z}$" for which $7 = 2x$.

S is the set of all even integers. We will look more at this set and related ones in Section 2.3.

The object in the set S was not the value substituted for the variable, but something based on that value. The following nonmathematical analogy may help with this "two-stage" use of the set-builder idea:

Analogy: "Children of Older Students" Club

Suppose a university decides to establish a club for the children of "older students"—say students over 40. There is a condition "x is over 40" involved in the membership process, but it is not the actual member who must be over 40. In sending out invitations to join, the university first looks at $\{x: x \text{ is a student and } x \text{ is over } 40\}$, and then invites the children of the people in that set. (Of course, some of those children might be over 40, and could even themselves be students at the university, but that wouldn't be why they were invited to join.)

The following example gives some more illustrations of this process:

Example 3 **Set-Builder Notation** List the elements (or give sample elements) for each of the following sets:

a) $A = \{x^2 + y^2 : x, y \in \mathbf{Z}\}$.
b) $B = \{Y \cup \{1\} : Y \in \mathcal{P}(\{2, 5, 9\})\}$.
c) $C = \{(x^2, y + 2) : (x, y) \in \{-1, 1, 2\} \times \{1, 5, 6\}\}$.
d) $D = \{|T| : T \subseteq \{2, 5, 9\}\}$.
e) $E = \{|x| : x \in \mathbf{R} \text{ and } x^2 + 5x - 6 = 0\}$.

Note: In d), "$|T|$" means the cardinality of T; in e), "$|x|$" means the absolute value of x. Although the notations are similar, the context makes clear which meaning is intended. In part d, the object between the vertical lines is a set, so we mean cardinality. In part e, it is a real number, so we mean absolute value.

SOLUTIONS

a) Some of the elements of A are:

13 (using $x = 2$, $y = 3$);
17 (using $x = 1$, $y = 4$);
0 (using $x = 0$, $y = 0$).

(The question of exactly which numbers are in A is an interesting problem in number theory.)

b) We are "throwing in 1" to each of the sets in the power set $\mathcal{P}(\{2,5,9\})$. Since $\mathcal{P}(\{2,5,9\}$ has eight elements, none of which includes 1, we get eight new sets, as follows:

$$\{1\}, \{1,2\}, \{1,5\}, \{1,9\}, \{1,2,5\}, \{1,2,9\}, \{1,5,9\}, \text{ and } \{1,2,5,9\}.$$

Each of these sets is an element of B.

c) Here we are "modifying" the elements of $\{-1,1,2\} \times \{1,5,6\}$, a Cartesian product with nine elements. For example, if we start with $(x,y) = (-1,5)$, we get that the element $(x^2, y+2) = (1,7)$ belongs to C. But we get the same element for C if we start with $(x,y) = (1,5)$. So C has fewer elements than the Cartesian product on which it is based. A case-by-case examination gives the following set for C:

$$C = \{(1,3),(1,7),(1,8),(4,3),(4,7),(4,8)\}.$$

d) As in b), we are doing something with the subsets of $\{2,5,9\}$. In this case, we are putting their *cardinalities* into a set. Subsets of $\{2,5,9\}$ come in four possible "sizes"—no elements, one element, two elements, or three elements. Thus, $D = \{0,1,2,3\}$. The following list gives, for each element of D, a possible choice for T which "causes" that element to belong to D:

0 ($T = \varnothing$);
1 ($T = \{9\}$);
2 ($T = \{2,9\}$);
3 ($T = \{2,5,9\}$).

e) The equation "$x^2 + 5x - 6 = 0$" has two real solutions, $x = 1$ and $x = -6$. Therefore, the set E contains $|1|$ (which equals 1), and $|-6|$ (which equals 6). Thus, $E = \{1,6\}$. □

The Algebra of Sets

Appendix B discusses several *operations* on sets (ways for "combining" sets to get other sets), and we have introduced two others here. Thus, for example, starting from two sets A and B, we can create $A \cup B$ (union), $A \cap B$ (intersection), $A - B$ (difference), A' (complement), $A \times B$ (Cartesian product), and $\mathcal{P}(A)$ (power set).

Working with these operations suggests an analogy with arithmetic, in which we have the operations of addition, subtraction, etc., for combining numbers. When you studied arithmetic, you learned some general principles concerning the basic operations, and the relationships among them. The next exploration is intended to pursue similar principles about sets and their operations.

Venn Diagrams

Before we send you off on this exploration, we want to introduce an important technique for thinking about sets. Figure 2.3 shows three sets, A, B, and C, each represented by a circle. The rectangle containing them represents the *universe*—the larger set which is

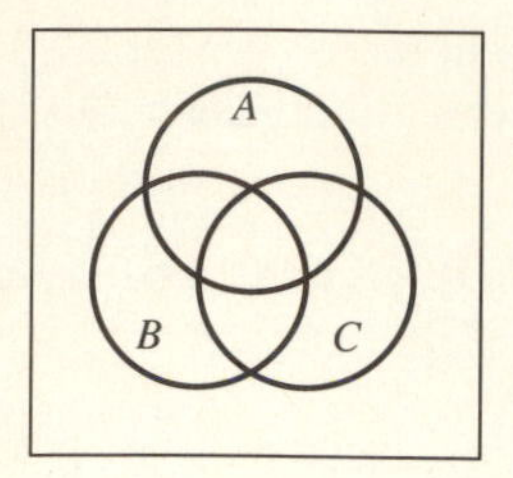

Figure 2.3 Each circle represents one of the sets A, B, and C. The overlaps represent intersections of the sets.

the context of the particular problem. Overlap between circles represents intersections of the corresponding sets. The space outside a particular circle is the complement of that set. This type of schematic representation of sets is known as a *Venn diagram*.

An element can be placed in any of eight "portions" of Figure 2.3 (shown numbered in Figure 2.4), depending on which of the sets A, B, and C it belongs to. For example, if an object belongs to A and C, but not to B, then it would be placed in the portion labeled "4" in Figure 2.4.

By shading certain portions of the diagram, we can indicate certain combinations of the sets. For example, the shaded area in Figure 2.5 schematically represents $(A \cup B) \cap C'$ (C' means the complement of the set C—the set of elements that are not in C; see Appendix B).

Caution: For particular sets A, B, and C, it may be that some of the portions are empty. For example, suppose $A = \{1, 2, 3, 5\}$, $B = \{2, 3, 6\}$, and $C = \{3, 4, 5\}$, with $\{1, 2, 3, \ldots, 10\}$ as the universe. The Venn diagram for A, B, and C is shown in Figure 2.6, and we see that $A' \cap B \cap C$ (the portion numbered "7" in Figure 2.4) is empty. (Remember that the numbers in Figure 2.4 are not elements of the sets, they are just labels for the portions.)

Get Your Hands Dirty 6

For small sets A, B, and C, we can make a Venn diagram showing the "location" of each element, i.e., placing it in the appropriate portion of Figure 2.3.

For each of the following conditions, choose specific sets $A, B, C \subseteq \{1, 2, 3, \ldots, 10\}$ which satisfy that condition (don't forget portion 8—the elements that belong to none of A, B, and C):

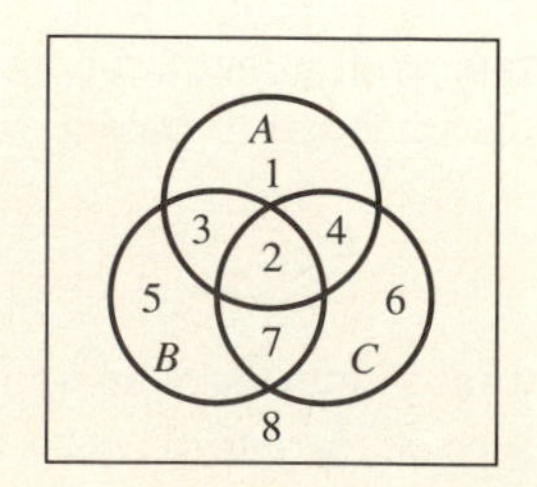

Figure 2.4 The numbers here are not elements of the sets—they are just labels for different portions of the overall universe.

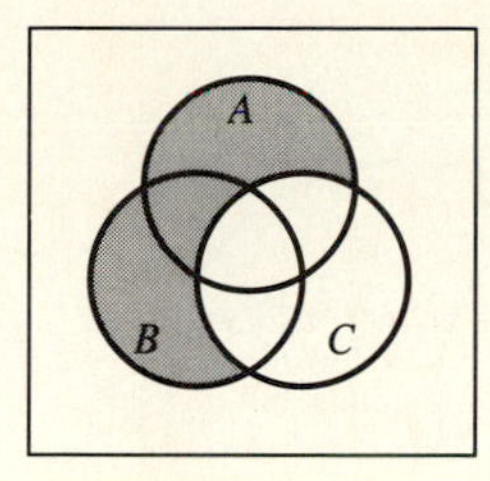

Figure 2.5 The shaded area is the set $(A \cup B) \cap C'$.

a) None of the portions is empty.
b) Exactly three of the portions are nonempty.
c) Exactly two of the portions are empty. □

There is another way in which Venn diagrams are used. Often we come across a problem in which various subsets of some universe are singled out, and we want to see how those subsets are related to each other. The following example illustrates how this is done.

Example 4 **Relations Among Specific Sets** Draw a single Venn diagram illustrating the subset relationships that exist among the following sets (**W** is the set of whole numbers, i.e., integers that are either zero or positive—see Appendix B):

$A = \{x \in \mathbf{N} : x \text{ is a prime}\}$
$B = \{2x : x \in \mathbf{N}\}$
$C = \{4x + 1 : x \in \mathbf{W}\}$
$D = \{2\}$
$E = \{6x + 1 : x \in \mathbf{W}\}$
$F = \{1\}$

Show where each of the integers from 1 through 60 belongs.

Try it!!

SOLUTION Figure 2.7 shows the various relationships. Note that $D = A \cap B$, $F \subseteq C \cap E$, and $B \cap (C \cup E) = \varnothing$. □

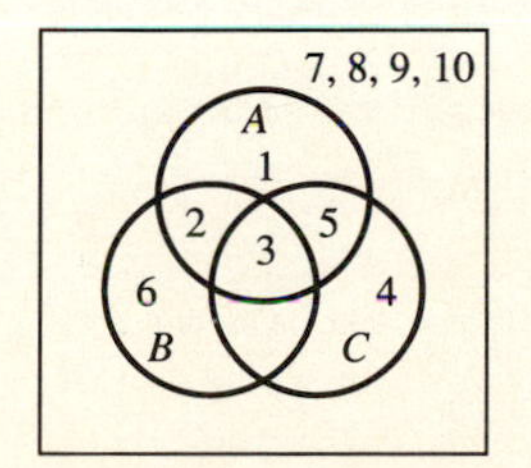

Figure 2.6 A Venn diagram for the sets $A = \{1, 2, 3, 5\}$, $B = \{2, 3, 6\}$, and $C = \{3, 4, 5\}$, with the universe $= \{1, 2, 3, \ldots, 10\}$.

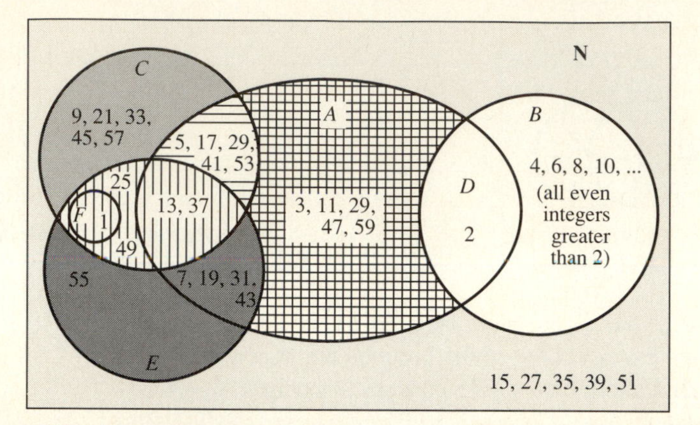

Figure 2.7 This complex Venn diagram shows the relationships among the sets of Example 4.

Comment: As indicated in Figure 2.7 by the different types of shading, the six sets $C \cap E$, $C \cap E' \cap A$, $C \cap E' \cap A'$, $C' \cap E$, $C' \cap E' \cap A \cap B'$, and $\mathbf{N} - (A \cup B \cup C \cup E)$ each have exactly five elements less than 60. Is this accidental? (B, of course, has exactly half of the elements from 1 to 60.)

Exploration 1: Set Operations—Analogies to Arithmetic

What general relationships or formulas can we find concerning the basic operations on sets? Let's begin with an analogy to arithmetic.

The operation of union of sets has some similarity to addition. In fact, addition of positive integers can be formally defined in terms of union of disjoint sets. Here are some general properties which hold for addition (x, y, and z represent real numbers). You may think of others. Which of them are valid formulas if the "+" sign is replaced by "$\cup$", and the variables representing numbers are replaced by variables representing sets?

1. $x + y = y + x$.
2. $(x + y) + z = x + (y + z)$.
3. $x + 0 = 0 + x = x$. (You may want to replace 0 by some analogous object for sets.)
4. If $x + z = y + z$, then $x = y$. (This is sometimes called the *cancellation property.*)

Similar properties exist for other arithmetic operations. Here are some of them:

5. $x(y + z) = xy + xz$.
6. If $x > y$, then $-y > -x$.
7. $x - (y + z) = x - y - z$.

You need to think about what the appropriate analogous operations or relations are. (">" and "<" are examples of relations—what corresponds to these for sets?) Think about how you might work the operation of intersection into some of the formulas.

HINTS

1. Use Venn diagrams frequently as a guide to what is happening.
2. Keep in mind that a conjecture that turns out false is not necessarily a lost cause. For example, one statement you might consider in this exploration is the following:

$$\text{If } A \cup C = B \cup C, \quad \text{then } A = B.$$

This statement is false. (Verify by a counterexample.) But that doesn't mean the statement has no interest. You can ask whether some variation of the statement is true. For example, is there some extra condition on the sets which would make it true? (What if the hypothesis also says that $C \subseteq A \cup B$ or that C is disjoint from $A \cup B$? Or what if you replace the "=" signs by "subset" signs?)

• E X P L O R E •

Now that you have done this exploration, we make two general observations about exploration techniques. Think about what methods were helpful in your work, and how the following ideas might help you in the future.

1. This exploration is a good illustration of the importance of using a *large variety of examples*. In order to test out various conjectures, you need to try different types of sets for the variables. You may want to try infinite as well as finite sets, sets that are disjoint as well as sets that overlap, empty as well as nonempty sets, and so on. The greater the variety in your choice of examples to explore, the more likely it is that the conjectures you make on the basis of your exploration will turn out to be correct.

2. Though a variety of examples is important, this exploration illustrates that some kinds of variety are *not important* in certain explorations. In this exploration, we are looking for general principles about sets. We aren't making any use of the "intrinsic" properties of the elements: for example, even if the set contains real numbers, we aren't involved with actually doing arithmetic with those numbers.

   Therefore, in choosing examples to try out various formulas involving sets, it makes no difference what the actual elements are. Though it does matter if two sets have elements in common, it won't matter what particular objects those elements are. You can choose any convenient set of objects, such as letters of the alphabet, positive integers, etc. (However, if you want to consider infinite sets, you won't be able to just use letters of the alphabet, and you should also keep in mind that some types of sets—such as intervals of real numbers—will lend themselves better to visual pictures than others.)

Cardinality

One of the most natural questions to ask about a set is "How many elements are in it?" or, put in more formal language, "What is its cardinality?" The last exploration of this section asks you to investigate what happens to cardinalities when new sets are formed from other sets.

Exploration 2: Cardinalities of Combinations of Sets

We have seen several ways to "create" new sets, based on other sets. If A, B, etc., are sets whose cardinalities are known, what will be the cardinalities of other sets expressed in terms of them, such as $A \cup B$, $A \times B$, $\mathcal{P}(A)$, etc.?

Assume that all sets being considered are finite.

HINTS

1.. Work with examples of small sets, so that you can actually list and count elements. Review GYHD 1 and 3.
2. As in Exploration 1, the choice of particular elements usually doesn't matter—what matters is how many there are, which sets they are in, etc.
3. Include the empty set as one of the possibilities.
4. Keep in mind that a formula which works in some special situation may need to be either adjusted to work in every situation, or stated in a way that specifies when it works.
5. In examining the question of $|\mathcal{P}(A)|$, consider the following approach:

    Suppose you have analyzed $\mathcal{P}(A)$ for some particular set A, and now you form a new set B, by putting *one new element* into A. Define $B = A \cup \{x\}$ (where x wasn't already in A). What subsets does B have that A doesn't? Which subsets of A are also subsets of B? How many are there of each kind?

    In other words, examine the collection of subsets of B *in relation to* the collection of subsets of A. This method of analysis, in which one example builds on the previous example, is closely connected with the method of mathematical induction, which will be discussed in Chapter 5.

•EXPLORE•

EXERCISES

Reminder: A complete list of symbols used in this book is provided following the appendices.

1. (GYHD 2) Is it possible for something to be both an element and a subset of the same set? In other words, can you create an example of two "things"—V and W—such that "$V \in W$" and "$V \subseteq W$" are both true statements? Think about the "type" of thing that V and W might each be.

2. (GYHD 4) Give five elements of the truth set for each of the following open sentences:

a) $3x - 7y = 12$ (universe $= \mathbf{R} \times \mathbf{R}$: make "$x$" the first coordinate).

b) $r + 2s + 3t = 17$ (universe $= \mathbf{N} \times \mathbf{N} \times \mathbf{N}$: make "$r$" the first coordinate and "s" the second coordinate).

c) $a \neq b$ (universe $= \{1,2,3,4\} \times \{2,3,4,5\}$: make "$a$" the first coordinate).

3. (GYHD 5) If A is some set (possibly empty), what are $A \times \varnothing$ and $\varnothing \times A$?

4. List explicitly all the elements of the truth set for each of the following open sentences, using the universe indicated.

a) $\{1\} \subseteq A \subseteq \{1,2,3,4\}$ [universe $= \mathcal{P}(Z)$].

b) $|S| = 3$ [universe $= \mathcal{P}(\{1,2,3,4,5\})$].

c) $A \subseteq B$ [universe $= \mathcal{P}(\{1,2\}) \times \mathcal{P}(\{1,2\})$: make "$A$" the first coordinate].

5. (Exploration 1) Examine each of the following formulas about sets and set operations. Show by means of a Venn diagram whether or not the formula is correct for all sets. If the formula is not correct, give a specific counterexample. (You can think of each formula as the conclusion to a conditional sentence that begins "If A, B, and C are sets, then . . .". Thus a counterexample consists of specific sets that, when substituted, give a false statement.)

Also, if the formula is not correct, find, if possible, an alternative expression to put on the right-hand side of the equation, to produce a correct formula.

a) $(A \cup B) \cap C = A \cup (B \cap C)$.

b) $(A \cap B)' = A' \cap B'$.

c) $A - (B \cup C) = (A - B) \cup (A - C)$.

d) $(A - B) - C = (A - C) - (B - C)$.

6. (Exploration 1) Examine each of the following formulas about sets and set operations. Decide by looking at examples whether the statement is true or false. (Venn diagrams are hard to use for problems involving power sets or Cartesian products.) If it is true, give an informal explanation. If it is false, give a specific counterexample.

a) $\mathcal{P}(A \cup B) = \mathcal{P}(A) \cup \mathcal{P}(B)$.

b) $\mathcal{P}(A \cap B) = \mathcal{P}(A) \cap \mathcal{P}(B)$.

c) $A \times (B \cup C) = (A \times B) \cup (A \times C)$.

d) $A \times (B \cap C) = (A \times B) \cap (A \times C)$.

e) $\mathcal{P}(A \times B) = \mathcal{P}(A) \times \mathcal{P}(B)$.

7. Make up three reasonable-looking formulas about sets and set operations that are actually false. (Do not use any already given in the text.)

8. Which of the following statements are true? Which are false? Which are vacuously true? Give informal explanations of your answers.

a) If $A \subset B$ and $B \subset A$, then $A = B$.

b) If $A \cap B \supseteq A$, then $A \subseteq B$.

c) If $A \cup B \subseteq A \cup C$, then $B \subseteq C$.

d) If $A \cap B \subseteq C \cap D$, then $A \subseteq C$ and $B \subseteq D$.

e) If $A \not\subseteq B$ and $B \not\subseteq C$, then $A \not\subseteq C$.

f) If $\mathcal{P}(A) = \varnothing$, then $A = \varnothing$.

g) If $A \in B$ and $B \in C$, then $A \in C$.

h) If $A \subseteq B$ and $B \in C$, then $A \in C$.

i) If $A \subseteq B$, then $B' \subseteq A'$.

j) If $A \times B \subseteq C \times D$, then $A \subseteq C$ and $B \subseteq D$.

k) If $A \times A \subseteq B \times B$, then $A \subseteq B$.
l) If $A \subseteq B$, then $A \cap B' = \emptyset$.

9. Make up three reasonable-sounding mathematical statements that are vacuously true.

10. List the elements of each of the following sets:
a) $\{y^2 : 1 \leq y \leq 7, y \in \mathbf{Z}\}$.
b) $\{|u| : -3 \leq u \leq 3, u \in \mathbf{Z}\}$.
c) $\{t^3 : t^6 - 3t^3 - 4 = 0, t \in \mathbf{R}\}$.
d) $\{s^2 : s^4 - 5s^2 - 6 = 0, s \in \mathbf{R}\}$ (careful!).
e) $\{5x + 3 : x \in \mathbf{Z}\}$ (give five sample elements).
f) $\{A \cup \{4\} : A \in \mathcal{P}(\{1,2,3\}) - \mathcal{P}(\{1,3,5\})\}$.
g) $\{(x^2, x^3) : x \in \{1,2,3\}\}$.

11. Let $U = \{x \in \mathbf{Z} : -20 \leq x \leq 20\}$. Consider the following subsets of U:

$$A = \{x \in U : x - 1 \text{ is even}\};$$
$$B = \{x \in U : x > 0\};$$
$$C = \{1, 5, 9, 13, 17\};$$
$$D = \{x \in U : 0 < 3x < 32\};$$
$$E = \{x \in U : x^2 - 3 \text{ is a multiple of } 4\};$$
$$F = \{x \in U : x^3 - 2x^2 - 15x = 0\}.$$

Draw a Venn diagram illustrating these sets, using U as the universe. Show where each element of U belongs.

12. (Exploration 2) Suppose A and B are finite sets, with $|A| = s$, $|B| = t$, and $|A \cap B| = u$. Express the following cardinalities in terms of s, t, and u:
a) $|A \cup B|$.
b) $|A \times B|$.
c) $|\mathcal{P}(A)|$.
d) $|A - B|$.

13. Figure 2.4 showed eight portions in the Venn diagram based on three sets A, B, and C. If there were n sets, how many portions would be created? (*Note:* The diagram becomes unwieldy beyond three sets, so you need to think about the number of "set combinations" possible for an element to belong to.)

14. There is another standard operation on sets known as *symmetric difference*. It is often represented by the symbol "$\triangle$" and is defined as follows:

$$A \triangle B = (A - B) \cup (B - A).$$

Thus, as Figure 2.8 illustrates, $A \triangle B$ consists of those elements which belong to either A or B but not both. It is called "symmetric" because $A \triangle B = B \triangle A$ (verify).

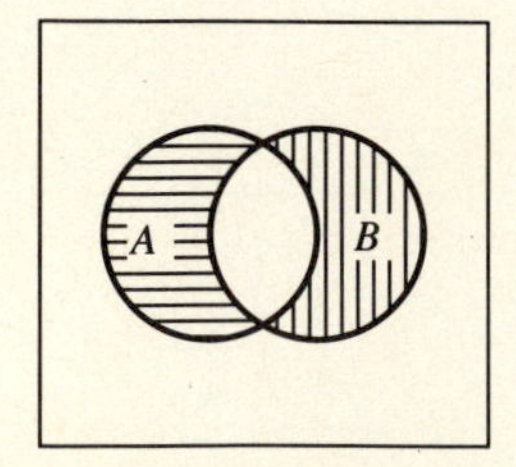

Figure 2.8 The two circles represent the sets A and B. The area with horizontal shading is $A - B$; the area with vertical shading is $B - A$. Their union, the total shaded area, is $A \triangle B$.

a) Determine whether each of the following formulas involving symmetric difference is correct or not. Justify each answer by means of Venn diagrams, a counterexample, or some other explanation. If the formula is incorrect, modify it if possible to produce a correct formula.
   i) $A \triangle (B \cup C) = (A \triangle B) \cup (A \triangle C)$.
   ii) $A \triangle (B \cap C) = (A \triangle B) \cap (A \triangle C)$.
   iii) $(A \triangle B) \triangle C = A \triangle (B \triangle C)$.
   iv) $A \triangle A = A$.

b) Suppose A and B are finite sets, with $|A| = s$, $|B| = t$, and $|A \cap B| = u$. Express $|A \triangle B|$ in terms of s, t, and u.

2.2 Proofs about Sets

In the last section, we asked you to examine and develop some general relationships concerning sets and the operations and relations on sets. In this section we will look at what is involved in proving some of these relationships. Some of the theorems we will examine are interesting because they state important results or will be needed later; others are of interest here primarily as illustrations of the proof process.

Sets and Subsets

Here is our first theorem and proof about sets:

Example 1 **A Subset of a Subset** This example illustrates how to prove that one set is a subset of another.

THEOREM 2.2. For sets A, B, and C,

$$\text{if } A \subseteq B \text{ and } B \subseteq C, \text{ then } A \subseteq C.$$

Proof: Assume, as in the hypothesis, that $A \subseteq B$ and $B \subseteq C$. Our goal is to show that $A \subseteq C$, i.e., that if $x \in A$, then $x \in C$.

In order to show this, suppose that x is an element of A. To prove the theorem, we need to show that $x \in C$.

Since $x \in A$ and $A \subseteq B$, we can conclude, by the definition of "subset", that $x \in B$. Similarly, since we now know that $x \in B$, and we also know that $B \subseteq C$, we can conclude that $x \in C$, as desired. ■

We can explain and summarize this proof as follows:

1. Since the theorem is an "if . . . , then . . ." statement, we began by assuming that its hypothesis—$A \subseteq B$ and $B \subseteq C$—was true. Our goal then became to prove from this assumption that the conclusion—$A \subseteq C$—was true.

2. We then looked at that goal: proving $A \subseteq C$. This statement itself can be understood as an "if . . . , then . . ." statement: "if $x \in A$, then $x \in C$"; so we assumed the hypothesis of this statement ("$x \in A$"), and had a new goal: deriving the conclusion of this statement ("$x \in C$").

3. We then started combining assumptions: we combined "$x \in A$" with "$A \subseteq B$" to get "$x \in B$", and combined this new fact, "$x \in B$", with "$B \subseteq C$" to get "$x \in C$". Since the last statement was the goal we set (see comment 2 above), that completed the proof.

4. You may be uncomfortable with a statement like "suppose x is an element of A", because A might be the empty set. But this "supposition" should be thought of as an assumption about x, not about A. We are not saying "suppose that A has an element", but rather "consider an element x, and consider the situation in which it is a member of A". We do not need to consider "$A = \varnothing$" as a special case in proofs like this, since any conclusion we arrive at will be in the form of a conditional sentence: "if $x \in A$, then . . .". Such a statement is always true if $A = \varnothing$.

Example 2 **Applying a Theorem** This example illustrates how to use a theorem that has already been proved.

THEOREM 2.3. If $U \subseteq W$, then $\mathcal{P}(U) \subseteq \mathcal{P}(W)$.

Proof: Assume that $U \subseteq W$. In order to prove that $\mathcal{P}(U) \subseteq \mathcal{P}(W)$, suppose that $T \in \mathcal{P}(U)$. We need to show $T \in \mathcal{P}(W)$.

By definition of "power set", we have $T \subseteq U$. But then the sets T, U, and W fit the hypothesis of Theorem 2.2; in other words, since $T \subseteq U$ and $U \subseteq W$, the substitution $A = T$, $B = U$, and $C = W$ makes "$A \subseteq B$ and $B \subseteq C$" true. Therefore, the conclusion "$A \subseteq C$" must be true, i.e., $T \subseteq W$, which means $T \in \mathcal{P}(W)$, as desired. ∎

COMMENTS

1. The above proof could have been done like Example 1, by referring to individual elements of the sets U and W, but it can also be done more efficiently by using Theorem 2.2.

2. We have discussed repeatedly that in order to *prove* an "if . . . , then . . ." statement, we assume its hypothesis, and then derive its conclusion. Example 2 illustrates how to *apply* an "if . . . , then . . ." statement that we have already proved (namely, Theorem 2.2): we prove its hypothesis, or, more precisely, we demonstrate that certain objects *make its hypothesis true,* and then we are guaranteed by the theorem that those certain objects make the conclusion true. Here we showed that T, U, and W fit the hypothesis of Theorem 2.2. We then applied that theorem, which lets us assert with certainty that those objects make the conclusion of that theorem true. Note that the letters in Theorem 2.3 are different from those in Theorem 2.2 (see comment 3 below).

The basic idea is that Theorem 2.2, like conditional statements generally, is an assertion about *all* objects that make its hypothesis true. In proving that theorem, we proved that any such sets will make its conclusion true. So, in particular, since the hypothesis is true when we replace A by T, B by U, and C by W, the conclusion is also true when we make this substitution.

3. The application of the theorem is clarified by the fact that Theorem 2.3 used completely different letters for its objects from those in Theorem 2.2. If this were not the case, the substitution corresponding to "$A = T$, $B = U$, and $C = W$" would be hard to state clearly.

4. We actually used a similar process twice in the proof of Theorem 2.2 itself. For example, we combined the statements "$x \in A$" and "$A \subseteq B$" to conclude that $x \in B$. The reasoning is as follows: "$A \subseteq B$" is an "if . . . , then . . ." statement (namely, "if $x \in A$, then $x \in B$"), and we are assuming that this statement is true; we also know that its hypothesis—"$x \in A$"—is true; therefore its conclusion—"$x \in B$"—is also true. This is a mathematical example of one of the classic logical forms called *syllogisms*. The most familiar type of syllogism looks like this:

   i) All men are mortal.
   ii) Socrates is a man.

   Therefore

   iii) Socrates is mortal.

   The statement "all men are mortal" can be rephrased in "if . . . , then . . ." form as "if x is a man, then x is mortal". If this conditional statement is true, and if, as ii) says, Socrates fits the hypothesis of this conditional statement, then we can conclude with certainty that Socrates fits the conclusion of the conditional statement, i.e., Socrates is mortal.

For emphasis, we state the principle involved in using the syllogism process:

In order to use a known conditional statement, show that certain objects satisfy the hypothesis of that statement. You are then guaranteed that those objects satisfy the conclusion of that statement.

Keep in mind that the "known conditional statement" involved here might come from some previously proved theorem, or might be part of the hypothesis being assumed in the proof (such as the "hidden conditional" statement "$A \subseteq B$" in Example 1), or might even be a result developed earlier in the proof.

Metaphor: Theorem as "Guarantee Producer"

You may like to think of proving a theorem as building a machine which produces guarantees of certain kinds of information. Of course, it will not produce such a guarantee without proper assurances—as provided in the hypothesis. But once you have proved the theorem, you can put objects into the machine, and then, if

> those objects provide the appropriate assurances (if they fit the hypothesis), you will get out a guarantee certificate — a statement that is 100% reliable and trustworthy — which says that those objects fit the conclusion of the theorem.

Thus, having proved Theorem 2.2, you now have a machine which, given three sets of the appropriate type — namely, one a subset of the second, and the second a subset of the third — will produce a certificate of guarantee that the first is a subset of the third.

The same image applies to the situation of a theorem whose hypothesis contains an "if . . . , then . . ." statement. For example, in Theorem 2.2, we are told in the hypothesis that $A \subseteq B$. We can think of this assumption as saying that there is a machine which will produce membership certificates in set B, whenever it is given an object which has a membership certificate in set A. We can then use that machine as part of the proof, which is essentially what we did in the third paragraph of that proof.

The proof of the following theorem is similar. You should do it now, to reinforce the ideas we have been discussing.

Get Your Hands Dirty 1

Prove the following:

If $A \subseteq B$ and $C \subseteq D$, then $A \cap C \subseteq B \cap D$. □

Proving Sets Equal

The previous examples involved showing that one set is a subset of another. What do you do if you want to prove that two sets are equal?

As noted in Appendix B, two sets are considered equal precisely when they consist of the same elements. In practice, we often prove that two sets are equal, i.e., that they have the same elements, by showing that each is a subset of the other. Using this method, a proof that $X = Y$ will have two parts: first show that $X \subseteq Y$, and then show $Y \subseteq X$.

The next example illustrates this method. The theorem expresses two important relationships involving union, intersection, and complementation. You may have discovered these as part of Exploration 1. We will give a proof only of the first equation — the proof of the second is similar, and is left for an exercise (Exercise 2).

Example 3 **Proving Sets Equal** This example illustrates how to prove that two sets are equal.

THEOREM 2.4. For any sets A and B:

i) $(A \cup B)' = A' \cap B'$

ii) $(A \cap B)' = A' \cup B'$

Proof of part i): To show $(A \cup B)' = A' \cap B'$, we will show $(A \cup B)' \subseteq A' \cap B'$ and $A' \cap B' \subseteq (A \cup B)'$.

To prove the first assertion, assume that $x \in (A \cup B)'$. In other words, x is not an element of $A \cup B$. By the definition of "union", this means that x is neither an element of A nor an element of B. But that says precisely that x is both in A' and in B', i.e., $x \in A' \cap B'$. Thus, $(A \cup B)' \subseteq A' \cap B'$.

To show the other half of part i, assume $x \in A' \cap B'$. Then x is in both A' and B'; i.e., $x \notin A$ and $x \notin B$. But if x is not in either A or B, then it is not in $A \cup B$. Thus $x \in (A \cup B)'$, as desired. Thus, $A' \cap B' \subseteq (A \cup B)'$. ■

COMMENTS

1. The two equations in this theorem are known as *De Morgan's Laws,* after the nineteenth century logician Augustus De Morgan (see Historical Note in Appendix A).

2. The proof itself involves little more than careful discussion of the meanings of the words "and", "or", and "not", combined together in different ways. This is natural, since the concepts involved—intersection, union, and complementation—are defined by means of these connecting words. Thus:

   $y \in S \cap T$ means "y is in S and y is in T".
   $y \in S \cup T$ means "y is in S or y is in T".
   $y \in S'$ means "y is not in S".

   We will see in Section 4.2 that there are analogous "equations" for combining open sentences.

3. A Venn diagram may aid intuitive understanding of a theorem like this. In Figure 2.9, the shaded portions of the diagrams show the sets indicated below each diagram. Since (ib) and (iic) have the same shaded area, this series of diagrams shows, pictorially, why $(A \cup B)'$ and $A' \cap B'$ are the same set.

COMMENT Does this series of diagrams constitute a proof that $(A \cup B)'$ is equal to $A' \cap B'$?

Our answer is "yes and no". A similar question is often posed concerning the use of diagrams in traditional high school geometry. The problem is that we need to be sure that the diagram really describes the situation accurately. There are many examples of *pseudo-proofs* in geometry—apparent proofs of false statements whose flaw comes from a misrepresentation in the diagram. For example, as noted in Section 2.1, some of the portions of a Venn diagram may actually represent the empty set, but may appear to be nonempty.

It is possible to prove a theorem that asserts that certain types of Venn diagram arguments are *complete*—that is, that a diagram such as Figure 2.9 really does completely analyze the situation. But such a *metatheorem*—a theorem about proofs—is beyond the scope of this book. Moreover, as you begin to learn to write proofs, it is important to be cautious, depending on diagrams only for intuition, and using more formal, verbal arguments for actual proofs.

So definitely use Venn diagrams to gain understanding, but then use the model of Example 3 when it comes to writing proofs.

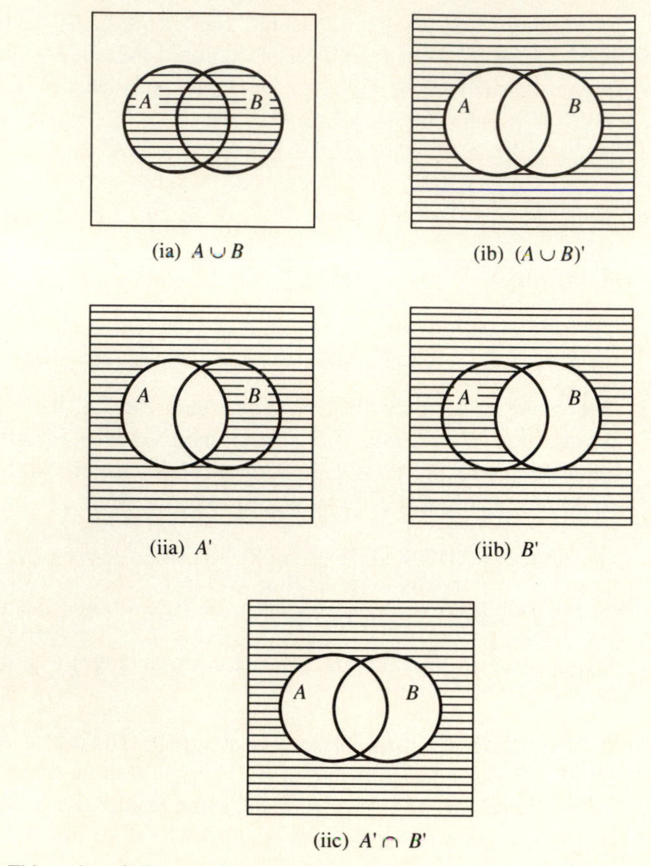

Figure 2.9 This series of diagrams illustrates the fact that $(A \cup B)' = A' \cap B'$.

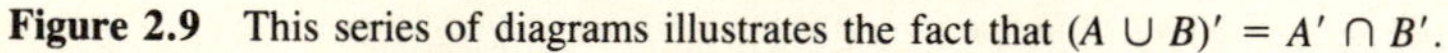

The result in the following "Get Your Hands Dirty" depends on the meaning of "not", "or", and "and", since set complement, union, and intersection are defined in terms of these words. A formal proof requires a technical definition of these everyday words, which is part of the development of formal logic. We will not be taking that approach, but we ask you to examine the theorem from an intuitive perspective.

Get Your Hands Dirty 2

Convince yourself of the following facts.

For any sets A and B:

a) $(A')' = A$.
b) $A \cup B = B \cup A$.
c) $A \cap B = B \cap A$. □

Proofs with Cartesian Product

The following examples deal with the issue of the "type" of elements in a set:

Example 4 **Products of subsets** This example illustrates how to work with Cartesian products.

THEOREM 2.5. If $A \subseteq B$ and $C \subseteq D$, then $A \times C \subseteq B \times D$.

Before reading the proof, you should think about this statement (as you should about every theorem). What's going on? Why is it true?

To help you with this process of analysis, we ask you to do the following:

Get Your Hands Dirty 3

Choose specific sets for A, B, C, and D, which fit the hypothesis. Make them small. Then actually write down the elements of both $A \times C$ and $B \times D$, and verify that the conclusion, "$A \times C \subseteq B \times D$", is satisfied. □

When you think that you understand why the theorem is true, read the proof.

Proof: Assume that A, B, C, and D are sets which make the hypothesis true, i.e., such that $A \subseteq B$ and $C \subseteq D$. We need to show $A \times C \subseteq B \times D$.

In order to prove this, assume that $w \in A \times C$. Then $w = (x, y)$ for some $x \in A$ and $y \in C$.

Since $x \in A$, and $A \subseteq B$, we know that $x \in B$. Similarly, since $y \in C$, and $C \subseteq D$, we know that $y \in D$.

Therefore $(x, y) \in B \times D$, which completes the proof. ■

COMMENTS

1. As always, when we want to show that one set is a subset of another, we suppose that we are given an object which is an element of the first set, and then need to show that this object is an element of the second set.

2. The key step in this proof occurs following the assumption that $w \in A \times C$. We need to realize that "being an element of $A \times C$" means "being an ordered pair, whose first component is from A and whose second is from C". That's what the sentence

   "Then $w = (x, y)$ for some $x \in A$ and $y \in C$"

   is saying. In other words, we are taking the assumption that $w \in A \times C$ and interpreting it, using the definition of Cartesian product. In order to do so, we need to introduce names for the two components of w. The single fact about w then becomes two individual facts about x and y: $x \in A$ and $y \in C$.

   Note: We have introduced the variable w in order to emphasize the role of the Cartesian product. In practice, many writers would omit w, and simply say "assume that $(x, y) \in A \times C$".

3. We then use the basic hypothesis of the theorem—that $A \subseteq B$ and $C \subseteq D$—to get further information about x and y, namely, that $x \in B$ and $y \in D$. Those two facts tell us that (x, y) (also known as w) is an element of $B \times D$, which completes the proof.

Example 5 **"Canceling" Set Factors** This example illustrates how to use the assumption that a set is nonempty.

THEOREM 2.6. If $A \times B \subseteq A \times C$, and $A \neq \varnothing$, then $B \subseteq C$.

As with the last theorem, we offer a task for you to get ready to read the proof:

Get Your Hands Dirty 4

a) Choose specific, small sets that satisfy the hypothesis. Can you come up with an example with $B \not\subseteq C$? Why not?

b) Find a pair of sets B and C satisfying $\varnothing \times B = \varnothing \times C$, with $B \not\subseteq C$. In other words, show that the condition $A \neq \varnothing$ is needed. □

Now read the proof:

Proof: Assume that $A \times B \subseteq A \times C$ and $A \neq \varnothing$. We need to show that $B \subseteq C$, so suppose that $x \in B$.

Since $A \neq \varnothing$, A must have at least one element. Choose such an element, and call it t. Then $(t, x) \in A \times B$, so we also have $(t, x) \in A \times C$ (since $A \times B \subseteq A \times C$). But that means x must be an element of C, as desired. ■

COMMENTS

1. It is very important here, as in Example 4, to distinguish between elements of the individual sets and elements of the Cartesian products.

2. The two objects x and t are brought into this proof in different ways: x is introduced, as an element of B, in order to *show* that $B \subseteq C$. The statement "$x \in B$" is *assumed*, because it is the hypothesis of the "if . . . , then . . ." form of this subset statement we are trying to prove. We are not assuming that there *are* elements in B, but only discussing what would happen *if* there were such an element, which we could call x. On the other hand, t is introduced by means of the hypothesis that $A \neq \varnothing$. We are guaranteed that some object must be an element of A, and we simply "ask A for one of its elements" and give it the name t.

3. Looking at the extreme case of the empty set can help clarify the difference between the roles of x and t. Let's see the difference in the effect on our theorem if either of the two sets they come from, B or A respectively, is the empty set. If $B = \varnothing$, the conclusion "$B \subseteq C$" is true (vacuously true), and the theorem holds. It doesn't matter if B is empty or not. However, if $A = \varnothing$, then $A \times B = \varnothing$ and $A \times C = \varnothing$, regardless of what sets B and C are. (Verify.) We can't conclude that $B \subseteq C$. If we remove "$A \neq \varnothing$" from the hypothesis, then the resulting statement is false. ($B = \{1, 2\}$ and $C = \{3\}$, together with $A = \varnothing$, would constitute a counterexample.)

We would be in serious trouble if our proof worked even when the statement was false. The proof doesn't work if $A = \varnothing$, because in that case we can't "ask for" the element t, and so we can't use the element x to form an ordered pair in $A \times B$. That makes it impossible to make use of the hypothesis that $A \times B \subseteq A \times C$.

This difference between A being empty and B being empty is reflected in the different ways in which x and t come into the proof.

Our next example is almost the same as Theorem 2.6. The way we deal with the difference illustrates an important idea about proofs.

Example 6 **Symmetry in the Hypothesis** This example illustrates how to use the fact that two variables appear similarly in the hypothesis.

COROLLARY 2.7. If $A \times B = A \times C$ and $A \neq \varnothing$, then $B = C$.

Compare with Theorem 2.6; what's different?

Proof: Assume $A \times B = A \times C$ and $A \neq \varnothing$. As discussed earlier, we need to show that $B \subseteq C$ and $C \subseteq B$.

Since $A \times B = A \times C$, we can be sure that $A \times B \subseteq A \times C$, so the hypothesis of Theorem 2.6 applies, and we can conclude that $B \subseteq C$.

But we also have $A \times C \subseteq A \times B$, so we can apply Theorem 2.6, with the roles of B and C reversed, to conclude that $C \subseteq B$.

Since $B \subseteq C$ and $C \subseteq B$, we have $B = C$. ■

COMMENTS

1. A *corollary* is a theorem which is proved easily using a previous theorem. The assertion in Example 6 follows directly from Theorem 2.6, and so is called a corollary. (The judgment about when to label a theorem as a corollary is a subjective one. The distinction does not affect the mathematical status of the result—a corollary is just as true as a theorem, and is used the same way.)

2. The hypothesis in this corollary is *symmetric in B and C*. In other words, if B and C were switched, the hypothesis would contain the same information as originally: the statement "$A \times B = A \times C$" is equivalent to the statement "$A \times C = A \times B$". (The hypothesis that $A \neq \varnothing$ doesn't involve either B or C.) Since the *hypothesis* is symmetric in this way, it follows that any *conclusion* we can reach from this hypothesis can also be symmetrized with regard to B and C—that is, we can draw a similar conclusion with B and C switched. In the second paragraph of the proof, we showed that "$B \subseteq C$" is a consequence of the hypothesis; the principle just stated tells us that "$C \subseteq B$" must also be a consequence of the hypotheses. When a proof of a second assertion is just like that of the first, except that the variables involved are different (or switched), mathematicians sometimes just say "similarly" or "by a parallel argument", and state the second conclusion without additional proof. For example, we could replace the entire third paragraph in the proof of Corollary 2.7 by saying "Similarly, $C \subseteq B$".

Warning: if you use this shortcut, be sure that the hypothesis on which it is based is really symmetric.

3. Notice that this switching of variables is hard to express by means of substitution, since Corollary 2.7 uses the same letters as Theorem 2.6.

Breaking Up a Proof into Cases

Sometimes, in the course of a proof, we are confronted with a situation where either of two things might happen. We want to be able to prove our conclusion regardless of which case occurs. The following example illustrates the method used in such a problem:

Example 7 **Proof by Cases** This example illustrates how to split a proof into "cases".

THEOREM 2.8. Suppose X, Y, and Z are sets such that:

i) $X \cap Y = X \cap Z$ and
ii) $X' \cap Y = X' \cap Z$.

Then $Y = Z$.

Proof: Assume that X, Y, and Z are sets satisfying i) and ii). As usual, to prove $Y = Z$, we need to show $Y \subseteq Z$ and $Z \subseteq Y$.

To prove the first, suppose $c \in Y$. We need to show $c \in Z$. We know, certainly, that either $c \in X$ or $c \notin X$. We look at the following two cases:

Case 1. $c \in X$. Then we have c in both X and Y, so $c \in X \cap Y$. By i), we must have $c \in X \cap Z$. In particular, this tells us that $c \in Z$, as desired.
Case 2. $c \notin X$. Then, by definition of complementation, we have $c \in X'$, so c is in $X' \cap Y$. By ii), we must have $c \in X' \cap Z$. In particular, this tells us that $c \in Z$, as desired.

Thus, in either case, we must have $c \in Z$. This completes the proof that $Y \subseteq Z$.
By a parallel argument, $Z \subseteq Y$. ∎

COMMENTS

1. Our hypothesis gave two conditions on X, Y, and Z. Which of these is helpful in our proof depends on whether the element c under consideration is in X or in X'. Since we can't tell which occurs, we have to prove that, no matter which case occurs, c can be shown to be in Z. Thus, in order to show $Y \subseteq Z$, we have to prove two separate conditional statements: "if c is in Y and c is in X, then c is in Z", and "if c is in Y and c is not in X, then c is in Z". Every element of Y satisfies the hypothesis of one of these two conditional statements. Since the two conditional statements have the same conclusion, "$c \in Z$", that conclusion must be true of every element of Y.

2. In this example, the two cases were *mutually exclusive*—that is, there was no possibility for an element c to fall into both cases. That isn't always the situation in proof

by cases, and it isn't important. What is crucial is that the cases be *exhaustive*—that is, that every possibility be included in at least one of the cases.

3. There is no reason why the number of cases must be two—sometimes there are more than two possibilities that must be considered.

4. Here, as in Corollary 2.7, the hypothesis has a symmetry to it. Y and Z are used in an identical way in conditions i) and ii). That symmetry allows us to prove $Y \subseteq Z$ in detail, and then just say "By a parallel argument, $Z \subseteq Y$".

The following Get Your Hands Dirty gives one other pair of formulas about sets that we will need to use as part of proofs later on.

Get Your Hands Dirty 5

Prove the following:

For any sets A, B, and C:

a) $A \cap (B \cup C) = (A \cap B) \cup (A \cap C)$.
b) $A \cup (B \cap C) = (A \cup B) \cap (A \cup C)$. □

EXERCISES

In doing proofs in this Exercise set, you may use any theorems proven in this section, but be sure to refer to such theorems clearly.

1. (GYHD 1) Prove the following:

If $A \subseteq B$ and $C \subseteq D$, then $A \cap C \subseteq B \cap D$.

2. a) Prove the second of De Morgan's Laws:

THEOREM 2.4. ii) For any sets A and B, $(A \cap B)' = A' \cup B'$.

b) Draw a Venn diagram (or series of diagrams) that shows why this formula holds.

3. Use a Venn diagram to explain Theorem 2.8. (*Hint:* Use the conditions given in the hypothesis to show that certain "portions" of the standard 3-set Venn diagram must be empty.)

4. (GYHD 5) Prove the following:
For any sets A, B, C:

a) $A \cap (B \cup C) = (A \cap B) \cup (A \cap C)$.
b) $A \cup (B \cap C) = (A \cup B) \cap (A \cup C)$.

(*Hint:* Use proof by cases.)

5. Prove or disprove each of the following formulas or assertions:

a) $(A \cap B)' = A' \cap B'$.
b) $(A - B)' = B' - A'$.
c) $(A \cup B) \cap C = A \cup (B \cap C)$.

d) $A - (B \cup C) = (A - B) \cup (A - C)$.
e) $(A - B) - C \subseteq (A - C) - (B - C)$.
f) $\mathcal{P}(A \cup B) = \mathcal{P}(A) \cup \mathcal{P}(B)$.
g) $\mathcal{P}(A \cap B) = \mathcal{P}(A) \cap \mathcal{P}(B)$.
h) If $\mathcal{P}(A) \subseteq \mathcal{P}(B)$, then $A \subseteq B$.
i) If $A \cup B \subseteq A \cup C$, then $B \subseteq C$.
j) If $A \cap B \subseteq C \cap D$, then $A \subseteq C$ and $B \subseteq D$.
k) If $A \not\subseteq B$ and $B \not\subseteq C$, then $A \not\subseteq C$.
l) $A \times (B \cup C) = (A \times B) \cup (A \times C)$.
m) $A \times (B - C) = (A \times B) - (A \times C)$.
n) $A \times (B \cap C) = (A \times B) \cap (A \times C)$.
o) If $A \subseteq B$, then $A \times A \subseteq B \times B$.
p) If $A \times A \subseteq B \times B$, then $A \subseteq B$.

6. PROOF EVALUATION (This type of exercise will appear occasionally): Each of the following is a proposed "proof" of a "theorem". However the "theorem" may not be a true statement, and even if it is, the "proof" may not really be a proof. You should read each "theorem" and "proof" carefully and decide and state whether or not the "theorem" is true. Then:

- If the "theorem" is false, find where the "proof" fails. (There has to be some error.)
- If the "theorem" is true, decide and state whether or not the "proof" is correct. If it is not correct, find where the "proof" fails.

a) "THEOREM": For any sets A, B, and C, if $A \cup B = A \cup C$, then $B = C$.
"Proof": Assume that $A \cup B = A \cup C$. We need to prove $B = C$. First we will prove that $B \subseteq C$.

Suppose x is in B. Then x is certainly in $A \cup B$. By assumption, therefore, x is also in $A \cup C$. But we did not assume x was in A, so x must be in C. Thus, we have shown $B \subseteq C$.

By a parallel argument, $C \subseteq B$.

b) "THEOREM": For any sets A and B, if $\mathcal{P}(A) \cap \mathcal{P}(B) \neq \emptyset$, then $A \cap B \neq \emptyset$.
"Proof": Assume $\mathcal{P}(A) \cap \mathcal{P}(B) \neq \emptyset$. Our goal is to show $A \cap B \neq \emptyset$. Since $\mathcal{P}(A) \cap \mathcal{P}(B)$ is nonempty, there is something in this intersection; call it X. So $X \subseteq A$ and $X \subseteq B$. Now let y be any element of X. So $y \in A \cap B$, and so $A \cap B \neq \emptyset$. This concludes the proof.

c) "THEOREM": For any sets A and B,

$$\mathcal{P}(A \cap B) \subseteq \mathcal{P}(A) \cap \mathcal{P}(B).$$

"Proof": Suppose $x \in \mathcal{P}(A \cap B)$. Then $x \in A$ and $x \in B$, so $x \in \mathcal{P}(A)$ and $x \in \mathcal{P}(B)$. Therefore $x \in \mathcal{P}(A) \cap \mathcal{P}(B)$.

d) "THEOREM": For any sets A and B,

$$\mathcal{P}(A \cap B) \subseteq \mathcal{P}(A) \cap \mathcal{P}(B).$$

"Proof": $A \cap B$ is a subset of A, since any object that belongs to both A and B must belong to A. Thus, by Theorem 2.3, $\mathcal{P}(A \cap B) \subseteq \mathcal{P}(A)$. Similarly, $\mathcal{P}(A \cap B) \subseteq \mathcal{P}(B)$. Therefore, $\mathcal{P}(A \cap B) \subseteq \mathcal{P}(A) \cap \mathcal{P}(B)$.

2.3 Exploring the Integers

In this section, we are working from an intuitive idea of what the integers are, using the basic assumptions about arithmetic described in Appendix B. We will look at some more formal aspects of the integers in Chapter 5.

Multiples and Divisors

Among the principal concepts involved in studying the integers are *multiple* and *divisor*. We begin with two definitions:

DEFINITION 2.4: For $n \in \mathbf{Z}$, the *set of multiples of n* is

$$\{n \cdot t : t \in \mathbf{Z}\}.$$

The set of multiples of n is generally abbreviated as $n\mathbf{Z}$. Each individual element of this set is called a *multiple of n*.

Example 1 Multiples

$$5\mathbf{Z} = \{\ldots, -15, -10, -5, 0, 5, 10, 15, \ldots\}.$$

30, −215, and 95 are each multiples of 5.

$$2\mathbf{Z} = \{\ldots, -6, -4, -2, 0, 2, 4, 6, \ldots\}.$$

26, −68, and 280 are each multiples of 2. (Multiples of 2 are generally called *even integers*—see Definition 2.7 below.)

Note that $0\mathbf{Z} = \{0\}$ and $1\mathbf{Z} = \mathbf{Z}$. Thus the only multiple of 0 is 0 itself, while every integer is a multiple of 1. Also note that, for any integer n, we have $0 \in n\mathbf{Z}$; i.e., 0 is a multiple of every integer. □

DEFINITION 2.5: For $x, y \in \mathbf{Z}$, with $x \neq 0$, "x is a *divisor* of y" (written "$x \mid y$") means: there is some integer, call it t, such that $x \cdot t = y$.

0 is not a divisor of any integer.

The notation "$x \mid y$" can also be read "x divides y". If x is not a divisor of y, we can write "$x \nmid y$". A divisor of an integer is also known as a *factor* of the integer.

Comparing the definitions of multiple and divisor, we can see that, for $x \neq 0$, the statements "x is a divisor of y" and "y is a multiple of x" are synonymous. In particular, every nonzero integer is a divisor of 0, and 1 is a divisor of every integer.

Example 2

$3 \mid 6$ (i.e., 3 is a divisor of 6), because $3 \times 2 = 6$.
$8 \mid 0$ (i.e., 8 is a divisor of 0), because $8 \times 0 = 0$.
$10 \mid -30$ (i.e., 10 is a divisor of -30), because $10 \times (-3) = -30$. □

Caution: "$3 \mid 6$" is a *statement* about 3 and 6. It is not a number. Do not confuse this statement with either of the fractions $\frac{3}{6}$ or $\frac{6}{3}$. Also do not confuse "$3 \mid 6$", which is true, with "$6 \mid 3$", which is false.

COMMENTS

1. You may wonder why we don't define "x is a divisor of y" to mean that, when we divide y by x, we get an integer for the quotient. After all, the equation $x \cdot t = y$ has $t = y/x$ as its solution—we simply want t to be an integer. Therefore, why not say "x is a divisor of y" means that $y/x \in \mathbf{Z}$? In fact, we could use that as the definition, but there are several advantages to the way we have presented the idea here. Here are two of them:

   a) Suppose we had only the set $\mathbf{Z}$ to work with—i.e., that the larger set $\mathbf{Q}$ of rational numbers weren't available, or that we simply wanted to limit our discussion strictly to the set of integers. Then y/x would not be defined in every case. It would be confusing to define divisibility based on whether y/x is an integer, when we aren't sure whether or not y/x even exists. Definition 2.5 avoids this problem by not actually referring to division at all. (There are actually many situations in abstract algebra in which the concept of divisibility presented in Definition 2.5 makes complete sense, even though the operation of division isn't defined.)
   b) Definition 2.5 focuses attention on the *existence* of that missing factor t (which is equal to the quotient y/x). Essentially, this definition forces us to ask whether y/x makes sense, and if it does, then we are forced to give a name to this quotient. The name is very helpful, because it makes it easier for us to work with this object algebraically. Proofs in the next section will illustrate the usefulness of this technique. We will examine other existence statements in later chapters.

2. Note that $x = 0$ is specifically excluded from being a divisor of any integer. This is consistent with the standard prohibition in algebra: you can't divide by 0. The intuitive thinking behind this prohibition depends on whether $y = 0$ or $y \neq 0$. In general, we would like y/x to be a number t satisfying the equation $x \cdot t = y$. Let's look at what happens to this equation if $x = 0$, i.e., if we try to solve $0 \cdot t = y$.

   a) If $y \neq 0$, then there is no number t (integer or not) such that $0 \cdot t = y$, so we can't define y/x in any way. Therefore, both "x is a divisor of y" and "y is a multiple of x" are considered false if $x = 0$ and $y \neq 0$.
   b) If $y = 0$, then the equation $0 \cdot t = y$ has infinitely many solutions—in fact, any number will work for t, integers as well as nonintegers. Thus the existence condition in Definition 2.5 is, in fact, satisfied. But in our usual understanding, divi-

sion of one number by another should have only one possible answer. There is no good way to choose just one numerical value for t and say "this is 0/0". We therefore want to consider 0/0 *undefined,* and to say that 0 is not a divisor of 0.

In order to do this, we have to specifically exclude $x = 0$ from the general definition of the phrase "x is a divisor of y". The existence condition ("there is some integer...") does not by itself make "$0 \mid 0$" false—such an integer does exist. Thus, "$0 \mid 0$" is false because we made an extra condition in the definition specifically to make it false. However, when x and y are both 0, we do want to say that y is a multiple of x, and Definition 2.4 allows this.

Prime Numbers

Before exploring divisibility, we introduce one other concept which is likely to be familiar to you:

> **DEFINITION 2.6:** For $t \in \mathbf{N}$, $t > 1$, "t is a *prime*" (or "*prime number*") means: the only positive integers which are divisors of t are t itself and 1. Otherwise, t is a *composite number.*

Note: This definition only refers to integers greater than 1. 1 itself is not considered either a prime number or a composite number. We will discuss the reason for this in Section 5.3 when we discuss the uniqueness of prime factorization.

The first several primes are 2, 3, 5, 7, 11, and 13. The first several composite numbers are 4, 6, 8, 9, 10, and 12.

Exploration 1: Properties of Divisibility

What general statements can you make about the concept of divisibility?

This is a very open-ended question. What we have in mind, roughly, is this: if you know something about certain numbers being divisors of others, what additional divisibility conditions can you deduce? We are looking for theorems that begin something like "If $a \mid b$ and $c \mid d$, then...", where the conclusion is also a divisibility statement.

• E X P L O R E •

With the concept of divisibility in mind, we ask you to reexamine Exploration 2 in Section 1.1, which raised the question of which subsets of $\mathbf{Z}$ are closed under addition or subtraction.

Exploration 2 (Section 1.1) Revisited: Sets Closed under Addition and Subtraction

How are the properties of divisibility that you found in Exploration 1 above reflected in sets which are closed under addition or subtraction? Does thinking about those sets suggest anything further about divisibility that you missed in doing Exploration 1?

• E X P L O R E •

One special type of divisibility is divisibility by 2. The next exploration asks about the numbers that have this property and those that do not. We preface this with a formal definition:

> **DEFINITION 2.7:** For $x \in \mathbf{Z}$:
>
> "x is *even*" means: there is an integer t such that $x = 2 \cdot t$.
> "x is *odd*" means: there is an integer y such that $x = 2 \cdot y + 1$.

Note: We are not defining "odd" as "not even". We will have more to say about the relationship between odd and even in the next section.

Exploration 2: Odd and Even Numbers

What general results can you state about odd and even numbers? Compare the type of questions you asked in Exploration 2 of Section 1.1 ("Sets Closed under Addition or Subtraction"). What about products of odd and even numbers? quotients? powers?

• E X P L O R E •

Linear Diophantine Equations

We conclude this section by posing an interesting and complex problem about the integers. The problem begins with a simple example.

Look at the following equation:

$$36x + 47y = 243\,. \qquad \text{(i)}$$

Are there *integers* x and y that make this equation true? (Keep in mind that x and y do not have to be positive.)

? ? ?

The area of mathematics called "Diophantine equations" explores the question of when certain equations have *integer* solutions. Since (i) is a linear equation in the variables x and y (i.e., its graph is a line), the above problem is called a *linear Diophantine equation*.

The equation (i), "$36x + 47y = 243$", is not of great interest in itself, but it raises a larger category of questions. Compare (i) with each of the following:

$$36x + 46y = 243\,. \qquad \text{(ii)}$$

$$36x + 44y = 242\,. \qquad \text{(iii)}$$

$$36x + 45y = 242\,. \qquad \text{(iv)}$$

> **Historical Note:** The term "Diophantine equation" comes from the name of a Greek mathematician, Diophantus of Alexandria, who probably lived in the fourth century C.E., and is credited with being one of the originators of the use of letters for unknown quantities in arithmetic. The problem posed in Exploration 3 below was studied by Diophantus, analyzed further by Hindu mathematicians in the seventh century, and completely resolved only in the 1860s by H. J. S. Smith (1826–1883).
>
> The most famous Diophantine-equation problem is *Fermat's Last Theorem,* which concerns integer solutions to the equation "$x^n + y^n = z^n$." The great French mathematician, Pierre de Fermat (1601–1665) wrote a note which was found after his death, in which he claimed to have proved that, for $n > 2$, this equation has no positive integer solutions. No proof was found, however, and it is generally believed that whatever Fermat thought was a proof was incorrect. The problem remains unsolved to this day, although many partial results have been obtained.

$$36x + 45y = 240. \qquad \text{(v)}$$

$$36x + 45y = 243. \qquad \text{(vi)}$$

How do these slight changes affect whether or not an integer solution exists? Are any of these easier to deal with than others? Is it easier to show something does not have a solution or to show that it does?

Exploration 3: Linear Diophantine Equations

Examine the following general problem:

For what integers a, b, and c will the equation

$$ax + by = c$$

have integral solutions for x and y?

• E X P L O R E •

EXERCISES

1. (Exploration 1) Which of the following statements are true? (a, b, and c are integers, not necessarily positive.) Give informal explanations of your answers.

- **a)** If $a \mid b$ and $b \mid c$, then $a \mid c$.
- **b)** If $a \mid b$ and $a \mid c$, then $a \mid (b + c)$.
- **c)** If $a \mid c$ and $b \mid c$, then $(ab) \mid c$.
- **d)** If $a \mid b$ and $c \mid d$, then $(ac) \mid (bd)$.
- **e)** If $a \mid b$ and $c \mid d$, then $(a + c) \mid (b + d)$.

f) If $a \mid (bc)$, then either $a \mid b$ or $a \mid c$.
g) If $a \mid b$ and $b \mid a$, then $a = b$.
h) If p and q are primes, and $p \mid q$, then $p = q$.

2. (Exploration 2)
   a) Which of the following statements are true? (a and b are integers.) Give informal explanations of your answers.
      i) If a and b are even, then ab is even.
      ii) If a and b are odd, then ab is odd.
      iii) If ab is odd, then a and b are both odd.
      iv) If ab is even, then a and b are both even.
   b) Formulate correct principles, similar in style to those of part i), for dealing with *sums* of odd and even integers.
3. (Exploration 3) Which of the equations (i) through (vi) (preceding Exploration 3) have solutions? Which do not? If you are not sure about a particular case, say so. Give informal explanations of your answers.
4. (Exploration 3) What happens with the linear Diophantine equation when $c = 0$? That is, for what integer values of a and b does the equation $ax + by = 0$ have an integer solution for x and y? Give an informal explanation of your answer. (*Hint:* Draw the graph of the equation.)
5. (Exploration 3) Suppose that the linear Diophantine equation $ax + by = c$ does have an integer solution for x and y. Will it necessarily have more than one solution? How many will it have? ("A solution" is a pair of integers for x and y that fit the equation.) Give informal explanations of your answer. (*Hint:* You may find your work on Exercise 4 helpful.)

2.4 Proofs about Integers

We will look at several proofs involving divisibility in this section, both to clarify and pin down the results themselves, and to illustrate some further ideas about proofs.

Basic Facts about Divisors

Example 1 **A Divisor of a Divisor** This example illustrates how to use the definition of divisibility in a proof.

THEOREM 2.9. For integers a, b, and c,

$$\text{if } a \mid b \text{ and } b \mid c, \text{ then } a \mid c.$$

Proof: Assume that a, b, and c are integers such that $a \mid b$ and $b \mid c$. We need to show that $a \mid c$.

Since $a \mid b$, there is an integer, call it x, such that

$$ax = b. \qquad \text{(i)}$$

Similarly, since $b \mid c$, there is an integer, call it y, with

$$by = c. \tag{ii}$$

We need to show that there is an integer, say z, such that

$$az = c. \tag{iii}$$

Define z to be the product xy. Because the product of two integers is an integer, we know that $z \in \mathbf{Z}$. We will show that this integer z "works", i.e., satisfies (iii). The proof is as follows: Using (i), we can replace b in (ii) by ax, to get $(ax)y = c$, or equivalently, $a(xy) = c$. This says precisely that $z = xy$ is a solution of (iii), which is exactly what we want. ■

COMMENTS

1. Since the definition of divisibility (Definition 2.5) involves the statement that something exists, we needed to find the "something" in order to prove this theorem. We did that by using the objects which the hypothesis told us had to exist. We built the object we wanted out of the objects we already had.
2. Be careful to verify that an object you create is in the set from which it's supposed to come—here, z must be an integer. Often proving that the object satisfies that restriction is the hard part of the proof.
3. Note the resemblance between this theorem and that of Example 1 in Section 2.2 ("A Subset of a Subset"). This common property of the subset relationship and the divisibility relationship is called *transitivity*. We will formally define this property in Section 6.1.

COROLLARY 2.10. Any multiple of an even number is even.

Proof: Suppose z is a multiple of some even number y. In the special case where $y = 0$, z must also be 0, so z is even, and the theorem holds. If $y \neq 0$, then our hypothesis (that y is even and z is a multiple of y) is the same as saying that $2 \mid y$ and $y \mid z$. Therefore this situation is simply a particular case of Theorem 2.9, with $a = 2$, $b = y$, and $c = z$. The conclusion we can draw from that theorem is that $2 \mid z$, i.e., z is even, which is precisely what we want here. ■

COMMENTS

1. Note once again the term "corollary"—this theorem is proved on the basis of the previous one.
2. The corollary was stated without reference to specific variables, but in order to apply Theorem 2.9, it was helpful to introduce variables. We chose different letters from those in that theorem, in order to be able to state the substitution clearly.

3. In applying Theorem 2.9, we showed that the objects 2, y, and z satisfied the hypothesis of that theorem, and thus were guaranteed (by a "guarantee-producing machine") that they satisfied the conclusion.

4. You should always be careful about special cases, such as situations involving zero or the empty set. They are usually easy to deal with individually, but sometimes they are actually exceptions to the theorem you are trying to prove. (For example, review Example 5 in Section 2.2.)

5. This corollary was stated in GYHD 1 back in Section 1.3.

The following result uses a similar approach.

Get Your Hands Dirty 1

Prove the following:
For $a, b, c \in \mathbf{Z}$, if $a \mid b$ and $a \mid c$, then $a \mid (b + c)$ and $a \mid (b - c)$. □

The result in GYHD 1 will be used later, so we restate it in different language, as a formal theorem. The form of the restatement may sound familiar:

THEOREM 2.11. If $a \in \mathbf{Z}$, then the set $a\mathbf{Z}$ is closed under both addition and subtraction.

Proof by Contradiction

We will now use the results of this section so far to get a partial answer to the question raised in Exploration 3—"Linear Diophantine Equations"—in Section 2.3. You may have noticed, for example, that the equation

$$36x + 46y = 243$$

does not have any integral solutions for x and y. We reason as follows:

Since 36 and 46 are both even integers, the expressions "$36x$" and "$46y$" will also both be even, no matter what integers we choose for x and y. And then the sum of the even integers $36x$ and $46y$ will again be even, so it can't equal 243.

We will prove the following simple theorem based on this reasoning:

THEOREM 2.12. For $a, b, c \in \mathbf{Z}$, if a and b are even, and c is not even, then the equation

$$ax + by = c$$

has no integral solution for x and y.

COMMENT The proof uses an important method known as *proof by contradiction*. The basic idea of the method is as follows: there is some statement that we are trying to prove. We show that it is impossible for that statement to be false, by showing that, if it were false, e.g., if there were a counterexample, then something contradictory or impossible would happen. Since we don't allow contradictions or impossible situations in mathematics, there can be no counterexample, and so we have proved our theorem.

Proof: Suppose that this theorem had a counterexample, i.e., suppose that there were integers a, b, and c, with a and b even, and c not even, for which the equation

$$ax + by = c$$

does have an integral solution for x and y. In other words, suppose that there are integers x and y with $ax + by = c$.

By Corollary 2.10, ax and by are both even (since they are multiples of the even numbers a and b). But then, by Theorem 1.1 (from Example 1 of Section 1.3), the sum $ax + by$ is also even, i.e., c is even (since c is equal to $ax + by$).

But we assumed c was not even, so c is both even and not even. This is impossible, and this impossible situation resulted from the supposition that the theorem had a counterexample. Therefore the theorem has no counterexample, i.e., the theorem is true. ■

COMMENTS

1. We repeat for emphasis the basic idea: we showed that, if there were a counterexample to the theorem, then both "$2 \mid c$" and "$2 \nmid c$" would be true statements. Since this is impossible, there must not be a counterexample, and the theorem must be true.

2. The methodology of a proof by contradiction is to assume that what you're trying to prove is false. We then use that assumption to deduce some sort of contradiction or impossible situation; often, we get a contradiction of some element of the hypothesis of the theorem itself. Having produced such a contradiction, we then conclude that the original statement must be true, which completes the proof.

3. The assertion that a particular statement is false is called the *negation* of that statement. The negation of any conditional statement is the assertion that there is a counterexample. Often the applicability of the method of proof by contradiction depends on how easy it is to express and work with the negation of the conclusion of the theorem under question. In our example, that negation was very easy to state, because the conclusion itself was a negative statement. The negation also was easy to work with because it provided us with elements (x and y) which fit a particular equation ($ax + by = c$). We could then combine that equation with our earlier results, giving us our contradiction. The concept of negation will be explored more thoroughly in Chapters 3 and 4, and proof by contradiction examined further in Section 4.2.

There is nothing special about divisibility by 2. The following theorem is a generalization of Theorem 2.12. Its proof, like its statement, is obtained from that of Theorem 2.12

by replacing the word "even" with "divisible by d" and using appropriate generalizations of Corollary 2.1 and Theorem 1.1 (from Example 1). (We leave details as Exercise 3.)

THEOREM 2.13. For $a, b, c \in \mathbf{Z}$ and $d \in \mathbf{Z} - \{0\}$, if a and b are divisible by d, and c is not divisible by d, then the equation

$$ax + by = c$$

has no integral solution for x and y.

Common Divisors

Theorem 2.13 deals with a situation where there is an integer d which is a divisor of both a and b. To help discuss this frequent situation, we introduce the following definition:

DEFINITION 2.8: For $x, y \in \mathbf{Z}$, "t is a *common divisor of x and y*" means: $t \mid x$ and $t \mid y$.

Example 2 Common Divisors

2 is a common divisor of 8 and 12.
-9 is a common divisor of 9 and 27.
1 is a common divisor of 6 and 22.
26 is a common divisor of 78 and 0. □

COMMENTS

1. 1 is a common divisor of any pair of integers, so any pair of integers has at least one common divisor.
2. A divisor of a positive integer cannot be larger than that integer (see Exercise 2). Therefore, if at least one of the two numbers x and y is not zero, then a common divisor cannot be larger than the absolute value of that nonzero number. Thus, any pair of integers which are not both zero can only have a finite number of common divisors, and so there must be a largest one.

Comment 2 suggests the following further definition:

DEFINITION 2.9: For $x, y \in \mathbf{Z}$, "t is the *greatest common divisor of x and y*" means:

i) t is a common divisor of x and y, and
ii) every common divisor of x and y is less than or equal to t.

The greatest common divisor of x and y is written GCD(x, y).

COMMENTS

1. By Comment 2 above, if at least one of x and y is nonzero, then there is a greatest common divisor for x and y. You should verify that a given pair of integers cannot have more than one greatest common divisor.
2. If x and y are both 0, then there is no greatest common divisor, since every nonzero integer is a common divisor.
3. If x and y are not both zero, then $\text{GCD}(x, y)$ is positive. (Why?)
4. We will see in Section 5.3 that every common divisor of x and y is actually a divisor of $\text{GCD}(x, y)$. Unfortunately, this is not an easy result to prove.

For many pairs of integers x and y, the only positive common divisor is 1, so $\text{GCD}(x, y) = 1$. This condition on x and y is related to many important results in number theory, and mathematicians have given it a name:

DEFINITION 2.10: For $x, y \in \mathbf{Z}$, "x and y are *relatively prime*" means:

$$\text{GCD}(x, y) = 1.$$

Example 3 Greatest Common Divisors

$\text{GCD}(8, -12) = 4.$
$\text{GCD}(27, 9) = 9.$
$\text{GCD}(-6, -22) = 2.$
$\text{GCD}(14, 5) = 1$ (so 14 and 5 are relatively prime).
$\text{GCD}(-18, 0) = 18.$ □

Get Your Hands Dirty 2

Find the following GCDs:

$\text{GCD}(10, -18)$
$\text{GCD}(7, 12)$
$\text{GCD}(24, 0)$
$\text{GCD}(8, -16)$ □

With these concepts in mind, we can state the following as a special case of Theorem 2.13:

COROLLARY 2.14. Suppose $a, b, c \in \mathbf{Z}$, with a and b not both 0, and $d = \text{GCD}(a, b)$. If $d \nmid c$, then the equation $ax + by = c$ has no integer solutions for x and y.

COMMENTS

1. This corollary only says that certain equations do not have a solution. It does not say that any equations do have a solution. You may have decided, in your exploration, that if $d \mid c$, then the equation has a solution. That assertion is true, but it is not part

of Corollary 2.14. The proof that the equation has a solution if $d \mid c$ is considerably more difficult than the proof of Corollary 2.14. We will prove that result in Section 5.3.

2. The first sentence in the statement of this theorem is really part of the hypothesis. It is stated as a separate "suppose", rather than as part of the "if . . . , then . . .", for readability. This is commonly done by mathematicians.

3. Comment 4 following the definition of the GCD implies that, if there is any common divisor d of a and b which does not divide c, then GCD(a, b) also does not divide c. Therefore, this special case of Theorem 2.13 is really "just as good as" the general case. By this we mean that, if there is any way of using Theorem 2.13 to rule out existence of a solution to "$ax + by = c$", then we can also use Corollary 2.14 to show that there is no solution.

4. You should clarify for yourself what role the assumption "a and b not both zero" played in the theorem.

5. You can think of this result as a corollary of the earlier theorems on divisibility. Notice how easily this theorem follows from those results.

6. Corollary 2.14 can be phrased in a more "positive" form, as follows:

   COROLLARY 2.14′. Suppose $a, b, c \in \mathbf{Z}$, with a and b not both 0, and $d =$ GCD(a, b). If the equation $ax + by = c$ has an integer solution for x and y, then $d \mid c$.

   In fact, the algebra of the proof given for Corollary 2.14 fits more closely with this variation. In Section 4.2, we will see that this variation on Corollary 2.14 is essentially its contrapositive, and that proving the contrapositive of a statement is equivalent to proving the statement itself.

Odd and Even Integers

In Definition 2.7, we defined an integer x to be *odd* if there is an integer t with $x = 2 \cdot t + 1$. We often think of "odd" as meaning simply "not even", but this is not what our definition says. We will now prove that an odd integer cannot be even. We begin with the special case $x = 1$ (verify that the following theorem says "1 is not even"):

THEOREM 2.15. If $u \in \mathbf{Z}$, then $2 \cdot u \neq 1$.

COMMENT The proof of this special case uses several of our "basic assumptions about arithmetic" (see Appendix B), especially the trichotomy property [item d) v)]. A special case of the trichotomy property says that every real number is either negative, zero, or positive, and that a real number can't be more than one of these. We will refer to others of these assumptions by item as needed.

Just to make sure we aren't overlooking the obvious, we will formally define 2 as $1 + 1$. Since $1 > 0$ [by f) i)], we have $1 + 1 > 1 + 0$ [by d) ii)], so $2 > 1$.

We now proceed with the proof of Theorem 2.15:

Proof of Theorem 2.15: Suppose u is an integer. Our proof is by cases: u is either negative, zero, or positive. We look at these one at a time:

Case 1: $u < 0$. Since $2 > 0$, we have $2 \cdot u < 0 \cdot u$ [by basic assumption d) iv), with $a = 2, b = 0, c = u$]. Since $0 \cdot u = 0$ [by b) ii)], this means that $2 \cdot u < 0$. But 1 is positive [by basic assumption f) i)], so $2 \cdot u \neq 1$ in this case. (The trichotomy property implies that a number cannot be both negative and positive.)

Case 2: $u = 0$. In this case, $2 \cdot u = 0$ [by basic assumption b) ii)]. Again, this tells us that $2 \cdot u \neq 1$, since $0 \neq 1$ (by the trichotomy property).

Case 3: $u > 0$. In this case, $u \geq 1$ [by basic assumption f) i)], so $2 \cdot u \geq 2$ [by basic assumption d) iii), with $a = u, b = 1, c = 2$]. But $2 > 1$, so we have $2 \cdot u > 1$, and so $2 \cdot u \neq 1$ (again, by the trichotomy property).

Thus, in all cases, we get $2 \cdot u \neq 1$, so the theorem is true. ■

With this special case taken care of, we can now prove the general statement:

THEOREM 2.16. For $x \in \mathbf{Z}$, if x is odd, then x is not even.

Proof: Suppose x is odd, and assume, by way of contradiction, that x is also even. By Definition 2.7 in Section 2.3, there are integers t and y with $x = 2 \cdot t$ and $x = 2 \cdot y + 1$. Therefore $2 \cdot t = 2 \cdot y + 1$. Subtracting $2 \cdot y$ from both sides of this equation, we get $2 \cdot (t - y) = 1$. Letting $u = t - y$, we have $u \in \mathbf{Z}$ and $2 \cdot u = 1$, i.e., we have a counterexample to Theorem 2.15.

But we proved that Theorem 2.15 is true, so it has no counterexamples. That is a contradiction, so Theorem 2.16 must be true. ■

COMMENTS

1. The contradiction here is that we demonstrated something which is contrary to a previously proved theorem. That is a very common form for proof by contradiction.
2. The phrase "assume, by way of contradiction" is often used as a short way of saying: "We are going to prove this theorem by the method of proof by contradiction. So assume that the conclusion is false, and we will show that something impossible happens. It will then follow that the conclusion must actually be true."

We can paraphrase Theorem 2.16 as follows:

THEOREM 2.16′. An integer cannot be both odd and even.

This paraphrased version simply says: Theorem 2.16 has no counterexamples.

COMMENT Theorem 2.16 does not say that "odd" and "not even" are synonymous, since it is still possible, based on our results, that a number could be neither odd nor even. The formal proof that this is impossible is based on a theorem known as the *Division Algorithm,* which we will prove in Section 5.2.

Common Multiples

There are concepts for multiples which are parallel to those for divisors that we have been discussing. We define "common multiple" here, and then refer you to Exercise 7 for more details:

DEFINITION 2.11: For $x, y \in \mathbf{Z}$, "t is a *common multiple of x and y*" means: t is a multiple of x and t is a multiple of y.

EXERCISES

1. (GYHD 1) Prove the following for $a, b, c \in \mathbf{Z}$:
   If $a \mid b$ and $a \mid c$, then $a \mid (b + c)$ and $a \mid (b - c)$.
2. Prove the following (using basic assumptions from Appendix B):
   If $a, x \in \mathbf{N}$ and $a \mid x$, then $a \leq x$.
3. Give details of the proof of Theorem 2.13.
4. Prove each of the following by contradiction. (You may use the fact that every integer must be either odd or even, and cannot be both odd and even.)
   - **a)** If $x \in \mathbf{Z}$ and x^2 is even, then x is even.
   - **b)** If $a, b \in \mathbf{Z}$ and ab is odd, then a and b are both odd.
   - **c)** If $a, b, c \in \mathbf{Z}$ and $a \nmid bc$, then a does not divide either b or c.
   - **d)** If S is a set and $S \times S \neq \emptyset$, then $S \neq \emptyset$.
5. Which of the following statements are true? If the statement is false, give a counterexample. If you can think of a way to "fix" a false statement, do so. (All variables are integers.) *Note:* Many proofs about GCD involve some very subtle concepts, so we are not asking you to write proofs of those statements that you think are true.
   - **a)** If $\text{GCD}(a, b) = 1$ and $t \neq 0$, then $\text{GCD}(at, bt) = t$.
   - **b)** If $\text{GCD}(a, b) = 1$ and $\text{GCD}(a, c) = 1$, then $\text{GCD}(a, bc) = 1$.
   - **c)** If $\text{GCD}(a, b) = x$ and $\text{GCD}(a, c) = y$, then $\text{GCD}(a, bc) = xy$.
6. If x is a prime number and $y \in \mathbf{Z}$, what are the possibilities for $\text{GCD}(x, y)$? Explain your answer.
7. There is a companion concept, related to greatest common divisor, called *least common multiple*. It is defined as follows:

   DEFINITION: For $x, y \in \mathbf{Z}$, both nonzero, "t is the *least common multiple of x and y*" means:

   i) t is a positive common multiple of x and y, and
   ii) every positive common multiple of x and y is greater than or equal to t.

The least common multiple of x and y is abbreviated LCM(x, y). *Note:* If either x or y is zero, LCM(x, y) is defined to be 0.

a) Prove that, if x and y are both nonzero integers, then there is at least one positive integer which is a multiple of both x and y. (*Hint:* First look at the case where x and y are both positive; there is a simple "generic" candidate.)
b) Explain why we require that both x and y be nonzero in the general definition of LCM.
c) Explore the relationship between the concepts of GCD and LCM. Consider the following questions (you may want to consider only positive x and y at first—try lots of examples):
  i) If GCD$(x, y) = 1$, what conclusion can you draw about LCM(x, y)?
  ii) If x is a prime number, what are the possibilities for LCM(x, y)?
  iii) What general equation can you write relating x, y, GCD(x, y), and LCM(x, y)?

8. Express the negation of each of these statements without just inserting "not" into the statement. That is, express the negation in "useful", "positive", or "direct" terms. (You may assume that the variables belong to some appropriate universe.) (*Example:* $A \subseteq B$. The negation could be stated "$A \not\subseteq B$", but it can be stated "positively" as "there is some object in $A - B$".)
a) $x < y$.
b) z is composite. (Assume that $z \in \mathbf{N}$.)
c) $t \in A \cup B$.

9. PROOF EVALUATION (see instructions for Exercise 6 in Section 2.2):

a) **"THEOREM":** For any sets A, B, and C, if $A \cup B = A \cup C$, then $B = C$.
"Proof": We will prove this by contradiction. Suppose that $A \cup B$ and $A \cup C$ are not equal. Then there is some object x that is in one and not the other. We proceed by looking at two cases:
First look at the case where $x \in A \cup B$ and not in $A \cup C$. Then x is not in A (because, if it were, it would also be in $A \cup C$). So x must be in B (otherwise it wouldn't be in $A \cup B$). Also, x is not in C (because, if it were, it would also be in $A \cup C$). Therefore x is in B and not in C, which contradicts the condition $B = C$.
The case where $x \in A \cup C$ and not in $A \cup B$ is done by a parallel argument.

b) **"THEOREM":** For any sets A, B, and C, if $A \cap B = \varnothing$ and $C - A = C - B$, then $C \subseteq A'$.
"Proof": We will prove this by contradiction. Suppose $A \cap B = \varnothing$ and $C - A = C - B$, but $C \not\subseteq A'$. Then there is an element, say x, which is in C but not in A'. Therefore x is in C and in A. But then x is not in $C - A$, so it is not in $C - B$ either. Since x is in C, it must also be in B (otherwise it would be in $C - B$). But then we have x in both A and B, which is a contradiction.

c) **"THEOREM":** For any sets A and B, if $\mathcal{P}(A) \cap \mathcal{P}(B) \neq \varnothing$, then $A \cap B \neq \varnothing$.
"Proof": We will prove this by contradiction. Assume A and B are sets, and to get a contradiction, assume A and B are *not* disjoint. Then there is some element x in $A \cap B$. So $\{x\} \in \mathcal{P}(A) \cap \mathcal{P}(B)$, which means that $\mathcal{P}(A) \cap \mathcal{P}(B)$ is not empty, and we are done.

d) **"THEOREM":** For $a, b, c \in \mathbf{N}$, if $a \mid b$ and $a \mid c$, then $a \mid (bc)$.
"Proof": Assume that the hypothesis is true. Then, by definition of divisibility, there

are integers x and y such that $b = ax$ and $c = ay$. Then $bc = (ax)(ay) = a(axy)$. Since $axy \in \mathbf{Z}$, this shows that $a \mid (bc)$.

e) **"THEOREM":** Suppose $a, b \in \mathbf{N}$, and $d = \text{GCD}(a, b)$. Then $d^2 = \text{GCD}(a^2, b^2)$.
"Proof": By hypothesis, we have that $d \mid a$ and $d \mid b$, so there are integers s and t with $a = ds$ and $b = dt$. Then $a^2 = d^2s^2$ and so $d^2 \mid a^2$. Similarly, $d^2 \mid b^2$. Thus d^2 is a common divisor of a^2 and b^2, as desired.

f) **"THEOREM":** For $a, b, c \in \mathbf{N}$, if $ac \mid bc$, then $a \mid b$.
"Proof": We will prove this by contradiction. Assume that $ac \mid bc$, but that "$a \mid b$" is false, i.e., that $b \mid a$. Then, for some $x \in \mathbf{Z}$, $a = bx$. Then $ac = bcx$, and so $bc \mid ac$, contrary to the opposite assumption.

3

The Language and Logic of Mathematics

Introduction

We have introduced the basic themes of this book—exploration and proof—and given some examples, in terms of sets and integers, of how these themes operate.

As we move through this book, we will be encountering "meatier" mathematical ideas, which require more complex forms of mathematical proof. Among the things that make "doing mathematics" difficult for some people are its special language and the complexity of its logic. Understanding an "if . . ., then . . .", or conditional, sentence requires the ability to analyze its hypothesis and its conclusion. Often these two components are themselves made up from other statements, so we have layers of complexity. We need to know how to use the assumptions provided by the hypothesis, and need to know exactly what is required from us in order to demonstrate its conclusion.

In this chapter, we will introduce you to the language and logic behind this type of analysis. We will illustrate how the basic terminology and concepts are used, and discuss their importance.

One of the keys to the language is the concept of a *quantifier,* which is a device for clarifying the usage and meaning of variables. There are two different quantifiers—*universal* and *existential*—and we will be examining each in detail.

The two primary ideas of formal mathematical logic that we will discuss are *implication* and *negation*. Implication can be combined with the universal quantifier to express the basic "if . . ., then . . ." statement. Negation provides a link between the two types of quantifiers, and is an essential element in understanding the connection between an "if . . ., then . . ." statement and a counterexample.

Section 3.1 will present an intuitive introduction to quantifiers and logic, including many examples, and will present the basic definitions for quantifiers, implication, and negation.

In Section 3.2, we will go further into the use of these ideas, illustrating them by the development of two separate mathematical topics. We will see how the language of quantifiers is used to give precise meaning, as well as concise expression, to complicated concepts.

3.1 Quantifiers, Implication, and Negation

One of the major tasks in understanding a mathematical theorem is analyzing what is required for its proof. Compare the following two theorems from Chapter 2:

THEOREM 2.2. For sets A, B, and C,

$$\text{if } A \subseteq B \text{ and } B \subseteq C, \text{ then } A \subseteq C.$$

THEOREM 2.9. For integers a, b, c,

$$\text{If } a|b \text{ and } b|c, \text{ then } a|c.$$

Although these two theorems appear to be very similar, their proofs follow significantly different paths. Let's look at what is needed in order to show each of their conclusions.

In order to show that the conclusion of Theorem 2.2—"$A \subseteq C$"—is true, we suppose that x is an element of A. We then are required to show that x is in C. Notice that the element x is *given to us,* and we have to *show something* about it.

In Theorem 2.9, on the other hand, in order to show "$a|c$", we have to *produce* an integer z which satisfies the condition $a \cdot z = c$. We *construct* z based on information in the hypothesis.

In understanding what's involved in proving these two theorems, we need to recognize that they have *hidden variables*. What the two proofs have in common is that each requires the introduction of new variables—variables referring to objects that are not specifically mentioned in either the hypothesis or the conclusion of the theorem. These variables are necessary because certain terminology in the theorem indirectly refers to them. Specifically, the definitions of both "subset" and "divides" are expressed in terms of "outside" variables.

Recall:

DEFINITION 2.2: For sets A and B, "A is a *subset* of B" means: if $u \in A$, then $u \in B$.

DEFINITION 2.5: For $x, y \in Z$, with $x \neq 0$, "x is a *divisor* of y" means: there is some integer, call it t, such that $x \cdot t = y$.

The subset relationship is defined by a statement that *all elements of* A have a certain property (namely, they belong to B). In order to prove a statement of this form, we need to introduce a variable to represent an arbitrary element of A.

The divisibility relationship is defined by a statement about the *existence of an integer* with a certain property (namely, it satisfies a particular equation). In order to prove a statement of this form, we have to define or construct an object which will represent this element.

Because mathematics builds upon itself, the terminology introduced in the above definitions will become a part of other, more complex, definitions. As the complexity grows, it becomes more difficult to trace through the levels of hidden and explicit variables, and understand exactly what can be assumed and what needs to be shown. The difference in form between the two definitions above is an important key in sorting through this complexity. One of the primary reasons for introducing quantifiers is to clarify this difference.

Universal and Existential Quantifiers

The subset relationship is an example of a property defined by a *universal condition,* that is, by a condition on all elements of a given type. The divisibility relationship is an example of a property defined by an *existential condition,* that is, by a statement that asserts the existence of an element of a given type.

This dichotomy is captured by the two quantifiers which we will examine. We will informally illustrate the use of both the universal and the existential quantifiers, and then give precise general definitions.

The Universal Quantifier

As we have noted, the statement "$A \subseteq B$" is an assertion about all elements of A. It is the statement that these elements all belong to B.

The symbol "$\forall$"—read "for all" or "for every"—is called the *universal quantifier.* It is generally followed by a variable name, a set restricting that variable, and an assertion or open sentence about the variable. We can illustrate the use of the universal quantifier by defining the phrase "$A \subseteq B$" by the following expression:

$$(\forall\ u \in A)(u \in B)\,. \tag{$*$}$$

This statement can be read: "For every element u in A, u is an element of B."

This is not really a statement about u, but rather is a statement about A and B. The variable u is only an incidental tool for the purpose of saying something about the relationship between the two sets. If the letter "u" were replaced by some other symbol in $(*)$, the meaning would be unchanged. Thus, "$(\forall\ w \in A)(w \in B)$" means exactly the same thing as "$(\forall\ u \in A)(u \in B)$". Both say "every element of A belongs to B".

The Existential Quantifier

The statement "$x \mid y$" is a different type of assertion. It is a statement about the existence of an object t with a certain property—namely, the object t satisfies the equation $x \cdot t = y$.

The symbol "∃"—read "there exists"—is called the *existential quantifier.* It is also generally followed by a variable name, a set restricting that variable, and an assertion or open sentence about the variable. We can illustrate the use of the existential quantifier by defining the phrase "$x \mid y$" by the following expression:

$$(\exists\, t \in \mathbf{Z})(x \cdot t = y)\,. \qquad (**)$$

This statement can be read: "There exists an element t in $\mathbf{Z}$ such that $x \cdot t = y$".

As with u in (*), the variable t here is only an incidental part of the statement. It could be replaced in (**) by any other symbol, and (**) is really a statement about x and y.

COMMENTS

1. You can remember the quantifier symbols by the fact that the universal quantifier is used for statements about *all* elements in a set, and the symbol is an upside-down letter A. The existential quantifier is used for statements about *existence* of elements in a set, and the symbol is a backwards letter E.
2. In reading a statement with "∃", it is important to insert the phrase "such that" for the statement to make sense. There is no corresponding phrase needed with "∀".
3. The set which occurs in the role played by A in (*) or by $\mathbf{Z}$ in (**) is sometimes referred to as a *replacement set*. With both universal and existential quantifiers, the replacement set is sometimes omitted when the context makes it clear. We shall see shortly that the replacement set can also be incorporated into the open sentence that follows it.

Free and Bound Variables

Each of statements (*) and (**) involves three variables: (*) has A, B, and u, and (**) has x, y, and t. As we noted, however, the statements are really *about* only two of the variables in each case. Variables that are used in incidental roles, such as u in (*) and t in (**), are called *bound variables, dummy variables,* or *quantified variables*. (These terms are used interchangeably.)

Generally speaking, a variable referred to by a quantifier is a bound variable. The role of a bound variable is circumscribed or defined by its quantifier. Often, a statement involving a bound variable can be made, perhaps clumsily, without explicitly naming the bound variable.

Thus, (*) could be stated: "every element of A belongs to B." (**) could be stated: "there is an integer which, when multiplied by x, gives y as the product."

A variable which is not bound is called a *free variable*. In (*), A and B are free variables, and (*) is really an open sentence about A and B. For certain pairs of sets A and B, (*) will be true; for other pairs of sets, it will be false. Similarly, x and y are free variables in (**), which is really an open sentence about x and y. It is true for some pairs of integers x and y, and false for other pairs. If we are looking for the truth set of a statement with both free and bound variables, we are asking what values for the free variable(s) will make it true.

We will often be working with statements that involve both free and bound variables. As long as there is at least one free variable, the statement will have the status of being an open sentence for its free variable(s). A statement whose variables are all bound is either true or false.

For example, "$(\forall\ t \in \mathbf{N})(w \mid t)$" has the bound variable t and the free variable w, and is an open sentence. We can express this open sentence in words as "w is a divisor of every natural number", and represent this open sentence as $p(w)$. This statement will be true for some values of w and false for others (see GYHD 2 on page 94). On the other hand, "$(\forall\ x \in \mathbf{R})\,(x^2 \geq 0)$" has only the bound variable x, and this is a true statement.

Similarly, "$(\exists\ v \in \mathbf{Z})(v^2 - 5v + r = 0)$" has the bound variable v and the free variable r, and is an open sentence which we can write as $q(r)$, while "$(\exists\ s \in \mathbf{Z})(s^2 = 7)$" has only the bound variable s, and is a false statement.

The following example examines in more detail the case of an existentially quantified statement with both a free and a bound variable:

Example 1. Truth Set of an Existentially Quantified Statement with Free and Bound Variables Consider the open sentence

$$(\exists\ x \in \mathbf{R})\,(x^2 = u) \qquad \text{with} \quad \mathbf{U} = \mathbf{R}.$$

This statement has "u" as a free variable (with universe $\mathbf{R}$) and "x" as a bound variable, and we can represent the statement as "$p(u)$". To understand what $p(u)$ is saying, we can try various real numbers for u, and see what the statement says:

$u = 3$: $p(3)$ says "$(\exists\ x \in \mathbf{R})(x^2 = 3)$". This is true, since $x = \sqrt{3}$ satisfies the equation. [$x = -\sqrt{3}$ also satisfies the equation, but all we need is one value for x to make $p(u)$ true.]

$u = 0$: $p(0)$ says "$(\exists\ x \in \mathbf{R})(x^2 = 0)$". This is true, since $x = 0$ satisfies the equation.

$u = -2$: $p(-2)$ says "$(\exists\ x \in \mathbf{R})(x^2 = -2)$". This is false, since there is no real number whose square is -2.

The statement $p(u)$ says "u has a real square root", and the truth set of $p(u)$ is $\mathbf{R}^{\geq 0}$ (the nonnegative real numbers). □

Get Your Hands Dirty 1

a) Give an example of a pair of real numbers x and y for which the statement "$(\exists\ t \in \mathbf{Z})(x + t = y)$" is true, and an example of a pair of real numbers x and y for which this statement is false.

b) Using $\mathbf{R} \times \mathbf{R}$ as the universe, describe the graph of the open sentence in part a. □

We will give an example using a universally quantified statement following the definition of implication (Definition 3.3). Meanwhile, we offer the following additional perspective on the distinction between free and bound variables.

Example 2. **Free and Bound Variables in Calculus** You may find it helpful to see the distinction between free and bound variables in another context. In calculus, we examine areas under curves by means of integrals. For example, the area under the curve $y = x^2$ between $x = 1$ and $x = 5$ is equal to

$$\int_1^5 x^2\,dx\,.$$

In this expression, x is a dummy variable—the above integral has the exact same meaning as

$$\int_1^5 u^2\,du\,.$$

We can introduce a free variable, by using a variable as one of the limits of integration. In the expression

$$\int_1^t u^2\,du$$

u is a dummy variable, and t is a free variable. This integral represents the area under the curve $y = x^2$ between $x = 1$ and $x = t$, and its value depends on t. □

Predicate of a Statement with Quantifiers

It is helpful to have a way of referring to the "underlying open sentence" involved in a statement with quantifiers. In "$(\forall\, t \in \mathbf{N})(w\,|\,t)$", the underlying open sentence is "$w\,|\,t$", which has two variables, w and t. In "$(\exists\, v \in \mathbf{Z})(v^2 - 5v + r = 0)$", the underlying open sentence is "$v^2 - 5v + r = 0$", which has two variables, v and r.

In any statement with its quantifiers at the beginning, the open sentence which comes after all the quantifier expressions will be called the *predicate* of the statement. Thus, the predicate of $p(w)$ is "$w\,|\,t$", while the predicate of $q(r)$ is "$v^2 - 5v + r = 0$".

Note: Variables may be free in the predicate itself, but then become bound when the quantifier is put in front of the predicate. For example, in the predicate, "$w\,|\,t$", both w and t are free variables. The variable t "becomes bound" when we write "$(\forall\, t \in \mathbf{N})$" in front of the predicate. Similarly, "$v^2 - 5v + r = 0$" has two free variables, r and v, but in the expression "$(\exists\, v \in \mathbf{Z})(v^2 - 5v + r = 0)$", v is bound and r is free.

Another Informal Example of Each Quantifier

Before turning to a formal definition of "$\forall$" and "$\exists$", it may be helpful to give another example of each.

Suppose that the symbol t is used to represent some integer greater than 1, and consider the statement "t is a prime".

How can we express this?

In Section 2.3, we defined this phrase as follows:

> **DEFINITION 2.6:** For $t \in \mathbf{N}$, $t > 1$, "t is a *prime*" means: the only positive integers which are divisors of t are t itself and 1.

The important consideration for our purposes here is that this definition expresses a condition about *all positive divisors of t*—namely, it says that these divisors are all either 1 or t.

There is a hidden variable in this definition—a variable which will represent such a divisor—and so this definition is an excellent candidate for the use of the universal quantifier. For convenience, let D represent the set of all positive divisors of t. We can then restate Definition 2.6 as follows:

> **DEFINITION 2.6′:** For $t \in \mathbf{N}$, $t > 1$, "t is a *prime*" means:
>
> $$(\forall\, x \in D)(x = 1 \text{ or } x = t)\,.$$

Notice that the use of the universal quantifier to define "t is a prime" makes the hidden variable explicit.

For comparison, consider the statement "u is a perfect square" (here, u is to be a nonnegative integer).

How can we express this?

We can define this phrase as follows:

> **DEFINITION 3.1:** For $u \in \mathbf{W}$, "u is a *square*" (or "*perfect square*") means: there is an integer whose square is u.

This is an existence statement—readily identified as such by the phrase "there is". This definition also has a hidden variable, referred to by the phrase "an integer". But in order to work with the idea of perfect squares, we will generally have to give that integer a name. And so, we have here a candidate for the use of the existential quantifier.

We can restate Definition 3.1 as follows:

DEFINITION 3.1′: For $u \in \mathbf{W}$, "u is a *square*" means:

$$(\exists\, y \in \mathbf{Z})(y^2 = u)\,.$$

Notice again that the use of the quantifier to define "u is a square" makes the hidden variable explicit.

In the quantifier version of the definition of "t is a prime" (Definition 2.6′), t is a free variable and x is a bound variable. The predicate is "$x = 1$ or $x = t$". The entire quantified statement "$(\forall\, x \in D)(x = 1$ or $x = t)$" is an open sentence about t, true for some values of t, false for others.

Note: the set D is itself defined in terms of t; specifically, $D = \{w \in \mathbf{N} : w \mid t\}$. We could make this description of D explicit by defining "t is a prime" as follows:

"t is a *prime*" means: $(\forall\, x \in \{w \in \mathbf{N} : w \mid t\})\,(x = 1 \text{ or } x = t)$,

but this is very clumsy, and introduces yet another variable, w. We will shortly show you a way to accomplish this goal more conveniently.

In the quantifier version of the definition of "u is a square", u is a free variable and y is a bound variable, and the predicate is "$y^2 = u$". The entire quantified statement "$(\exists\, y \in \mathbf{Z})\,(y^2 = u)$" is an open sentence about u, true for some values of u, false for others.

Formal Definitions of "∀" and "∃"

With these examples in mind, we now give formal definitions of the use of the symbols "∀" and "∃":

DEFINITION 3.2: For an open sentence $p(x)$ and a set S:

"$(\forall\, x \in S)\,(p(x))$" means:

every element of S makes $p(x)$ true;

"$(\exists\, x \in S)\,(p(x))$" means:

there is an element of S which makes $p(x)$ true.

More concisely, using P to represent the truth set of $p(x)$, we can write

"$(\forall\, x \in S)\,(p(x))$" means: $S \subseteq P$,

and

"$(\exists\, x \in S)\,(p(x))$" means: $S \cap P \neq \varnothing$.

The open sentence $p(x)$ used in these definitions may actually involve other variables in addition to the variable x. This has been the case in most of the examples discussed so far.

The set S itself can be considered a variable in these quantified statements. If x is the only variable in the predicate $p(x)$, then each of the statements "$(\forall\, x \in S)\,(p(x))$" and "$(\exists\, x \in S)\,(p(x))$" is really an assertion about S.

For example, consider the statement "$(\forall\, x \in S)\,(x^2 \geq 0)$", in which $p(x)$ is the open sentence "$x^2 \geq 0$". We can eliminate the variable x by rephrasing the quantified statement as "every element of S has a nonnegative square". The truth value of this statement depends on S, so it is an open sentence with S as a free variable. For example, the statement is true if $S = \mathbf{R}$ and it is false if $S = \mathbf{C}$ (the complex numbers—see Appendix B).

Universal Quantifiers and Implication

A "universally quantified" statement can be thought of as just a conditional statement in disguise. For example, the statement "$(\forall\, x \in A)\,(x \in B)$" means exactly the same thing as "if $x \in A$, then $x \in B$". Both express the condition "$A \subseteq B$".

In general, the statement

$$(\forall\, x \in S)\,(p(x))$$

is just another way to say

$$\text{"if } x \in S, \text{ then } p(x)\text{"}.$$

Both of these express the relationship "$S \subseteq P$" [where P is the truth set for $p(x)$].

We want to introduce another way to state this which retains the flavor of the "if . . . , then . . .", or conditional, statement, but which incorporates the use of the universal quantifier.

Defining "Implication"

Intuitively, a conditional sentence suggests a causal relationship between its hypothesis and its conclusion. Our goal is to capture this idea by the symbol "$\Rightarrow$", which is called *implication* and is read "implies".

The general form in which this symbol is to be used is the following:

$$(\forall\, x \in \mathbf{U})\,(p(x) \Rightarrow q(x))\,,$$

which is read "for all x in $\mathbf{U}$, $p(x)$ implies $q(x)$" (here, $\mathbf{U}$ is some "universe of discourse").

We want to define "$\Rightarrow$" so that this statement will have exactly the same meaning as "if $p(x)$, then $q(x)$".

However, there is a technical problem we need to deal with. Whatever follows "$(\forall\, x \in \mathbf{U})$" should be an open sentence with x as a variable, so we have to formally define "$p(x) \Rightarrow q(x)$" as an open sentence. That is, we have to know how to determine the truth value of a particular statement "$p(c) \Rightarrow q(c)$", which results from substituting c for x.

In order to do so, we ask the following:

QUESTION 1 How can we define "$p(x) \Rightarrow q(x)$" as an open sentence so that the complete expression "$(\forall\, x \in \mathbf{U})\,(p(x) \Rightarrow q(x))$" will have exactly the same meaning as the conditional sentence "if $p(x)$, then $q(x)$"? What do we need to say about "all x"? Under what circumstances should "$p(c) \Rightarrow q(c)$" be defined as true?

Examine the principles by which we defined truth and falsehood for conditional sentences in Section 1.3.

We said that a conditional sentence would be considered true if every substitution for its variables produced one of the following combinations of truth values for the hypothesis and conclusion:

- hypothesis true, conclusion true;
- hypothesis false, conclusion true;
- hypothesis false, conclusion false.

A conditional sentence is considered false only if there is a substitution of variables which makes the hypothesis true and the conclusion false, that is, if there is a counterexample.

We will simply adapt this definition for defining implication:

DEFINITION 3.3: For open sentences $p(x)$ and $q(x)$, and an element $c \in \mathbf{U}$,

"$p(x) \Rightarrow q(x)$"

is an open sentence, for which the truth value for a substitution $x = c$ is defined as follows:

"$p(c) \Rightarrow q(c)$" is true in each of the following cases:

- if $p(c)$ is true and $q(c)$ is true,
- if $p(c)$ is false and $q(c)$ is true,
- if $p(c)$ is false and $q(c)$ is false.

"$p(c) \Rightarrow q(c)$" is false:

- if $p(c)$ is true and $q(c)$ is false.

Thus, according to Definition 3.3, "$p(c) \Rightarrow q(c)$" is true unless c is a counterexample to the conditional sentence "if $p(x)$, then $q(x)$"; in that case, "$p(c) \Rightarrow q(c)$" is false. That is, we have defined implication so that the statement "$p(c) \Rightarrow q(c)$" means "c is not a counterexample to the conditional sentence 'if $p(x)$, then $q(x)$' ".

Therefore, the assertion that the conditional sentence is true has exactly the same meaning as the assertion that "$p(x) \Rightarrow q(x)$" is true *for all* x. Both say that there is no value of c which makes $p(c)$ true and $q(c)$ false.

In other words, we have defined implication in such a way that the following theorem holds:

THEOREM 3.1. For open sentences $p(x)$ and $q(x)$, the two statements

$$\text{“if } p(x), \quad \text{then } q(x)\text{”}$$

and

$$\text{“}(\forall\, x \in \mathbf{U})\,(p(x) \Rightarrow q(x))\text{”}$$

mean the same thing.

COMMENTS

1. Though there is a distinction between "if $p(x)$, then $q(x)$" and "$p(x) \Rightarrow q(x)$", mathematicians don't always observe this distinction carefully. You may often see the phrase "$p(x)$ implies $q(x)$", or its symbolic form "$p(x) \Rightarrow q(x)$", used without a universal quantifier to mean "if $p(x)$, then $q(x)$". In the context of "working mathematics", this doesn't cause any confusion.

2. Since any universally quantified statement has the same meaning as its corresponding conditional statement, we will use the terminology of "hypothesis" and "conclusion" for universally quantified statements as well. Specifically, for the implication "$p(x) \Rightarrow q(x)$", or its quantified form "$(\forall\, x \in \mathbf{U})\,(p(x) \Rightarrow q(x))$", we will refer to "$p(x)$" as the hypothesis and "$q(x)$" as the conclusion. Similarly, we will use the word "counterexample" to mean an element of $\mathbf{U}$ that makes $p(x)$ true and $q(x)$ false.

3. The intuitive connection between implication and the concept of a conditional sentence is the reason why mathematical logic defines implication as it does. For example, the conditional sentence "if $x + 3 = 7$, then $2x = 8$" is true. Since $x = 1$ is not a counterexample to this conditional sentence, we also need to treat the statement "if $1 + 3 = 7$, then $2 \times 1 = 8$" as true. Out of context, that may not make much sense, but it fits the larger picture.

4. We rarely want to make a theorem out of the statement that there is something which is not a counterexample to a particular conditional sentence. Therefore, we almost never prove statements of the form "$(\exists\, x \in \mathbf{U})\,(p(x) \Rightarrow q(x))$". If some value $x = c$ is not a counterexample, then either $p(c)$ is false or $q(c)$ is true. It makes more sense to prove either "$(\exists\, x \in \mathbf{U})$('$p(x)$' is false)" or "$(\exists\, x \in \mathbf{U})\,(q(x))$" (depending on which is so) than to prove "$(\exists\, x \in \mathbf{U})\,(p(x) \Rightarrow q(x))$".

With Definition 3.3 in mind, we now look at the problem of finding the truth set for a universally quantified implication statement which has both free and bound variables:

Example 3 **Truth Set of a Universally Quantified Statement with Free and Bound Variables** Consider the open sentence

$$(\forall\, v \in \mathbf{R})\,(v < t \Rightarrow v^2 > 10) \qquad \text{with} \quad \mathbf{U} = \mathbf{R}\,.$$

This statement has "t" as a free variable (with universe **R**) and "v" as a bound variable, and we can represent the statement as "$q(t)$". To understand what $q(t)$ is saying, we can try various real numbers for t, and see what the statement says:

$t = 3$: $q(3)$ says "$(\forall\, v \in \mathbf{R})\,(v < 3 \Rightarrow v^2 > 10)$". This is false, since the substitution $v = 1$ makes the hypothesis "$v < 3$" true but makes the conclusion "$v^2 > 10$" false.

$t = -5$: $q(-5)$ says "$(\forall\, v \in \mathbf{R})\,(v < -5 \Rightarrow v^2 > 10)$". This is true: if v is a real number less than -5, then v^2 must be greater than 10.

$t = -\sqrt{10}$: $q(-5)$ says "$(\forall\, v \in \mathbf{R})\,(v < -\sqrt{10} \Rightarrow v^2 > 10)$". This is true: if v is a real number less than $-\sqrt{10}$, then v^2 must be greater than 10.

The statement $q(t)$ says "all of the numbers less than t have squares greater than 10". and the truth set of $q(t)$ is the interval $(-\infty, -\sqrt{10}]$ (i.e., $\{t \in \mathbf{R} : t \leq -\sqrt{10}\}$). □

Get Your Hands Dirty 2

Identify the free and bound variables and find the truth sets for each of these open sentences:

a) $(\forall\, t \in \mathbf{N})\,(w \mid t)$.
b) $(\exists\, v \in \mathbf{Z})\,(v^2 - 5v + r = 0)$.

Use **N** as the universe for both w and r. □

Working with Implication and "∀"

Every use of "∀" can be thought of as expressing an implication. Consider the quantified expression "$(\forall\, x \in S)\,(p(x))$".

How can this be restated using implication? (*Hint:* Try stating it in "if..., then..." form.)

We simply write "$(\forall\, x \in \mathbf{U})\,(x \in S \Rightarrow p(x))$". Both this and "$(\forall\, x \in S)\,(p(x))$" have the same meaning as "if $x \in S$, then $p(x)$".

COMMENTS

1. Since the statement "$(\forall\, x \in S)\,(p(x))$" can be interpreted as an abbreviation for "$(\forall\, x \in \mathbf{U})\,(x \in S \Rightarrow p(x))$", we will consider "$x \in S$" the hypothesis of this statement and "$p(x)$" its conclusion.

2. In a statement of the form "$(\forall\, x \in \mathbf{U})\,(x \in S \Rightarrow p(x))$", we end up assuming both "$x \in S$" and "$x \in \mathbf{U}$", and so the condition "$x \in \mathbf{U}$" becomes redundant (since $S \subseteq \mathbf{U}$). We therefore sometimes just write "$(\forall\, x)\,(x \in S \Rightarrow p(x))$", with the understanding that the universe is some set that contains S.

3. We have made the replacement set part of the predicate for a universally quantified statement by using it in the hypothesis of an implication. When working with an existential quantifier, we can make the replacement set part of the predicate by writing "$(\exists\, x \in S)\,(p(x))$" as "$(\exists\, x)\,(x \in S$ and $p(x))$". (*Note:* We use "and", not "⇒", with the existential quantifier.)

Get Your Hands Dirty 3

For each of the statements i)–iii) below, do the following:

a) Rewrite the statement using "$\Rightarrow$".
b) Rewrite the statement in "if . . . , then . . ." form using the given variables.
c) Rewrite the statement in words without explicitly naming the bound variables.
i) $(\forall\, u \in A)(u \in B)$.
ii) $(\forall\, x \in \mathbf{R})(x^2 \geq 0)$.
iii) $(\forall\, t \in \mathbf{N})(w \mid t)$. □

We can use implication to improve the quantifier version of the definition of "t is a prime". Recall that definition:

DEFINITION 2.6′: For $t \in \mathbf{N}$, $t > 1$, "t is a *prime*" means:

$$(\forall\, x \in D)(x = 1 \text{ or } x = t),$$

where D is the set of all positive divisors of t.

This formulation has the disadvantage that the quantified expression itself does not include an explicit definition of D. Simply replacing D by "$\{w \in \mathbf{N} : w \mid t\}$" leads to an awkward expression with the additional variable w.

How can we use implication to solve this dilemma? Try to formulate the definition in "if . . . , then . . ." form.

The definition of prime can be phrased as a conditional sentence as follows:

If x is a divisor of t, then $x = 1$ or $x = t$.

This essentially provides our answer. We simply use "x is a divisor of t" as the hypothesis and "$x = 1$ or $x = t$" as the conclusion. The natural choice for the universe in this context is $\mathbf{N}$. The new version is the following:

DEFINITION 2.6″: For $t \in \mathbf{N}$, $t > 1$, "t is a *prime*" means:

$$(\forall\, x \in \mathbf{N})(x \mid t \Rightarrow [x = 1 \text{ or } x = t]).$$

This is probably the most useful way in which to work with the concept of "prime". It makes the hidden variable explicit, it eliminates the set D, and it provides a hypothesis and conclusion to use in the context of later proofs.

COMMENTS

1. In Chapter 1, we discussed the use of the "if . . . , then . . ." form of a sentence as a way of *stating theorems*. In the discussion above, we are using "if . . . , then . . ." as a

way of *expressing a property*. We are *defining* the phrase "t is a prime" by the conditional statement "if $x \mid t$, then $x = 1$ or $x = t$". This conditional statement is not a theorem; it is a statement about t. It is an open sentence, neither true nor false. Its truth value depends on the substitution for t. When we then use the condition "t is a prime" in the hypothesis or conclusion of a theorem, we end up with "if . . . , then . . ."s inside "if . . . , then . . ."s. We will examine that sort of complication carefully in Section 4.1.

2. Our discussion shows that we can express the general subset condition "$A \subseteq B$" either as "$(\forall\, x \in A)\,(x \in B)$" or as "$(\forall\, x)\,(x \in A \Rightarrow x \in B)$". Though these two quantifier expressions have the identical meaning, each has a distinctive "flavor". The first can be thought of as a declarative statement about elements of A: they belong to B. The second feels more like a "relationship" between the two conditions "$x \in A$" and "$x \in B$". If you want to emphasize the "causal" nature of the statement, then the implication may be more appropriate. If you simply want to make a statement about the elements of A, then the other form may be better.

Though this distinction has no mathematical meaning, it may affect how you think about the concept of subset.

Definitions with Quantifiers

We began this section by comparing the statement "$A \subseteq B$" with the statement "$x \mid y$". Though the two may appear to be very similar, we saw that the first is basically a *universal* property: it can be expressed by using a universal quantifier to make a hidden variable explicit. The second, on the other hand, is basically an *existential* property: it can be expressed using an existential quantifier to make a hidden variable explicit.

Similarly, we have seen that the statement "t is a prime" expresses a universal property, while "u is a square" describes an existential property.

We would like next to give you some more sample definitions which use each of the two quantifiers.

Definitions using "∀"

The following are some typical definitions expressed using the universal quantifier:

DEFINITION 3.4: For a set $S \subseteq \mathbf{R}$, "*S is closed under subtraction*" means:

$$(\forall\, x, y \in S)\,(x - y \in S)$$

(In this quantified statement, "S" is a free variable, and "x" and "y" are dummy variables. The statement is "about" S.)

COMMENT In Definition 3.4 (and in later definitions), we have two variables quantified by the same quantifier. This is common practice, although it would also be correct to write

$$\text{"}(\forall\, x \in S)(\forall\, y \in S)(\ldots)\text{"}.$$

If the variables have different replacement sets, this must be made clear. For example, "$(\forall\, x \in \mathbf{R}, y \in \mathbf{Z})$" means: "for every real number x and integer y".

DEFINITION 3.5: For integers a, b, and c, "$\text{GCD}(a, b) = c$" means:

$$c \mid a \text{ and } c \mid b \text{ and } (\forall\, z \in \mathbf{N})([z \mid a \text{ and } z \mid b] \Rightarrow z \leq c)\,.$$

(In this quantified statement, "a", "b", and "c" are free variables, and "z" is a dummy variable. The statement is "about" a, b, and c.) [Verify that Definition 3.5 agrees with the earlier definition of GCD (Definition 2.9 in Section 2.4).]

DEFINITION 3.6: For a function f with domain $\mathbf{R}$, "f is *strictly increasing*" means:

$$(\forall\, x, y \in \mathbf{R})\,(x < y \Rightarrow f(x) < f(y))$$

(In this quantified statement, "f" is a free variable, and "x" and "y" are dummy variables. The statement is "about" f.)

Note: We will define the term "function" precisely in Section 3.2 (see Definition 3.14). For now, your intuition about functions should suffice.

Get Your Hands Dirty 4

For each of the Definitions 3.4–3.6 with "$\forall$":

a) express the definition in words without explicitly naming the bound variables;
b) find a substitution for each of the free variables that makes the quantified statement true;
c) find a substitution for each of the free variables that makes the quantified statement false. □

Get Your Hands Dirty 5

Write each of these statements using "$\forall$":

a) S is closed under addition. (Here, S is a subset of $\mathbf{Z}$.)
b) Any set with n elements has 2^n subsets. (The set should be described as a subset of some universe $\mathbf{U}$.)
c) The sum of three consecutive integers is always divisible by 3. □

Definitions using "∃"

The following are some typical definitions expressed using the existential quantifier:

> **DEFINITION 3.7:** For $x \in \mathbf{N}$, "x has a *proper divisor*" means:
> $$(\exists\, y \in \mathbf{N})(y \mid x \text{ and } 1 < y < x).$$

(In this quantified statement, "x" is a free variable, and "y" is a dummy variable. The statement is "about" x.)

Note: Often the word "has", as in the expression "x has a proper divisor", is a hint that an existential quantifier is called for.

> **DEFINITION 3.8:** For a function f with domain $\mathbf{R}$, "z is in the *image set of* f" means:
> $$(\exists\, x \in \mathbf{R})(f(x) = z).$$

(In this quantified statement, "z" and "f" are free variables, and "x" is a dummy variable. The statement is "about" z and f.)

See Definition 3.14 in Section 3.2 for the formal definition of "function".

> **DEFINITION 3.9:** For $t \in \mathbf{N}$, "t is *composite*" means:
> $$(\exists\, u, v \in \mathbf{N}) \quad (u > 1 \text{ and } v > 1 \text{ and } t = u \cdot v).$$

(In this quantified statement, "t" is a free variable, and "u" and "v" are dummy variables. The statement is "about" t.)

> **DEFINITION 3.10:** For $x \in \mathbf{R}$, "x is *rational*" means:
> $$(\exists\, p \in \mathbf{Z}, q \in \mathbf{N})\left(x = \frac{p}{q}\right).$$

(In this quantified statement, "x" is a free variable, and "p" and "q" are dummy variables. The statement is "about" x.)

COMMENT In each of the last two definitions, as in Definition 3.6, there are two variables that are governed by the same quantifier. In Definition 3.10, the two variables have different replacement sets.

Get Your Hands Dirty 6

For each of Definitions 3.7–3.10, with "∃":

a) express the definition in words without explicitly naming the bound variables;
b) find a substitution for each of the free variables that makes the quantified statement true;
c) find a substitution for each of the free variables that makes the quantified statement false. □

Get Your Hands Dirty 7

Write each of these statements using "∃":

a) A is not empty. (Here, A is a set.)
b) x is a power of 2. (Here, x is an integer.)
c) The equation "$x^2 = 2$" has a real solution. □

Quantifiers in Theorems

Much of the time, a theorem will have parallel use of quantifiers in its hypothesis and conclusion.

For example, Theorem 2.3 states: "if $U \subseteq W$, then $\mathcal{P}(U) \subseteq \mathcal{P}(W)$". As we have discussed, the subset symbol expresses a universal property. Thus, in this theorem, both the hypothesis and the conclusion involve universal quantifiers.

Similarly, Theorem 2.9 states: "if $a \mid b$ and $b \mid c$, then $a \mid c$". We have seen that the divisibility relationship expresses an existential condition. Thus, in this theorem, both the hypothesis and the conclusion involve existential quantifiers.

Example 4 **Stating a Theorem with Quantifiers** Consider the simple conditional sentence "if x^2 is even, then x is even" (with universe = $\mathbf{Z}$). We can write this symbolically using quantifiers as follows:

$$(\forall\, x \in \mathbf{Z})([(\exists\, t \in \mathbf{Z})(x^2 = 2t)] \Rightarrow [(\exists\, s \in \mathbf{Z})(x = 2s)])\,.$$

The first statement in brackets, "$(\exists\, t \in \mathbf{Z})(x^2 = 2t)$", simply says "$x^2$ is even". This is the hypothesis of the implication. The second statement in brackets, "$(\exists\, s \in \mathbf{Z})(x = 2s)$", says "$x$ is even". This is the conclusion of the implication.

We can read this complicated quantified statement as follows:

> For any integer x, if there is an integer t such that $x^2 = 2t$, then there is an integer s such that $x = 2s$.

Notice that the integers t and s here are dummy variables. Technically, we could have used the same letter in both places, but that might have mistakenly led to the expectation that the same integer was involved. The use of different letters suggests that they need not be the same (although, if $x = 0$, they are the same). □

As a general rule, existential information is easier to use than universal information. Having some existential condition in the hypothesis can be a powerful assumption, because it provides a specific object to work with. For example, we can manipulate the given object in equations or inequalities, to get further conclusions. Therefore, it is often fairly straightforward to prove theorems whose hypothesis involves an existential quantifier. We will illustrate this technique, which we call "name it—then use it", in Section 4.1.

The use of a hypothesis with a universal quantifier is more difficult to master, and involves a method known as *specialization*. We will also see several examples of specialization in Section 4.1.

In Section 5.3, we will prove a powerful theorem about the integers (Corollary 5.29) which says "if $\mathrm{GCD}(a, b) = 1$, then there are integers x and y with $ax + by = 1$". As Definition 3.5 shows, the hypothesis of this theorem expresses a universal property. The usefulness of this theorem comes in part from the fact that it has an existential conclusion. The ability of this theorem to convert universal information to existential information makes it an extraordinarily valuable tool in studying the integers.

Get Your Hands Dirty 8

For each of the following open sentences:

a) identify the free variable(s);
b) find a substitution for the free variables (using the given universe) which makes the sentence true;
c) find a substitution for the free variables (using the given universe) which makes the sentence false.

i) $(\forall\, u \in \mathbf{R})\,(u \le x \Rightarrow u \le x^2)$, $\mathbf{U} = \mathbf{R}$.
ii) $(\forall\, x \in C)\,(x \in A \text{ or } x \in B)$, $\mathbf{U} = \mathcal{P}(\{1, 2, 3, 4\})$ for each free variable.
iii) $(\exists\, t \in \mathbf{N})\,(c \mid t^2 \text{ and } c \nmid t)$, $\mathbf{U} = \mathbf{N}$. □

Why Bother with Formal Quantifiers?

You may wonder why mathematicians introduce the artificial notation of quantifiers, when these symbols express notions that can be stated in ordinary language.

One of the major reasons is that ordinary language is filled with ambiguities. Consider the following two sentences:

i) A horse is in my house.
ii) A horse is a mammal.

The syntax of these two statements is very similar, but the phrase "a horse" is used in two entirely different ways.

In statement i), the phrase is used *existentially*—that is, we mean "*there is* a horse that is in my house". In ii), the phrase is used *universally*—we mean "*every* horse is a mammal". Use of formal quantifiers makes it easier for us to make this distinction explicit.

Here is another example of confusing use of language:

iii) If there is a solution to the equation $x^2 - 13x + 32 = 0$, then it is positive.

Though the phrase "there is" makes this statement sound existential, the actual meaning of this assertion is universal. It says that *every* solution to the given equation is positive, so it can be expressed more formally as:

$$(\forall\, x \in \mathbf{R})\,(x^2 - 13x + 32 = 0 \Rightarrow x > 0)\,. \tag{*}$$

Here "x" is a bound variable, and there are no free variables. This is not an open sentence, but is instead a true statement. (Verify that this is true, using the quadratic formula.)

It may be tempting to write iii) as

$$[(\exists\, x \in \mathbf{R})\,(x^2 - 13x + 32 = 0)] \Rightarrow x > 0 \tag{**}$$

or

$$(\exists\, x \in \mathbf{R})\,(x^2 - 13x + 32 = 0 \Rightarrow x > 0)\,, \tag{***}$$

but neither of these is a correct "translation".

The problem with (**) is an ambiguity of free vs. bound variables. The hypothesis here, "$(\exists\, x \in \mathbf{R})\,(x^2 - 13x + 32 = 0)$", is a true statement, with a bound variable x. But the conclusion, "$x > 0$", is an open sentence, neither true nor false. There is no quantifier here, and the x here is not the same x that the hypothesis says exists. (**) says "if the equation has a solution, then $x > 0$", but it loses track of what "x" is in the process. What we want to say is: "if x is a solution to the equation, then it is positive", which is precisely what the universally quantified expression (*) says.

The difficulty with (***) is different. As we discussed earlier (see comment 4 following Theorem 3.1), it is rarely correct (and never good mathematics) to write an expression of the form "$(\exists\, x \in S)\,(p(x) \Rightarrow q(x))$". Recall that "$p(c) \Rightarrow q(c)$" means that c is not a counterexample to the conditional sentence "if $p(x)$, then $q(x)$". Therefore, "$(\exists\, x \in S)\,(p(x) \Rightarrow q(x))$" simply means that there is some object which is not a counterexample. That's not much of a statement.

In our example, the value $x = -4$ is not a counterexample because it is not a solution to the equation "$x^2 - 13x + 32 = 0$", i.e., it makes the hypothesis false. Similarly, $x = 10$ is not a counterexample, because it makes the conclusion, "$x > 0$", true. Either one of these facts by itself makes (***) true, since all (***) says is that there is at least one number which is not a counterexample. So (***) is certainly not what is intended by iii).

Negating Quantified Statements

The assertion that a particular statement is false is called its *negation*. The symbol "$\sim$" is used to represent this idea: that is, "$\sim p(x)$" means "$p(x)$ is false". The negation of a conditional sentence is the assertion that the conditional sentence has a counterexample. Because of the fundamental role played by counterexamples in our understanding of conditional sentences, it is essential that we be able to state, simplify, and understand the negation of complex statements.

So let us examine what negation means in the context of quantifiers. What does it mean for either a universally or an existentially quantified statement to be false? Here is an example of each:

Example 5 **Negation of a Universal Statement** Consider the following simple statement, which could be either true or false, depending on what the sets A and B are:

$$(\forall\, x \in A)\,(x \in B)\,. \qquad (*)$$

(This is just the formal way of saying that A is a subset of B.) What does it mean to say that $(*)$ is false?

Think about it—express the negation of $(*)$ in terms of quantifier language.

If we treat $(*)$ as an implication—"$(\forall\, x)\,(x \in A \Rightarrow x \in B)$"—then for this to be false means that there is a counterexample; in other words, it means that "there is an element of A which is not in B". Thus, the negation of the statement $(*)$ is an existential statement, namely, $(\exists\, x \in A)\,(x \notin B)$.

In brief,

$$\text{"}\sim[(\forall\, x \in A)\,(x \in B)]\text{"}$$

has the same meaning as

$$\text{"}(\exists\, x \in A)\,(x \notin B)\text{"}\,.$$ □

Now let's look at what happens if we start with an existential statement:

Example 6 **Negation of an Existential Statement** Consider the following statement (use **R** as the universe):

$$(\exists\, x \in A)\,(x^2 < 1)\,. \qquad (**)$$

In ordinary language, this says that some element of A satisfies the inequality $x^2 < 1$. Again, this could be either true or false, depending on what the set A is.

What does it mean to say that $(**)$ is false? [Suggestion: suppose A were the finite set $\{1, 2, 3, 4\}$. How would you prove that $(**)$ is false?]

Think about it—express the negation of $(**)$ in terms of quantifier language.

Working with the suggestion, you could prove $(**)$ false by trying each of the elements in A, and showing each of them has a square that is at least 1. In other words, to say that there are no elements of A that "fit" the inequality is the same as saying that every element of A "fails to fit" the inequality. Thus, the negation of $(**)$ is a universal statement, namely, $(\forall\, x \in A)\,(x^2 \not< 1)$. We can simplify "$x^2 \not< 1$" by writing it as "$x^2 \geq 1$".

In brief,

$$\text{“}\sim[(\exists\, x \in A)\,(x^2 < 1)]\text{”}$$

has the same meaning as

$$\text{“}(\forall\, x \in A)\,(x^2 \geq 1)\text{”}. \qquad \square$$

We can summarize the principles behind these two examples as follows:

i) For it to be false that every element "works" means that some element "fails".
ii) For it to be false that some element "works" means that every element "fails".

In other words, negation interchanges the two quantifiers, turning universal statements into existential statements, and vice versa.

We can express these principles by the following general theorem:

THEOREM 3.2. For any open sentence $p(x)$ and set S:

i) "$\sim[(\forall\, x \in S)\,(p(x))]$" has the same meaning as "$(\exists\, x \in S)\,(\sim p(x))$", and
ii) "$\sim[(\exists\, x \in S)\,(p(x))]$" has the same meaning as "$(\forall\, x \in S)\,(\sim p(x))$".

(This can be proved formally by expressing each statement in terms of truth sets.)

Of course, finding out what "$\sim p(x)$" means for a particular open sentence may be fairly complicated. Here's a chance to get some practice:

Get Your Hands Dirty 9

For each of the following open sentences i)–iii) (from GYHD 8):

a) express the negation using quantifiers, without using the negation symbol (you may use symbols such as $\nmid$, $\notin$, etc.), and simplify your answer as much as possible;
b) find a substitution for the free variables (using the given universe) which makes the negation true (i.e., makes the original sentence false).

i) $(\forall\, u \in \mathbf{R})\,(u \leq x \Rightarrow u \leq x^2)$, $\mathbf{U} = \mathbf{R}$.
ii) $(\forall\, x \in C)\,(x \in A \text{ or } x \in B)$, $\mathbf{U} = \mathcal{P}(\{1, 2, 3, 4\})$ for each free variable.
iii) $(\exists\, t \in \mathbf{N})\,(c \mid t^2 \text{ and } c \nmid t)$, $\mathbf{U} = \mathbf{N}$. $\square$

Exploration 1: General Principles about Quantifers and Logic

It's time to do some general exploration with quantifiers and logic. We have stated some general theorems—Theorem 3.1 about implication, Theorem 3.2 about negation—but there are many others. Here are some avenues to pursue:

a) *Combine open sentences with "or", "and", and "not":* What are the truth sets for statements like "$p(x)$ and $q(x)$", "$p(x)$ or $q(x)$", "$\sim[p(x)$ and $q(x)]$", etc.? How can they be expressed in terms of the truth sets of $p(x)$ and $q(x)$? How can De Morgan's Laws (Theorem 2.4) be used to simplify the answers? How can De Morgan's Laws be interpreted in terms of logic to develop rules for equivalent sentences—statements which say the same thing?

b) *Combine* "$\exists$" *with "and":* Consider the statement

$$\text{“}(\exists\, x \in S)\,(p(x) \text{ and } q(x))\text{”}.$$

This statement says that there is an object x in S which makes both $p(x)$ and $q(x)$ true. How does this statement compare with the combined statement

$$\text{“}(\exists\, x \in S)\,(p(x)) \text{ and } (\exists\, x \in S)\,(q(x))\text{”?}$$

Can the original statement be true and the combined statement false? Or vice versa? Try some examples with specific open sentences, and look for general principles.

c) Try the same idea as b), using "$\forall$".

d) Try b) and c), using "or" instead of "and".

• E X P L O R E •

EXERCISES

1. (GYHD 3) For each of statements i)–iii) below, do the following:

a) Rewrite the statement using "$\Rightarrow$".

b) Rewrite the statement in "if . . . , then . . ." form using the given variables. Where possible, use words rather than symbols.

c) Rewrite the statement in words without explicitly naming the bound variables.

i) $(\forall\, u \in A)\,(u \in B)$.

ii) $(\forall\, x \in \mathbf{R})\,(x^2 \geq 0)$.

iii) $(\forall\, t \in \mathbf{N})\,(w \mid t)$.

2. **a)** List explicitly the elements of the truth set for statement i) of Exercise 1, using $\mathcal{P}(\{1,2\})$ as the universe for each of A and B. [Each element will be an ordered pair of the form (A, B), with $(A, B) \in \mathcal{P}(\{1,2\} \times \mathcal{P}(\{1,2\}$.]

b) List explicitly the elements of the truth set for statement iii) of Exercise 1, using $\mathbf{Z}$ as the universe for w.

3. (GYHD 5) Write each of these statements using "$\forall$":

a) S is closed under addition. (Here, S is a subset of $\mathbf{Z}$.)

b) Any set with n elements has 2^n subsets. (The set should be described as a subset of some universe $\mathbf{U}$.)

c) The sum of three consecutive integers is always divisible by 3.

4. (GYHD 7) Write each of these statements using "$\exists$":

a) A is not empty. (Here, A is a set.)

b) x is a power of 2. (Here, x is an integer.)

c) The equation "$x^2 = 2$" has a real solution.

5. (GYHD 8) For each of the following open sentences:

a) identify the free variable(s);

b) find a substitution for the free variable(s) (using the given universe) which makes the sentence true;

c) find a substitution for the free variable(s) (using the given universe) which makes the sentence false.

i) $(\forall\, u \in \mathbf{R})\,(u \leq x \Rightarrow u \leq x^2)$, $\mathbf{U} = \mathbf{R}$.

ii) $(\forall\, x \in C)\,(x \in A \text{ or } x \in B)$, $\mathbf{U} = \mathcal{P}(\{1, 2, 3, 4\})$ for each free variable.

iii) $(\exists\, t \in \mathbf{N})\,(c \mid t^2 \text{ and } c \nmid t)$, $\mathbf{U} = \mathbf{N}$.

6. Express the negation of each of the following open sentences using quantifier notation, but without using the negation symbol. (You may, however, use symbols such as $\nmid$ or $\neq$.) Simplify the predicates as much as possible. [Items d) and e) are from GYHD 8.]

a) $(\forall\, a \in \mathbf{N})\,(b \mid a \text{ and } c \mid a \Rightarrow (bc) \mid a)$.

b) $(\forall\, x, y \in \mathbf{N})\,(a \mid (xy) \Rightarrow a \mid x \text{ or } a \mid y)$.

c) $(\exists\, x, y \in \mathbf{Z})\,(ax + by = c)$.

d) $(\forall\, u \in \mathbf{R})\,(u \leq x \Rightarrow u \leq x^2)$.

e) $(\exists\, t \in \mathbf{N})\,(c \mid t^2 \text{ and } c \nmid t)$.

7. For each statement in Exercise 6, find a substitution for the free variable(s) which will make the statement true, and a substitution which will make the statement false. Give informal explanations of your answers. Use the universes given below:

a) $b, c \in \mathbf{N} - \{1\}$.

b) $a \in \mathbf{N} - \{1\}$.

c) $a, b, c \in \mathbf{Z}$.

d) $x \in \mathbf{R}$.

e) $c \in \mathbf{N}$.

8. Based on your work in Exercise 7, and further exploration as needed, make a conjecture as to what the truth set is for each of the open sentences of Exercise 6. (*Note:* These are not easy open sentences to analyze. Remember to *explore*, and use your intuition.)

9. Read and think about the meaning of each of the following quantified statements. Express the meaning in natural English. Do not just translate the quantifiers. As much as possible, you should use terminology defined so far in this book, and other standard terminology, to make your statements concise. Where possible, avoid referring to the bound variables. The word "not" may be helpful in expressing some of these statements.

a) $(\forall\, S \subseteq \mathbf{U})\,(\forall\, n \in \mathbf{W})\,(|S| = n \Rightarrow |\mathcal{P}(S)| = 2^n)$.

b) $(\exists\, x, y \in \mathbf{R})\,(x < y \text{ and } f(x) \geq f(y))$ (here, f is a function from $\mathbf{R}$ to $\mathbf{R}$).

c) $(\forall\, x \in A)\,(f(x) \neq t)$ (here, f is a function from A to B and t is an element of B).

d) $(\forall\, y \in \mathbf{N})\,(a \cdot y \neq b)$ (here, a and b are positive integers).

10. For each statement of Exercise 9, identify the free variables, and then, if possible:

a) find at least one substitution for the free variables which make the statement true, and

b) find at least one substitution for the free variables which make the statement false.

11. (Exploration 1) For each of the following statements:

- if the statement is true:
  - **i)** give a brief explanation of why it is true;
  - **ii)** if a "better" true statement can be made, give such a statement, and indicate how it is better;
- if the statement is false, open or does not make sense:
  - **i)** give a brief explanation of why it is false, open, or does not make sense;
  - **ii)** write a true statement, if possible, using the same predicate(s) (e.g., change quantifiers, rearrange parentheses, replace "and" by "or", etc.).

a) $(\exists\, x \in \mathbf{N})\,(x \text{ is even} \Rightarrow 3\,|\,x)$.
b) $(\forall\, x \in \mathbf{N})\,(x \text{ is even})$ or $(\forall\, x \in \mathbf{N})\,(x \text{ is odd})$.
c) $(\exists\, x \in \mathbf{N})\,(x \text{ is a prime})$ and $(\exists\, x \in \mathbf{N})\,(x \text{ is even})$.
d) $(\exists\, x \in \mathbf{R})\,(x^2 = y \Rightarrow y \geq 0)$.
e) $(\forall\, y \in \mathbf{N})\,(x\,|\,y \Rightarrow x \leq y)$.

3.2 Working with Quantifiers

One of the important uses for quantifiers and other precise mathematical language is to express concepts for which we have some intuitive feeling, but which are hard to explain.

The relationship between intuition and the formal definition is a two-way street. Definitions should be seen as tools for assisting our understanding. Over the centuries, mathematicians have often adjusted their definitions to better reflect their intuition, and to broaden the applicability of the underlying concepts.

Generally, we begin with an intuitive feeling for a concept. After some experimentation, we develop a formal definition, often using quantifiers, which describes what we want the concept to mean. Sometimes we introduce new terminology to express the ideas more concisely.

Then we use the new terminology as a kind of shorthand for the precise, often complex, formal definition. In order to explore the concepts, and make conjectures and develop theorems about them, we generally revert to our intuitive pictures. Our intuition is more likely to give us insights than the formalism. However, intuition can only be relied on for direction, not certainty. When it comes to proving theorems, mathematics has to get back to the formalism. In the final analysis, the actual definitions are the arbiter of what something means.

In this section, we will examine two important mathematical topics. We will see how quantifiers and other special mathematical language and notation can be used to clarify the important concepts. We will begin, in each case, with an intuitive description of the relevant ideas. We will try to capture these intuitive ideas by means of formal definitions. In Section 4.1, we will use the new terminology to state some basic theorems, and prove them by referring to the actual definitions.

Topic 1: Upper Bounds and Sets Bounded Above

Throughout the discussion of this topic, our universe will be the set of real numbers. Lowercase variables will represent real numbers, and uppercase variables will represent sets of real numbers.

We begin with the open sentence "$x \geq y$". This is a statement about x and y. If we replace each of these variables by a specific real number, we get a statement which is either true or false.

Now let's consider the situation where we want to compare x with many objects, not just y. Suppose we have a set S of real numbers, i.e., $S \subseteq \mathbf{R}$, and we want to say that x is greater than or equal to "all of S". This is sometimes expressed by saying "x is an upper bound for S".

QUESTION 1 What exactly do we mean by saying "x is an upper bound for S"? How can this statement be expressed in terms of quantifiers?

Try to define this phrase precisely.

Our formal definition says, but in more precise terms, that "x is at least as big as everything in S".

> **DEFINITION 3.11:** For $x \in \mathbf{R}$ and $S \subseteq \mathbf{R}$, "x is an *upper bound* for S" means:
>
> $$(\forall\, y \in S)(x \geq y).$$

COMMENTS

1. This is a statement about x and S. y is now a dummy variable, since it has been quantified.
2. An upper bound for S can be, but does not have to be, an element of S. We made no restriction that $x \in S$ in the definition. If some element of S is an upper bound for S, we generally call that element the *maximum* (or *greatest*) element of S.
3. We can make an analogous definition for "lower bound" by replacing "$\geq$" with "$\leq$". The theorems on this topic all have "lower bound" analogs.
4. Another way to express the concept of upper bound would be to say that nothing in S is bigger than x. That is a "nonexistence" statement, which could be written in quantifier notation as "$\sim(\exists\, y \in S)(y > x)$". Theorem 3.2 can be used to show formally that this is equivalent to the condition in the above definition.
5. To understand what is going on, you may wish to picture S as a set along a number line, as shown in Figure 3.1. The statement "x is an upper bound for S" says, intuitively, "x is to the right of S".

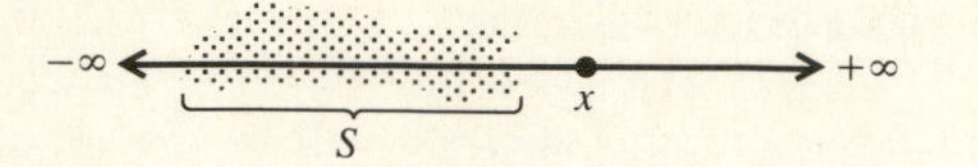

Figure 3.1. The set S is some set of points on the number line, and x is "to the right" of S, and so x is an upper bound for S

Whenever you see a new definition, you should ask yourself if it fits your intuitive idea of what the terminology should mean. One way to test if we have made a "good" definition is to try some examples. Before reading Example 1, play around with Definition 3.11—consider various sets and real numbers, and ask if the given number is an upper bound for the given set. Come up with several examples of pairs of sets and real numbers in which the real number is an upper bound for the set, and several where it is not.

? ? ?

The following example is intended to give some further guidance in exploring this idea:

Example 1 **Is x an Upper Bound for S?** In each of the following cases, decide if the given value for x is an upper bound for the particular set S, according to Definition 3.11.

i) $x = 5$, $S = \mathbf{R}$.
ii) $x = 10$, $S = \mathbf{R}^-$ (the negative real numbers).
iii) $x = 0$, $S = \mathbf{R}^{\leq 0}$ (the "nonpositive" real numbers).
iv) $x = 135$, $S = 2\mathbf{Z}$.
v) $x = 1$, $S = [-2, 3]$.
vi) $x = -6$, $S = \varnothing$ (think carefully about this one).

Draw pictures. Reread the definition. You also need to think about what "x is *not* an upper bound for S" means.

SOLUTIONS The negation of the quantified statement, "$(\forall\, y \in S)\,(x \geq y)$", in Definition 3.11 can be written as

$$\text{“}(\exists\, y \in S)\,(y > x)\text{”}. \qquad (*)$$

Thus, to show that x is not an upper bound, we must find an element of y which is greater than x.

With this in mind, we can look at the individual examples:

i) 5 is not an upper bound for $\mathbf{R}$. $(*)$ is true in this case: choose $y = 6$ (for example).
ii) 10 is an upper bound for $\mathbf{R}^-$. 10 is actually greater than every negative real number .
iii) 0 is an upper bound for $\mathbf{R}^{\leq 0}$. It is important to note that an element of S can be an upper bound for S, because Definition 3.11 has the predicate "$x \geq y$", not "$x > y$".
iv) 135 is not an upper bound for $2\mathbf{Z}$: choose, for example, $y = 140$, to see that $(*)$ is true in this case.
v) 1 is not an upper bound for $[-2, 3]$: choose, for example, $y = 2$, to see that $(*)$ is true in this case.

vi) -6 is an upper bound for $\varnothing$. In fact, any real number is an upper bound for $\varnothing$, because, no matter what number is substituted for x, the statement "$(\forall\, y \in \varnothing)\,(x \geq y)$" will be vacuously true. □

COMMENTS

1. In cases i), iv), and v), we showed that the existential statement $(*)$ was true by identifying a specific number y in S which made "$y > x$" true. This is typical of the proof of an existential statement.

2. The statement "$(\forall\, y \in S)\,(x \geq y)$" is a formal way of writing the conditional sentence "if y is an element of S, then $x \geq y$". The negation of this—the statement $(*)$—simply asserts that there is a counterexample to the conditional sentence.

Exploration 1: Looking for Theorems about Upper Bounds

Examine Definition 3.11, Example 1, your own examples, and the comments above. What else can you say about the concept of upper bound? What general theorems do you think are true? What conjectures can you make?

Look specifically for theorems that have a hypothesis about some real number being an upper bound for some set, and a conclusion about the same or some related number being an upper bound for the same or some related set. Consider combinations of sets, such as union, intersections, differences, etc.

• E X P L O R E •

Sets Bounded Above

Our discussion so far shows that some sets have many upper bounds, but suggests that others might not have any upper bound at all. For example, **R** itself has no upper bound, since, no matter what real number we choose for x, there will always be a larger one.

How would you express, using the language and notation of quantifiers, that a particular set "is bounded above", i.e., has an upper bound?

Write out a formal definition for "S is bounded above".

We define this concept as follows:

DEFINITION 3.12: For $S \subseteq \mathbf{R}$, "S is *bounded above*" means:

$$(\exists\, x \in \mathbf{R})\,(\forall\, y \in S)\,(x \geq y)$$

COMMENTS

1. This is our first quantified statement that combines the two quantifiers, "$\exists$" and "$\forall$". We read this: "there exists a real number x such that, for every element y in S, x is greater than or equal to y."

2. The quantified statement in Definition 3.12 is a statement about S. Both x and y are dummy variables.
3. "S is bounded below" is defined analogously.

Example 2 **Is S Bounded Above?** Which of the sets used in Example 1 have an upper bound? What does it mean to say "S is not bounded above"?

? ? ?

ANSWER We begin with the second question. It may be easiest to start by thinking of "S is bounded above" as saying "there is an x which is an upper bound". If this is false, it means "every x fails to be an upper bound", i.e., "$(\forall x \in \mathbf{R})$ (x is not an upper bound for S)".

What does it mean to say "x is not an upper bound for S"?

? ? ?

We saw in Example 1 that it means that there is some element of S which is bigger than x. Thus, "S is not bounded above" means:

$$(\forall x \in \mathbf{R})(\exists y \in S)(y > x). \qquad (**)$$

This is read: "for every real number x, there exists an element y in S such that y is greater than x".

Note: The statement (**) could also be obtained by starting with the formal negation of "S is bounded above", i.e., the statement "$\sim(\exists x \in \mathbf{R})(\forall y \in S)(x \geq y)$". If we apply Theorem 3.2 to this twice (once for each quantifier), and simplify "$\sim(x \geq y)$" as "$y > x$", we get precisely (**). We essentially did the same process on an intuitive level.

Example 1 showed that $\mathbf{R}^-$, $\mathbf{R}^{\leq 0}$, and $\varnothing$ were bounded above, because the example actually gave an upper bound for each of these sets. The interval $[-2, 3]$ *is* bounded above, even though the particular value for x given in Example 1 was not an upper bound. For example, 7 is an upper bound for $[-2, 3]$. $\mathbf{R}$ and $2\mathbf{Z}$ are not bounded above. □

Get Your Hands Dirty 1

Give an intuitive explanation for the assertion that the sets $\mathbf{R}$ and $2\mathbf{Z}$ are not bounded above. [*Hint:* verify (**) by describing how to find an appropriate y in terms of a "generic" x.] □

Least Upper Bound

It is easy to see that a set can have many upper bounds. For example, both 5 and 6 are upper bounds for the set $[-2, 3]$. We will prove a simple theorem in Section 4.1 that shows that sets which are bounded above actually have infinitely many upper bounds. We conclude Topic 1 by asking the following:

QUESTION 2 If a particular set S is bounded above, is there one of its upper bounds which is the smallest of all of them? Think about this in the case of the sets $\mathbf{R}^-$, $\mathbf{R}^{\leq 0}$, and $[-2, 3]$, which, we already saw, are bounded above.

? ? ?

In all three cases, there is such an object: $x = 0$ works for the first two cases, and $x = 3$ for the third. It is an important and profound property of the set **R** that any nonempty subset of **R** which has an upper bound has a smallest one. This is known as the *completeness property* of the real numbers. Its proof requires a careful definition of the set **R**. This is usually done in courses in real analysis. Our interest here is in making the concept of "least upper bound" precise. How would you define "least upper bound" for a set S?

Try it—Remember that the "least upper bound" must itself be an upper bound.

In addition to being an upper bound, the least upper bound must be "smaller than all the others". We use both of these criteria in the following definition:

DEFINITION 3.13: For $x \in \mathbf{R}$ and $S \subseteq \mathbf{R}$, "x is the *least upper bound* for S" means:

a) x is an upper bound for S, and
b) $(\forall\, z \in \mathbf{R})(z \text{ is an upper bound for } S \Rightarrow z \geq x)$

The phrase "least upper bound" is often abbreviated "lub". In the case of lower bounds (see comment 2 following Definition 3.11), the analogous object is called the "greatest lower bound", and is abbreviated "glb".

COMMENTS

1. The completeness property described above says that every nonempty subset of **R** which is bounded above has a least upper bound in **R**. This property is not necessarily true if **R** is replaced by some other set— see Exercise 5.

2. We spoke of "an" upper bound in Definition 3.11, but here we speak of "the" least upper bound. A set cannot have more than one "least upper bound". (Why not? See Exercise 2 of Section 4.1.)

3. Both condition a) and condition b) are statements about x and S; in condition b), z is a dummy variable.

4. Both conditions must hold in order for x to be the least upper bound for S (see Exercise 2).

5. In condition b), the object x itself fits the hypothesis of "being an upper bound for S". Though x is supposed to be "less than the *other* upper bounds", we don't want the definition to say that x is less than itself. Therefore the implication has "$z \geq x$" as its conclusion instead of "$z > x$".

6. In condition b), the hypothesis that z is an upper bound for S is itself a statement involving a quantifier, namely, "$(\forall\, y \in S)\,(z \geq y)$". Thus, b) altogether says:

$$(\forall\, z \in \mathbf{R})\,([(\forall\, y \in S)\,(z \geq y)] \Rightarrow z \geq x)\,.$$

Perhaps surprisingly, we cannot rearrange the parentheses and write this as:

$$(\forall\, z \in \mathbf{R})\,(\forall\, y \in S)\,(z \geq y \Rightarrow z \geq x)\,.$$

For example, the latter statement is false if $S = [-2, 3]$ and $x = 3$ (verify!), even though 3 is the least upper bound for $[-2, 3]$. This situation is explored further in Exercise 3.

Topic 2: Functions

The concept of a "function" is used throughout mathematics. It is often introduced in an intuitive way in the early grades of elementary school, using the metaphor of an "In–Out Machine". That is, we think of a function as a "machine" in which one number goes in, and some other number comes out, usually computed according to some formula or equation.

In algebra and calculus, we begin to work more precisely with functions. We talk about the *graph* of a function, and combine functions using an operation called *composition* (as well as using addition, multiplication, etc.).

Also, in calculus, we often study functions which cannot be expressed by a simple rule. In other areas of mathematics, such as abstract algebra, we are interested in functions involving things other than numbers. Thus, we find ourselves stretching the function concept as we go, and so it becomes more and more important to decide what exactly a function is.

Developing a Definition

Suppose A and B are any two sets, of any kind. We want to define precisely what we mean by "a function from A to B". What should this mean?

Formulate a tentative definition.

One of the metaphors that is used most commonly in talking about functions is that of "matching" or "pairing" elements from one set with elements of the other. This suggests the idea of thinking of a function as a *set of ordered pairs*. More specifically, a function from A to B can be thought of as a certain type of subset of $A \times B$. This idea goes well with the concept of the *graph* of a function, which actually does consist of various ordered pairs, representing points in a coordinate system.

We then have to decide what kinds of subsets will be allowed. In algebra, you may have learned the "vertical-line test" for deciding if a graph represents a function. This test says that if there is a vertical line that goes through a graph in more than one point, then the graph is not the graph of a function (see Figure 3.2).

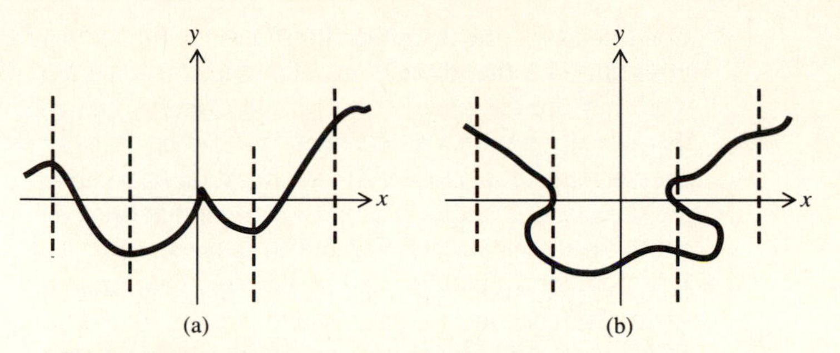

Figure 3.2. (a) This graph "passes" the vertical line test. No vertical line crosses the graph more than once, so this graph represents a function. (b) This graph "fails" the vertical line test. Some of the vertical lines do cross the graph more than once, so this graph does not represent a function.

With these ideas in mind, try to come up with a precise, formal definition for what we mean by "a function from A to B".

Try to write down a definition for "function" using ordered pairs and the language of quantifiers.

As indicated already, such a function will be some kind of subset of $A \times B$. The restriction on what subsets are considered functions can be expressed in various equivalent ways. The most concise is essentially the following:

Every element of A must be used in exactly one of the ordered pairs of the subset.

This is often expressed as two separate conditions:

i) every element of A is used in at least one pair, and
ii) no element of A is used in two different pairs.

We will use the following as our formal definition:

DEFINITION 3.14: For any sets A and B, a *function from A to B* is a set $S \subseteq A \times B$ which satisfies the following conditions:

a) $(\forall\, u \in A)(\exists\, v \in B)((u, v) \in S)$, and
b) $(\forall\, u \in A)(\forall\, v, w \in B)([(u, v) \in S \text{ and } (u, w) \in S] \Rightarrow v = w)$.

The set A is called the *domain* of the function, and the set B is called the *codomain* of the function.

COMMENTS

1. Condition a) in Definition 3.14 may not fit your experience with functions in calculus. For example, you may have seen something like $f(x) = 1/(x^2 - 9)$, which is not defined when $x = \pm 3$, described as a function from $\mathbf{R}$ to $\mathbf{R}$. However, strictly speaking, this formula expresses a function from $\mathbf{R} - \{-3, 3\}$ to $\mathbf{R}$.

2. Condition b) in the definition illustrates a common way in which mathematicians express the idea that there is at most one of something. They say: "if there are two, then they are the same". You should examine this condition until it is clear to you that it really says that no element of A can be "used twice" or paired with two different elements of B. Notice that the negation of condition b) is the statement that there are objects u, v, and w, with $v \neq w$, such that (u, v) and (u, w) are both in S. In other words, the negation of condition b) is the statement (in a context more general than $\mathbf{R} \times \mathbf{R}$) that S fails the vertical-line test. You may like to think of condition b) as saying "a function cannot be schizophrenic"—that is, it cannot give one value some of the time and a different value at other times.

3. Notice that there is no restriction on how often an element of B can be used. An extreme illustration of this is a "constant function", in which one element of the codomain is used as the second element in every pair of the function. To be specific, let $A = \mathbf{R}$ and $B = \mathbf{R}$, and let S be the set $\mathbf{R} \times \{2\}$, i.e., $S = \{(x, 2) : x \in \mathbf{R}\}$. This set fits the two conditions in Definition 3.14, and is therefore considered a function.

4. There is a third quantifier, called the "unique existence" quantifier, which can be used to combine the two conditions into one. This is examined in Exercise 8.

To reconcile Definition 3.14 with the common, everyday function notation, we make the following definition, in which we use the letter "f" for the set of ordered pairs, replacing "S", and introduce the usual "$f(x)$" notation:

DEFINITION 3.15: For sets A and B, "$f: A \rightarrow B$" means: f is a function from A to B. If f is a function, then "$f(u) = v$" means: $(u, v) \in f$.

Most of the time we will use this familiar notation, and think of a function intuitively as a way of getting from an element in A to an element in B. We will rely on the formal definition of a function as a set of ordered pairs in difficult cases, such as deciding what "function" means when A or B is the empty set.

COMMENTS

1. The use of the standard notation has built into it implicitly the fact that a function is "not schizophrenic" (see Comment 2 above). When we write "$f(u)$", we assume that this has a single meaning, and cannot represent more than one value. We can sometimes eliminate variables using this notation, writing "$(u, f(u))$" instead of "(u, v)" for an element of f.

2. We can often describe functions by means of equations, using phraseology like

   $$\text{"}f : A \rightarrow B, \text{ with } f(x) = 2x - 1\text{"},$$

   which means f is the function from A to B consisting of the set

   $$\{(x, y) \in A \times B : y = 2x - 1\}\text{"}.$$

If f is a function from A to B, and u and v are elements of A and B respectively, there are various phrases by which mathematicians commonly express the condition "$f(u) = v$":

i) v is the image of u under f.
ii) v is the value of f at u.
iii) u is a preimage of v under f.
iv) u is mapped to v.

The element u is sometimes referred to as the *argument* of f.

DEFINITION 3.16: For a function $f: A \to B$, the *image set* of f [written "Im(f)" or "$f(A)$"] is

$$\{f(t) : t \in A\}.$$

This is the set of all those elements of B which are images of some element of A. (Verify that this terminology is consistent with Definition 3.8, given in Section 3.1.)

COMMENTS

1. If f is a function from A to B, then f can also be thought of as a function from A to Im(f), or to any set between its image set and B. In other words, the codomain of a function is not inherent in the function considered as a set of ordered pairs. The same set of pairs can be viewed as a function with different codomains. In many situations, it does not matter which set is thought of as the codomain. However, we will introduce an important concept shortly in which the codomain plays an important role.

2. The image set is sometimes also referred to as the *range* of the function. However, the word "range" is used in some texts to mean what we have called the codomain. To avoid confusion, we will not use "range".

In Chapter 6, we will see that the concept of a function can be seen as a special case of a more general concept called a relation.

Example 3. **Sample Functions** In each of the following examples, S is a function with the given set A as the domain and the given set B as the codomain.

i) $A = \{1, 2, 3\}$, $B = \{1, 2, 3\}$: $S = \{(1, 2), (2, 3), (3, 2)\}$. The image set for this function is $\{2, 3\}$.
ii) $A = \mathbf{R}$, $B = \mathbf{R}$: $S = \{(x, y) \in \mathbf{R} \times \mathbf{R} : y = x^2\}$. The image set for this function is $\mathbf{R}^{\geq 0}$ (the nonnegative real numbers).
iii) $A = \mathbf{Z} - \{0\}$, $B = \{-1, 1\}$: $S = \{(x, x/|x|) : x \in \mathbf{Z}\}$. The image set for this function is B.
iv) $A = \mathbf{Q}$, $B = \mathbf{Q}$: $S = \{(x, 2x) : x \in \mathbf{Q}\}$. The image set for this function is $\mathbf{Q}$.

Verify that, in each case, the set S satisfies the two conditions in Definition 3.14. □

Finite Functions

If the domain is a small, finite set, it is common to specify a function using an *arrow diagram* instead of ordered-pair notation. For example, function i) of Example 3 might be written by naming the function, say calling it f, and writing

$$\begin{aligned} &1 \to 2 \\ f:&2 \to 3 \\ &3 \to 2 \end{aligned}$$

If we have each of the elements of the domain listed once in the left-hand column of such an arrow diagram, and the objects in the right-hand column are all elements of the codomain, then we automatically have a function. This notation provides a convenient way to analyze all of the functions from one small set to another. For example, if $A = \{1,2\}$ and $B = \{3,4\}$, then there are exactly four distinct functions from A to B, and they can be listed as follows:

$$f_1: \begin{matrix} 1 \to 3 \\ 2 \to 3 \end{matrix} \qquad f_2: \begin{matrix} 1 \to 3 \\ 2 \to 4 \end{matrix} \qquad f_3: \begin{matrix} 1 \to 4 \\ 2 \to 3 \end{matrix} \qquad f_4: \begin{matrix} 1 \to 4 \\ 2 \to 4 \end{matrix}$$

We will ask you later (Exploration 3) to examine how many functions there are in general for finite sets A and B, and further to explore how many there are that fit certain special conditions we will be defining.

Meanwhile, we ask you to look at the most extreme case of finite sets:

Get Your Hands Dirty 2

Suppose either A or B is the empty set. What does that mean about possible functions from A to B? Are there any? If so, what are they like? □

Surjective Functions

In examples iii) and iv) of Example 3, the image set was equal to the entire codomain B. In the first two examples, it was not. It turns out the distinction between these situations represents an important concept. A function whose image set is equal to its codomain is called *onto* or *surjective*.

How can this concept be defined directly in terms of the elements of A and B, in the style of Definition 3.14?

Try to write a formal definition. We are saying something about every element of *B*.

We want to say that every element of B is the image of at least one element of A; that is, it is used in at least one ordered pair of the function. This is exactly the same as the first condition defining "function" except that the roles of A and B have been reversed. The formal definition is as follows:

DEFINITION 3.17: For a function $f: A \to B$, "f is *onto*" (or "*surjective*") means:

$$(\forall\, v \in B)(\exists\, u \in A)(f(u) = v).$$

The condition "$(\exists\, u \in A)(f(u) = v)$" says precisely "$v \in \mathrm{Im}(f)$", so the condition "$f$ is onto" can be stated concisely as $B \subseteq \mathrm{Im}(f)$, or simply $\mathrm{Im}(f) = B$.

COMMENTS

1. We pointed out following the definition of "image set" that a given set of ordered pairs can sometimes be interpreted with different codomains. Whenever we use the terms "onto" or "surjective", however, it is essential that we know what the codomain is. Every function is "onto" when its image set is used as the codomain.

2. The two terms, "onto" and "surjective", are used quite commonly, though "surjective" is the term generally used in advanced mathematical writing. We will use them both, to help you get used to the variation in language. A similar comment applies to Definition 3.18 below.

If we use an arrow diagram to describe a function f for finite sets A and B, then we can tell if f is surjective simply by looking down the right-hand column: if every element of B appears, then the function is surjective.

For example, if $A = \{1, 2, 3, 4\}$ and $B = \{3, 4, 5\}$, then the following function is surjective:

$$g_1: \begin{array}{l} 1 \to 3 \\ 2 \to 4 \\ 3 \to 3 \\ 4 \to 5 \end{array}$$

but this one is not:

$$g_2: \begin{array}{l} 1 \to 3 \\ 2 \to 4 \\ 3 \to 3 \\ 4 \to 3 \end{array}$$

(but, as comment 1 above points out, g_2 is surjective when considered as a function from A to $\{3, 4\}$).

Here is an example of how to use this concept for infinite sets, where the function is defined by a formula:

Example 4. Proving That a Function Is Onto

Proof that $S = \{(x, 3x + 5) : x \in \mathbf{R}\}$ ***is onto as a function from R to R:*** Assume that $v \in \mathbf{R}$. We need to show that there is a real number s such that $(s, v) \in S$.

Let $s = (v - 5)/3$. By definition of S, $(s, 3s + 5) \in S$. But $3s + 5 = v$, so we have $(s, v) \in S$, as desired. ■

COMMENT We leave it to you to figure out how we found the right s. (*Hint:* Exploration.)

Get Your Hands Dirty 3

Prove in detail that the function $f: \mathbf{R} \to \mathbf{R}$ given by the equation $f(x) = 2x^3 - 5$ is surjective. □

Injective Functions

As we noted, the condition in the definition of "onto" is the reverse of the first condition in that of "function", and describes those functions in which each element of the codomain is used in at least one ordered pair. We get a second important category of functions by reversing the roles of A and B in the second condition of the definition of "function". These are the functions in which no element of B is used in more than one pair.

DEFINITION 3.18: For a function $f: A \to B$, "f is *one-to-one*" (or *injective*) means:

$$(\forall\, v \in B)(\forall\, u, w \in A)([(u, v) \in f \text{ and } (w, v) \in f] \Rightarrow u = w).$$

COMMENTS

1. As in condition b) of the definition of "function", this quantified statement expresses "uniqueness": it says that, for a particular element of B, there is at most one element of A which is mapped to it. We will look at how this type of condition is used in a proof in Example 5 below (see comment 4 following that proof).
2. Definition 3.18 has been stated in such a way as to emphasize its similarity to condition b) of the definition of "function". In fact, it can be expressed more succinctly, using standard function notation to eliminate v, as follows:

DEFINITION 3.18′: For any function $f: A \to B$, "f is *one-to-one*" (or *injective*) means:

$$(\forall\, u, w \in A)(f(u) = f(w) \Rightarrow u = w).$$

In practice, we will generally use this version. (Verify that the two versions of the definition mean the same thing.)

As with "surjective", the question of whether or not a function is injective is easy when we have a function described by an arrow diagram: if the objects in the right-hand column are all different, then the function is injective.

For example, if $A = \{1, 2, 3\}$ and $B = \{3, 4, 5, 6\}$, then the following function is injective:

$$\begin{aligned} &1 \to 3 \\ h_1 : &2 \to 4 \\ &3 \to 6 \end{aligned}$$

but this one is not:

$$\begin{aligned} &1 \to 5 \\ h_2 : &2 \to 4 \\ &3 \to 5 \end{aligned}$$

For infinite functions, the issue is not so simple, and as our definitions get more complex, it may become harder to figure out what their negations are. So we ask the following:

What does it mean for a function to *not* be one-to-one? (Use Definition 3.18′)

Here the predicate of our quantified statement is an implication, and so its negation is the statement that there is a counterexample, i.e., there are values of u and v for which the hypothesis "$f(u) = f(w)$" is true but the conclusion "$u = w$" is false. In other words, the negation of the condition in Definition 3.18′ is

$$(\exists\, u, w \in A)\,(f(u) = f(w) \text{ and } u \neq w)\,. \qquad (***)$$

Get Your Hands Dirty 4

Prove that each of functions i), ii), and iii) of Example 3 is not one-to-one; i.e., find specific values of u and w which make the predicate of (***) true. □

However, function iv) of Example 3 is one-to-one, as we show below in Example 5. (Since this function is expressed formally as a set of pairs, we use that notation in the proof. An alternate proof is given in comment 3 below.)

Example 5. Proving That a Function is One-to-One

Proof that the function $S = \{(x, 2x) : x \in Q\}$ is one-to-one: Assume that $v \in \mathbf{Q}$ and that $u, w \in \mathbf{Q}$, with $(u, v) \in S$ and $(w, v) \in S$. We need to show that $u = w$.

Since $(u, v) \in S$, we must have $v = 2u$. Similarly, $v = 2w$. Thus $2u = 2w$, so $u = w$, as desired. ■

COMMENTS

1. If we were to try to mimic this proof for function ii) of Example 3 [$S = \{(x, y) \in \mathbf{R} \times \mathbf{R} : y = x^2\}$], the proof would break down at the last step. Using the same notation as above, we would get $u^2 = w^2$. But we cannot conclude from this equation that $u = w$, so we cannot prove that the function is one-to-one.

2. In the case where the domain and codomain are the same finite set, a function which is one-to-one must also be onto, and vice versa. (Exercise 10 asks you to explain intuitively why this is so.) But you may find it surprising to learn that for infinite sets this is not the case. If A is any infinite set, there are functions from A to A which are one-to-one but not onto, and functions which are onto but not one-to-one (see Exercise 11).

3. If we were to use "standard" function notation, we might write function iv) as $f : \mathbf{Q} \to \mathbf{Q}$, with $f(x) = 2x$, instead of writing it explicitly as a set of pairs. The proof would then read:

   > Assume that $u, w \in \mathbf{Q}$, with $f(u) = f(w)$. We need to show that $u = w$. Since $f(u) = 2u$ and $f(w) = 2w$, we have $2u = 2w$, and so $u = w$, as desired.

4. This is a typical proof of "uniqueness": in order to show that there is a unique ordered pair in S with v as its second coordinate, we assume that there are two such pairs, (u, v) and (w, v), and show that they are the same, by showing $u = w$.

There is a term for functions which are both injective (one-to-one) and surjective (onto).

DEFINITION 3.19: For any function $f : A \to B$, "f is *bijective*" means: f is both injective and surjective.

A bijective function is also known as a *one-to-one correspondence*.

There is an important connection between bijective functions and the concept of cardinality. Intuitively, if there is a bijective function from A to B, then the elements of A and B can be "matched up", and so the sets "have the same size". We will use this idea in Chapter 6, when we prove a theorem about the cardinality of power sets, and use it in a more theoretical way in Section 8.1 when we discuss cardinality more formally.

For now, we will assume without proof that, for finite sets A and B, if there is a bijective function from A to B, then $|A| = |B|$.

Composition

We include here one more important definition concerning functions—the concept of *composition*. If we have two functions, $f : A \to B$ and $g : B \to C$, then we can put the two functions together to create a third function, written $g \circ f$, from A to C. As in calculus, we want $(g \circ f)(x)$ to be equal to $g(f(x))$. The diagram in Figure 3.3 shows how this is done.

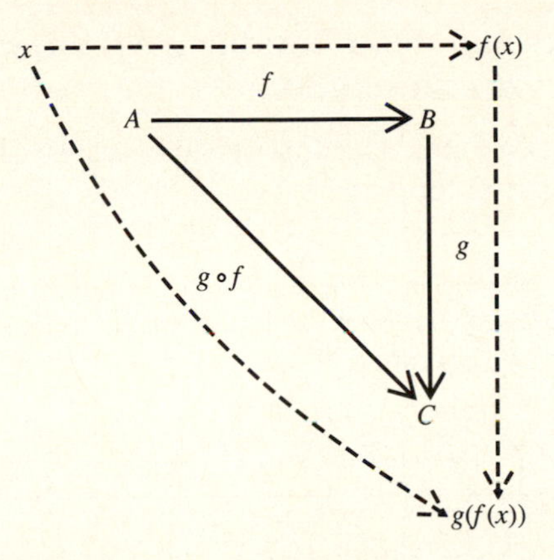

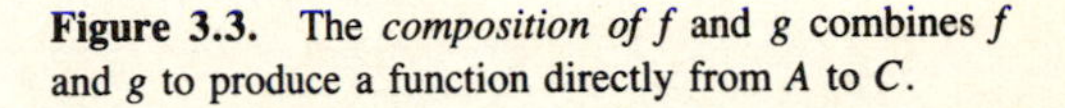
Figure 3.3. The *composition of* f and g combines f and g to produce a function directly from A to C.

The image in Figure 3.3 fits very well for functions defined in terms of arrow diagrams. For example, suppose $A = \{1,2,3\}$, $B = \{3,5,9,10\}$, and $C = \{1,4,9\}$. Let $f: A \to B$ and $g: B \to C$ be given by

$$f: \begin{array}{l} 1 \to \ \ 3 \\ 2 \to 10 \\ 3 \to \ \ 9 \end{array} \quad \text{and} \quad g: \begin{array}{r} 3 \to 4 \\ 5 \to 1 \\ 9 \to 4 \\ 10 \to 9 \end{array}$$

We can find the composition $g \circ f$ by connecting elements in the right-hand column of f with the same elements in the left-hand column of g, and then following the arrows, as in Figure 3.4.

Thus, 1 is mapped to 3, using f [i.e., $f(1) = 3$], and we then follow 3 over to the diagram for g, and see that 3 is mapped to 4 using g [i.e., $g(3) = 4$]. Altogether, this tells us that $(g \circ f)(1) = g(f(1)) = 4$. Similarly, we have $(g \circ f)(2) = 9$ and $(g \circ f)(3) = 4$. Thus, we have

$$g \circ f: \begin{array}{l} 1 \to 4 \\ 2 \to 9 \\ 3 \to 4 \end{array}$$

We still need a formal definition for composition in general. How can we define composition based on the notion of a function as a set of ordered pairs?

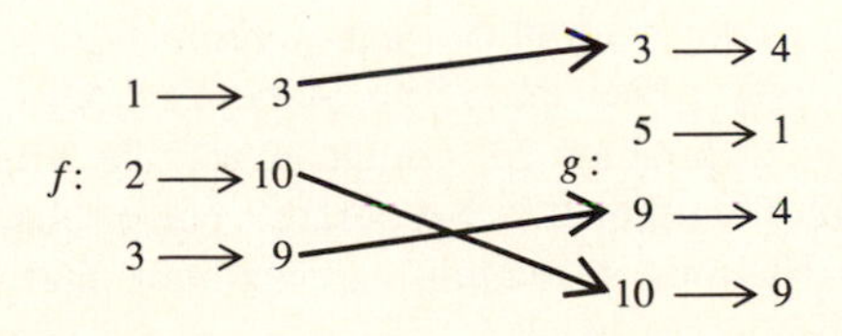

Figure 3.4. We can calculate the composition of f and g by following the arrows.

Try to write down a formal definition for $g \circ f$, where f and g are functions.

We will give the definition in two forms. Our first version uses the ordered-pair notation for f and g themselves, as well as for $g \circ f$:

> **DEFINITION 3.20:** For functions $f: A \to B$ and $g: B \to C$, the *composition of f and g* (written $g \circ f$, and read "g composition f" or "g of f") is the subset of $A \times C$ given by
>
> $$g \circ f = \{(u, w) \in A \times C : (\exists\ v \in B)((u, v) \in f \text{ and } (v, w) \in g)\}.$$

Intuitively, a pair (u, w) belongs to g precisely when there is a "middle element" v that "connects them" via f and g.

We can use standard function notation for f and g to eliminate the need for v in the definition:

> **DEFINITION 3.20′:** For functions $f: A \to B$ and $g: B \to C$, the *composition of f and g* (written $g \circ f$, and read "g composition f" or "g of f") is the subset of $A \times C$ given by
>
> $$g \circ f = \{(u, w) \in A \times C : g(f(u)) = w\}.$$
>
> In other words, "$(u, w) \in g \circ f$" means $w = g(f(u))$.

Important note about notation: In the composition $g \circ f$, the function f is "done first"—that is, we first find the image of u under f, i.e., $f(u)$, and then find the image of $f(u)$ under g. The image of u under $g \circ f$ is $g(f(u))$.

COMMENT The question of notation for composition has created an unsolvable dilemma for mathematicians. Because we write the image of x under f as $f(x)$, we express the image of x under the composition as $g(f(x))$. Thus, in the composition $g \circ f$, the functions appear in the same order as in the expression $g(f(x))$. That is perfectly natural.

However, it would also be natural to write the function that is "done first" on the left, since we read most mathematics from left to right. In fact, we are doing the opposite: in the composition $g \circ f$, the function f, which is "done first", is on the right.

The only way to reconcile these two contradictory requirements is to do something like write the image of x under f as xf rather than $f(x)$. Some textbooks do this. Others write $f(x)$, but use the reverse of our notation for composition. The approach we have chosen here is the one most commonly used. It is not perfect, but it seems to be the best choice available. You should be cautious in reading any textbook that discusses functions to be clear on what notation is being used.

We will look at some theorems involving the concepts of "injective", "surjective", and "composition" in Section 4.1. For now, we conclude Topic 2 with two explorations:

Exploration 2: How Many Functions Are There from A to B?

Suppose A and B are both finite sets, with $|A| = r$ and $|B| = s$. How many functions are there from A to B? Does your answer make sense in the case where r or s is zero? (*Hint:* Look at your results for GYHD 2. Also look at specific examples where r and s are small, such as $r = 2$, $s = 3$, or $r = 5$, $s = 2$. Then try to explain your results.)

Harder: How many of the functions from A to B are one-to-one? How many are onto? How many are one-to-one and onto? (*Note:* The general question of how many surjective functions there are does not have a simple answer. You should be satisfied if you get partial results, such as the case where $|B| = 2$.)

• EXPLORE •

Exploration 3: Composition with One-to-One and Onto Functions

Suppose f and g are functions with $f: A \to B$ and $g: B \to C$. Is there any connection between whether either of the functions f or g is one-to-one or onto and whether the composition $g \circ f$ is one-to-one or onto?

• EXPLORE •

EXERCISES

Note: The following exercises focus on intuitive understanding of the terminology introduced in this section and the use of quantifiers. Section 4.1 contains exercises on the topics of this section that focus on proof.

Topic 1: Upper Bounds and Sets Bounded Above

1. For a set $S \subseteq \mathbf{R}$, the statement "S is bounded above" is defined by the quantifier expression

$$\text{“}(\exists\, x \in \mathbf{R})\,(\forall\, y \in S)\,(x \geq y)\text{”}.$$

Consider the statement "$(\forall\, y \in S)\,(\exists\, x \in \mathbf{R})\,(x \geq y)$", in which the order of the quantified variables has been reversed.

a) In intuitive terms, what does this new statement mean?

b) Are there sets that are bounded above for which this new statement is false?

c) Are there sets that are not bounded above for which this new statement is true?

2. For the set $S = [-2, 3]$, find a real number x, different from 3, which satisfies the following condition:

$$\text{“}(\forall\, z \in \mathbf{R})\,(z \text{ an upper bound for } S \Rightarrow z \geq x)\text{”}.$$

[This is condition b) of the definition of "least upper bound" (Definition 3.13).]

3. Consider the two statements:

i) $(\forall\, z \in \mathbf{R})(\forall\, y \in S)(z \geq y \Rightarrow z \geq x)$.

ii) $(\forall\, z \in \mathbf{R})([(\forall\, y \in S)(z \geq y)] \Rightarrow z \geq x)$.

Suppose $x = 3$ and $S = [-2, 3]$. Explain why statement i) is false and statement ii) is true.

4. Write the statement "x is not the least upper bound for S" using quantifiers, without using the negation symbol.

5. We can generalize the concept of completeness by making the following definition:

DEFINITION: For $S \subseteq \mathbf{R}$, "S has the *completeness property*" means:

a) every nonempty subset T of S which is bounded above by an element of $\mathbf{R}$ has a least upper bound in S, and

b) every nonempty subset T of S which is bounded below by an element of $\mathbf{R}$ has a greatest lower bound in S.

[It is an important part of this definition that the lub or glb be elements of S, though they need not be in T. Also note that the upper or lower bound for T need not be in S. It is possible for a set to have property a) but not b), or vice versa. $\mathbf{R}$ itself has both properties, and for $\mathbf{R}$, b) can proved from a).]

Which of the following sets have the completeness property? Justify your answers with either an informal explanation or a counterexample:

a) $\mathbf{Q}$

b) $[0, 1]$

c) $(0, 1)$

d) $[0, 1)$

e) $[0, \infty)$

Topic 2: Functions

6. (GYHD 2)

a) Describe the set of all functions from A to B if $A = \varnothing$ and $B \neq \varnothing$.

b) Describe the set of all functions from A to B if $B = \varnothing$ and $A \neq \varnothing$.

c) Describe the set of all functions from $\varnothing$ to $\varnothing$.

7. For a function $f: A \to B$, the statement "f is surjective" is defined by the quantifier expression

$$\text{“}(\forall\, v \in B)(\exists\, u \in A)(f(u) = v)\text{”}.$$

Consider the statement "$(\exists\, u \in A)(\forall\, v \in B)(f(u) = v)$", in which the order of the quantified variables has been reversed.

a) In intuitive terms, what does this new statement mean?

b) Are there functions for which this new statement is true? If so, are they all surjective?

c) Are there surjective functions f for which this new statement is false?

8. There is a third quantifier, called the "unique existence" quantifier, which is used to abbreviate statements like the definition of "function" (Definition 3.14). The symbol for this quantifier is "$\exists!$", the existential quantifier followed by an exclamation point. It is defined as follows:

DEFINITION: For an open sentence $p(x)$ and a set S, "$(\exists!\, x \in S)(p(x))$" (read "there is a unique x in S such that $p(x)$") means:

$$(\exists\, x \in S)(p(x)) \quad \text{and} \quad (\forall\, u, v \in S)([p(u) \text{ and } p(v)] \Rightarrow u = v).$$

a) Explain, in intuitive terms, how "$(\exists!\, x \in S)(p(x))$" says that there is exactly one element from S in the truth set of $p(x)$.

b) Rewrite Definition 3.14 using the unique existential quantifier.

c) Suppose $f: A \to B$. Rephrase the quantifier statement

$$\text{“}(\forall\, v \in B)(\exists!\, u \in A)(f(u) = v)\text{”}$$

using terminology for functions introduced in this section.

d) Express "$\sim(\exists!\, x \in S)(p(x))$" without using negation except directly next to the predicate $p(x)$ (careful! this is not easy).

e) Explain, in intuitive terms, why "$(\exists!\, x \in S)(p(x))$" means the same thing as

$$\text{“}(\exists\, x \in S)[p(x) \text{ and } ([\forall\, u \in S][p(u) \Rightarrow u = x])]\text{”}.$$

9. Let f be the function from $\mathbf{R}$ to $\mathbf{R}$ with $f(x) = 3x - 5$.

a) Prove in detail that f is one-to-one.

b) Prove in detail that f is onto.

10. Explain in intuitive terms why any function from a finite set A to itself which is either one-to-one or onto must be both one-to-one and onto.

11. (*Hint* on this problem: Use arrow diagrams for ideas.)

a) Find a specific example of a function $f: \mathbf{N} \to \mathbf{N}$ which is one-to-one but not onto.

b) Find a specific example of a function $f: \mathbf{Z} \to \mathbf{Z}$ which is one-to-one but not onto.

c) Find a specific example of a function $f: \mathbf{N} \to \mathbf{N}$ which is onto but not one-to-one.

d) Find a specific example of a function $f: \mathbf{Z} \to \mathbf{Z}$ which is onto but not one-to-one.

12. (Exploration 2)

a) How many functions are there from $\{1, 2\}$ to $\{1, 2, 3\}$?

b) How many functions are there from $\{1, 2, 3, 4, 5\}$ to $\{1, 2\}$?

c) Suppose that A and B are finite sets, with $|A| = r$ and $|B| = s$. Express the number of functions from A to B as a general formula in terms of r and s.

13. (Exploration 2)

a) How many one-to-one functions are there from $\{1, 2\}$ to $\{1, 2, 3\}$? How many onto functions?

b) How many one-to-one functions are there from $\{1, 2, 3, 4, 5\}$ to $\{1, 2\}$? How many onto functions?

c) Suppose that A and B are finite sets, with $|A| = r$ and $|B| = s$. Express the number of one-to-one functions from A to B as a general formula in terms of r and s. [*Suggestion:* You should consider more examples than those in parts a) and b) before deciding on your answer.]

d) Suppose that A is a finite set, with $|A| = r$. Express the number of onto functions from A to $\{1, 2\}$ as a general formula in terms of r.

14. (Exploration 3) In the following problem, $f: A \to B$ and $g: B \to C$. Decide whether each of the following statements is true or false. If it is false, give a counterexample; if it is true, give an intuitive explanation. (*Hint:* Use arrow diagrams to explore.)

a) If f and g are both injective, then $g \circ f$ is injective.

b) If f and g are both surjective, then $g \circ f$ is surjective.

c) If $g \circ f$ is injective, then f must be injective.

d) If $g \circ f$ is injective, then g must be injective.

e) If $g \circ f$ is surjective, then f must be surjective.

f) If $g \circ f$ is surjective, then g must be surjective.

g) If f is injective, then $g \circ f$ is injective.

h) If g is surjective, then $g \circ f$ is surjective.

Other Exercises

15. Consider the following definition:

DEFINITION: For $S \subseteq \mathbf{R}$, "S is an *interval*" means:

$$(\forall u, v \in S)(\forall x \in \mathbf{R})(u < x < v \Rightarrow x \in S).$$

(*Note:* This may not exactly fit your intuitive idea of what an interval should be, but you should use this definition in this exercise nevertheless.)

a) Is $\varnothing$ an interval? Explain your answer.
b) Is $\{1\}$ an interval? Explain your answer.
c) Can a finite set S with $|S| > 1$ be an interval?

16. (Use the definition from Exercise 15 for this exercise.) Decide whether each of the following statements is true or false. If it is false, give a counterexample; if it is true, give an intuitive explanation:

a) If S and T are intervals, then $S \cup T$ is an interval.
b) If S and T are intervals, then $S \cap T$ is an interval.
c) If $S \cup T$ is an interval, then S and T are intervals.
d) If $S \cap T$ is an interval, then S and T are intervals.

17. In each of the following pairs of statements, the order of the two quantified variables has been reversed. Compare the meaning of the two statements in each pair. Is each true or false? (Compare this with Exercises 1 and 7.)

a) "$(\forall y \in \mathbf{Z})(\exists x \in \mathbf{Z})(x + y = 0)$" and "$(\exists x \in \mathbf{Z})(\forall y \in \mathbf{Z})(x + y = 0)$".
b) "$(\forall A \subseteq \mathbf{R})(\exists B \subseteq \mathbf{R})(A \cup B = \mathbf{R}$ and $A \cap B = \varnothing)$" and "$(\exists B \subseteq \mathbf{R})(\forall A \subseteq \mathbf{R})(A \cup B = \mathbf{R}$ and $A \cap B = \varnothing)$".

18. Read and think about the meaning of each of the following quantified statements. Express the meaning in natural English. Do not just translate the quantifiers. As much as possible, you should use terminology defined so far in this book, and other standard terminology, to make your statements concise. The word "not" may be helpful in expressing some of these statements. [*Warning:* Pay careful attention to grouping symbols—parentheses and brackets—in interpreting statements d) through g).]

a) $(\forall x \in \mathbf{R})(\exists y \in S)(y > x)$ (here, S is a variable representing a subset of $\mathbf{R}$).
b) $(\forall x, y \in A,\ u, v \in B)((x, u) \in f$ and $(y, v) \in f \Rightarrow u = v)$ (here, f is a variable representing a function from A to B).
c) $(\exists w \in B)(\forall r \in A)(f(r) \neq w)$ (here, f is a variable representing a function from A to B).
d) $(\forall a, b, c, d \in \mathbf{R})(a \neq 0 \Rightarrow [(\exists t \in \mathbf{R})(at^3 + bt^2 + ct + d = 0)])$.
e) $(\forall x \in \mathbf{Z})([(\exists t \in \mathbf{Z})(x^2 = 2t)] \Rightarrow [(\exists s \in \mathbf{Z})(x = 2s)])$.
f) $(\forall u \in \mathbf{Q})([(\forall x \in S)(u \geq x)] \Rightarrow [(\exists v \in \mathbf{Q})([(\forall y \in S)(v \geq y)]$ and $v < u)])$ (here, S is a variable representing a subset of $\mathbf{Q}$).
g) $(\forall n \in \mathbf{N})([n > 2$ and $(\forall u \in \mathbf{N})(u \mid n \Rightarrow u = 1$ or $u = n)] \Rightarrow (\exists t \in \mathbf{N})(n = 2t + 1))$.

19. For each of the statements a), b), c), and f) of Exercise 18, identify the free variables, and then:

a) find at least one substitution for the free variables which makes the statement true, and
b) find at least one substitution for the free variables which makes the statement false.

20. Write each of the following statements using quantifiers as fully as possible. Use $\mathbf{U}$ or an appropriate specific set as your universe.

a) There is no largest integer.
b) If x^2 is odd, then x is odd. (Here, x is a variable representing an integer.)
c) t can be written as the sum of the squares of four integers. (Here, t is a variable representing a positive integer.)
d) The sum of an integer and its square is always even.
e) Every prime greater than 2 is odd.
f) There are primes over a million.
g) Any nonempty set has more than one subset. (Use the notation for cardinality.)
h) If f is strictly increasing, then f is injective. (Let **U** be the set of all functions from **R** to **R**.)
i) Every nonempty subset of S has a maximal element. (Here, S is a variable representing a subset of **R**.)
j) Either n is a square or it has two distinct divisors other than 1 and itself. (Here, n is a variable representing a positive integer.)
k) Some surjective functions from A to B are not injective. (Here A and B are variables representing two sets. Let **U** be the set of all functions from A to B.)
l) All functions from S to S are either injective or surjective. (Here, S is a variable representing a set. Let **U** be the set of all functions from S to S.)

21. For statements c), i), j), k), and l) of Exercise 20, find at least one substitution for the free variable(s) which makes the statement true. For statements i), j), k), and l) of Exercise 20, find at least one substitution for the free variables which makes the statement false.

22. Express the negations of each of the following statements using quantifier notation as fully as possible, but without using the negation symbol (you may use symbols such as $\nmid$, $\neq$, etc.):
a) x is an upper bound for S. (Here, S is a subset of **R** and x is a real number.)
b) t is a prime. (Here, t is an integer greater than 1.)
c) u is rational. (Here, u is a real number.)
d) d is the GCD of a and b. (Here, a, b, and d are positive integers.) (Careful!)
e) R is a function from A to B. (Here, R is a subset of $A \times B$.)
f) f is strictly increasing. (Here, f is a function from **R** to **R**.)
g) f is injective. (Here, f is a function from A to B.)
h) f is surjective. (Here, f is a function from A to B.)
i) S is an interval. (Here, S is a subset of **R**. Use the definition of "interval" in Exercise 15.)

23. For a real number r, the notation $[r]$ is often used to represent the largest integer that is less than or equal to r. For example, $[3.1] = 3$, $[-2.7] = -3$, and $[6] = 6$. For $r \in \mathbf{R}$ and $n \in \mathbf{N}$, define the condition "$n = [r]$" in terms of quantifiers.

4

Basic Methods of Proof and Exploration

Introduction

In Chapter 3, we introduced and developed the language of quantifiers and logic, and used that language to express some important mathematical ideas. In this chapter, we turn from the expression of ideas to the exploration and proof of theorems.

We will begin Section 4.1 by looking at the stage of exploration in which we examine potential theorems more carefully (see Section 1.2), including looking at how to make hidden quantifiers explicit.

We will then examine the process of developing and writing proofs. We will see that the form of a mathematical theorem—especially the presence of quantified variables (hidden or explicit) in the hypothesis and conclusion of a potential theorem—can provide valuable clues to the mathematician. These clues suggest how to explore or investigate the truth of a particular statement, as well as how to set up and carry through a formal proof of a statement believed to be true. We will see that the process of developing a proof involves a different style of thinking from that used in actually presenting a proof.

In Section 4.2, we will examine ways in which statements can be rewritten to get simpler or more usable statements with the same meaning. In particular, we will expand our earlier discussion (in Section 2.4) of proof by contradiction, and discuss a related method—proof using the contrapositive.

4.1 Exploration and Proof with Quantifiers

Exploring a Universal Statement

Consider the following statement (for the universe **N**):

If a and b both divide c, then ab divides c.

As an "if . . . , then . . ." statement, this can be translated into a universally quantified implication:

$$(\forall\, a, b, c \in \mathbf{N})([a \mid c \text{ and } b \mid c] \Rightarrow (ab) \mid c)\,. \qquad (*)$$

Is the statement (*) true? How would you go about deciding? How can you investigate this assertion?

? ? ?

Since the predicate has the form of an implication, you can "experiment", by choosing some objects that make the hypothesis true. (The hypothesis here includes the fact that a, b, and c are natural numbers, in addition to the divisibility conditions "$a \mid c$" and "$b \mid c$".) Which objects? This is a key question in the exploration process.

What guidelines are there for choosing some values for a, b, and c? Remember that you are exploring, not proving.

There are no specific "right" or "wrong" choices for a, b, and c. We want to get a sense of whether the conclusion is true for all objects that fit the hypothesis or not. We can experiment by trying *some* sets of three positive integers. If we make the choices "diverse" in some way, we can hope to get a fair reading of the situation.

So if you haven't already done so, find several triples of positive integers for a, b, and c which fit the divisibility hypotheses, and find out whether those triples fit the conclusion or not.

> **Hint:** Look for diversity. For example, in the condition "$a \mid c$", there are the extreme cases where $a = 1$ and where $a = c$. Try these extremes as well as cases in between. Also, the condition on a is the same as that on b. Try cases where $a = b$ and where $a \neq b$.

(Don't read on until you are fairly certain whether (*) is true or false.)

You should have reached the conclusion that (*) is false. You should have come to this conclusion by finding a counterexample. If you followed the guidelines for diversity, you would at some point try the case where both a and b were equal to c (unless you found a counterexample sooner). For example, if $a = 4$, $b = 4$, and $c = 4$, then the hypothesis of (*) is true and the conclusion is false.

COMMENTS

1. Since we found a specific counterexample to (*), there is no need to discuss the question of proof. A counterexample is a definitive result. It completely settles the question. The statement is false.

2. That doesn't end our interest in (*). You should ask yourself if there are some reasonable restrictions on a, b, and c which will make (*) true (see Exercise 11).

Exploring an Existential Statement

Our next example involves both the universal and existential quantifiers. As we shall see, the order in which they appear has a major significance for the meaning of the statement.

Consider the following statement:

Every pair of positive integers has a positive common multiple.

This statement doesn't have "if . . . , then . . ." form, so we need to do a little work to express this in terms of quantifiers.

Where do we start?

It is a statement about "every pair of positive integers", and so we can write it as "$(\forall\, a, b \in \mathbf{N})(\ldots)$".

What do we want to say about every such pair of positive integers?

?　　?　　?

We are saying "they *have* a common multiple". As we remarked in Section 3.1, the word "have" is a clue that we are talking about an existential quantifier: "there exists an element which is a common multiple for a and b".

So our statement becomes:

$$(\forall\, a, b \in \mathbf{N})(\exists\, c \in \mathbf{N})(a \mid c \text{ and } b \mid c)\,. \qquad (**)$$

As with our exploration of a universal statement, we ask: is it true? How would you investigate it?

?　　?　　?

It may help to think of "$(\exists\, c \in \mathbf{N})$ $(a \mid c$ and $b \mid c)$" as an open sentence $p(a, b)$ for the free variables a and b, and view (**) as saying

$$(\forall\, a, b \in \mathbf{N})(p(a, b)).$$

Since this is a universally quantified statement, you can explore it by trying out specific objects. For example, suppose $a = 6$ and $b = 8$. What does $p(6, 8)$ say?

Write it out.

Plugging in the values for a and b, we get

$$(\exists\, c \in \mathbf{N})(6 \mid c \text{ and } 8 \mid c)\,.$$

This simply says there is a positive integer which is divisible by both 6 and 8. Is that true?

?　　?　　?

Of course. There are many, but all we need is one, and $c = 24$ will do.

If we made a different substitution for a and b, we would get a slightly different statement for $p(a, b)$. [If you're not sure what's going on, try this. Choose several pairs of numbers for a and b, and in each case, write out the resulting statement $p(a, b)$.]

Will there always be a "good" value for c? How can we describe it "generically"?

? ? ?

Looking at diverse examples should help. Since the subject here is divisibility, it may be a good idea to try various extremes in that regard. For example, if $a \mid b$, then b itself will be a common multiple of a and b, and we can choose $c = b$. But that won't work in general. Another extreme is the case where the integer a "totally doesn't divide" b, i.e., $\text{GCD}(a, b) = 1$. For example, if $a = 5$ and $b = 7$, then probably the first choice that comes to mind is $c = 35$.

Why 35?

Because $5 \times 7 = 35$. One way to get a common multiple is to just multiply a and b together. And that ought to work all the time.

This all may seem very elementary, but the point is that our investigation provides a roadmap to how we will prove the assertion. As you try different values of a and b, you get a *method* for finding a "good" c. Writing a proof of the assertion in this example will require us to *construct* or *define* c in terms of generic positive integers a and b, and then prove that c has the required property.

One more preparatory detail: What is the "required property" just mentioned? What exactly does the requirement on c mean?

? ? ?

We want a value of c such that "$a \mid c$ and $b \mid c$". These divisibility conditions are themselves existential statements. So, to work out all the details of a proof, we will have to find integers u and v (again, in terms of a and b) such that $c = u \cdot a$ and $c = v \cdot b$.

Proving That Something Exists

With all of this done, we can write a proof very easily, since we know what we're looking for.

THEOREM 4.1. Every pair of positive integers has a common multiple.

Proof: Assume that a and b are elements of $\mathbf{N}$.

We need to prove that a and b have a common multiple. In other words, we must show

$$(\exists\, c \in \mathbf{N})(a \mid c \text{ and } b \mid c)\,.$$

Let $c = ab$. Then $c \in \mathbf{N}$ (since the product of positive integers is a positive integer).

Further $a \mid c$ and $b \mid c$, for the following reason: "$a \mid c$" means (using Definition 2.5 in Section 2.3) "$(\exists\, u \in \mathbf{Z})(a \cdot u = c)$". And the latter statement is true, since "$a \cdot u = c$" is true when $u = b$; i.e., $u = b$ "works". Similarly, $b \mid c$.

Thus we have found an object which fits the desired conditions. This completes the proof. ■

DISCUSSION What does it mean, in general, to prove a statement of the form "$(\exists\ t)(q(t))$"? We have seen that there are two basic steps:

1. Identify, define, or construct the object you wish to use for t.
2. Prove that the object you identified is in the truth set for $q(t)$.

In our present theorem, we did step 1 by giving a *formula*—an equation—for c, in terms of other variables, namely, $c = ab$. We had already seen, at least intuitively, that choosing $c = ab$ would "work".

In other cases, we might use some other way to "identify, define, or construct", or otherwise "point to", the object that we claim will "work". Sometimes we can use a previously proved existence theorem. In Chapter 5, we will look at an important axiom, as well as some existence theorems, which can be used to "point to" the element we want, without requiring a specific formula or procedure for calculating the object.

Often we can construct or describe the "something" we are looking for in terms of other variables. In this proof, we described c in terms of a and b. We were allowed to do this because the quantifier for c comes after the quantifiers for a and b.

When other variables are introduced prior to the one we are looking for, you can often use exploration to figure out how to do step 1. Ask yourself what you would choose in a particular case. Actually try some specific values for the previous variables (in this case, a and b), and see what choice would "work" for the desired object (in this case, c). How might you generalize that choice? Is there a formula or equation that will help identify the object? Is there some earlier theorem or something in the main hypothesis that says something similar about "something existing"?

All of these questions may give clues for your proof.

COMMENTS

1. Although there are actually infinitely many values for c that will "work", all we needed to do was give one. The one we gave may not be the smallest, but that wasn't expected or required, and so we chose the one that was easiest to express and work with.

2. The work involved in exploration does not appear in the proof itself. All we see is the statement "let $c = ab$". In writing proofs, mathematicians usually don't bother to explain what led them to the particular choice that they give. They just state it, and then prove that it "works".

3. If you want to be especially fussy, you can justify the "similarly, $b \mid c$" statement by pointing out that ab and ba are equal, so that the symmetry of the problem is not disturbed by defining c as ab rather than ba.

Summary

The form of this proof is straightforward. We have to show that "something exists". As already discussed, the first stage of such a proof involves identifying a candidate for the required object, and the second stage is to show that the identified object has the necessary property.

We want to emphasize the important role which exploration plays in developing this type of proof. In the case of Theorem 4.1, our exploration led us to expect that the product ab would "work" for c. Though the investigation that led to this expectation is behind the scenes, it is nevertheless an essential part of the overall proof process. In reading existence proofs, ask yourself how the writer chose the object which is being shown to satisfy the condition.

Importance of Order of Quantifiers

In our last example, it was very important that the existential quantifier in this example came after the universal quantifiers. This told us that the object c could be chosen in terms of a and b.

Look at the following statement, which has the quantifiers in a different order:

$$(\exists\, c \in \mathbf{N})(\forall\, a, b \in \mathbf{N})(a \mid c \text{ and } b \mid c).$$

Think about the portion of this which follows the existential quantifier: $(\forall\, a, b \in \mathbf{N})(a \mid c$ and $b \mid c)$. This is an open sentence for the variable c. Call this statement $q(c)$.

Try some values for c. What does $q(c)$ say? For what values of c is it true?

The open sentence $q(c)$ says "every positive integer a and every positive integer b is a divisor of c". There is no positive integer c for which this is true. No matter what positive integer is chosen for c, there will be values of a and b which are not divisors of it. Thus, by moving the existential quantifier and its variable to the front, we have changed a true statement into a false one.

Changing the Universe

Here is another example of the importance of order of quantifiers. In this example, the universe plays an important role.

Compare the following pair of statements in the two cases $S = \mathbf{R}$ and $S = \mathbf{R}^{\geq 0}$ (nonnegative reals):

i) $(\forall\, x \in S)(\exists\, y \in S)(y \leq x)$, and
ii) $(\exists\, y \in S)(\forall\, x \in S)(y \leq x)$.

Are these statements true or false for each S?

Statement i) begins as a universally quantified statement. It says something about every x in S. What it says about every such x is "$(\exists\, y \in S)(y \leq x)$", which is an open sentence with x as a free variable. This open sentence simply says "there is an element of S which is less than or equal to x". This is certainly true for every element x in S, since x itself is "an element of S which is less than or equal to x". So statement i) is true, no matter what set S is.

Statement ii) begins as an existentially quantified statement. It says "there is an element y in S satisfying a certain condition". The condition which y must satisfy is expressed as "$(\forall x \in S)(y \leq x)$", which is an open sentence with y as a free variable. This open sentence says "y is less than or equal to every element of S". Since $y \in S$, this means that y is the smallest element of S. Thus, we can paraphrase ii) as saying: "S has a smallest element". The truth value of this statement depends on S: if $S = \mathbf{R}$, "S has a smallest element" is false. [Verify: Choose some values for y, and look at the statement "$(\forall x \in S)(y \leq x)$".] If $S = \mathbf{R}^{\geq 0}$, then "S has a smallest element" is true. (Verify: What's the element?).

To summarize, we have seen that i) and ii) have the same truth value if $S = \mathbf{R}^{\geq 0}$ (both are true), while they have different truth values if $S = \mathbf{R}$ [i) is true, ii) is false].

Exploration 1: Order of Quantifiers

What general statement, if any, can be made about the relationship between a statement of the form

$$(\forall x \in S)(\exists y \in S)(r(x,y))$$

and the similar statement

$$(\exists y \in S)(\forall x \in S)(r(x,y))\,?$$

Examine the examples just discussed, and look at Exercises 1, 7, and 17 in Section 3.2. Compare the two statements for various predicates $r(x,y)$. Try to find other examples where one of these statements is true and the other is false.

• EXPLORE •

Quantifiers in the Hypothesis—Introduction

We have seen that certain mathematical terminology is defined in terms of quantified statements involving bound variables. Thus, the statement "$a \mid b$" involves an existentially quantified hidden variable, representing a missing factor; the statement "$A \subseteq B$" involves a universally quantified variable representing an element of A.

We now examine in more detail how the nature of these quantified variables affects the structure of a proof. As we noted before, theorems with hidden quantified variables as part of their hypothesis will often have hidden variables in their conclusion, usually with the same type of quantifier. This is natural, since the same terminology is likely to appear in both parts of the theorem.

Thus, for example, we might have a theorem about upper bounds, in which the hypothesis says that one set has an upper bound, and the conclusion says that another set has an upper bound. In such a theorem, both the hypothesis and the conclusion would have hidden "$\exists$" quantifiers, which refer to hidden variables that name those upper bounds. Similarly, theorems about subsets are likely to have hidden "$\forall$" quantifiers in both hypothesis and conclusion.

The presence of these hidden, quantified variables has a major impact on the way a theorem is proved. When there are universally quantified variables in the hypothesis of a theorem, we often use a technique known as *specialization*. When there are existentially quantified variables in the hypothesis of a theorem, we can generally use a technique that we call "name it—then use it".

We begin with the universal quantifier.

"∀" in the Hypothesis: Specialization

We will give two proofs that illustrate how to use a universally quantified statement in the hypothesis of a theorem. The first example of the specialization technique states a very simple result about upper bounds, which should seem intuitively clear. The theorem itself is not all that interesting, but the simplicity of the result allows us to concentrate on understanding the technique. Our second example is an important result about injective functions that is part of Corollary 4.7, an important theorem which will be used in Chapter 10.

A Theorem about Upper Bounds

Our first example of specialization uses the condition of being an upper bound. For convenience, we restate the definition:

DEFINITION 3.11: For $x \in \mathbf{R}$ and $S \subseteq \mathbf{R}$, "x is an *upper bound* for S" means:

$$(\forall\, y \in S)(x \geq y)\,.$$

Thus, the statement "x is an upper bound for S" involves a universally quantified hidden variable. It may help to think of the quantified statement as an implication:

$$\text{“}(\forall y)(y \in S \Rightarrow x \geq y)\text{”},$$

with hypothesis "$y \in S$" and conclusion "$x \geq y$".

The content of the following theorem is fairly straightforward, and you should think about what is going on before you read the proof. We will make comments on the proof as we go.

THEOREM 4.2. For $S \subseteq \mathbf{R}$ and $x, z \in \mathbf{R}$, if x is an upper bound for S, and $z \geq x$, then z is also an upper bound for S.

Think about this. Does it seem reasonable? Draw a picture. (We have one later.)

Proof: Assume that $S \subseteq \mathbf{R}$, that x is an upper bound for S, and that $z \geq x$. We need to show that z is an upper bound for S.

What does this mean?

| Thus, we need to show $(\forall\, t \in S)(z \geq t)$.

Comment: The statement "z is an upper bound for S" has a hidden variable, which we are making explicit here. We've used t instead of y for this hidden variable, for clarity in our explanation later on. Since y was a dummy variable, this doesn't change the meaning of the statement.

| So suppose $t \in S$. We need to show $z \geq t$.

Comment: We're now proving "if $t \in S$, then $z \geq t$" so we assume the hypothesis "$t \in S$" of this statement, and must prove its conclusion, "$z \geq t$".

How can you use the fact that x is an upper bound for S? What does that mean?

| Since x is an upper bound for S, and $t \in S$, we are guaranteed that $x \geq t$.

Comment: The reasoning in the step just completed is called *specialization,* and is explained below.

| But we also know that $z \geq x$ (from the hypothesis of the theorem). Therefore $z \geq t$, as desired. That completes the proof. ■

DISCUSSION — SPECIALIZATION The key step in this proof is the one that says "since x is an upper bound for S, and $t \in S$, we are guaranteed that $x \geq t$". What is crucial is the way this step uses the hypothesis "x is an upper bound".

What the proof does, without specifically saying so, is to interpret that assumption as the universally quantified statement "$(\forall\, y \in S)(x \geq y)$", and then apply that assumption to the particular element t in S. That is, since we are assuming that "$x \geq y$" is true for all elements y of S, we are guaranteed that it is true in the particular case $y = t$ (since $t \in S$).

As we noted before the proof, it may help to think of "$(\forall\, y \in S)(x \geq y)$" as an "if..., then..." statement: "if $y \in S$, then $x \geq y$". Since the hypothesis of this statement, "$y \in S$", is true when $y = t$, we are guaranteed that its conclusion, "$x \geq y$", is also true when $y = t$.

This key step is called *specialization*. The idea of specialization involves the same principle as "applying a theorem", as discussed in Section 2.2 (following Theorem 2.3). We either know (from a previous theorem) or are assuming (because it's in the hypothesis of the current theorem) that some universally quantified statement is true. We can therefore apply that statement to a particular case that happens to be relevant to the problem at hand.

When a universally quantified statement is in the hypothesis of a theorem, it is like a theorem that has previously been proved. It is a "guarantee producer", using our metaphor from Section 2.2. We are entitled to use it to produce appropriate guarantees. In order to use it, we need to provide it with objects that fit its conditions, i.e., its hypothesis. Often we will get precisely such objects as the result of assuming the "hypothesis of the conclusion". That is exactly what happened in this proof.

It may be helpful to outline the proof, and paraphrase the universally quantified statements and inequalities. The hypothesis has two parts:

i) Every element of S is $\leq x$.
ii) $x \leq z$.

The conclusion says:

iii) Every element of S is $\leq z$.

This conclusion can be thought of as the statement "if $t \in S$, then $t \leq z$". To prove this statement, we need to assume its hypothesis, namely, that we have an element of S, which we call t. Since $t \in S$, we can apply hypothesis i) to t. We are guaranteed by this hypothesis that $t \leq x$. That is the key step of the proof. (We then combine the inequality $t \leq x$ with $x \leq z$ to get $t \leq z$, as needed.)

COMMENT If you haven't already done so, you should think about what this theorem says, on an intuitive level. It should be "obvious" to you that the statement is true, and it should also be "clear" why it is true. You can imagine S as some set on a number line. The hypothesis tells us that "x is to the right of S" and that "z is to the right of x". The theorem concludes that "z is to the right of S". (See Figure 4.1.)

As we stated earlier, the content of this theorem is not very profound. We hope it has allowed you to follow the use of specialization without worrying about content.

A Schematic View of Specialization

At the risk of overexplaining, we give a schematic diagram of how specialization works. The theorem we are proving has a hypothesis and a conclusion. Since the theorem involves terminology with hidden, universally quantified, variables, both the hypothesis and the conclusion themselves can be thought of as "if . . . , then . . ." statements. So the overall picture is something like this:

$$\begin{array}{ccc} \text{MAIN HYPOTHESIS} & \Rightarrow & \text{MAIN CONCLUSION} \\ [\text{hypothesis}_1 \Rightarrow \text{conclusion}_1] & \Rightarrow & [\text{hypothesis}_2 \Rightarrow \text{conclusion}_2] \end{array}$$

We assume that MAIN HYPOTHESIS, and set out to prove the MAIN CONCLUSION. But that MAIN CONCLUSION is an "if . . . , then . . ." statement, so in order to prove it, we need to assume *its* hypothesis (labeled "hypothesis$_2$" here). Our goal becomes to derive the *conclusion of the MAIN CONCLUSION* ("conclusion$_2$").

In assuming "the hypothesis of the MAIN CONCLUSION", we are given an object w that satisfies this hypothesis. In the proof, we now try to show that this object, or some object u based on it, satisfies "the hypothesis of the MAIN HYPOTHESIS" ("hypothesis$_1$"). If u satisfies "the hypothesis of the MAIN HYPOTHESIS", then u must also

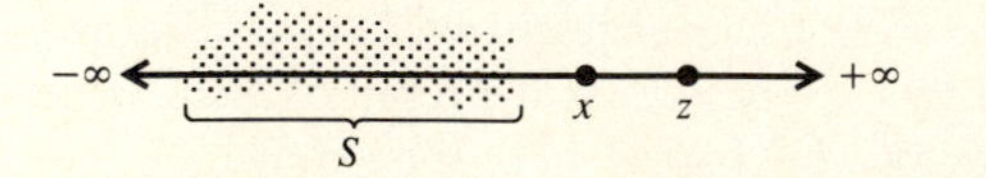

Figure 4.1. The number x is an upper bound for S, so x is "to the right of S". Since $z \geq x$, z is even further "to the right", so it is also an upper bound for S.

satisfy "the conclusion of the MAIN HYPOTHESIS" ("conclusion$_1$"), since we are assuming that the MAIN HYPOTHESIS is true.

We then try to use the fact that u satisfies "the conclusion of the MAIN HYPOTHESIS" to show that our earlier object w satisfies "the conclusion of the MAIN CONCLUSION".

In Theorem 4.2, "the hypothesis of the MAIN CONCLUSION" is the statement "$t \in S$", which is also "the hypothesis of the MAIN HYPOTHESIS". (The fact that these are identical makes Theorem 4.2 a simplified illustration of specialization.) We therefore can conclude that t satisfies "the conclusion of the MAIN HYPOTHESIS", namely, $x \geq t$. We finish the proof by combining this inequality with the other part of the MAIN HYPOTHESIS, namely, the inequality $z \geq x$, to get $z \geq t$, which is "the conclusion of the MAIN CONCLUSION", and we're done.

We want to emphasize that, for many students, the idea of specialization is one of the hardest proof techniques to understand.

Composition of Injective Functions

Our second example of specialization involves the concept of a injective function. We repeat the definition here for convenience:

> **DEFINITION 3.18′:** For any function $f:A \to B$, "f is *injective*" means:
>
> $$(\forall\, u, w \in A)\,(f(u) = f(w) \Rightarrow u = w)\,.$$

Again, the terminology involves a universal quantifier and implication.

The theorem we prove here is deeper than Theorem 4.2.

THEOREM 4.3. If $g:X \to Y$ and $h:Y \to Z$ are both injective functions, then $h \circ g:X \to Z$ is an injective function.

Think about this. Try some examples. Try to construct a counterexample and see why it's impossible to do so.

We begin thinking about the proof by examining the definition of the term "injective", which appears in both the hypothesis and the conclusion of Theorem 4.3. As Definition 3.18′ shows, this term involves hidden variables that are universally quantified.

We will begin the proof by assuming the MAIN HYPOTHESIS (that g and h are injective) and try to show the MAIN CONCLUSION (that $h \circ g$ is injective). To do so, we will assume that we have objects that satisfy "the hypothesis of the MAIN CONCLUSION"—that is, the hypothesis of the universally quantified implication statement that says that $h \circ g$ is injective. In other words, we will have objects that satisfy a statement of the same form as "$f(u) = f(w)$", as in Definition 3.18′. These objects will belong to the domain of $h \circ g$.

But we will then be able to show, fairly easily, that these objects, or objects based on them, satisfy "the hypotheses of the MAIN HYPOTHESIS"—the hypotheses of the implication statements that assert that g and h are individually injective. These will again be statements of the form "$f(u) = f(w)$". This stage of the proof is very simple in terms of mechanics, because the statements we are showing are so similar to the one we are assuming.

We will then be guaranteed the truth of certain statements based on "the conclusions of the MAIN HYPOTHESIS", which will be statements like "$u = w$". And these in turn will give us "the conclusion of the MAIN CONCLUSION", which is what we wanted. Whew!

With this introduction, try to write a complete proof.

Here's one way to write a complete proof. We give some brief explanatory comments:

Proof: Assume that g and h are injective. In order to show that $h \circ g$ is injective, suppose that r and s belong to A, with $(h \circ g)(r) = (h \circ g)(s)$. We need to show that $r = s$.

Comment: We have assumed the MAIN HYPOTHESIS, and the hypothesis of the MAIN CONCLUSION. We need to derive the conclusion of the MAIN CONCLUSION.

We can rewrite $(h \circ g)(r) = (h \circ g)(s)$ as $h(g(r)) = h(g(s))$. Thus $g(r)$ and $g(s)$ are elements of B whose images under h are equal.

Comment: One part of the MAIN HYPOTHESIS—the part that says h is injective—says

$$\text{“}(\forall\, u, w \in B)(h(u) = h(w) \Rightarrow u = w)\text{”}.$$

We are saying that $g(r)$ and $g(s)$ satisfy the hypothesis of this implication, i.e., if $u = g(r)$ and $w = g(s)$, then the statement "$h(u) = h(w)$" is true. We are therefore guaranteed that the substitution $u = g(r)$ and $w = g(s)$ makes the conclusion of this implication true. That's what our next step says.

Since h is injective, this tells us that $g(r) = g(s)$.

Comment: Next, we'll go through a similar process using the fact that g is injective.

Thus, r and s are elements of A whose images under g are equal.

Comment: In other words, the substitution $u = r$ and $w = s$ satisfies the hypothesis of the implication:

$$(\forall\, u, w \in A)\,(g(u) = g(w) \Rightarrow u = w)\,,$$

and so we can conclude that this substitution satisfies the conclusion of this implication. That's what comes next.

| Since g is injective, this tells us that $r = s$.

Comment: That's what we wanted.

| This completes the proof. ■

COMMENT You may want to reread the proof itself now, without stopping to read the commentary. If it still doesn't make sense, go over it again.

The following GYHD will give you practice in the technique of specialization:

Get Your Hands Dirty 1

a) Prove the following, identifying the use of specialization:

If x is an upper bound for S and for T, then x is an upper bound for $S \cup T$.

b) Make up a "stronger" theorem for $S \cap T$, and prove it. □

"∃" in the Hypothesis: "Name It—Then Use It"

We will give three illustrations of proofs in which the hypothesis involves an existential condition.

The first is very simple both in terms of its content—perfect squares—and in terms of its logic. The term "perfect square", which appears in both the hypothesis and the conclusion, represents a straightforward "$(\exists\, x)(p(x))$" statement.

Our second example involves sets that are bounded above, and resembles Theorem 4.2. The intuitive idea is as simple as that result, but the reasoning is a bit more complicated, since the basic definition has the form

$$\text{"}(\exists\, x \in U)(\forall\, y \in U)(p(x,y))\text{"}.$$

This quantified expression appears (hidden) in the conclusion, and so we have to show that something exists, and then prove that it has a certain universal property. To do this, we use the fact that the same type of expression appears in the hypothesis. This guarantees the existence of "something with a universal property". Because the terminology of this theorem involves both universal and existential quantifiers, the proof uses specialization as well as "name it—then use it".

The third example concerns the concept of surjective functions. Here the basic definition has the form

$$\text{"}(\forall\, x \in U)(\exists\, y \in U)(p(x,y))\text{"}.$$

Thus, the object which is said to exist depends on some other variable, which is universally quantified. In Theorem 4.1 ("every pair of positive integers has a common multiple"), this was the form of the whole theorem. Here we have this type of quantified statement both in the hypothesis and in the conclusion. Once again, we use specialization in combination with "name it—then use it".

We conclude this section by combining our third example with its counterpart for injective functions, Theorem 4.3, to give an important result that is used in Chapter 10.

Product of Perfect Squares

Here is our first example:

THEOREM 4.4. If a and b are perfect squares, then ab is a perfect square.

Proof: We assume that a and b are perfect squares, so there are integers r and s such that $r^2 = a$ and $s^2 = b$. We need to find an integer t such that $t^2 = ab$.

The integer rs has this property: $(rs)^2 = r^2s^2 = ab$. That completes the proof. ■

COMMENTS

1. The hypothesis of this theorem says that "something exists"—actually two "somethings"—since the definition of "perfect square" involves the existential quantifier. The conclusion requires us to *show* that "something exists". In each case the "something" satisfies an appropriate condition.

   The basic approach of this proof is to *name* those particular objects which we are told exist, and then *use* them to show that there is an element which fits the condition required by the conclusion. This two-step outline—name it, then use it—is typical of theorems in which the existential quantifier appears in the hypothesis and the conclusion.

2. The definition of "x is a perfect square" reads: "$(\exists\ w \in \mathbf{Z})(w^2 = x)$". Often, when we are using a definition involving "$\exists$", we simply use the letter in the definition as the name for the object that it says exists. But that letter is just a dummy variable, and the "object that exists" can be given any name. It's important here that the object which the hypothesis "a is a perfect square" says must exist is not necessarily the same integer as the object which the hypothesis "b is a perfect square" says must exist, so we can't name them both "w". Therefore, we avoid confusion by choosing to give new names, r and s, to these two objects.

3. In this case, finding the right t was easy. In other theorems, you may need to do some substantial exploration before you figure out how to use the given object(s) to create the required object(s). In this context, you may want to reread the discussion and comments about Theorem 4.1.

Sets Bounded Above

Our next example is more complicated. As with Theorem 4.2, you should create a mental picture of the situation in order to understand the proof intuitively as well as formally.

THEOREM 4.5. For $S \subseteq \mathbf{R}$:
If S is bounded above, and $T \subseteq S$, then T is bounded above.

Draw a picture.

Comment: Compare this to Theorem 4.3. There we "replaced" x by something bigger. Here we "replace" S by something smaller.

Proof: Assume that S is bounded above. We need to show that T is bounded above.

What does this hypothesis tell us? What does the conclusion require from us?

Since S is bounded above, there is some real number, call it c, which is an upper bound for S. We need to show that T also has an upper bound.

Comment: So we need to identify some real number which will work. What should we choose? That step is based on our intuitive understanding of the situation.

We will show that c itself is an upper bound for T, i.e., that $(\forall\, x \in T)(c \geq x)$. So suppose $x \in T$. We need to show that $c \geq x$.

How will we do that? What do we already know about c?

Since $x \in T$, and $T \subseteq S$, we have $x \in S$. But c is an upper bound for S, so we have "$(\forall\, y \in S)(c \geq y)$". Therefore, $c \geq x$, completing the proof.

Comment: The last step uses specialization twice. First, we showed that $x \in S$, by specializing the statement "$T \subseteq S$" (which says "if $w \in T$, then $w \in S$"). We then used the fact that $x \in S$ to specialize the statement "$(\forall\, y \in S)(c \geq y)$", and conclude that $c \geq x$. ■

COMMENTS

1. There are three key elements involved in proving this theorem.

The first is recognizing and making use of the fact that the hypothesis "S is bounded above" is an existence statement. We use that hypothesis by *giving a name* to the object which it says exists.

The second key step is *identifying the object* which we have to show exists. We have to find some real number which is an upper bound for T. In order to do so, we use the object we already have. In this case, the object we want is the same as the object we have, but in other proofs, we might identify the object we want *in terms of* the object (or objects) we have, by an equation or formula. (The process of identifying the desired object is a step that requires insight. By exploring familiar examples, we get a sense of what object is likely to work. Much of this work is done "behind the scenes", and all that appears in the written proof is the result of the exploration.)

The third step is to justify the insights developed in the exploration. That is, what remains is to *prove* that the object we have identified really works, i.e., is an upper bound for T. In this theorem, we did this using specialization—that is, we used the property of the given object to get the appropriate property for the constructed object.

The last two steps are the steps described after Theorem 4.1, for "proving that something exists".

2. Once the object c was specified as the "desired object", we were proving the following:

   If c is an upper bound for S, and $T \subseteq S$, then c is an upper bound for T.

   In a sense, this assertion is stronger than Theorem 4.5. It not only says that T is bounded above, but gives us a specific upper bound. This assertion is in standard form for the use of specialization.

3. For clarity, we wrote the definition for "c is an upper bound for T" using a different letter for the quantified variable from that used in stating the assumed condition that "c is an upper bound for S".

4. As in Theorem 4.1, finding a successful candidate for the required object was easy, and was based on an intuitive understanding of the theorem. Figure 4.2 gives the general picture of what's happening in this theorem.

Get Your Hands Dirty 2

Prove:

If S and T are bounded above, then $S \cup T$ is bounded above. □

Composition of Surjective Functions

Our last illustration of "name it—then use it" involves the concept of a surjective (onto) function:

THEOREM 4.6. If $g: X \to Y$ and $h: Y \to Z$ are both surjective functions, then $h \circ g: X \to Z$ is surjective.

DISCUSSION Recall the definition of "surjective" in Section 3.2:

DEFINITION 3.17: For a function $f: A \to B$, "f is *surjective*" means: $(\forall\ v \in B)(\exists\ u \in A)(f(u) = v)$.

According to this definition, proving that $h \circ g$ is surjective will entail assuming that we have an element, say t, in its codomain Z, and showing that something exists, say r, in its domain X, such that $(h \circ g)(r) = t$. To find r, we will use the assumption that g and

Figure 4.2 Since x is an upper bound for S, it is "to the right of S". But T is "part of" S, so x is also "to the right of T"; i.e., x is an upper bound for T.

h are each surjective, which tells us that, for each element in *their* codomains, something exists in *their* domains, with the appropriate property.

Based on this introduction, try to write a complete proof of this theorem. Begin by exploring. Take a particular X, Y, Z, g, h, and t, and decide how you would find the appropriate r.

? ? ?

Proof: Assume that g and h are surjective, and suppose that $t \in Z$. We need to show that there is an element r of X such that $(h \circ g)(r) = t$.

How can you use the hypothesis that g and h are surjective?

Since t is in Z (the codomain of h), and h is surjective, there is an element, call it s, in Y (the domain of h) such that $h(s) = t$.

Now what? Remember that Y is not only the domain of h; it is also the codomain of g.

Since s is in the codomain of g, and g is surjective, there is an element, call it r, in X (the domain of g) such that $g(r) = s$.

Comment: It's not an accident that we labeled this element r. It's the object we're looking for.

Then $(h \circ g)(r) = h(g(r)) = h(s) = t$. This completes the proof. ■

COMMENT Proving that a function is surjective, or proving any statement of the form "$(\forall x)(\exists y)(p(x, y))$", can be thought of as a "challenge" process: "someone" gives you an x, and you have to find an appropriate y. There may be more than one possibility for a successful y, and you are allowed to express y in terms of x, but you then have to prove that your y works for that specific x.

Fortunately, the hypothesis of Theorem 4.6 gives us a tool for meeting this challenge, because it also uses this combination of quantifiers. In Theorem 4.6, we couldn't actually create a formula for the object r we were looking for, but we used the information in the hypothesis to "locate" such an element. You may find the following metaphor a useful way to think about such a hypothesis:

Metaphor: Universal–existential hypothesis as "element finder"

The "$(\forall x \in A)(\exists y \in B)(p(x, y))$" type of hypothesis doesn't give us a specific element. Instead, it guarantees the *existence* of an element, which we call y, for any given $x \in A$, such that the pair (x, y) satisfies a certain condition ("$p(x, y)$").

Just as a "$(\forall x)(q(x))$" statement in the hypothesis is a "guarantee producer", so a "$(\forall x)(\exists y)$" statement in the hypothesis can be thought of as an "element finder", which is triggered by being given an element x. Whether such a statement is in the hypothesis of a theorem, or is a theorem already proved, we can use it to help us locate or identify an element with certain properties, as long as we give it an appropriate x. The statement also provides us with a guarantee that $p(x, y)$ is true.

Theorem 4.6 tells us that the composition of surjective functions is surjective. Earlier, Theorem 4.3 stated that the composition of injective functions is injective. Together, these theorems combine to give the following result, which we will be using in Section 10.1:

COROLLARY 4.7. If $g:X \to Y$ and $h:Y \to Z$ are both bijective functions, then $h \circ g:X \to Z$ is bijective.

EXERCISES

1. Prove that the sets **R** and 2**Z** are not bounded above. (GYHD 1 of Section 3.2 asked you to explain this intuitively.)
2. Prove that a subset of **R** cannot have more than one least upper bound. (*Hint:* Assume that there are two, and prove that they are equal.)
3. Consider the following definition:

DEFINITION: For $S \subseteq \mathbf{R}$ and $y \in \mathbf{R}$, "y is the *maximum element* of S" means:

$$y \in S \text{ and } y \text{ is an upper bound for } S.$$

Prove:

- **a)** If $S \subseteq \mathbf{R}$ and y is the maximum element of S, then y is the least upper bound for S.
- **b)** A subset of **R** cannot have more than one maximum element.

Also:

- **c)** Give an example of a subset of **R** which has a least upper bound but not a maximum element.

4. (GYHD 1)
   - **a)** Suppose S and T are subsets of **R**, and that x is a real number which is an upper bound for both S and T. Prove that x is an upper bound for $S \cup T$.
   - **b)** Make up a "stronger" theorem for $S \cap T$, and prove it.
5. (For this exercise, you may assume that every nonempty subset of **R** which is bounded above has a least upper bound.)
   - **a)** Prove that every nonempty subset of the interval $[0, 1]$ has a least upper bound in $[0, 1]$.
   - **b)** Prove that every nonempty subset of **R** which is bounded below has a greatest lower bound in **R**. (*Hint:* Start with some set S. To find a glb for S, form a related set T, and use the lub of T to produce a glb for S.)
6. [For this exercise, you may assume that every nonempty subset of **R** has a least upper bound, and you may use the result of Exercise 5b).] We introduced the following terminology in Exercise 15 of Section 3.2:

DEFINITION: For $S \subseteq \mathbf{R}$, "S is an *interval*" means:

$$(\forall\, u, v \in S)(\forall\, x \in \mathbf{R})(u < x < v \Rightarrow x \in S).$$

Suppose that T is an interval with at least two elements which is bounded above and below. Prove that T is a "standard" interval, i.e., for some $a, b \in \mathbf{R}$, with $a < b$, T is equal to one of the sets (a, b), $[a, b)$, $(a, b]$, $[a, b]$.

7. Prove the following:

For $f: A \to B$ and $g: B \to C$, if $g \circ f$ is surjective and g is injective, then f is surjective.

8. Prove the following:

For $f: \mathbf{R} \to \mathbf{R}$, if f is strictly increasing, then f is injective.

(Use Definition 3.6 in Section 3.1.)

9. Consider the following definition:

DEFINITION: For sets A and B, "$\overline{\overline{A}} \leq \overline{\overline{B}}$" means:
there is an injective function $f: A \to B$.

a) Give an informal argument to show that the condition "$\overline{\overline{A}} \leq \overline{\overline{B}}$" expresses the intuitive idea that the cardinality of A is at most that of B.

b) Which of the following statements are true? Give a proof or counterexample for each.

i) If $\overline{\overline{A}} \leq \overline{\overline{B}}$ and $\overline{\overline{B}} \leq \overline{\overline{C}}$, then $\overline{\overline{A}} \leq \overline{\overline{C}}$.

ii) For any set A, $\overline{\overline{A}} \leq \overline{\overline{A}}$.

iii) If $\overline{\overline{A}} \leq \overline{\overline{B}}$, and C is any set, then $\overline{\overline{A \cup C}} \leq \overline{\overline{B \cup C}}$.

iv) If $\overline{\overline{A}} \leq \overline{\overline{B}}$, and C is any set, then $\overline{\overline{A \cap C}} \leq \overline{\overline{B \cap C}}$.

v) If $A \subseteq B$, and $\overline{\overline{B}} \leq \overline{\overline{A}}$, then $A = B$.

10. Consider the following definitions:

DEFINITION: For $f: \mathbf{R} \to \mathbf{R}$, "$f$ is *increasing*" means:

$$(\forall\, x, y \in \mathbf{R})\,(x \leq y \Rightarrow f(x) \leq f(y)).$$

DEFINITION: For $f: \mathbf{R} \to \mathbf{R}$ and $g: \mathbf{R} \to \mathbf{R}$:

$f + g$ is the function from $\mathbf{R}$ to $\mathbf{R}$ whose value at a real number u is $f(u) + g(u)$.
$f \cdot g$ is the function from $\mathbf{R}$ to $\mathbf{R}$ whose value at a real number u is $f(u) \cdot g(u)$.
$f - g$ is the function from $\mathbf{R}$ to $\mathbf{R}$ whose value at a real number u is $f(u) - g(u)$.

Which of the following statements are true? Give a proof or counterexample for each.

a) If f and g are both increasing functions from $\mathbf{R}$ to $\mathbf{R}$, then $g \circ f$ is increasing.

b) If f and g are both increasing functions from $\mathbf{R}$ to $\mathbf{R}$, then $f + g$ is increasing.

c) If f and g are both increasing functions from $\mathbf{R}$ to $\mathbf{R}$, then $f \cdot g$ is increasing.

d) If f and g are both increasing functions from $\mathbf{R}$ to $\mathbf{R}$, then $f - g$ is increasing.

11. We saw at the beginning of this section that the statement "if a divides c and b divides c, then ab divides c" is false. We found a counterexample in which a, b, and c were all equal.

a) Does every substitution from $\mathbf{N}$ in which a, b, and c are equal make the implication "$[a \mid c$ and $b \mid c] \Rightarrow (ab) \mid c$" false? Explain your answer with a proof or counterexample.

b) Are there counterexamples to "$[a \mid c$ and $b \mid c] \Rightarrow (ab) \mid c$)" in which a, b, and c are not equal? Explain your answer.

c) Can you find some additional hypothesis on a and b which will give us a true statement? In other words, is there some reasonably simple open sentence $p(a, b)$ for which

"$(\forall\, a, b, c \in \mathbf{N})\,([a \mid c$ and $b \mid c$ and $p(a, b)] \Rightarrow (ab) \mid c)$"

is true? (You do not need to prove your answer.)

12. Give three examples of predicates $p(x, y)$ in two variables for which the statement "$(\forall\, x)\,(\exists\, y)\,(p(x, y))$" is true, but the statement "$(\exists\, y)\,(\forall\, x)\,(p(x, y))$" is false. Make your predicates mathematical in nature, and do not use those discussed thus far in the text.

13. PROOF EVALUATION: (see instructions for Exercise 6 in Section 2.2)

a) **"THEOREM":** For sets $S, T \subseteq \mathbf{R}$, if $S \cap T$ is bounded above, then either S or T is bounded above.

"Proof": Assume that $S \cap T$ is bounded above, and so there is an element u which is an upper bound for $S \cap T$.

By way of contradiction, assume that neither S nor T is bounded above. Expressed in terms of quantifiers, we have

$$(\forall\, x \in \mathbf{R})(\exists\, y \in S)(y > x)$$

and

$$(\forall\, x \in \mathbf{R})(\exists\, y \in T)(y > x).$$

Specializing these statements, by letting $x = u$, we get

$$(\exists\, y \in S)(y > u) \quad \text{and} \quad (\exists\, y \in T)(y > u).$$

Thus, we have an element y in both S and T, with $y > u$. But u was an upper bound for $S \cap T$, so this is a contradiction. Therefore, our assumption that neither S nor T is bounded above must have been false; i.e., one of these sets must be bounded above.

b) **"THEOREM":** For $x, y \in \mathbf{N}$: if $x \mid y$, then $x \leq y$.

"Proof": Assume that $x, y \in \mathbf{N}$, with $x \mid y$, so there is an element $w \in \mathbf{Z}$ such that $x \cdot w = y$.

We first show that w must be positive. By basic assumption d) v), w must be positive, negative, or 0. If $w = 0$, then $x \cdot w = 0$, by basic assumption b) ii). Thus $y = 0$, contrary to the fact that y is positive. [By basic assumption d) v), y can't be both positive and 0.] If $w < 0$, then $x \cdot w < 0$, by basic assumption d) iv), using $b = 0$. Again, this is contrary to the fact that y is positive.

Since $w > 0$, it follows from basic assumption f) i) that $w \geq 1$, and so $x \cdot w \geq x \cdot 1$ [using basic assumption d) iii)]. Also $x \cdot 1 = x$ [by basic assumption b) iii)]. Thus $x \cdot w \geq x$, i.e., $y \geq x$, as desired.

c) **"THEOREM":** For $f : A \to B$ and $g : B \to C$, if $g \circ f$ is surjective, then f is surjective.

"Proof": Assume that $g \circ f$ is surjective. To show that f is surjective, let t be an element of B, so we need to find an element r in A such that $f(r) = t$.

Since $t \in B$, we have $g(t) \in C$. Since $g \circ f$ is surjective, there is an element r in A such that $(g \circ f)(r) = g(t)$. But if $g(f(r)) = g(t)$, then $f(r) = t$. Therefore, we have found the desired element r.

4.2 "What to Prove Instead", and Other Equivalences

Up until now, with the exception of Theorem 2.12 and Theorem 2.16, our proofs have begun by assuming that the hypothesis is true, and then using that assumption to show that the conclusion must also be true. This is known as a *direct proof*.

In proving Theorem 2.12 and Theorem 2.16, we used a method called *proof by contradiction,* sometimes called *indirect proof*. In this section, we will examine proof by contradiction further, and look at some other methods of proof, including the method of contrapositive. Our purpose is to show that often one can prove a particular theorem by proving something *logically equivalent* — that is, something that can be shown *by virtue of its form* to have the same truth value as the original. Often this alternative may be easier to prove, or may suggest a method of proof that the original does not. (We formally define "logically equivalent" in Appendix A.)

Next, we will look at ways in which conditional sentences are phrased, besides the usual "if . . . , then . . ." form. We will also examine the concept of the converse of a conditional sentence, which is often confused with the sentence itself. We conclude the section with some other general principles of proof.

Proof by Contradiction

The basic idea of the method of *proof by contradiction* is to prove a statement by showing that, if it were false, we could derive some consequence which we know is impossible. In the case of a conditional sentence, we show that, if there were a counterexample, then something would happen that we know cannot be so.

Thus, in Theorem 2.16, we showed that, if an odd integer x were also even, then the integer 1 would be even, contrary to a previous theorem (Theorem 2.15). In Theorem 2.12, the hypothesis says that a and b are even, and that c is not even, and the conclusion says that the equation "$ax + by = c$" has no integer solutions for x and y. We saw that, if there were a counterexample, then c *would be* even, contradicting part of the hypothesis.

The general form of a proof by contradiction is to begin with the assumption that the statement is false. In the case of a conditional sentence, this is done in terms of the variables that appear in the hypothesis and conclusion; that is, we assume that we have objects that provide a counterexample, i.e., objects that can be substituted for the variables to make the hypothesis true and the conclusion false. We then show that these assumptions lead to a contradiction.

A contradiction of what?

? ? ?

That's often the hard part of developing the proof. The contradiction can be any statement that can't be true or any combination of statements that cannot all be true. For example, the statement "$2 = 0$" is, by itself, a contradiction; the pair of statements "x is even" and "x is odd" together are a contradiction; a statement that some previously proved theorem has a counterexample is also a contradiction.

To further illustrate this type of proof, we will re-prove an earlier theorem using the method of proof by contradiction:

THEOREM 2.6. For sets A, B, C, if $A \times B \subseteq A \times C$ and $A \neq \varnothing$, then $B \subseteq C$.

Proof: Suppose there is a counterexample; i.e., assume that there are sets A, B, and C, with $A \neq \varnothing$, and $A \times B \subseteq A \times C$, such that $B \not\subseteq C$.

Since $B \not\subseteq C$, there is an element y with $y \in B$ and $y \notin C$. Also, since $A \not\subseteq \varnothing$, there is an element $x \in A$.

Then $(x, y) \in A \times B$, but $(x, y) \notin A \times C$. This element (x, y) is a contradiction to the assumption that $A \times B \subseteq A \times C$.
Therefore the theorem is true. ■

What is the contradiction here? We found—or more accurately, we showed that there must exist—an object which belongs to $A \times B$ but does not belong to $A \times C$. That contradicted the assumption, from the hypothesis of the theorem, that $A \times B \subseteq A \times C$.

It is a common approach in a proof by contradiction to show that some part of the hypothesis is false. Since our basic assumption is that the hypothesis is true, that's a contradiction.

It may be helpful to give a schematic comparison of direct proof with proof by contradiction. Let's say our original theorem says "if $p(x)$, then $q(x)$". In a direct proof, we assume that $p(x)$ is true, and set out to prove $q(x)$. In a proof by contradiction, we assume that $p(x)$ is true and that $q(x)$ is false; that is, we assume that the conditional statement has a counterexample. Our goal is to derive a contradiction—that is, to derive from our assumptions some impossible conclusion, or two contradictory conclusions. We often represent a contradiction generically as "$r(x)$ and $\sim r(x)$".

Thus, we have:

	Assume	*Show*
Direct proof:	$p(x)$	$q(x)$
Contradiction:	$p(x)$ and $\sim q(x)$	$r(x)$ and $\sim r(x)$ (a contradiction)

Proof by Contrapositive

We look next at a method of proof which can be thought of as a special case of proof by contradiction. We begin with the following definition:

> **DEFINITION 4.1:** For open sentences $p(x)$ and $q(x)$:
> the *contrapositive* of the conditional sentence "if $p(x)$, then $q(x)$" is the conditional sentence "if $\sim q(x)$, then $\sim p(x)$".
> the *contrapositive* of the implication "$p(x) \Rightarrow q(x)$" is the implication "$\sim q(x) \Rightarrow \sim p(x)$".

For example, the contrapositive of the statement "if $A \subseteq B$, then $\mathcal{P}(A) \subseteq \mathcal{P}(B)$" is the statement "if $\mathcal{P}(A) \not\subseteq \mathcal{P}(B)$, then $A \not\subseteq B$".

Notice that "the contrapositive of the contrapositive" is the original conditional sentence (since "the negation of the negation" of any open sentence is that open sentence itself).

Get Your Hands Dirty 1

What would you need in order to have a counterexample to the statement "if $A \subseteq B$, then $\mathcal{P}(A) \subseteq \mathcal{P}(B)$"? What would you need in order to have a counterexample to its contrapositive, "if $\mathcal{P}(A) \not\subseteq \mathcal{P}(B)$, then $A \not\subseteq B$"? Compare the two. □

The importance of the concept of contrapositive is that the contrapositive of a conditional sentence always has the same truth value as the conditional sentence itself. Or, put another way, if we can prove either the original conditional sentence or the contrapositive, then we have really proved both. We give two different explanations for this assertion:

Method 1. Using proof by contradiction. We can reason as follows: Suppose we've proved the contrapositive of some statement—i.e., we know that, if $q(x)$ is false, then $p(x)$ is false. We could use that result to prove the original conditional sentence, by contradiction, as follows:

Assume that $p(x)$ is true, and suppose, by way of contradiction, that $q(x)$ is false, i.e., that "$\sim q(x)$" is true. Since we know that the contrapositive is true, we are guaranteed that "$\sim p(x)$" is true, i.e., $p(x)$ is false. But that's a contradiction, since $p(x)$ was assumed true. Therefore, $q(x)$ can't be false, so it must be true.

Similarly, if we've proved the original conditional sentence, we could use that to prove the contrapositive.

Method 2. Comparing counterexamples. (Re-examine GYHD 1 before reading this.) What would a counterexample be for the conditional statement "if $p(x)$, then $q(x)$" or its contrapositive "if $\sim q(x)$, then $\sim p(x)$"? A counterexample for any conditional sentence is an object that makes its hypothesis true and its conclusion false. For "if $p(x)$, then $q(x)$", a counterexample is an object that makes $p(x)$ true and $q(x)$ false. For the contrapositive, "if $\sim q(x)$, then $\sim p(x)$", a counterexample is an object that makes $\sim q(x)$ true and $\sim p(x)$ false. But an object that makes $\sim q(x)$ true and $\sim p(x)$ false is exactly the same thing as an object that makes $p(x)$ true and $q(x)$ false. And so, if we can show that either one of the two conditional sentences is true, then it has no counterexamples, and so neither does the other, and it must also be true.

The methods described above both work without any regard to what the particular open sentences $p(x)$ and $q(x)$ are. Therefore, we have the following general principle:

THEOREM 4.8. For any open sentences $p(x)$ and $q(x)$, if either "if $p(x)$, then $q(x)$" or its contrapositive "if $\sim q(x)$, then $\sim p(x)$" is true, then so is the other.

We can express this more formally as follows:

THEOREM 4.8′. For any open sentences $p(x)$ and $q(x)$, the statement "if $p(x)$, then $q(x)$" and its contrapositive "if $\sim q(x)$, then $\sim p(x)$" are logically equivalent.

COMMENT This theorem uses the technical phrase "are logically equivalent". Intuitively, this means that the two statements are expressing the same assertion *by virtue of their form*. That is, we can tell that the statements are either both true or both false without any consideration of what the *content* of the open sentences might be. A proof of Theorem 4.8′ requires a formal development of logic which is not really in the intuitive spirit of this book.

For those who want a taste of this material, we have included some of this formalism in Appendix A. That appendix discusses the "algebra" of combining propositions and open sentences to make more complex ones, the notion of a *truth table*, and the use of the algebra of propositions and open sentences to define and prove theorems about logical equivalence.

In the body of this book, we will work on building an intuitive skill for recognizing logically equivalent statements—that is, statements that are essentially "saying the same thing" by virtue of their form—and using these logical equivalences to prove theorems with interesting mathematical content. There will be other theorems like Theorem 4.8′ in this section. The basis for their proofs is discussed in Appendix A.

Theorem 4.8 has tremendous practical significance. It says that, if you are trying to prove an "if . . . , then . . ." statement, you can instead prove its contrapositive, and be guaranteed that the original "if . . . , then . . ." statement will also be true. Mathematicians use this idea all the time, often without explicit mention of the process. They will simply start a proof with a phrase like "assume that the conclusion is false", then show, using that assumption, that the hypothesis is false, and consider the proof complete. We will call this method of proving a theorem *proof by contrapositive*. The outline of the method is simple: to prove "if $p(x)$, then $q(x)$" by contrapositive, we assume that $q(x)$ is false, and show that $p(x)$ is false.

Example 1. **Statements and Their Contrapositives** Here are some other examples of conditional statements and their contrapositives:

Statement	*Contrapositive*
If $S = \emptyset$, then $S \times S = \emptyset$.	If $S \times S \neq \emptyset$, then $S \neq \emptyset$.
If $x > 0$, then $x^3 > 0$.	If $x^3 \leq 0$, then $x \leq 0$.
Assume $A \neq \emptyset$. If $A \times B \subseteq A \times C$, then $B \subseteq C$.	Assume $A \neq \emptyset$. If $B \not\subseteq C$, then $A \times B \not\subseteq A \times C$.

□

Why Bother with Contrapositives?

You may still be skeptical of all the fuss, and ask why we should care about the contrapositive of a statement. When would we want to use "proof by contrapositive"? What would be the advantage?

The answer is that, depending on the form of the hypothesis and conclusion of the conditional statement, the contrapositive may be easier to work with.

For example, if the hypothesis and conclusion involve universal quantifiers, then their negations will involve existential quantifiers. We saw in Section 4.1 that "existential information" is often easy to work with: we can use the "name it—then use it" approach. On the other hand, "universal information" can be difficult to use: it often requires specialization of some kind.

The use of the contrapositive of a theorem as an equivalent way of saying the same thing often allows us to choose the type of information we get to use. The following proof, of the first statement in Example 1, illustrates the usefulness of the contrapositive:

EXAMPLE 2. **Proof by Contrapositive** We will prove the following by means of contrapositive:

THEOREM 4.9. If $S = \varnothing$, then $S \times S = \varnothing$.

Proof: Assume that $S \times S \neq \varnothing$. Then there is an element $t \in S \times S$. Such an element t must be of the form (x, y), where $x \in S$ and $y \in S$. Since $x \in S$, we have $S \neq \varnothing$.

This completes the proof. ■

COMMENTS

1. We didn't explicitly say in the proof that we were using "proof by contrapositive". Mathematicians often don't.
2. It is very difficult to use the hypothesis "$S = \varnothing$" directly. It contains "negative information" about "all elements"—it says "$(\forall\, x)(x \notin S)$". Working with the contrapositive gave us "positive information" about one of the elements of $S \times S$, which was easy to use.
3. We should point out that proof by contrapositive and proof by contradiction are not always easier than direct proof. With experience, you will learn to recognize when it is advantageous to assume the negation of the conclusion as part of your proof process.
4. Don't let the fact that $t \in S \times S$ mislead you into thinking that t has the form (x, x). The two components of t need not be the same, even if they both must belong to the same set S.

Contrapositive and Contradiction

Proof by contradiction and proof by contrapositive are very similar. In both cases, we assume that $q(x)$ is false. In proof by contradiction, we also assume that $p(x)$ is true, and derive a contradiction. In proof by contrapositive, we do not make any additional assumption, and we show that $p(x)$ is false.

We can compare the two methods symbolically as follows:

	Assume	*Show*
Contradiction:	$p(x)$ and $\sim q(x)$	contradiction
Contrapositive:	$\sim q(x)$	$\sim p(x)$

Any proof by contrapositive can be reformulated as a proof by contradiction as follows: instead of assuming only that $q(x)$ is false, we also assume that $p(x)$ is true. We then proceed as in the proof by contrapositive, using the assumption that $q(x)$ is false to show that $p(x)$ is false. We thus have a contradiction, because we have $p(x)$ both true and false. We can represent this as follows:

	Assume	*Show*
Modified contrapositive:	$p(x)$ and $\sim q(x)$	$\sim p(x)$ [which contradicts $p(x)$]

This is just a special case of proof by contradiction.

Each method—contradiction and contrapositive—has its advantages. In a proof by contradiction, you get to assume more—you assume both that the hypothesis is true and that the conclusion is false. You then can prove the theorem by producing *any* contradiction. In a proof by contrapositive, you start with only the assumption that the conclusion is false, and you must contradict the hypothesis.

The advantage of the proof by contrapositive is that its logic is less complicated. After some experience using Theorem 4.8, you will find yourself taking this equivalence for granted.

Alternative Language for Conditional Sentences

We have been discussing alternative ways to *prove* a conditional sentence of the form "if $p(x)$, then $q(x)$". We switch gears for a moment to talk about alternative ways to *state* that conditional sentence. Mathematicians have developed a wide variety of ways to express the basic "if . . . , then . . ." relationship between two open sentences. This is the focus of the next example.

Example 3. **Ways to Say "If . . . , Then . . ."** All of the following phrases are sometimes used to mean "if $p(x)$, then $q(x)$":

i) $p(x)$ implies $q(x)$.
ii) $q(x)$ is implied by $p(x)$.
iii) $p(x)$ is a sufficient condition for $q(x)$.
iv) $q(x)$ is a necessary condition for $p(x)$.
v) $p(x)$ is a stronger condition than $q(x)$.
vi) $q(x)$ is a weaker condition than $p(x)$.
vii) $p(x)$ only if $q(x)$.
viii) $q(x)$ if $p(x)$.
ix) Whenever $p(x)$, $q(x)$.
x) $q(x)$ whenever $p(x)$.
xi) $\sim p(x)$ unless $q(x)$.
xii) $q(x)$ unless $\sim p(x)$. □

DISCUSSION

i) was discussed following the definition of implication and Theorem 3.1. We just remind you here that this is used colloquially as if it had "for all x" in front of it.

ii) is just a variation of i).

iii) means, intuitively, that it is enough to know $p(x)$ in order to be certain of $q(x)$.

iv) means, intuitively, that $q(x)$ is a requirement in order for $p(x)$ to be true. Another way to say this would be that you can't have $p(x)$ without $q(x)$; i.e., if you don't have $q(x)$, then you don't have $p(x)$. The latter statement is essentially the contrapositive of "if $p(x)$, then $q(x)$".

v) means, intuitively, that knowing $p(x)$ is better than knowing $q(x)$—it gives you more information. For the "if . . . , then . . ." statement to be true means that knowing $p(x)$ guarantees the truth of $q(x)$ as well.

vi) is another way of stating v).

vii) is like iv): it says you can't have $p(x)$ without $q(x)$.

viii) is just a syntactical variation of the original "if . . . , then . . .".

ix) says: any time $p(x)$ is true, $q(x)$ will be also. This uses a time metaphor to express the idea of substituting different values for the variable.

x) is a variation of ix).

xi) and xii) are tricky. "Unless" is essentially synonymous with "if not". So "$\sim p(x)$ unless $q(x)$" means "$p(x)$ will be false if $q(x)$ is not true". Formally this becomes "$\sim p(x)$ if $\sim q(x)$", which is essentially the contrapositive of the original "if . . . , then . . ." statement. Similarly, "$q(x)$ unless $\sim p(x)$" means "$q(x)$ will be true if $\sim p(x)$ is not true", i.e., "$q(x)$ if $\sim(\sim p(x))$". The latter statement is essentially the original "if . . . , then . . .".

However, "unless" also conveys an impression of *equivalence*. Consider the following real-life example: "I won't mow the lawn unless you pay me". This says unequivocally that if payment is not given, then the lawn won't get mowed. But it also broadly suggests that if payment is given, the lawn will get mowed. Because of this ambiguity in colloquial English, the "unless" phraseology is not a very good one to use in a mathematical context.

Converse

After this discussion about different statements that essentially mean the same thing as a given conditional sentence, it is important to clarify a common misconception. We begin with a definition, similar to that of "contrapositive" (Definition 4.1):

> **DEFINITION 4.2:** For open sentences $p(x)$ and $q(x)$,
> the *converse* of "if $p(x)$, then $q(x)$" is "if $q(x)$, then $p(x)$",
> the *converse* of "$p(x) \Rightarrow q(x)$" is "$q(x) \Rightarrow p(x)$".

For example (with $\mathbf{Z}$ as the universe), the converse of the statement "if x is composite, then $2x$ is composite" is the statement "if $2x$ is composite, then x is composite."

Get Your Hands Dirty 2

What would you need in order to have a counterexample to the statement "if x is composite, then $2x$ is composite"? What would you need in order to have a counterexample to its converse, "if $2x$ is composite, then x is composite"? Compare the two. □

Perhaps the most important point to make about the converse of a statement is that, unlike the contrapositive, *the converse of a statement does not mean the same thing as the statement itself.* It is possible for the converse to have a counterexample, even if the original statement is true. In our example, the original statement, "if x is composite, then $2x$ is composite", is true, but its converse, "if $2x$ is composite, then x is composite", is false (e.g., $x = 5$ is a counterexample).

"If . . . , then . . ." statements are often confused with their converses. Or, more precisely, it is often thought that, when a conditional statement is true, its converse must also be true. This confusion arises in part from the fact that, in everyday speech, "if . . . , then . . ." is often used differently from the way mathematicians use it. For example, the statement "if I finish doing my work, then I will go to the movies" is commonly interpreted to also mean "if I don't finish my work, then I won't go to the movies". In other words, we interpret the statement to mean that "finishing my work" and "going to the movies" will either both happen or neither will happen. "If . . . , then . . ." is often used to mean that the hypothesis and the conclusion are either both true or both false.

This is not how "if . . . , then . . ." is used in mathematics (and also not always the way it is used in everyday speech). In mathematics, the "if . . . , then . . ." language allows for the possibility of a false hypothesis and a true conclusion. If we only want to allow "both true" or "both false", we use the phrase "if and only if" (more on that later).

There are many theorems—true statements which can be proved—whose converses are not true statements. Example 4 gives some instances of this.

Example 4. **Theorems Whose Converses Are False** In each of the following pairs of statements, the first statement is a theorem, and the second statement, which is the converse of the first, is false.

i) (x is an integer.)
  a) If $x > 0$, then $x^2 > 0$.
  b) If $x^2 > 0$, then $x > 0$.
ii) (ABC and DEF are triangles.)
  a) If ABC and DEF are congruent, then they are similar.
  b) If ABC and DEF are similar, then they are congruent.
iii) (A, B, and C are sets.)
  a) If $A \subseteq B$, then $A \cup C \subseteq B \cup C$.
  b) If $A \cup C \subseteq B \cup C$, then $A \subseteq B$. □

The relationship between a conditional statement and its converse may be clarified by looking at the truth sets involved. Suppose we are considering open sentences $p(x)$ and $q(x)$, with their respective truth sets P and Q.

The statement "if $p(x)$, then $q(x)$" means that $P \subseteq Q$. The converse "if $q(x)$, then $p(x)$" means that $Q \subseteq P$. If P is a subset of Q, that certainly doesn't mean that Q is a subset of P.

The converse of a statement is not the same as the negation of the original statement either. If "$P \subseteq Q$" is false, that does not necessarily guarantee that "$Q \subseteq P$" is true. The two sets may be unrelated to each other, and neither a subset of the other.

The following example shows some true conditional statements whose converses are also true:

Example 5. **Theorems Whose Converses Are True** The following are pairs of (true) theorems in which the second is the converse of the first.

i) (x is an integer.)
  a) If x is odd, then x^2 is odd.
  b) If x^2 is odd, then x is odd.
ii) (ABC is a triangle.)
  a) If ABC has two equal sides, then it has two equal angles.
  b) If ABC has two equal angles, then it has two equal sides.
iii) (A and B are sets.)
  a) If $A \subseteq B$, then $\mathcal{P}(A) \subseteq \mathcal{P}(B)$.
  b) If $\mathcal{P}(A) \subseteq \mathcal{P}(B)$, then $A \subseteq B$. □

"If and Only If"

As we have just seen, sometimes a conditional statement "if $p(x)$, then $q(x)$" and its converse are both true. When this happens, we can express that fact by the shorthand phrase "$p(x)$ *if and only if* $q(x)$". The phrase "if and only if" is sometimes abbreviated "*iff*".

The statement "$p(x)$ if and only if $q(x)$" is also expressed by the phrase "$p(x)$ and $q(x)$ are equivalent" (see comment below, following Definition 4.3). Our discussion of the relationship between "if . . . , then . . ." statements and the truth sets of their hypothesis and conclusion gives us the following result:

THEOREM 4.10. For open sentences $p(x)$ and $q(x)$, $p(x)$ and $q(x)$ are equivalent if and only if they have the same truth set.

It is perhaps worth noting that, in the phrase "$p(x)$ if and only if $q(x)$", the "$p(x)$ if $q(x)$" part says "if $q(x)$, then $p(x)$" and the "$p(x)$ only if $q(x)$" part says "if $p(x)$, then $q(x)$".

"If and Only If" vs. "Logically Equivalent"

We stated in Theorem 4.8′ that a conditional statement and its contrapositive are "logically equivalent"—that is, they essentially say the same thing. The reason for this logical equivalence was based on the form of the statements: it had nothing to do with the specific content of $p(x)$ and $q(x)$.

The word "equivalent" as used in Theorem 4.10 is different. For example, we can say that "$A \subseteq B$" and "$\mathcal{P}(A) \subseteq \mathcal{P}(B)$" are equivalent (as open sentences with variables A

and B). These two open sentences *mean* different things, but they have the same truth set; that is, they are true for the same combinations of sets A and B. We can prove, *based on the meaning of* "$\subseteq$", that each of these conditions implies the other. (We will prove this shortly as an example.)

Just as the implication symbol is used to capture the idea of "if . . . , then . . .", so we have a symbol for open sentences that embodies the idea of "if and only if".

> **DEFINITION 4.3:** For open sentences $p(x)$ and $q(x)$, "$p(x) \Leftrightarrow q(x)$" means:
>
> "$p(x) \Rightarrow q(x)$ and $q(x) \Rightarrow p(x)$".

"$\Leftrightarrow$" is called *double implication.*

COMMENT The quantified statement "$(\forall\, x)\,(p(x) \Leftrightarrow q(x))$" means the same thing as "$p(x)$ if and only if $q(x)$". We have noted that the distinction between "if . . . , then . . ." and "$\Rightarrow$" is often blurred. Similarly, the distinction between "if and only if" and "$\Leftrightarrow$" is often ignored. The notation "$p(x) \Leftrightarrow q(x)$" is often read "$p(x)$ is equivalent to $q(x)$".

Just as we can say "if $p(x)$, then $q(x)$" in a variety of ways, so there are also several ways to express the statement "$p(x)$ if and only if $q(x)$".

Example 6. **Ways to Say "If and Only If"** The following are alternative ways to say "$p(x)$ if and only if $q(x)$".

i) $p(x)$ iff $q(x)$.
ii) $p(x) \Leftrightarrow q(x)$.
iii) $p(x)$ is equivalent to $q(x)$.
iv) $p(x)$ is necessary and sufficient for $q(x)$.
v) $p(x)$ precisely when $q(x)$. □

As we noted, i) is just an abbreviation, and ii) is like the blurring of the distinction between "if . . . , then . . ." and implication.

We have also discussed "equivalent". The intuitive feeling of iii) is that the two conditions "are equally strong" [compare v) and vi) of the alternative statements for "if $p(x)$, then $q(x)$"].

Statement iv) is a combination of two statements. As discussed earlier, "$p(x)$ is sufficient for $q(x)$" is the same as saying "if $p(x)$, then $q(x)$"; "$p(x)$ is necessary for $q(x)$" is the same as saying "if $q(x)$, then $p(x)$". Together they make up an "if and only if" statement.

In statement v), the phrase "precisely when" can be thought of as short for "when, and only when": it uses a time metaphor to express the idea that $p(x)$ and $q(x)$ are true for the same set of values.

Using "If and Only If" in Definitions

The phrase "if and only if" is commonly used in definitions, since, as we pointed out in Section 1.3, definitions are always two-directional. For example, we might define the concept of set intersection by saying

$$x \in A \cap B \quad \text{if and only if} \quad x \in A \text{ and } x \in B .$$

This statement tells us two things: it says that, for x to belong to $A \cap B$, it must belong to both A and B; and it also says that, if x does belong to both A and B, then it is a member of $A \cap B$.

In other words (using language discussed in Example 6), the requirement "$x \in A$ and $x \in B$" is both a necessary and a sufficient condition for x to belong to $A \cap B$.

Proving "If and Only If"

The proof of an "if and only if" statement is generally written as a two-part proof, i.e., as separate proofs for each direction. Here is a simple illustration:

Example 7. Proving an "If and Only If" Statement We will prove the following:

THEOREM 4.11. For sets A and B,

$$A \subseteq B \quad \text{if and only if} \quad \mathcal{P}(A) \subseteq \mathcal{P}(B) .$$

Proof:

$\Rightarrow$: Suppose $A \subseteq B$ and $S \in \mathcal{P}(A)$. We will show $S \in \mathcal{P}(B)$. Since $S \in \mathcal{P}(A)$, we have $S \subseteq A$. But also $A \subseteq B$, so it follows (by Theorem 2.2) that $S \subseteq B$, and so $S \in \mathcal{P}(B)$, as desired.

$\Leftarrow$: Suppose $\mathcal{P}(A) \subseteq \mathcal{P}(B)$, and we will show that $A \subseteq B$. Since $A \in \mathcal{P}(A)$, and $\mathcal{P}(A) \subseteq \mathcal{P}(B)$, it follows that $A \in \mathcal{P}(B)$. Therefore, $A \subseteq B$, as desired. ∎

COMMENTS

1. Rather than write "First we will prove that, if $A \subseteq B$, then $\mathcal{P}(A) \subseteq \mathcal{P}(B)$", we simply labeled the first part of the proof as "$\Rightarrow$:". This means that we are showing that the condition on the left of the "if and only if" implies the condition on the right.

2. You might see a theorem like this stated simply:

$$A \subseteq B \quad \Leftrightarrow \quad \mathcal{P}(A) \subseteq \mathcal{P}(B)$$

The use of the arrow symbols as described here and in comment 1 is common, and reflects the blurring of the distinction between implication and "if . . . , then . . .".

Sometimes there is a collection of several conditions which we want to show are equivalent. The next example illustrates a common format for stating and proving a theorem of this type.

Example 8. Proving Several Statements Equivalent We will prove the following:

THEOREM 4.12. For sets A and B, the following are equivalent:

a) $A \subseteq B$
b) $\mathcal{P}(A) \subseteq \mathcal{P}(B)$
c) $A \cap B' = \varnothing$

Proof:

a) ⇒ b): Suppose $A \subseteq B$ and $S \in \mathcal{P}(A)$. We will show $S \in \mathcal{P}(B)$. Since $S \in \mathcal{P}(A)$, we have $S \subseteq A$. But also $A \subseteq B$, so it follows (by Theorem 2.2) that $S \subseteq B$, and so $S \in \mathcal{P}(B)$, as desired.

b) ⇒ c): Suppose $\mathcal{P}(A) \subseteq \mathcal{P}(B)$. We will prove that $A \cap B' = \varnothing$ by contradiction. To do so, we suppose that there is some element x in $A \cap B'$. Since $x \in A$, we have $\{x\} \in \mathcal{P}(A)$. But $\mathcal{P}(A) \subseteq \mathcal{P}(B)$, so we must also have $\{x\} \in \mathcal{P}(B)$. But this means $\{x\} \subseteq B$, so that $x \in B$. This contradicts our assumption that $x \in B'$.

c) ⇒ a): Suppose $A \cap B' = \varnothing$. To show $A \subseteq B$, we will suppose that some object x is in A, and show that $x \in B$. Since $x \in A$, and $A \cap B' = \varnothing$, we have $x \notin B'$, and so $x \in B$, as desired. ■

COMMENTS

1. The content of this proof is rather unexciting, but it nicely illustrates the "circular" method of proving more than two conditions equivalent. Here we have three conditions, and so there are six "if . . . , then . . ." statements which must hold if these conditions are equivalent. But we only showed three of them. The other three are immediate consequences of the ones we proved. For example, since a) implies b) and b) implies c), we are automatically guaranteed that a) implies c). (The formal basis of this idea is discussed in Appendix A.) As long as a complete chain of implications can be proved, which includes each of the conditions and gets back to where it started, all the conditions must be equivalent.

2. The phrase "the following are equivalent" occurs frequently in mathematical writing, and is sometimes abbreviated "TFAE".

Get Your Hands Dirty 3

Suppose p, q, r, and s are four propositions, and suppose we have proved each link in the "chain" of implications, $p \Rightarrow q$, $q \Rightarrow r$, $r \Rightarrow s$, and $s \Rightarrow p$.

Explain, in intuitive terms, why this guarantees that each of the four propositions implies the other three. □

"And"s, "Or"s, and "Not"s

One of the essential elements of working with proof by contradiction is being able to state the negation of an assertion clearly. Expressing negations clearly is also important in proving that some property does *not* hold in a given situation—e.g., showing that a specific function is not surjective, or that a particular subset of **R** does not have an upper bound.

One area where the process of negation can be difficult is with sentences involving the words "and" and "or". (In Exploration 1—"General Principles about Quantifiers and Logic"—at the end of Section 3.1, we asked you to look at how "and", "or", and "not" fit together in open sentences. You may wish to review your work on that exploration now.)

Let's look at a specific example:

Example 9. **"Not the GCD"** In Definition 3.5 (Section 3.1), we defined GCD(a, b) as follows (for $a, b, c \in \mathbf{N}$): GCD$(a, b) = c$ means

$$c \mid a \text{ and } c \mid b \text{ and } (\forall\, z \in \mathbf{N})([z \mid a \text{ and } z \mid b] \Rightarrow z \leq c)\,.$$

In other words, in order for c to be the GCD of a and b, c must satisfy three conditions:

i) c must divide a;
ii) c must divide b;
iii) every positive integer that divides both a and b must be less than or equal to c.

What does it mean for a positive integer *not* to be the GCD of a and b?

In other words, how can the statement "GCD$(a,b) \neq c$" be simplified or analyzed?

Consider the case $a = 8$, $b = 12$. Which of the conditions i), ii), and iii) are satisfied by each of the following four values of $c : c = 6$; $c = 8$; $c = 5$; $c = 2$?

Work it out.

By division, we see that the value $c = 6$ satisfies condition ii) and not condition i). Since the only positive integers that divide both 8 and 12 are 1, 2, and 4, $c = 6$ also satisfies condition iii).

In a similar way, we find that $c = 8$ satisfies i) and iii), but not ii); that $c = 5$ satisfied iii), but not i) or ii); and that $c = 2$ satisfies both i) and ii), but not iii).

Therefore, none of these four values is the GCD of 8 and 12.

Why not?

The answer may seem trivial, but it is an important point. In order to be the GCD, a positive integer has to satisfy all three criteria. If it fails to satisfy even one of the conditions (as did the three values tested), then it is not the GCD.

In other words, the statement "GCD$(a, b) \neq c$" means:

$$\text{"either } c \nmid a \text{ or } c \nmid b \text{ or } (\exists\, z \in \mathbf{N})(z \mid a \text{ and } z \mid b \text{ and } z > c)\text{"}.$$

[We are using the fact that the negation of condition iii)—"$(\forall\, z \in \mathbf{N})([z \mid a \text{ and } z \mid b] \Rightarrow z \leq c)$"—is the statement "$(\exists\, z \in \mathbf{N})(z \mid a \text{ and } z \mid b \text{ and } z > c)$"; i.e., the negation of an implication is a counterexample.] □

Important note: In mathematics, the word "or" is always used in the "inclusive" sense—that is, it allows the possibility that more than one of the alternatives under consideration are true. For example, suppose $p(x)$ is the statement "x is even or x is greater than 5". When $x = 10$, both components of $p(x)$ are true, and so $p(10)$ is considered a true statement. Of course, $p(x)$ is also considered true if only one of its components is true. For example, when $x = 4$, x is even, but is not greater than 5. When $x = 7$, x is not even, but is greater than 5. Both $p(4)$ and $p(7)$ are considered true, since, in each case, at least one of the components connected by "or" is true.

In ordinary English, the word "or" is often used in the "exclusive" sense, i.e., "one or the other but not both". For example: "you will study hard or you will fail the course". People hearing this are likely to take it to mean that, if they do study hard, they will not fail the course; that is, "studying hard" and "failing the course" cannot both happen. Writers sometimes use "and/or" if they want to clarify that they intend the "inclusive" meaning of "or"; e.g., "x is even and/or greater than 5". We will occasionally write "either... or... or both" to emphasize the inclusive nature of mathematical "or", but the phrase "or both" is technically unnecessary.

COMMENT This use of "or" is consistent with the way we define union of sets: "$x \in S \cup T$" means: "$x \in S$ or $x \in T$". Objects that belong to both sets are included in $S \cup T$.

In Example 9, there were three conditions that needed to be satisfied in order for c to be the GCD of a and b. We can symbolically represent the statement "GCD$(a, b) = c$" as "$p(x)$ and $q(x)$ and $r(x)$". The negation is then represented as "$\sim p(x)$ or $\sim q(x)$ or $\sim r(x)$"—a combination of three statements using "or". If one or more of these negations is true—i.e., if c fails one or more of the original conditions—then it is not the GCD. Of the cases examined in Example 9, $c = 6$, $c = 8$, and $c = 2$ each made one of the negations true, while $c = 5$ made two of them true. Exercise 9 asks you to examine making other combinations of these negations true.

Example 9 illustrates the following general principle:

THEOREM 4.13. For open sentences $p(x)$ and $q(x)$, "$\sim[p(x)$ and $q(x)]$" is logically equivalent to "$\sim p(x)$ or $\sim q(x)$".

Although Theorem 4.13 is stated in terms of combining *two* open sentences, this result can be generalized to apply to the negation of any number of open sentences that are combined by "and". For instance, in Example 9, we saw that negating a set of three conditions combined by "and" gave us the negations of the individual conditions combined by "or".

Theorem 4.13 is intimately connected with one of De Morgan's Laws about sets:

$$(P \cap Q)' = P' \cup Q' .$$

Like Theorem 4.13, De Morgan's Law can be generalized to give the complement of any intersection of sets as the union of their complements.

The formal connection between De Morgan's Law and the principle of logic embodied in Theorem 4.13 is based on working with truth sets of the open sentences. This is discussed further in Appendix A. Theorem 4.13 is sometimes called a "De Morgan's Law for Logic".

Example 10. **A Model for "De Morgan's Law for Logic"** It may be helpful to have a simple, nonmathematical example of the use of this logical version of De Morgan's Law. We offer the following model:

We can define the word "mother" to mean "female and parent". That is, in order for someone to fit the condition of being a mother, that person must satisfy two conditions: i) be female, and ii) be a parent.

What does it mean to *not be* a mother? How does one qualify for that designation?

? ? ?

There are two causes for being labeled a nonmother: all males are nonmothers, and anyone who is not a parent is a nonmother. In other words, the statement "x is not a mother" means the same thing as "either x is not a female or x is not a parent". (Or both—the set of males and the set of nonparents are certainly not disjoint, and male nonparents are not mothers.)

We can use this example to illustrate De Morgan's Law for sets as well: let M = the set of mothers, F = the set of females, and P = the set of parents. Our definition of "mother" says: $M = F \cap P$. What we have shown is that $M' = F' \cup P'$. □

There is a principle of logic that corresponds to the other De Morgan's Law as well. Recall this principle about sets:

$$(P \cup Q)' = P' \cap Q' .$$

How would you translate this rule about sets into a theorem analogous to Theorem 4.13?

We'll look at an example before stating the general principle:

Example 11. **"Not" with "Or"** Consider the following statement (for $x \in \mathbf{R}$):

$$\text{"if } x^2 - 3x + 2 > 0, \text{ then } x > 2 \text{ or } x < 1\text{"}. \qquad (*)$$

Suppose you wanted to prove this by using the contrapositive. How would you simplify the negation of its conclusion? What would you assume?

Simplify "$\sim(x > 2$ or $x < 1)$".

If the conclusion "$x > 2$ or $x < 1$" is false, then both parts must be false. In other words, "$\sim(x > 2$ or $x < 1)$" means the same thing as "$\sim(x > 2)$ and $\sim(x < 1)$". We can simplify this in turn to "$x \leq 2$ and $x \geq 1$", or simply "$1 \leq x \leq 2$".

Thus, the contrapositive of $(*)$ is

$$\text{"if } 1 \leq x \leq 2, \quad \text{then } x^2 - 3x + 2 \leq 0\text{"}. \qquad \square$$

The important point of this example is that the negation of an "or" statement is an "and" statement. More precisely, we have the following general principle:

THEOREM 4.14. For open sentences $p(x)$ and $q(x)$, "$\sim[p(x)$ or $q(x)]$" is logically equivalent to "$\sim p(x)$ and $\sim q(x)$".

Get Your Hands Dirty 4

State the negation of each of the following, simplifying your answer:

i) f is surjective and injective (here, f is a function).
ii) x is even or $x \geq 3$ (here, x is a positive integer). □

What to Assume? What to Prove?

A major goal in this section has been to point out that there are alternatives to the basic "assume $p(x)$, prove $q(x)$" approach to proving theorems. The following examples expand on that theme:

Example 12. **"Or" in the Conclusion** Consider the following conditional sentence, whose conclusion has the form "$p(x)$ or $q(x)$":

If p is a prime and $p \mid ab$, then either $p \mid a$ or $p \mid b$.

(This is a true statement, which we will prove in Section 5.3. The proof is not easy. Our interest here is in the general plan of the logic, not the details of the mathematical content.)

How do you prove an "either . . . or" statement? You can't necessarily show that $p \mid a$, nor can you prove that $p \mid b$. You don't know which one is true—it depends on exactly which p, a, and b you started with, and you don't know that.

So what do you do? What do we assume? What do we have to prove?

? ? ?

It may have occurred to you to use either proof by contradiction or "proof by cases". These methods both work, but there is another approach that is often used and that is generally considered simpler.

What we show, in practice, is that if p does not divide a, then it must divide b. We don't know if p divides a or not. If it does, then we are done. So we can act as if we need only worry about the case when p does not divide a. That becomes an additional assumption: $p \nmid a$. The goal is then to show: $p \mid b$.

Thus, what we actually prove is the following:

If p is a prime, and $p \mid ab$, and $p \nmid a$, then $p \mid b$. □

It is important to note that the discussion in Example 12 had absolutely nothing to do with primes or divisibility. The process by which we reformulated the theorem was generic—it could have been done with any "if . . . , then . . ." statement which had open sentences connected by "or" in the conclusion. We can therefore generalize this discussion to the following principle, which we state without proof:

THEOREM 4.15. For open sentences $p(x)$, $q(x)$, and $r(x)$, if either the statement "if $p(x)$, then $q(x)$ or $r(x)$" or the statement "if $p(x)$ and $\sim q(x)$, then $r(x)$" is true, then so is the other.

This theorem, like those that follow the next two examples, can be explained intuitively by describing what a counterexample would be for each of the two statements. As with our discussion of contrapositive, we could show that the two statements in Theorem 4.15 have the same counterexamples (if any). A more formal approach is described in Appendix A.

Example 13. **"If..., Then..." in the Conclusion** Consider the following conditional sentence (the universe is $\mathbf{Z}$):

If x is odd, then any divisor of x must also be odd.

The conclusion here has a hidden variable, and means "if u is a divisor of x, then u is odd". In other words, the conclusion itself has the form "if . . . , then . . ." (although it is an open sentence, with free variable x).

What do we assume here? What do we have to prove?

?　　?　　?

As usual, we start by assuming the hypothesis, "x is odd". But having done that, we look at what we are trying to prove "if $u \mid x$, then u is odd."

As usual, when we want to prove an "if . . . , then . . ." statement, we assume its hypothesis. Here we assume the hypothesis "$u \mid x$". We then have a new goal: to prove "u is odd".

Thus, what we actually prove is the following:

If x is odd and $u \mid x$, then u is odd. □

As with Example 12, our discussion in this example can be formulated into a general principle:

THEOREM 4.16. For open sentences $p(x)$, $q(x)$, and $r(x)$, if either the statement "if $p(x)$, then: if $q(x)$, then $r(x)$" or the statement "if $p(x)$ and $q(x)$, then $r(x)$" is true, then so is the other.

The combinations of "if . . . , then . . ."s in Theorem 4.16 leads to somewhat clumsy language. Here is an alternative formulation:

THEOREM 4.16′. For open sentences $p(x)$, $q(x)$, and $r(x)$, the open sentences "$p(x) \Rightarrow [q(x) \Rightarrow r(x)]$" and "$[p(x)$ and $q(x)] \Rightarrow r(x)$" are logically equivalent.

We give one more example of this reformulation process:

Example 14. **"Or" in the Hypothesis** Consider the following conditional sentence:

If either A or B is the empty set, then $A \times B = \varnothing$.

The assumption here is that either A or B is the empty set. But how can you use that assumption? You don't know that $A = \varnothing$ and you don't know that $B = \varnothing$. You just know that (at least) one of these two sets is empty.

So what do we assume?

? ? ?

As in Example 12, proof by contradiction is a possibility. But here the standard approach is proof by cases. From that perspective, we need to prove that, in either the case "$A = \varnothing$" or the case "$B = \varnothing$", the conclusion "$A \times B = \varnothing$" is correct. We essentially have to do two separate proofs: "if $A = \varnothing$, then $A \times B = \varnothing$" and "if $B = \varnothing$, then $A \times B = \varnothing$". □

The following theorem states the general principle represented by Example 14:

THEOREM 4.17. For open sentences $p(x)$, $q(x)$, and $r(x)$, if either the statement "if $p(x)$ or $q(x)$, then $r(x)$" or the statement "[if $p(x)$, then $r(x)$] and [if $q(x)$, then $r(x)$]" is true, then so is the other.

We conclude this chapter with an exploration:

Exploration 1: Developing Useful Equivalences

We have described various pairs of "if . . . , then . . ." statements as "logically equivalent". Our list so far includes the following examples:

Original	*Logical Equivalent*
If $p(x)$, then $q(x)$.	If $\sim q(x)$, then $\sim p(x)$.
If $p(x)$, then $q(x)$ or $r(x)$.	If $p(x)$ and $\sim q(x)$, then $r(x)$.
If $p(x)$, then: if $q(x)$, then $r(x)$.	If $p(x)$ and $q(x)$, then $r(x)$.
If $p(x)$ or $q(x)$, then $r(x)$.	If $p(x)$, then $r(x)$, and if $q(x)$, then $r(x)$.

What other logical equivalences can you think of? If you come up with a candidate, test it out by looking at specific predicates. Make a list of both "true" logical equivalences and "pseudo" logical equivalences—that is, pairs of statements that appear to be related but are not logically equivalent.

• E X P L O R E •

EXERCISES

1. (GYHD 4) State the negation of each of the following, simplifying your answer (you do not need to express these negations in terms of quantifiers):

a) f is surjective and injective (here, f is a function).
b) x is even or $x \geq 3$ (here, x is a positive integer).

2. State the contrapositive and the converse of each of the following conditional sentences; state whether the original statement and its contrapositive are true or false; and state whether the converse is true or false:
a) $(x, y \in \mathbf{R})$ If $x > y$, then $x^2 > y^2$.
b) $(A, B, C$ sets$)$ If $B \subseteq C$, then $A \times B \subseteq A \times C$.
c) $(a, b, c \in \mathbf{N})$ If $a \mid b$ and $b \mid c$, then $a \mid c$.

3. Prove each of the following, either by contradiction or by proving the contrapositive:
a) If $x \in \mathbf{R}$, and $x^2 \leq x$, then $x \leq 1$.
b) Suppose a, b, and c are integers. If $\text{GCD}(a, bc) = 1$, then $\text{GCD}(a, b) = 1$ and $\text{GCD}(a, c) = 1$.
c) If $X \cap Y = \varnothing$ and $X' \cap Y = \varnothing$, then $Y = \varnothing$.
d) If either A or B is the empty set, then $A \times B = \varnothing$.
e) If x is an odd integer, then any divisor of x must be odd.

4. Give three examples of theorems not presented thus far in the text whose converses are true, and three examples of theorems whose converses are false.

5. For each of the pairs of statements i)–iv) below, do the appropriate one of the following:
a) give an explanation why the two statements are logically equivalent; or
b) give an explanation why one statement implies the other, and give a pair of predicates for $p(x)$ and $q(x)$ which shows that the two statements are not logically equivalent; or
c) give two pairs of predicates for $p(x)$ and $q(x)$ which together show that neither statement implies the other.
i) "$\sim(p(x)$ and $q(x))$" and "$\sim p(x)$ and $\sim q(x)$".
ii) "$(\exists\, x)(p(x)$ and $q(x))$" and "$(\exists\, x)(p(x))$ and $(\exists\, x)(q(x))$".
iii) "$(\exists\, x)(p(x)$ or $q(x))$" and "$(\forall\, x)(\sim p(x)) \Rightarrow (\exists\, x)(q(x))$".
iv) $(\forall\, x)(\exists\, y)(p(x, y))$" and "$(\exists\, y)(\forall\, x)(p(x, y))$".

6. Using the examples in Exercise 5 as inspiration, look for other potential pairs of logically equivalent statements. Keep track of the pairs that actually turn out to be logically equivalent, as well as those that do not. For each pair of statements, look for reasons why they are logically equivalent or examples demonstrating that they are not. If a pair of statements are not logically equivalent, does one imply the other? Can one of them be modified so that they will be logically equivalent?

7. For each of the following statements:
- write the statement in "if . . . , then . . ." form;
- write the statement using quantifiers as fully as possible (use $\mathbf{U}$ or an appropriate specific set as your universe).

a) A complex number has a negative square only if it isn't real.
b) For a positive integer to be a prime, it is necessary that it be odd or 2.
c) A function is surjective whenever its codomain has exactly one element and its domain is nonempty.
d) "$x > 1$" is a stronger condition than "$x^2 > 1$".
e) There is at least one function from A to B unless B is empty.

8. Prove "if $x^2 - 3x + 2 > 0$, then $x > 2$ or $x < 1$" by contrapositive.

9. In Example 9, we pointed out that the statement "$\text{GCD}(a, b) = c$" means that c satisfies three conditions:

i) c divides a;
ii) c divides b;
iii) every positive integer that divides both a and b is less than or equal to c.

For each of cases a)–c) below, and using $a = 8$, $b = 12$, find, if possible, a positive integer for c that satisfies the stated combination of conditions. If no such value exists, explain why not, and find a different substitution for a and b for which such a value for c does exist, or explain why *that* is impossible.

a) A value of c which satisfies i), but neither ii) nor iii).
b) A value of c which satisfies ii), but neither i) nor iii).
c) A value of c which fails all three criteria.

10. Give an example of open sentences $p(x)$ and $q(x)$ such that either $p(x)$ or $q(x)$ can be true, for an appropriate value of x, but they cannot both be true for the same value of x. (For such open sentences, the distinction between "inclusive or" and "exclusive or" is irrelevant.)

11. PROOF EVALUATION: (see instructions for Exercise 6 in Section 2.2):

a) "THEOREM": If x is a non-zero integer, then x is both positive and negative.
"Proof": By contradiction. Assume that the hypothesis is true and the conclusion is false; i.e., suppose that x is a nonzero integer, and that x is not positive and not negative. The trichotomy property states that, for any two real numbers a and b, exactly one of the following holds: "$a < b$", "$a = b$", or "$a > b$". Applying this with $a = x$, $b = 0$, we have that either $x < 0$, $x = 0$, or $x > 0$. We are assuming that "$x < 0$" is false and that "$x > 0$" is false. Therefore "$x = 0$" must be true. This contradicts our assumption that x is nonzero.

b) "THEOREM": For $x, y \in \mathbf{Z}$, if xy is even, then x and y are even.
"Proof": Let x and y be integers, and assume that x and y are not even. Then there are integers m and n such that $x = 2m + 1$ and $y = 2n + 1$. Therefore

$$xy = (2m + 1)(2n + 1) = 2(2mn + m + n) + 1,$$

so xy is odd. By Theorem 2.16, this means xy cannot be even. Thus we have proved the contrapositive of this theorem, so the theorem is true.

c) "THEOREM": For $x, y \in \mathbf{Z}$, if xy is even, then x and y are even.
"Proof": Let x and y be integers. If x and y are even, then there are integers m and n such that $x = 2m$ and $y = 2n$. Therefore

$$xy = (2m)(2n) = 2(2mn)$$

so xy is even, as desired.

d) "THEOREM": For sets A and B, the following are equivalent:
i) $A \subseteq B$,
ii) $A \cap B = A$,
iii) $A \cup B = B$.

"Proof": First, we note the following facts, for any sets A and B:

$$A \cap B \subseteq A \qquad (*)$$
$$B \subseteq A \cup B. \qquad (**)$$

(These conditions follow immediately from the definitions of intersection and union.) We use these facts to prove the equivalence, as follows:

i) ⇒ ii): Assume $A \subseteq B$. We know, by $(*)$, that $A \cap B \subseteq A$. To prove the inclusion in the reverse direction (i.e., that $A \subseteq A \cap B$), suppose $x \in A$.

Then we have $x \in B$, since $A \subseteq B$. Therefore x is in both A and B, i.e., $x \in A \cap B$, as desired.

ii) ⇒ iii): Assume $A \cap B = A$. We know, by (**), that $B \subseteq A \cup B$. To prove the reverse direction, suppose $x \in A \cup B$. Therefore $x \in (A \cap B) \cup B$ (since $A \cap B = A$). But $A \cap B \subseteq B$, so $(A \cap B) \cup B \subseteq B \cup B$. Also, $B \cup B = B$. Therefore $x \in B$, as desired.

i) ⇒ iii): Assume $A \subseteq B$. As above, we know, by (**), that $B \subseteq A \cup B$. To prove the reverse direction, suppose $x \in A \cup B$. Since $A \subseteq B$, we have $A \cup B \subseteq B \cup B$. Also, $B \cup B = B$. Therefore $x \in B$, as desired.

e) **"THEOREM":** For $a, b \in \mathbf{Z}$, if $\mathrm{GCD}(a, b) = 1$, then there are integers x and y such that $ax + by = 1$.

"Proof": Suppose, on the contrary, that a and b are integers with $\mathrm{GCD}(a, b) \neq 1$. Then there is some integer $d > 1$ such that $d \mid a$ and $d \mid b$, and so there are integers u and v with $a = d \cdot u$ and $b = d \cdot v$.

We will show that there cannot be integers x and y, such that $ax + by = 1$. For suppose there were. Then we would have:

$$1 = ax + by = (du)x + (dv)y = d(ux + vy).$$

But this would mean $d \mid 1$, contrary to the assumption that $d > 1$, and thus proving the theorem.

PART II

Core Mathematics

5

Induction and the Integers

Introduction

In this chapter we will look at some fundamental properties of one of the most important sets in mathematics — the integers. Actually, we will begin with a set that is even more basic — the natural numbers, or positive integers, which is represented by the symbol **N**.

The positive integers are a natural place to start studying mathematics, because they are the first system of numbers we learn as children, and the first system developed historically. Indeed, there is a famous saying attributed to the 19th century mathematician Leopold Kronecker: "The natural numbers are the work of God; all the rest is the work of man." Somehow, the natural numbers seem just that — *natural* — as if they always existed, whereas entities like negative, irrational, or complex numbers had to be invented.

But what exactly are the natural numbers? Because they form an infinite set, we can't list them all, or write down all the properties they have. Many people would respond to this question with an answer like: "$1, 2, 3$ and so on." They probably assume that everyone understands what "and so on" means in this context.

Mathematicians are somewhat uncomfortable with that kind of vagueness, since it might lure them into accepting false statements as theorems. They have attempted to capture the essence of this "and so on" property of the natural numbers in several ways. The Principle of Induction is the most familiar of these, and we will begin with that in Section 5.1. This principle is the basis for an important method of proof, which we illustrate by several examples. In Section 5.2, we will look at two other ways to capture the essential idea of the natural numbers, and use methods based on these alternatives to prove two important theorems about arithmetic in **N**.

Finally, in Section 5.3, we will use ideas and methods developed earlier in this chapter, including the theorems about **N** from Section 5.2, to examine several other important properties of the integers, including the uniqueness of prime factorization and a solution to the Linear-Diophantine-Equation problem discussed in Exploration 3 in Section 2.3.

Reminder: We are assuming certain basic facts about arithmetic — see Appendix B. To prove those statements, we would need a formal definition of the arithmetic operations, in terms of sets. That approach is possible, but is the subject for a different book.

5.1 The Principle of Induction

Inductive Sets

The natural numbers begin with 1. After 1 comes 2. After 2 comes 3. After 3 comes 4. And so on.

The essence of the "and so on" here is that we can get the "next" number each time by the simple process of "adding 1". One of the basic features of **N** is that, if you add 1 to a natural number, the result is another natural number.

This feature of **N** is important, although it doesn't tell us all we need to know. But before going further, we will give a name to this property.

DEFINITION 5.1: A set of numbers S is called *inductive* iff

$$(\forall\, t)(t \in S \Rightarrow t + 1 \in S).$$

Another way to describe such a set is that it is *closed under adding* 1.

COMMENT The statement "S is inductive" involves a hidden variable, given by t in the above definition, which is used in a universally quantified implication. The nature of this definition plays an important role in the way this concept is used in proofs.

The set **N** is not the only inductive set. Others include:

$\{6, 7, 8, 9, 10, \ldots\}$,
$\{-3.2, -2.2, -1.2, -0.2, 0.8, 1.8, 2.8, \ldots\}$,
R,
Z.

Examples of noninductive sets are

$$\{1\}, \quad \{\tfrac{1}{2}, \tfrac{2}{3}, \tfrac{3}{4}, \ldots\}, \quad \text{and} \quad \{2, 4, 6, 8, \ldots\}.$$

Get Your Hands Dirty 1

a) Give two more examples of inductive sets and two more examples of noninductive sets.

b) Are there any finite sets that are inductive? What about the empty set? □

Another feature of the set **N** is that it contains the number 1. But the inductive sets **R** and **Z** also contain the number 1, and these sets are different from **N**.

We would like to find some description of **N**, related to this idea of inductiveness, which *completely characterizes* **N**—that is, which describes **N** but no other set. We can then say exactly what **N** "is"—it is *the* set with that particular property.

Before reading on, think about this question. What exactly is it about **N** that captures the essence of this set?

? ? ?

Recall where we began—we said that the natural numbers begin with 1, and then there's 2, then 3, then 4, and so on. This feels like a good intuitive description of **N** because this process produces all of **N** (and nothing else). We need to capture the idea that the process of "starting with 1" and "adding 1 repeatedly" does, in fact, result in the entire set of natural numbers, that we don't leave anything out in this process. Although **R** contains 1, and is inductive, it also contains a lot of numbers that are not natural numbers. The same is true of **Z**, or of $\{-3, -2, -1, 0, 1, 2, 3, 4, \ldots\}$.

The essential fact is this:

N is the smallest inductive set that contains 1. (∗)

Unfortunately, we can't prove this statement, because we don't have a formal definition of what the natural numbers are. But deep in our hearts, we know that this statement should be true about the set **N**.

The Need for an Axiom

One way to resolve this dilemma is to make the statement (∗) into an axiom. (We are taking this approach.) An *axiom* is a basic assumption, used to express properties and results which mathematicians feel are the most fundamental. Euclid, in his classic treatment of geometry, chose to use axioms to express some basic facts about points and lines, because he recognized that one must make some assumptions in order to make use of definitions. Once a set of axioms is selected, everything else is based on definitions and theorems.

The axiom we will use about **N** is known as the *Principle of Induction* (or the Principle of Mathematical Induction). It says the following:

AXIOM 5.1 (Principle of Induction). Any set which contains the number 1 and which is inductive must contain all the natural numbers.

We repeat: the Principle of Induction is an *assumption*. Many of the basic theorems about **N** (and about **Z**) depend on this assumption. We will adopt the Principle of Induction as an axiom, and simply recognize that our future results may depend on it. From now on we will be working under this assumption. Any time we come across a set S which contains 1 and which is inductive, we will consider ourselves justified in concluding that **N** is a subset of S. The Principle of Induction does guarantee that **N** is "the *smallest* inductive set that contains 1."

COMMENT In Section 5.2, we will look at two other fundamental properties of **N** which could be used as alternative axioms. We will prove each of these properties using the Principle of Induction, and we will also show that the Principle of Induction can itself be proved if we assume either one of these other properties. (In the latter proofs, we need to temporarily drop the assumption that the Principle of Induction is true.)

Proofs by Induction

The Principle of Induction is a formidable tool for proving certain types of theorems about **N**. For convenience, we restate the Principle of Induction in a form more useful for proofs than that given above in Axiom 5.1:

AXIOM 5.1′ (Principle of Induction). For any set S, if $1 \in S$ and S is inductive, then $\mathbf{N} \subseteq S$.

How do we use this principle in proving real theorems? For example, suppose we want to prove the following (we will prove this shortly):

THEOREM 5.2. If n is a natural number, then $n < 2^n$.

There is no set S in this theorem. What is the connection between this theorem and the Principle of Induction?

? ? ?

As a first step, we reformulate the theorem as a universally quantified statement:

$$\text{“}(\forall\, n \in \mathbf{N})\,(n < 2^n)\text{”}.$$

Where do we get the set S for the Principle of Induction?

How can we reformulate "$(\forall\, n \in \mathbf{N})(n < 2^n)$" as a statement about some set?

The key is to look at the truth set of the open sentence "$n < 2^n$". If S is the truth set of this open sentence, then Theorem 5.2 says simply: "$\mathbf{N} \subseteq S$". [Recall, from Definition 3.2 of "$\forall$" (in Section 3.1), that if P is the truth set of some open sentence $p(n)$, then the statement "$(\forall\, n \in \mathbf{N})\,(p(n))$" means precisely "$\mathbf{N} \subseteq P$".]

"$\mathbf{N} \subseteq S$" is precisely the type of conclusion that the Principle of Induction (as reformulated in Axiom 5.1′) is designed to give. Thus, to prove a statement of the form "$(\forall\, n \in \mathbf{N})\,(p(n))$", we need only prove that the truth set P of $p(n)$ satisfies the following two conditions:

i) $1 \in P$, and
ii) P is inductive.

Metaphor: The Induction Genie

We can think of our axiom of the Principle of Induction as a genie that issues "Guarantee Certificates" for proofs of theorems. We bring the genie a proof that a certain set contains 1 and is inductive, and the genie issues us a certificate that our set contains all of $\mathbf{N}$.

The mathematical world trusts this genie—because it "believes" the Principle of Induction. If you are trying to prove that a certain open sentence is true for all natural numbers, the mathematical world will accept, *as a proof of that assertion,* a proof that the open sentence's truth set contains 1 and is inductive.

In most cases, proving the first condition will be straightforward. We only need to show that $p(1)$ is a true statement. The proof of the second condition, that P is inductive, often takes some imagination. But the form of the proof will always be the same. Since the definition of "inductiveness" involves a universally quantified implication—$(\forall\, t)\,(t \in P \Rightarrow t + 1 \in P)$—we will assume the hypothesis of that implication and prove its conclusion. That is, we will assume that some symbol t represents an element of P, and prove, assuming nothing about t except that it is in P, that $t + 1$ is in P. The proof of the second condition is often called the *induction step* of the proof, and the assumption that $t \in P$ is called the *induction hypothesis*. The universally quantified variable used in the original open sentence (n, in our prototype above) is called the *induction variable,* and the proof of the entire theorem is often described by the phrase "by induction on n".

It is time for some examples.

Inequalities and Formulas

Inequalities and other formulas concerning the natural numbers are typical candidates for proof by induction. We will give several examples of this. Our first example is Theorem 5.2 as stated above:

THEOREM 5.2. If n is a natural number, then $n < 2^n$.

Proof: As suggested in our discussion above, we begin by letting $p(n)$ be the open sentence "$n < 2^n$" (for the universe $\mathbf{N}$), and let P be its truth set. We will show:

i) $1 \in P$, and
ii) P is inductive.

By the principle of induction, that will complete the proof of the theorem.

First, $p(1)$ says "$1 < 2^1$". Since $2^1 = 2$, and "$1 < 2$" is true, we have that $1 \in P$.

Next, we show that P is inductive. To do so, we assume that some object t is an element of P, i.e., we assume that $p(t)$ is true. To complete the proof, we only need to show that $t + 1 \in P$, i.e., that $p(t + 1)$ is true.

Since $t \in \mathbf{N}$ ($\mathbf{N}$ is our universe), we have $1 \leq t$, so, adding t to both sides, we get

$$t + 1 \leq 2t. \tag{a}$$

Since we are assuming $t \in P$, we know

$$t < 2^t. \tag{b}$$

Multiplying both sides of (b) by 2 gives

$$2t < 2 \times 2^t. \tag{c}$$

Also

$$2 \times 2^t = 2^{t+1}. \tag{d}$$

Combining (a), (c), and (d), we get

$$t + 1 \leq 2t < 2 \times 2^t = 2^{t+1}. \tag{e}$$

Thus, we have

$$t + 1 < 2^{t+1}. \tag{f}$$

But this is precisely the statement $p(t + 1)$, so $t + 1 \in P$. This completes the proof. ■

COMMENTS

1. In proving the induction step, we assumed the fact that $t < 2^t$. This is the induction hypothesis, and this assumption is perfectly legitimate, since the induction step is an "if . . . , then . . ." statement, and "$t < 2^t$" is its hypothesis.

2. Though the statement "$t < 2^t$" resembles the statement of the theorem itself, it is in fact quite different. The statement of the theorem is "$(\forall\, n \in \mathbf{N})(p(n))$". In proving the induction step, we are assuming that t is a *particular* number such that $p(t)$ is true. This assumption is the hypothesis of the implication which defines the statement "P is inductive".

3. The expression "$p(n)$" is an open sentence. It is not a function, and should not be confused with the two expressions "n" and "2^n" which are on either side of the inequality in Theorem 5.2.

4. Notice how helpful it is to use a different variable in the induction step from that used in initially stating the theorem. In the notation used above, n is the variable of the theorem itself. The condition "P is inductive" involves a hidden variable, in this case t.

Often no reference is made to the truth set P. Instead, the proof can be written directly in terms of the open sentence $p(n)$. To use this method, we can state the Principle of Induction as follows:

AXIOM 5.1″ (Principle of Induction). For any open sentence $p(n)$, if "$p(1)$" and "$(\forall\, t \in \mathbf{N})\,(p(t) \Rightarrow p(t+1))$" are both true, then "$(\forall\, n \in \mathbf{N})\,(p(n))$" is true.

(Again, notice the use of two different variables: t in the induction step and n in the final conclusion.)

Here's an intuitive view of the use of the application of this principle. The induction step, "$(\forall\, t \in \mathbf{N})\,(p(t) \Rightarrow p(t+1))$", is a universally quantified implication, and so we can *specialize* it; i.e., we can substitute specific values for t, and get that all of the following are true:

$$p(1) \Rightarrow p(2)$$

$$p(2) \Rightarrow p(3)$$

$$p(3) \Rightarrow p(4)$$

But we also prove that $p(1)$ is true, and we can combine $p(1)$ with the first implication to get $p(2)$, next combine $p(2)$ with the second implication to get $p(3)$, then combine $p(3)$ with the third implication to get $p(4)$, and so on. In this way we get $p(n)$ for all n.

The following metaphor may help you understand this process:

Metaphor: Induction as a Chain of Dominoes

Imagine the statements $p(1), p(2), p(3), \ldots$ as a set of dominoes, standing on end, one after the other. When we prove the induction step, we are proving that, when any domino falls, it knocks down the next one. When we prove $p(1)$, we are knocking down the first domino in the chain. If we can prove both $p(1)$ and the induction step, we have knocked down all the dominoes—i.e., we have proved "$(\forall\, n \in \mathbf{N})\,(p(n))$".

Perhaps the most interesting aspect of the proof of Theorem 5.2 is invisible. We are referring to the thinking behind the proof that P is inductive. In trying to show "$t + 1 < 2^{t+1}$", once we've assumed "$t < 2^t$", how do we know to compare $t + 1$ with $2t$, for example, or to multiply the inequality $t < 2^t$ by 2?

The answer, as you may realize by this stage in the book, lies in exploration. Before you set out to do any proof, you should ask yourself why the statement seems to be true. Here, you should ask yourself why the truth of the inequality "$t < 2^t$" should guarantee the truth of "$t + 1 < 2^{t+1}$". The "behind the scenes" reasoning might go something like this:

In order to explore the question of whether P is inductive, we remind ourselves that we will be assuming the hypothesis $p(t)$ and trying to prove the conclusion $p(t + 1)$. How can we relate the inequality $p(t + 1)$ to the inequality $p(t)$? If we want to examine whether $t + 1$ really is less than 2^{t+1}, we need to use what we know, namely, the induction hypothesis "$t < 2^t$". We must try to relate the expressions $t + 1$ and 2^{t+1}, on each side of the inequality we're exploring, to the expressions t and 2^t, from each side of the inequality that we're assuming.

How are these expressions related? The expression $t + 1$ is 1 more than t. The expression 2^{t+1} is twice 2^t. In other words, if we begin with the known inequality $t < 2^t$, then we want to add 1 to the left side and double the right side. Do these two steps "preserve the inequality"? That is, will the new left side be less than the new right side?

Certainly, if we were to double both sides of "$t < 2^t$", the result would be a valid inequality. But we only want to add 1 to the left side, instead of doubling it. Since $t + 1 \leq 2t$, adding 1 to the left side of "$t < 2^t$" increases it, at most, by the same amount that doubling it would. Thus, the two steps of "adding 1 to the left side" and "doubling the right side", applied to the inequality "$t < 2^t$", will give us a valid inequality.

By this type of analysis, or something similar, we get an understanding of the situation, and a clue to the proof. The actual proof simply summarizes the essential ideas in the reasoning, and expresses them succinctly.

The key to exploring the induction step, as in any exploration of a conditional statement, is to ask how the conclusion is related to the hypothesis. In proving an inductive step, we ask how the statement $p(t + 1)$ is related to the statement $p(t)$—that is, how knowing that $p(t)$ is true will help you show that $p(t + 1)$ is true. If you can come to an understanding of the relationship between the two assertions, then you will be on the way toward an idea of how to organize the proof.

Here is another theorem which can be proved by induction (we will outline the basic idea, and leave the details for Exercise 4):

THEOREM 5.3. For any $n \in \mathbf{N}$,

$$1^3 + 2^3 + 3^3 + \cdots + n^3 = \left(\frac{n(n+1)}{2}\right)^2.$$

The equation in Theorem 5.3 is an open sentence using the variable n, and we will represent this open sentence by $p(n)$. As usual, we need to prove $(\forall\, n \in \mathbf{N})\,(p(n))$.

Let's look at the induction step. We can think of $p(n)$ as giving us a short way of adding up the first n cubes. How does that idea help us to relate the statement $p(t + 1)$ to the statement $p(t)$?

How does knowing the sum of the first t cubes help you to add the first $t + 1$ cubes?

The sum of the first $t + 1$ cubes is $(t + 1)^3$ more than the sum of the first t cubes. To prove $p(t + 1)$, using $p(t)$, we need to add $(t + 1)^3$ to the expression for the sum of the first t cubes given to us by $p(t)$, and see if the result is equal to the expression for the sum of the first $t + 1$ cubes which $p(t + 1)$ says it should be equal to.

In other words, we need to show the following:

LEMMA 5.4. For any $t \in \mathbf{N}$,

$$\left(\frac{t(t+1)}{2}\right)^2 + (t+1)^3 = \left(\frac{(t+1)[(t+1)+1]}{2}\right)^2.$$

The right side of the equation in this assertion is obtained by substituting $t + 1$ for n in the expression $(n(n + 1)/2)^2$. The proof of Lemma 5.4 and the details of the proof of Theorem 5.3 are left to the reader as Exercise 4.

COMMENTS

1. A *lemma* is a minor, often technical, result whose primary importance is its use in proving some other theorem. The word is from a Greek root, meaning "assumption". The distinction between a lemma and a theorem is a subjective one, and has no formal mathematical significance.

2. We repeat here the first comment we made following Theorem 5.2, because it is a source of such anxiety for students. The induction hypothesis, which in this case says

$$1^3 + 2^3 + 3^3 + \cdots + t^3 = \left(\frac{t(t+1)}{2}\right)^2,$$

is a thoroughly legitimate assumption. It is an essential part of proving that the truth set of $p(n)$ is an inductive set. The assertion that a set is inductive is, itself, an "if . . . , then . . ." statement. To prove it, you assume its hypothesis and derive its conclusion.

3. There are some theorems which can be proved by induction, but are perhaps better explained in some other manner. A simple example is the formula "$1 + 2 + \cdots + n = n(n + 1)/2$". Exercise 8a) asks you to prove this by induction, as part of a more general exploration, but there are alternative ways to prove it. The following piece of mathematical lore suggests the method:

> When the great German mathematician Karl Friedrich Gauss was a child, his class was once punished for misbehaving, and the children were told to add the numbers from 1 to 1000. The teacher expected this task to keep them occupied for quite some time, but the ten-year-old Gauss turned in his answer after just a moment. The rest of the class struggled on for an hour, eventually turning in lengthy computations, all filled with errors. Gauss's paper consisted of a single number—the correct answer. He had mentally combined $1 + 1000$, $2 + 999$, $3 + 998$, etc., getting 500 separate sums, each totaling 1001. It was then just a matter of multiplying 1001 by 500 to get the answer.

The Generalized Principle of Induction

We have looked at inductive sets that contain 1, in order to prove that certain open sentences are true for all elements of $\mathbf{N}$. What can we do about open sentences that are true

Historical note: Gauss (1777–1855) has been called "the Prince of Mathematics", and is considered by many to have been the greatest mathematician of all time. He made major contributions to the physical sciences as well as to many aspects of mathematics. Among his better known achievements are the Prime-Number Theorem, which describes the frequency with which primes occur among the natural numbers, and the invention (more or less simultaneously with Bolyai and Lobachevsky) of hyperbolic geometry, which resolved the centuries-old question about the need for Euclid's Parallel Postulate.

only for the natural numbers above a certain value? For example, as we will prove shortly, the inequality "$(n + 1)^2 \leq 2^n$" is true if $n \geq 6$, but is false if $n = 1, 2, 3, 4$, or 5. As n increases beyond 6, the inequality gets stronger; i.e., the right side grows faster than the left side. (You should do some arithmetic to verify this.)

Can we modify the method of proof by induction to prove a theorem of form "$(\forall\, n \geq k)\,(p(n))$"?

Is there some variation of the Principle of Induction for inductive sets containing some arbitrary integer *k*?

The answer is yes. For convenience, for any $k \in \mathbf{Z}$, we define the set $\mathbf{N}_{[k,\infty)} = \{n \in \mathbf{Z} : n \geq k\}$. (This notation is used, even when k is negative, to emphasize the similarity in structure between this set and the set $\mathbf{N}$.) We have the following theorem, called the *Generalized Principle of Induction:*

THEOREM 5.5 (Generalized Principle of Induction). For any set S and any $k \in \mathbf{Z}$, if $k \in S$ and S is inductive, then $\mathbf{N}_{[k,\infty)} \subseteq S$.

The proof is left to the reader as Exercise 5.

COMMENTS

1. Theorem 5.5 is a *theorem*. We can prove it on the basis of the assumption of the Principle of Induction. (The method of proof outlined in Exercise 5 requires separate arguments for the cases $k > 0$, $k = 0$, and $k < 0$.)

2. The Generalized Principle of Induction is usually applied with $k = 0$ or with some integer $k > 1$. However, the value of k in this theorem can, in fact, be negative. For example, if an inductive set S contains the number -3, then it must also contain $-2, -1, 0, 1, 2, 3, 4, \ldots$.

3. $\mathbf{N}_{[1,\infty)}$ is just the set $\mathbf{N}$, and in this case, Theorem 5.5 is just the "ordinary" Principle of Induction. Also note that $\mathbf{N}_{[0,\infty)} = \mathbf{W}$, the set of whole numbers. This case of Theorem 5.5 is often used to prove theorems.

4. The alternative axioms for $\mathbf{N}$ which we will discuss in the next section also have "generalized" forms that apply to $\mathbf{N}_{[k,\infty)}$. In practice, mathematicians often don't

make a distinction between the "ordinary" form of the principle and the "generalized" form. For ease of discussion, when the distinction is not important, we will sometimes follow that practice.

The Generalized Principle of Induction is applied in a proof using the same two-step process that we use for the (ordinary) Principle of Induction. First, we prove $p(k)$, and then we prove the induction step, $(\forall\, t \in \mathbf{N}_{[k,\infty)})\,(p(t) \Rightarrow p(t + 1))$. This method is illustrated by the following proof:

THEOREM 5.6. For $n \in \mathbf{N}$, $n \geq 6$,

$$(n + 1)^2 \leq 2^n.$$

Proof: Let $p(n)$ represent the open sentence $(n + 1)^2 \leq 2^n$ for the universe $\mathbf{N}_{[6,\infty)}$, and let P be its truth set; i.e., $P = \{n \in \mathbf{N}_{[6,\infty)} : (n + 1)^2 \leq 2^n\}$. The proof will consist of showing two things:

i) $6 \in P$, and
ii) P is inductive.

The proof of i) is easy: $p(6)$ says "$(6 + 1)^2 \leq 2^6$". The left side of this inequality is 49; the right side is 64; and 49 is less than or equal to 64. Thus $p(6)$ is true, so $6 \in P$.

To prove ii), assume $t \in P$, i.e.,

$$(t + 1)^2 \leq 2^t. \qquad (*)$$

We need to prove $t + 1 \in P$, i.e.,

$$[(t + 1) + 1]^2 \leq 2^{t+1}. \qquad (**)$$

Comment: The right side of (**) is exactly twice the right side of (*). If we can show that the left side of (**) is at most twice the left side of (*), that will allow us to put together inequalities to prove (**). In other words, we would like to show that, for $t \geq 6$:

$$(t + 2)^2 \leq 2 \cdot (t + 1)^2. \qquad (***)$$

(Remember that $t \geq 6$, since $t \in \mathbf{N}_{[6,\infty)}$.)

Our proof will put this detail into a lemma.

We need the following lemma, whose proof is left to the reader as Exercise 6:

LEMMA 5.7. For $t \in \mathbf{Z}$, if $t \geq 6$, then $(t + 2)^2 \leq 2 \cdot (t + 1)^2$.

By Lemma 5.7 and the induction hypothesis (*), we have

$$(t + 2)^2 \leq 2 \cdot (t + 1)^2 \leq 2 \times 2^t = 2^{t+1}.$$

Thus, $(t + 2)^2 \leq 2^{t+1}$, which is the statement $p(t + 1)$. This proves that $t + 1 \in P$, completing the induction step and the proof of the theorem.

Summation Notation—A Digression

Theorem 5.3 provided a formula for the sum

$$1^3 + 2^3 + \cdots + n^3.$$

Both in induction problems and elsewhere, we often deal with sums like this. More generally, we study sums of the form

$$f(1) + f(2) + f(3) + \cdots + f(n), \qquad (*)$$

where f is some function whose domain is $\mathbf{N}$.

There is a standard notation which is used to abbreviate this type of expression. The sum in $(*)$ is represented by the following symbolism:

$$\sum_{t=1}^{n} f(t) \quad \text{or} \quad \sum_{t=1}^{n} f(t).$$

This is called a summation expression, and is read "the summation, from t equals 1 to n, of $f(t)$."

The symbol "Σ" is the Greek capital letter sigma, and is referred to in this notation as a "summation sign". It tells us that we are representing a sum. (There is a similar notation using "Π", the Greek capital letter pi, for representing products.)

The letter "t" here plays the role of a dummy variable (similar to that of a bound variable with quantifiers). The number "1" below the summation sign tells us to begin by substituting $t = 1$; i.e., the first term of the sum is $f(1)$. We then increase t by 1 repeatedly, getting terms $f(2)$, $f(3)$, etc. The variable "n" above the summation sign tells us that the last term of the sum is $f(n)$.

Just as we can vary the starting number in an induction proof, so we can also adjust the starting number in a summation expression. Suppose v and w are any integers (possibly negative) with $v \leq w$, and that f is a function whose domain includes all integers t with $v \leq t \leq w$. Then the notation

$$\sum_{t=v}^{w} f(t) \qquad (**)$$

represents the sum

$$f(v) + f(v + 1) + \cdots + f(w).$$

[If $v = w$, this sum has only one term, $f(v)$, and $\sum_{t=v}^{w} f(t)$ is equal to $f(v)$.]

In the expression $(**)$, the integer v is called the *lower limit*, and the integer w is called the *upper limit*.

By convention, any summation expression whose lower limit is greater than its upper limit [i.e., with $v > w$ in $(**)$] is defined to be 0.

Example 1. **Using "Σ" Notation** The following equations show the meaning of each Σ expression as a sum:

a) $\Sigma_{t=1}^{5} t^2 = 1^2 + 2^2 + 3^2 + 4^2 + 5^2$.

b) $\Sigma_{y=1}^{4} \dfrac{y}{y+1} = \dfrac{1}{2} + \dfrac{2}{3} + \dfrac{3}{4} + \dfrac{4}{5}$.

c) $\Sigma_{x=-2}^{3} 2^x = 2^{-2} + 2^{-1} + 2^0 + 2^1 + 2^2 + 2^3$.

d) $\Sigma_{u=2}^{2} (u^3 + 4) = 2^3 + 4$.

e) $\Sigma_{s=7}^{4} (s^2 + 1) = 0$. □

Get Your Hands Dirty 2

Express each of the following sums using "Σ" notation (there may be more than one way to do so):

a) $4 + 5 + 6 + 7$.

b) $\dfrac{1}{2} + \dfrac{1}{4} + \dfrac{1}{6} + \cdots + \dfrac{1}{20}$.

c) $5 + 9 + 13 + 17 + 21 + 25 + 29$. □

Here is an example of an induction proof, written using summation notation:

THEOREM 5.8. For $n \in \mathbf{N}$,

$$\sum_{t=1}^{n} t^2 = \frac{n(n+1)(2n+1)}{6}.$$

Proof: Let

$$S = \left\{ n \in \mathbf{N} : \sum_{t=1}^{n} t^2 = \frac{n(n+1)(2n+1)}{6} \right\}.$$

We need to show $S = \mathbf{N}$.

When $n = 1$, $\Sigma_{t=1}^{n} t^2$ is equal to $\Sigma_{t=1}^{1} t^2$, or 1^2, which is 1. The value of $n(n+1) \times (2n+1)/6$ is equal to $1 \times (1+1) \times (2 \times 1 + 1)/6$, which is also 1. Since the two are equal, $1 \in S$.

Now suppose that $k \in S$. We need to show that $k + 1 \in S$, so we look at $\Sigma_{t=1}^{k+1} t^2$. By the definition of the summation notation, this sum is equal to $(\Sigma_{t=1}^{k} t^2) + (k+1)^2$. Since $k \in S$, we know that $\Sigma_{t=1}^{k} t^2$ is equal to $k(k+1)(2k+1)/6$. Therefore

$$\sum_{t=1}^{k+1} t^2 = \frac{k(k+1)(2k+1)}{6} + (k+1)^2.$$

We now do some algebra:

$$\sum_{t=1}^{k+1} t^2 = \frac{k(k+1)(2k+1)}{6} + (k+1)^2 \qquad \text{(as just noted)}$$

$$= \frac{k(k+1)(2k+1)}{6} + \frac{6(k+1)^2}{6}$$

$$= \frac{(k+1)[k(2k+1) + 6(k+1)]}{6}$$

$$= \frac{(k+1)[2k^2 + 7k + 6]}{6}$$

$$= \frac{(k+1)(k+2)(2k+3)}{6}$$

$$= \frac{(k+1)[(k+1)+1][2(k+1)+1]}{6}.$$

The final expression is precisely the value of $n(n+1)(2n+1)/6$ for $n = k+1$, so the series of equations shows that $k + 1 \in S$. That completes the proof. ■

COMMENT It's amazing how the algebra worked out, isn't it? No, not really. It worked out because the theorem was true. If we had been trying to prove the wrong formula, the algebra would not have come out to the expression we needed.

Get Your Hands Dirty 3

a) Try to prove that, for every $n \in \mathbf{N}$, $\sum_{v=1}^{n}(2v - 1) = n^2$. (First test it for $n = 1$ and $n = 2$.)
b) Try to prove that for every $n \in \mathbf{N}$, $\sum_{x=1}^{n} x^2 = n^2 + n - 1$. (First test it for $n = 1$ and $n = 2$.) □

Cardinality for Finite Power Sets—Proof by Induction

Our next illustration of the use of induction comes from a different area—the study of cardinality of sets. In Exploration 2 in Section 2.1, you may have discovered a formula for finding the cardinality of a power set $\mathcal{P}(A)$ in terms of the cardinality of A itself. (If not, you should explore that idea now before reading on.) We here state such a formula as a theorem:

THEOREM 5.9. For any finite set A, $|\mathcal{P}(A)| = 2^{|A|}$.

This may not look like an appropriate theorem for the use of induction, because there is no induction variable in sight. The first task in working with this theorem is to rephrase it to make that hidden variable explicit.

Where might such a variable come from?

? ? ?

It comes from the expression $|A|$. Since A is a finite set, its cardinality is some integer n. Now actually, we don't know if n is positive or not; n might be 0. Therefore we will use the Generalized Principle of Induction (Theorem 5.5), with $k = 0$, instead of the ordinary Principle of Induction, and refer to the set **W** (the whole numbers, or nonnegative integers), instead of **N**. But the general idea is the same.

Our proof will include lots of detail, and be interrupted by numerous comments that explain the purpose of each step. Afterward, we will give a bare-bones summary of the proof. Before beginning the proof, we will restate the theorem to make the induction variable explicit:

THEOREM 5.9′. For any $n \in \mathbf{W}$,

$$\text{if } |A| = n, \text{ then } |\mathcal{P}(A)| = 2^n .$$

Comment: We need to think of the "if . . . , then . . ." statement in the theorem as an open sentence with the variable n (and A as a universally quantified variable). In other words, $p(n)$ will be the statement

$$\text{“}(\forall A)(|A| = n \Rightarrow |\mathcal{P}(A)| = 2^n)\text{”} .$$

For a particular value of n, this statement might be either true or false. We want to show that this statement is true for all $n \in \mathbf{W}$.

Proof: Let $p(n)$ be the open sentence "$(\forall A)(|A| = n \Rightarrow |\mathcal{P}(A)| = 2^n)$". We need to prove "$(\forall n \in \mathbf{W})(p(n))$".

We begin by showing that $p(0)$ is true. To do so, suppose $|A| = 0$. Then $A = \varnothing$, and so $\mathcal{P}(A) = \{\varnothing\}$. Thus $|\mathcal{P}(A)| = 1$, which is equal to 2^0. This proves $p(0)$.

Now for the induction step.

Comment: We are trying to prove

$$(\forall t \in \mathbf{W})(p(t) \Rightarrow p(t + 1)) .$$

Since each statement $p(t)$ and $p(t + 1)$ is itself a universally quantified implication, the discussion about specialization in Section 4.1 is relevant to this proof. You may wish to reexamine the schematic diagram provided there.

Let $t \in \mathbf{W}$ and suppose that $p(t)$ is true. In other words, we assume the following:

$$(\forall A)(|A| = t \Rightarrow |\mathcal{P}(A)| = 2^t) .$$

Comment: We will use this induction hypothesis by means of specialization; that is, we will find a set A which satisfies the "hypothesis of the hypothesis", namely, "$|A| = t$", and then be guaranteed that A also satisfies the "conclusion of the hypothesis", namely, "$|\mathcal{P}(A)| = 2^t$".

What we need to do is prove $p(t + 1)$, which is the universally quantified statement "$(\forall B)(|B| = t + 1 \Rightarrow |\mathcal{P}(B)| = 2^{t+1})$". To start the proof of this, suppose we have a set B with $|B| = t + 1$.

Comment: First, note that, in stating $p(t + 1)$, we used a variable different from A for the set that's involved. That will help keep things clear.

Second, because $p(t + 1)$ is a universally quantified implication statement, the "induction conclusion", $p(t + 1)$, has its own hypothesis, "$|B| = t + 1$". We assume this "hypothesis of the conclusion", and we need to prove the "conclusion of the conclusion", namely, $|\mathcal{P}(B)| = 2^{t+1}$.

What we need to show now is that $|\mathcal{P}(B)| = 2^{t+1}$. To make use of the induction hypothesis, we will form a set of cardinality t as follows: Let x be any element of B. (Since $|B| = t + 1$ and $t \geq 0$, we know that $B \neq \varnothing$.) Define the set C to be $B - \{x\}$, so C is a set with exactly t elements, and therefore C fits the "hypothesis of the induction hypothesis". We are thus guaranteed that C satisfies the "conclusion of the induction hypothesis", that is, we know that $|\mathcal{P}(C)| = 2^t$.

Comment: So far, we have been primarily concerned with the form of this proof. Now we get down to the substance. The crucial question is this: *How is the cardinality of B related to the cardinality of C?*

To answer this question, we need to know which subsets of B are also subsets of C, which are not, and how many there are of each kind. (For example, compare the power set of $\{a, b, c, d\}$ with the power set of $\{a, b, c\}$.)

? ? ?

The key idea is that the subsets of B which *are* subsets of C are precisely those which do not contain x, and these are exactly half of the subsets of B. Put another way, we will show that the number of subsets of B which *are* subsets of C is equal to the number of subsets of B which are *not* subsets of C. We can summarize this assertion as

$$|\mathcal{P}(C)| = |\mathcal{P}(B) - \mathcal{P}(C)|.$$

We will prove that the sets $\mathcal{P}(C)$ and $\mathcal{P}(B) - \mathcal{P}(C)$ have the same cardinality by setting up a one-to-one, onto function from one to the other, as follows. (Recall our comment following the definition of "bijective" (Definition 3.19 in Section 3.2): if there is a bijective function from one set to another, then the sets have the same cardinality.)

Consider the function $f: \mathcal{P}(C) \to \mathcal{P}(B) - \mathcal{P}(C)$, defined by $f(S) = S \cup \{x\}$. That is, for every subset S of C [i.e., for $S \in \mathcal{P}(C)$], define $f(S)$ to be the set $S \cup \{x\}$, which is a subset of B that is *not* a subset of C [i.e., $S \cup \{x\} \in \mathcal{P}(B) - \mathcal{P}(C)$]. We will show that this function f is one-to-one and onto, thereby proving that $|\mathcal{P}(C)| = |\mathcal{P}(B) - \mathcal{P}(C)|$.

To prove that f is one-to-one, suppose that S and T are elements of $\mathcal{P}(C)$ such that $f(S) = f(T)$. We need to show $S = T$.

Since $f(S) = f(T)$, we have $S \cup \{x\} = T \cup \{x\}$. But x belongs to neither S nor T, since S and T are subsets of C, and $C = B - \{x\}$. Therefore, $S \cup \{x\} - \{x\} = S$ and $T \cup \{x\} - \{x\} = T$. So we have $S = S \cup \{x\} - \{x\} = T \cup \{x\} - \{x\} = T$, and $S = T$, as desired.

To prove that f is onto, suppose $U \in \mathcal{P}(B) - \mathcal{P}(C)$, so that x is in U. We need to find a set $W \in \mathcal{P}(C)$ such that $f(W) = U$.

Let $W = U - \{x\}$. Since $x \in U$, we have $W \cup \{x\} = U$. Also, since $U \subseteq B$, and $x \notin W$, we have $W \subseteq C$, i.e., $W \in \mathcal{P}(C)$. Finally, we have $f(W) = W \cup \{x\} = U$, as desired.

Thus we have that f is both one-to-one and onto, and so $|\mathcal{P}(C)| = |\mathcal{P}(B) - \mathcal{P}(C)|$.

Now, since every subset of B is either in $\mathcal{P}(C)$ or in $\mathcal{P}(B) - \mathcal{P}(C)$, and can't be in both, we have $|\mathcal{P}(B)| = |\mathcal{P}(C)| + |\mathcal{P}(B) - \mathcal{P}(C)|$, and so $|\mathcal{P}(B)| = 2 \cdot |\mathcal{P}(C)|$. Since $|\mathcal{P}(C)| = 2^t$, this shows $|\mathcal{P}(B)| = 2^{t+1}$, which completes the proof. ■

COMMENTS

1. If you are finding this complicated, that's because it is. An induction proof in which the induction hypothesis is itself an "if . . . , then . . ." statement will, inevitably, require a rather complex series of steps to set up. Proofs of theorems in which the hypothesis involves an implication statement will generally involve the use of specialization. In this proof, we used specialization when we applied the induction hypothesis to the set C. In preparation for this, we had to first create the set C from B. Recall that the set B itself came from the "hypothesis of the induction conclusion". Combining the difficult idea of specialization with the method of proof by induction can make for a rather intricate overall theorem. You may wish to reread the discussion of specialization in Section 4.1 to help sort out the logic of this proof.

2. As in any induction proof, the crucial substantive element is relating the induction conclusion $p(t + 1)$ to the induction hypothesis $p(t)$. In the present case, this involved relating the cardinality of $\mathcal{P}(B)$ to the cardinality of $\mathcal{P}(C)$, where C was a set of size t which we created for the purpose of making this comparison.

3. This proof has the added complication that both the domain and codomain of the function f are sets of sets. Working with such "set-valued" functions can be confusing until one gains experience with them.

Since this proof was interrupted by so many comments and explanatory notes, we give here a bare-bones version, such as might appear in a text for a set-theory course:

THEOREM 5.9′. For any $n \in \mathbf{W}$,

$$\text{if} \quad |A| = n, \quad \text{then} \quad |\mathcal{P}(A)| = 2^n.$$

Proof (Bare-bones Version): When $n = 0$, we have $A = \varnothing$, so the conclusion follows immediately.

Assume that the statement is true when $n = t$, and let B be a set of cardinality $t + 1$. Since $|B| > 0$, we can let x be any element of B. Then let $C = B - \{x\}$, so that $|C| = t$.

Then $\mathcal{P}(B)$ is the disjoint union of $\mathcal{P}(C)$ and $\mathcal{P}(B) - \mathcal{P}(C)$, so $|\mathcal{P}(B)| = |\mathcal{P}(C)| + |\mathcal{P}(B) - \mathcal{P}(C)|$. If we can show that $|\mathcal{P}(B) - \mathcal{P}(C)| = |\mathcal{P}(C)|$, we will be done, since we know by the induction hypothesis that $|\mathcal{P}(C)| = 2^t$.

But $f: \mathcal{P}(C) \to \mathcal{P}(B) - \mathcal{P}(C)$, given by $f(S) = S \cup \{x\}$, can be shown to be a one-to-one, onto function. This completes the proof. ■

COMMENT Even in the bare-bones version, this proof may not be the best way to understand or prove this theorem. But this approach is a good illustration of an important type of induction proof.

The following is a sketch of another approach to understanding and proving the formula for cardinality of a finite power set:

> Suppose A is a set with n elements, and we're trying to make a list of all the subsets of A. As we go to create each subset, we have to decide, element by element, whether to put a given element of A in the subset or leave it out. That gives two choices for each element. Since there are n elements, we have to make such an "in or out" choice n times. That gives 2^n choices altogether for how to make a subset of A.

A proof based on this approach is suggested in Exercise 13.

Generalizing Formulas from the Case $n = 2$

There are many operations that are defined for combining two objects. We have defined composition of functions, union of sets, and GCD of integers, all as binary operations in which we start with two objects and produce a third. (We will discuss binary operations more formally in Section 7.1.)

The Principle of Induction can be used to prove generalizations of equations that involve binary operations to results that apply to more than two objects. We give here an illustration using the Law of Exponents, $a^x \cdot a^y = a^{x+y}$. We will assume that this formula has been proved valid whenever a is a positive real number and $x, y \in \mathbf{N}$ (see Exercise 15), and prove a generalization involving a product of an arbitrary number of exponential expressions.

THEOREM 5.10. Suppose $n \geq 2$, that $a \in \mathbf{R}^+$, and that $x_1, x_2, \ldots, x_n$ are positive integers. Then

$$a^{x_1} \cdot a^{x_2} \cdot \cdots \cdot a^{x_n} = a^{x_1+x_2+\cdots+x_n} \qquad (*)$$

Proof: We have assumed that the case $n = 2$ has already been proved, so we turn to the induction step. Suppose the formula $(*)$ is true whenever the number of factors is t, and suppose we have an expression of the form $a^{x_1} \cdot a^{x_2} \cdot \cdots \cdot a^{x_{t+1}}$, with $t + 1$ factors.

We can write $a^{x_1} \cdot a^{x_2} \cdot \cdots \cdot a^{x_{t+1}}$ as $(a^{x_1} \cdot a^{x_2} \cdot \cdots \cdot a^{x_t}) \cdot a^{x_{t+1}}$. The expression in parentheses is a product with t factors, so our induction hypothesis tells us that it is equal to $a^{x_1+x_2+\cdots+x_t}$. But the product $a^{x_1+x_2+\cdots+x_t} \cdot a^{x_{t+1}}$ is a product with only two factors, and so $(*)$ is valid for this product as well. Therefore, we can write this product as $a^{(x_1+x_2+\cdots+x_t)+x_{t+1}}$, which is equal to the right side of $(*)$ for $n = t + 1$.

This completes the proof. ∎

COMMENT The universes selected for the x_i's and for a do not affect the above proof, as long as the case $n = 2$ has been proved for the particular universes.

Fallacious Proofs by Induction

Because proof by induction can be so complex, it is often helpful to see examples of what *not* to do. The next two examples illustrate areas of potential error:

Example 2. **Forgetting to Get Started** This example points out the importance of the initial step of an induction proof:

NONTHEOREM 5.11. For $n \in \mathbf{N}$,

$$1 + 2 + \cdots + n = \frac{n^2 + n + 2}{2}.$$

"Proof": Let $p(n)$ be the open sentence $1 + 2 + \cdots + n = (n^2 + n + 2)/2$, and assume that $p(t)$ is true; i.e., assume

$$1 + 2 + \cdots + t = \frac{t^2 + t + 2}{2}. \qquad (*)$$

We want to prove $p(t + 1)$, i.e., we want to show

$$1 + 2 + \cdots + t + 1 = \frac{(t + 1)^2 + (t + 1) + 2}{2}. \qquad (**)$$

The left side of $(**)$ is $t + 1$ more than the left side of $(*)$, so it is also equal to $t + 1$ more than the right side of $(*)$. Thus we have [adding $t + 1$ to both sides of $(*)$]

$$1 + 2 + \cdots + t + 1 = \frac{t^2 + t + 2}{2} + t + 1. \qquad (***)$$

But the right side of $(***)$ can be combined as follows:

$$\begin{aligned}\frac{t^2 + t + 2}{2} + t + 1 &= \frac{t^2 + t + 2 + 2(t + 1)}{2}\\ &= \frac{(t^2 + 2t + 1) + (t + 1) + 2}{2}\\ &= \frac{(t + 1)^2 + (t + 1) + 2}{2}.\end{aligned}$$

This proves $p(t + 1)$, and completes the proof. ■

COMMENTS

1. The fallacy in this proof is that we didn't prove $p(1)$. In fact, $p(1)$ is false (verify!), and the formula is incorrect. (Try a few values. Compare the left and right sides.)

2. The above "proof" *is* a legitimate proof of the statement "$(\forall t \in \mathbf{N})(p(t) \Rightarrow p(t+1))$". In other words, the truth set of $p(n)$ is, in fact, an inductive set. But as it happens, the hypothesis of the implication here is false for every t, because the truth set of $p(n)$ is actually the empty set. [The empty set is inductive, since the statement "$(\forall t)(t \in S \Rightarrow t + 1 \in S)$" is vacuously true if $S = \varnothing$.]

3. It may help to look at the above proof of the statement "$(\forall t \in \mathbf{N})(p(t) \Rightarrow p(t+1))$" in concrete terms. Suppose $t = 100$. The statement $p(100)$ says "$1 + \cdots + 100 = 5051$" [since for $t = 100$, the expression $(t^2 + t + 2)/2$ comes out to 5051]. The above proof simply adds 101 to both sides, and concludes that $1 + \cdots + 101 = 5152$, which is the statement $p(101)$ [since for $t = 101$, the expression $(t^2 + t + 2)/2$ comes out to 5152]. In essence, we are proving the statement "if $1 + \cdots + 100$ is equal to 5051, then $1 + \cdots + 101$ is equal to 5152". That statement makes some intuitive sense, even though both hypothesis and conclusion are false, since we can add 101 to both sides of the hypothesis to get the conclusion. And from the way we defined implication, the open sentence $p(t) \Rightarrow p(t+1)$ is considered true when $t = 100$.

Example 3. **"All Horses Have the Same Color"** The following "proof" is a commonly used example of abuse of the Principle of Induction. The error here, however, is much more subtle than that in Example 2. See if you can find it as you read along.

NONTHEOREM 5.12. For $n \in \mathbf{N}$, if S is a set of n horses, then all the horses in S have the same color.

Comment: This in effect says that all horses have the same color. We have stated this nontheorem in such a way as to make the induction variable explicit. This is similar to the restating of Theorem 5.9 as Theorem 5.9′. There is no error in this. This "theorem" also resembles Theorem 5.9 in that the open sentence $p(n)$ is itself a universally quantified implication, here with the variable S. The structure of this proof will be quite similar to that one.

Proof: Let $p(n)$ represent the open sentence "If S is a set of n horses, then all the horses in S have the same color".

We begin with the case $n = 1$. If S is any set with only 1 horse in it, then it is certainly true that all the horses in S have the same color.

Next, for the induction step, assume the induction hypothesis $p(t)$. We need to prove $p(t+1)$, so we assume *its* hypothesis, namely, that S is a set with $t + 1$ horses. We need to prove the conclusion of $p(t+1)$, namely, that all horses in S have the same color.

Let x and y be any two of those horses, with $x \neq y$. (We know that $t + 1 \geq 2$.) We need to show that x and y have the same color. Define two sets A and B by letting $A = S - \{x\}$ and $B = S - \{y\}$. These are each sets with t horses. Choose any $z \in A \cap B$.

Then both y and z are in A. Since A is a set with t horses, the induction hypothesis $p(t)$ tells us that y and z must have the same color. Similarly, x and z are in B, so x and

z have the same color. Since x and y both have the same color as z, x and y must have the same color as each other. This completes the proof. ■

Where is the flaw in this proof?

COMMENT Of course, the assertion of Nontheorem 5.12 is false, and so the "proof" must not be correct, but the flaw is well hidden. We leave it as Exercise 16 for you to locate the flaw. There is only one invalid step in the "proof".

Definition "by Induction"

The induction process of "starting with 1" and then "doing the next step repeatedly" can be used in making definitions, as well as in proving theorems. Known more properly as *definition by recursion,* the process is closely parallel to that of proof by induction.

Typically, the process is used to define a function f whose domain is $\mathbf{N}$. The codomain can be any set A. We first define $f(1)$ (this is called the *initial condition*), and then state, in general terms, how to find $f(t + 1)$ in terms of $f(t)$ and t. This general rule is called the *recursion step* or *recursion equation*. Thus $f(2)$ is defined in terms of $f(1)$, and then $f(3)$ in terms of $f(2)$, and so on.

(The validity of this method as a means of defining functions is based on a fairly complicated result, called the *Recursion Theorem,* whose proof is based on the Principle of Induction. Though the individual details of the proof are not very difficult, the overall proof is quite long and technical, and will not be done in this book.)

COMMENT The recursion method can be used just as well for a domain such as $\mathbf{W}$ or $\{-3, -2, -1, 0, 1, 2, \ldots\}$. The proof of the more general result uses the Generalized Principle of Induction.

Example 4. **Definition by Recursion—Factorials** There is an important function with domain $\mathbf{N}$, known as the *factorial* function, which is used in many "counting problems". Its value for a given natural number n is called "n factorial", and is written as "$n!$" ("n" followed by an exclamation point). This function can be defined recursively as follows:

$$1! = 1$$

and

$$(t + 1)! = (t + 1) \cdot (t!) \qquad \text{for} \quad t \geq 1\,.$$

(This second equation is the recursion equation.)

Thus, since we know the value of $1!$, we can find the value of $2!$ by applying the recursion equation with $t = 1$. The recursion equation tells us that $2! = 2 \times (1!) = 2 \times 1 = 2$.

Similarly,

$$3! = 3 \times (2!) = 3 \times 2 = 6$$

$$4! = 4 \times (3!) = 4 \times 6 = 24$$

$$5! = 5 \times (4!) = 5 \times 24 = 120$$

and so on.

Each value of n factorial is defined as a product. If we don't actually carry out the multiplication, what we get is an increasingly long sequence of factors, with an additional factor at each stage. The product for $(t + 1)!$ is just $t + 1$ times the product for $t!$. Writing it out this way gives

$$1! = 1$$
$$2! = 2 \times (1!) = 2 \times 1$$
$$3! = 3 \times (2!) = 3 \times 2 \times 1$$
$$4! = 4 \times (3!) = 4 \times 3 \times 2 \times 1$$
$$5! = 5 \times (4!) = 5 \times 4 \times 3 \times 2 \times 1$$

and, in general,

$$n! = n \times (n - 1) \times (n - 2) \times \cdots \times 3 \times 2 \times 1.$$ □

COMMENTS

1. There are some important advantages we want to point out in the use of the recursive definition. One is that it avoids the potential for ambiguity or confusion in the "$\cdots$" notation. Second, in some other situations, it is difficult to write an explicit expression for $f(n)$, and so the recursive definition becomes the only method available. Our next example will illustrate that. Finally, the recursive definition is perfectly suited for proofs by induction, as our next example will also illustrate.

2. In some contexts, it is helpful to define the factorial function for 0 as well. In order for the recursion equation "$(t + 1)! = (t + 1) \cdot t!$" to be valid for $t = 0$, and still have $1! = 1$, we define $0!$ to be equal to 1.

We leave the proof of the following result as Exercise 17:

THEOREM 5.13. For $n \in \mathbf{N}$, and sets A and B, if $|A| = |B| = n$, then the number of bijective functions from A to B is $n!$.

The case when $A = B$ is of particular interest, and will be mentioned again in Section 10.1. It turns out that it is easier to prove Theorem 5.13, and derive the result for $A = B$ as a special case, than to prove the result for the case $A = B$ by itself [see part *b*) of Exercise 17].

Example 5. **The Fibonacci Sequence** The recursion process can be varied to define the next value, not just in terms of the immediately preceding value, but in terms of some other combination of values already defined.

A sequence well known in number theory, called the *Fibonacci numbers,* is defined by recursion based on the two preceding terms. In order to do this, we need to give the first two values explicitly, before turning to a recursion equation. The Fibonacci numbers, usually written $F_1, F_2, F_3, \ldots$, are defined as follows:

$$F_1 = 1$$

$$F_2 = 1$$

and

$$F_{t+2} = F_{t+1} + F_t \qquad \text{for} \quad t \geq 1\,.$$

Thus, we can find F_3 by using $t = 1$ in the recursion equation, to get

$$F_3 = F_2 + F_1 = 1 + 1 = 2\,.$$

Similarly,

$$F_4 = F_3 + F_2 = 2 + 1 = 3$$

$$F_5 = F_4 + F_3 = 3 + 2 = 5$$

$$F_6 = F_5 + F_4 = 5 + 3 = 8$$

and so on. □

COMMENT It is a rather sophisticated problem to develop an expression for F_n directly in terms of n. This is one reason for using the recursive definition. More importantly, the recursive definition captures the spirit of most applications of the Fibonacci sequence, and the method of proof by induction is well suited to proofs concerning recursive definitions. The following theorem illustrates this:

THEOREM 5.14. Let F_n represent the nth Fibonacci number. Then the following relationship holds for all $n \in \mathbf{N}$:

$$F_1 + F_2 + \cdots + F_n = F_{n+2} - 1\,. \qquad (*)$$

Proof: When $n = 1$, the assertion $(*)$ says "$F_1 = F_3 - 1$". Since $F_1 = 1$ and $F_3 = 2$, this assertion is true.

Now assume that $(*)$ is true when $n = r$. In other words, assume that the equation

$$F_1 + F_2 + \cdots + F_r = F_{r+2} - 1\,. \qquad \text{(i)}$$

is correct. We need to prove that $(*)$ is true when $n = r + 1$; i.e., we need to verify the equation

$$F_1 + F_2 + \cdots + F_{r+1} = F_{(r+1)+2} - 1\,. \qquad \text{(ii)}$$

If we simply add F_{r+1} to both sides of (i), we get

$$F_1 + F_2 + \cdots + F_{r+1} = F_{r+1} + F_{r+2} - 1\,. \qquad \text{(iii)}$$

But, applying the recursion equation (with $t = r + 1$), we have that $F_{r+1} + F_{r+2} = F_{r+3}$.

Thus the right side of (iii) is equal to the right side of (ii). This completes the proof. ■

COMMENT As usual, we use a different variable in the induction step (r, here) from that used in the recursion equation (t) to make it easier to refer to a specific case of the recursion equation. Thus, we were able to get the equation $F_{r+1} + F_{r+2} = F_{r+3}$ by substituting $r + 1$ for t in the general equation $F_t + F_{t+1} = F_{t+2}$ (which is a rewriting of the recursion equation).

EXERCISES

1. [GYHD 1b)] Are there any finite sets that are inductive? What about the empty set?

2. (GYHD 2) Express each of the following sums using "Σ" notation (there may be more than one way to do so):

a) $4 + 5 + 6 + 7$.

b) $\frac{1}{2} + \frac{1}{4} + \frac{1}{6} + \cdots + \frac{1}{20}$.

c) $5 + 9 + 13 + 17 + 21 + 25 + 29$.

3. (GYHD 3)

a) Try to prove that, for every $n \in \mathbf{N}$, $\Sigma_{v=1}^{n}(2v - 1) = n^2$. (First test it for $n = 1$ and $n = 2$.)

b) Try to prove that, for every $n \in \mathbf{N}$,. $\Sigma_{x=1}^{n} x^2 = n^2 + n - 1$. (First test it for $n = 1$ and $n = 2$.)

4. **a)** Prove:

LEMMA 5.4. For $t \in \mathbf{N}$,

$$\left(\frac{t(t+1)}{2}\right)^2 + (t+1)^3 = \left(\frac{(t+1)[(t+1)+1]}{2}\right)^2.$$

(This proof does not require induction.)

b) Use part a) to give a detailed proof of the following:

THEOREM 5.3. For any $n \in \mathbf{N}$:

$$1^3 + 2^3 + 3^3 + \cdots + n^3 = \left(\frac{n(n+1)}{2}\right)^2.$$

5. The Generalized Principle of Induction says the following:

THEOREM 5.5. For any set S and any $k \in \mathbf{Z}$, if $k \in S$ and S is inductive, then $\mathbf{N}_{[k,\infty)} \subseteq S$. (Recall: $\mathbf{N}_{[k,\infty)} = \{t \in \mathbf{Z} : t \geq k\}$.)

a) Prove this theorem in the case $k \in \mathbf{N}$. (*Hint:* Apply the Principle of Induction to the set $T = S \cup \{1, 2, \ldots, k - 1\}$.)

b) Prove this theorem in the case $k = 0$. (*Hint:* First show that $1 \in S$, and apply the Principle of Induction.)

c) Prove this theorem for $k < 0$. [*Hint:* Write $k = -n$, with $n \in \mathbf{N}$, and let $p(n)$ be the open sentence "if $-n \in S$ and S is inductive, then $\mathbf{N}_{[-n,\infty)} \subseteq S$". Prove $p(n)$ by induction on n. You may use part b) to help with the case $n = 1$.]

6. Prove:

LEMMA 5.7. For $t \in \mathbf{Z}$,

$$\text{if } t \geq 6, \text{ then } (t+2)^2 \leq 2(t+1)^2.$$

(Induction is not needed for this proof.)

7. Prove each of the following, for every $n \in \mathbf{N}$, using the Principle of Induction:

a) $1 + 2 + 4 + \cdots + 2^n = 2^{n+1} - 1$.

b) $1 \times (1!) + 2 \times (2!) + \cdots + n \times (n!) = (n + 1)! - 1$.

c) $8^n \mid (4n)!$.

d) $2^{n-1} \leq n!$.

e) $(1 + x)^n \geq 1 + nx$ for all $x \in \mathbf{R}^+$.

f) $F_1 + F_3 + \cdots + F_{2n-1} = F_{2n}$ (where F_i is the ith Fibonacci number).

8. **a)** Prove by induction:

$$\sum_{t=1}^{n} t = \frac{n(n + 1)}{2} \qquad (\text{for all} \quad n \in \mathbf{N}).$$

(*Note:* There are other, perhaps better, ways to prove this formula, but you should do it here by induction.)

b) Prove by induction:

$$\sum_{t=2}^{n} t = \frac{(n - 1)(n + 2)}{2} \qquad (\text{for} \quad n \geq 2).$$

c) Find, and prove by induction, a similar formula for evaluating the sum $\sum_{t=5}^{n} t$. (Your formula, and proof, should be for $n \geq 5$.)

d) Find, and prove by induction, a general formula for evaluating the sum $\sum_{t=m}^{n} t$ (for $n \geq m$). (*Note:* You do not need to use induction on m in your proof. For a generic m, prove this by induction for $n \geq m$.)

9. Assume the trigonometric formulas:

$$\cos(x + y) = \cos x \cos y - \sin x \sin y,$$
$$\sin(x + y) = \sin x \cos y + \cos x \sin y.$$

Prove, by induction,

$$(\cos x + i \sin x)^n = \cos(nx) + i \sin(nx) \qquad (\text{where} \quad i = \sqrt{-1})$$

(This is called *De Moivre's Theorem.*)

10. Prove each of the following for the subset of $\mathbf{N}$ indicated:

a) $2^{n+1} \leq 3^n$ (for $n \geq 2$).

b) $2^{n+2} \leq n!$ (for $n \geq 6$).

c) $n^4 \leq 2^n$ (for $n \geq 16$).

11. Assume that the sum of the interior angles of a triangle is 180°. Sketch a proof by induction, using intuitive geometric ideas, that, for $n \geq 3$, the sum of the interior angles of an n-sided polygon is $180(n - 2)°$.

12. Recall that, if A and B are sets, the Cartesian product $A \times B$ is defined as the set of all ordered pairs (x, y) such that $x \in A$ and $y \in B$. As part of Exploration 2 or Exercise 12b) in Section 2.1, you perhaps discovered the cardinality relationship $|A \times B| = |A| \cdot |B|$, for finite sets A and B. (If you haven't seen this equation before, experiment until you are convinced of it.)

a) Assuming this formula for two sets, prove by induction that $|A_1 \times A_2 \times \cdots \times A_n| = |A_1| \cdot |A_2| \cdot \cdots \cdot |A_n|$, where each A_i is a finite set. (*Hint:* Use the number of sets as your induction variable, and follow the model of Theorem 5.10.) [*Note:* To be strictly correct, we should make a distinction in discussing Cartesian products between a set like $A \times B \times C$ and the similar set $(A \times B) \times C$. The former is a set of triples of the form (a, b, c); the latter is a set of pairs $((a, b), c)$ whose first member is also a pair. In this exercise, we are ignoring that distinction, as mathematicians generally do.]

b) For any set S, define S^n to be the Cartesian product $S \times S \times \cdots \times S$ with n factors. Use part a) to explain the formula $|S^n| = |S|^n$.

13. The following is an expansion of the alternative approach to Theorem 5.9 introduced in the comment following the bare-bones version of the proof:

Suppose A is a set with n elements, say $A = \{x_1, x_2, x_3, \ldots, x_n\}$. Let X be the set $\{0, 1\}$, so the cardinality of X^n is 2^n [see part b) of Exercise 12]. Define a function $f: \mathcal{P}(A) \to X^n$ as follows: for $S \in \mathcal{P}(A)$, let $f(S)$ be the n-tuple whose ith entry is 0 if $x_i \notin S$ and 1 if $x_i \in S$. If we can show that f is one-to-one and onto, we will have proved that $|\mathcal{P}(A)| = 2^n$.

Prove that the function f given here is one-to-one and onto.

14. In each of the following, assume that the statement is true for $n = 2$, and prove that it is true for all $n \geq 2$.

a) If $x_1, x_2, \ldots, x_n \in \mathbf{R}$, then

$$|x_1 + x_2 + \cdots + x_n| \leq |x_1| + |x_2| + \cdots + |x_n|.$$

(The case $n = 2$ is known as the *Triangle Inequality,* and is discussed in Section 9.2.)

b) If p is a prime, and $x_1, x_2, \ldots, x_n \in \mathbf{N}$, and $p \mid x_1 \cdot x_2 \cdot \cdots \cdot x_n$, then $p \mid x_i$ for some i. (The case $n = 2$ is known as *Euclid's Lemma* and is proved in Section 5.3.)

15. **a)** Give a recursive definition for expressions of the type a^n, where $a \in \mathbf{R}^+$ and $n \in \mathbf{N}$, in which you start defining a^1 to be equal to a, and then define a^{t+1} in terms of a^t.

b) Use your definition to prove

$$(\forall\, a \in \mathbf{R}^+)(\forall\, m, n \in \mathbf{N})(a^m \cdot a^n = a^{m+n}).$$

(*Hint:* Let a be any element of $\mathbf{R}^+$, and let m be any natural number. Then prove the formula by induction on n. Use your recursive definition for the case $n = 1$.)

16. Find the flaw in the proof of Nontheorem 5.12 ("All horses have the same color").

17. **a)** Prove the following by induction:

THEOREM 5.13. For $n \in \mathbf{N}$, and sets A and B, if $|A| = |B| = n$, then the number of bijective functions from A to B is $n!$.

b) Explain why it is easier to prove Theorem 5.13 as stated, rather than prove directly that if $|A| = n$, then the number of functions from A to A is $n!$.

18. **PROOF EVALUATION:** (see instruction for Exercise 6 in Section 2.2)

a) **"THEOREM":** For any $n \in \mathbf{N}$,

$$1 + 2 + 4 + \cdots + 2^n = 2^{n+1} - 1. \qquad (*)$$

"Proof": When $n = 1$, the equation says "$1 + 2 = 2^{1+1} - 1$". Since both sides of the equation are equal to 3, this is a true statement.

Now suppose the equation $(*)$ holds for $n = t$, i.e., assume the following:

$$1 + 2 + 4 + \cdots + 2^t = 2^{t+1} - 1.$$

We now replace t by $t + 1$, giving

$$1 + 2 + 4 + \cdots + 2^{t+1} = 2^{(t+1)+1} - 1.$$

Since this is the equation $(*)$ for $n = t + 1$, we have proved the theorem by induction.

b) **"THEOREM":** For any $n \in \mathbf{N}$,

$$1 + 3 + 5 + \cdots + (2n - 1) = n^2 + 1. \qquad (**)$$

"Proof": Suppose the equation (**) is true when $n = t$; i.e., assume the following:

$$1 + 3 + 5 + \cdots + (2t - 1) = t^2 + 1\,.$$

We now add $2t + 1$ to both sides, giving

$$[1 + 3 + 5 + \cdots + (2t - 1)] + (2t + 1) = [t^2 + 1] + (2t + 1)\,. \qquad (***)$$

The left side of (***) is equal to

$$1 + 3 + 5 + \cdots + (2[t + 1] - 1)\,,$$

which is the left side of (**) for $n = t + 1$. The right side of (***) is equal to $(t^2 + 2t + 1) + 1$, which is equal to $(t + 1)^2 + 1$. Thus, this is the right side of (**) for $n = t + 1$.

Thus, we have shown that (**) holds when $n = t + 1$, proving the theorem by induction.

c) **"THEOREM":** For any $n \in \mathbf{N}_{[4,\infty)}$,

$$2^n \geq n^2\,.$$

"Proof": When $n = 4$, $2^n = 16$ and $n^2 = 16$, so the formula holds when $n = 4$.

To prove the induction step, assume that $2^t \geq t^2$. We need to show that $2^{t+1} \geq (t + 1)^2$.

Since $t \geq 4$, we have $1/(t + 1) \leq \frac{1}{5}$. Therefore,

$$\frac{t}{t + 1} = 1 - \frac{1}{t + 1} \geq \frac{4}{5}\,,$$

and so

$$\left(\frac{t}{t + 1}\right)^2 \geq \frac{16}{25} > \frac{1}{2}\,.$$

Therefore, we have the following sequence of inequalities (using the induction hypothesis for the second step):

$$2^{t+1} = 2 \times 2^t \geq 2 \times t^2 = 2 \times (t + 1)^2 \times \left(\frac{t}{t + 1}\right)^2$$

$$> 2 \times (t + 1)^2 \times \frac{1}{2} = (t + 1)^2.$$

which shows that $2^{t+1} \geq (t + 1)^2$, completing the proof.

d) **"THEOREM":** For any $n \in \mathbf{N}$, if $a_1, a_2, \ldots, a_n$, and c are elements of $\mathbf{N}$ such that $c \mid (a_1 \cdot a_2 \cdot \cdots \cdot a_n)$, then, for some value i, $c \mid a_i$.

"Proof": Let $p(n)$ be the assertion "if $a_1, a_2, \ldots, a_n$, and c are elements of $\mathbf{N}$ such that $c \mid (a_1 \cdot a_2 \cdot \cdots \cdot a_n)$, then $c \mid a_i$ for some i with $1 \leq i \leq n$."

(Notice that this statement is itself a conditional sentence.)

The assertion $p(1)$ simply says, "if a_1 and c are elements of $\mathbf{N}$ such that $c \mid a_1$, then $c \mid a_1$". This statement is true, since its conclusion is the same as its hypothesis.

Now assume that the statement $p(t)$ is true. We need to prove that $p(t + 1)$ is true. Since $p(t + 1)$ is a conditional statement, we assume its hypothesis, that $a_1, a_2, \ldots, a_{t+1}$, and c are elements of $\mathbf{N}$ such that $c \mid (a_1 \cdot a_2 \cdot \cdots \cdot a_{t+1})$. We need to show that $c \mid a_i$ for some i with $1 \leq i \leq t + 1$.

Let $b_1 = a_1, b_2 = a_2, \ldots, b_{t-1} = a_{t-1}$, and let $b_t = a_t \cdot a_{t+1}$. Then $b_1 \cdot b_2 \cdot \cdots \cdot b_t = a_1 \cdot a_2 \cdot \cdots \cdot a_{t+1}$, so $c \mid (b_1 \cdot b_2 \cdot \cdots \cdot b_t)$. Therefore, the hypothesis of $p(t)$ is satisfied. Since our induction hypothesis tells us that $p(t)$ is true, we can conclude that $c \mid b_i$ for some i with $1 \le i \le t$.

If $i < t$, then we have $c \mid a_i$ (since $b_i = a_i$), as desired. If $i = t$, we have $c \mid (a_t \cdot a_{t+1})$. In this case, let $d_1 = a_t$ and $d_2 = a_{t+1}$, so $c \mid (d_1 \cdot d_2)$. We now apply $p(2)$ [we know $t \ge 2$, so we've already proved $p(2)$ from $p(1)$], and conclude that $c \mid d_1$ or $c \mid d_2$, i.e., $c \mid a_t$ or $c \mid a_{t+1}$.

Thus, in any case, we have $c \mid a_i$ for some i with $1 \le i \le t + 1$, which completes the proof.

Section 5.2 Alternative Axioms and Two Important Theorems about N

As we indicated in the introduction to this chapter, there are two other basic principles about **N** which are sometimes used as alternatives to the Principle of Induction as the basic axiom. As with the Principle of Induction, each of these principles can be a valuable tool in proving theorems. As we shall see, there are theorems which would be extremely difficult to prove directly from the Principle of Induction, but which are easy to prove using one or the other alternative. Thus you need to be familiar with all three principles.

We will prove each of these alternative principles as theorems, based on our axiom of the Principle of Induction. We will also show that, if we were to assume either of these alternative principles as our axiom, we could prove the Principle of Induction itself.

In the course of this section, we will prove two very important theorems about arithmetic in **N**, whose proofs specifically lend themselves to the use of these alternative principles.

The Second Principle of Induction

Our first alternative principle for **N** is called the *Second Principle of Induction*. It resembles the Principle of Induction, but uses a concept which is a variation on the notion of an inductive set. It is used when the proof of the open sentence $p(t + 1)$ does not use $p(t)$, but instead requires statements $p(x)$ for $x < t$.

For convenience, we will use the notation $\mathbf{N}_{[1,t]}$ to represent the "interval" of **N** consisting of natural numbers up to t; i.e., $\mathbf{N}_{[1,t]} = \{x \in \mathbf{N} : x \le t\}$. (Intuitively, $\mathbf{N}_{[1,t]} = \{1, 2, \ldots, t\}$; we use $\{x \in \mathbf{N} : x \le t\}$ rather than $\{1, 2, \ldots, t\}$ as our definition of $\mathbf{N}_{[1,t]}$ because part of the purpose of working with the Principle of Induction is to avoid possible ambiguity in the use of the "..." notation.)

Notice that, for $t = 0$, this formal definition gives the empty set, so, according to our definition, we have $\mathbf{N}_{[1,0]} = \varnothing$. This makes sense, since we expect $\mathbf{N}_{[1,t]}$ to have t elements.

Using this notation, we make the following key definition:

DEFINITION 5.2: For $S \subseteq \mathbf{N}$, "S is *weakly inductive*" means:

$$(\forall\, t \in \mathbf{N})\,[\mathbf{N}_{[1,t]} \subseteq S \Rightarrow t + 1 \in S]\,.$$

Get Your Hands Dirty 1

Give two examples of subsets of **N** that are weakly inductive, and two examples of subsets that are not. □

COMMENTS

1. Any subset of **N** that is inductive must also be weakly inductive. (Verify—see Exercise 1.)

2. The converse of comment 1 is false. For example, $\{2\}$ is not an inductive set, but it is weakly inductive, since the hypothesis of the implication in Definition 5.2 is false for every $t \in \mathbf{N}$. In fact, any subset of **N** which does not contain 1 must be weakly inductive.

3. There is no standard mathematical terminology for the concept we are calling "weakly inductive". As we discussed in Example 3 of Section 4.2, the term "weaker" is often used in mathematics to compare two properties. If a statement of the form $(\forall\, x)\,(p(x) \Rightarrow q(x))$ is true, then we call the condition $q(x)$ "weaker" then $p(x)$ [and call $p(x)$ "stronger" than $q(x)$]. Intuitively, you can "get more out of" the stronger condition than you can from the weaker condition, since if the stronger condition holds, then the weaker one also holds, but not vice versa.

Thus, comment 1 states that the condition we are calling "weakly inductive" is a weaker condition on subsets of **N** than "inductive". That relationship between the two conditions is the motivation for our terminology.

4. Though we defined "inductive" to apply to arbitrary sets of numbers, this definition of "weakly inductive" only applies to subsets of **N**. This is because we need to be able to refer to an "initial segment" for the set, which does not make sense for arbitrary sets of numbers. It is possible, however, to generalize the concept of "weakly inductive" to allow subsets of any $\mathbf{N}_{[k,\infty)}$. (Recall $\mathbf{N}_{[k,\infty)} = \{n \in \mathbf{Z} : n \geq k\}$.) This is discussed briefly preceding Theorem 5.16.

Though not every weakly inductive set is inductive, those weakly inductive sets that contain 1 are inductive. In fact, we can prove more, using the principle of induction:

THEOREM 5.15 (Second Principle of Induction). Any subset of **N** which contains the number 1 and which is weakly inductive must be equal to **N**.

We will prove this shortly, but first we make a few observations.

COMMENTS

1. Notice how closely this theorem resembles the Principle of Induction. Theorem 5.15 tells us that, with regard to subsets of **N** that contain 1, the concepts of "inductive" and "weakly inductive" are equivalent. The only subset of **N** containing 1 with either of these properties is **N** itself.

2. Because every inductive set is weakly inductive, the Principle of Induction can be proved as an immediate consequence of the Second Principle of Induction (see Exercise 2). Therefore the Second Principle of Induction is sometimes called the *Principle of Strong Induction,* and a proof using this principle is sometimes called a proof "by strong induction". This is a slight variation on the use of the word "strong" as described in comment 3 following the definition of "weakly inductive". Here we are using it with regard to a theorem or statement rather than to a condition. One theorem or statement is called "stronger" than a second if the two have the same conclusion, but the first uses a weaker hypothesis, i.e., it assumes less. It is thus stronger because it accomplishes just as much, but using less. In the case of induction, both the Principle of Induction and the Principle of Strong Induction have the same conclusion, "$S = \mathbf{N}$" (assuming $S \subseteq \mathbf{N}$), and both have the statement "$1 \in S$" as part of the hypothesis. But the Principle of Induction also includes the statement "S is inductive" in its hypothesis, while the Principle of Strong Induction instead has only the weaker assumption "S is weakly inductive". (A "stronger" theorem can also have the same hypothesis as the other, but have a stronger conclusion.)

3. When we want to emphasize the distinction between the principles, we will refer to our original Principle of Induction as the *First Principle of Induction.*

4. You can picture the logic behind the Second Principle of Induction as follows: Suppose S is a subset of **N** which contains 1 and is weakly inductive. Let $p(t)$ represent the open sentence "$\mathbf{N}_{[1,t]} \subseteq S \Rightarrow t + 1 \in S$", which is true for every $t \in \mathbf{N}$, since S is weakly inductive. We reason as follows:

   $1 \in S$, so $\{1\} \subseteq S$.
   Now, since $\{1\} \subseteq S$, and $p(1)$ is true, we have $2 \in S$, so $\{1, 2\} \subseteq S$.
   Now, since $\{1, 2\} \subseteq S$, and $p(2)$ is true, we have $3 \in S$, so $\{1, 2, 3\} \subseteq S$.
   Now, since $\{1, 2, 3\} \subseteq S$, and $p(3)$ is true, we have $4 \in S$, so $\{1, 2, 3, 4\} \subseteq S$.
   Etc.

   The formal proof of Theorem 5.15 involves clarifying the "etc." here, which is done using the First Principle of Induction. We will assume that S is a subset of **N** which contains 1 and is weakly inductive, and prove by induction that, for every $n \in \mathbf{N}$, $\mathbf{N}_{[1,n]}$ is a subset of S.

5. Application of the Second Principle of Induction follows the same sort of two-step outline as proofs by the First Principle of Induction. We first prove that $1 \in S$. We

then assume that $\mathbf{N}_{[1,t]} \subseteq S$, and use that assumption to prove that $t + 1 \in S$. We will still refer to this second step as the *induction step,* and in the context of such a proof, we will refer to the statement "$\mathbf{N}_{[1,t]} \subseteq S$" as the *weak-induction hypothesis*.

Proof of Theorem 5.15: In order to prove the Second Principle of Induction, suppose that S is a subset of $\mathbf{N}$ which contains 1 and is weakly inductive. We need to show that $S = \mathbf{N}$.

Let $p(n)$ be the open sentence "$\mathbf{N}_{[1,n]} \subseteq S$". We will prove, by induction, that $p(n)$ is true for all $n \in \mathbf{N}$. Since $n \in \mathbf{N}_{[1,n]}$, this will show that $n \in S$ for all $n \in \mathbf{N}$, which will complete the proof.

First, $p(1)$ (the statement "$\mathbf{N}_{[1,1]} \subseteq S$") is true, since $\mathbf{N}_{[1,1]} = \{1\}$, and $1 \in S$.

What remains is to prove the induction step, so suppose that $p(t)$ is true, i.e., $\mathbf{N}_{[1,t]} \subseteq S$. Since S is weakly inductive, we therefore have that $t + 1 \in S$. But $\mathbf{N}_{[1,t+1]} = \mathbf{N}_{[1,t]} \cup \{t + 1\}$, so it follows that $\mathbf{N}_{[1,t+1]} \subseteq S$; i.e., $p(t + 1)$ is true.

This completes the induction step, so the proof of the theorem is complete. ∎

Generalized Second Principle of Induction

Like the First Principle of Induction, the Second Principle of Induction can be generalized to subsets of any $\mathbf{N}_{[k,\infty)}$. Unfortunately, to do so requires generalizing the concept of "weakly inductive", which is somewhat complicated. Intuitively, a subset S of $\mathbf{N}_{[k,\infty)}$ is called *weakly inductive with respect to* k if, for any $t \geq k$, knowing that all the integers from k up to t are in S guarantees that $t + 1$ is in S. (The "ordinary" meaning of weakly inductive is just the case $k = 1$.) The Generalized Second Principle of Induction, which we state without proof, then says:

THEOREM 5.16 (Generalized Second Principle of Induction). For any $k \in \mathbf{Z}$, any subset of $\mathbf{N}_{[k,\infty)}$ which contains k and which is weakly inductive with respect to k must be equal to $\mathbf{N}_{[k,\infty)}$.

COMMENTS

1. The proof is essentially identical to that of the ordinary case $k = 1$, but uses the Generalized First Principle of Induction instead of the ordinary First Principle of Induction.

2. Mathematicians don't always bother to distinguish among the different principles of induction—second vs. first, generalized vs. ordinary. We will make these distinctions when they have pedagogical value, but otherwise we will use the phrases "Principle of Induction" or "by induction" to refer to any one of them.

Existence of Prime Factorization: Using the Second Principle of Induction

As we stated earlier, the Second Principle of Induction is a useful tool for proving theorems of the form "$(\forall\ n \in \mathbf{N})\,(p(n))$", in situations where the truth of $p(t + 1)$ does not depend directly on the truth of $p(t)$, but instead involves statements $p(x)$ for $x < t$. The following is an important theorem about **N** for which the Second Principle of Induction provides the most appropriate method of proof. The proof will follow a brief discussion.

THEOREM 5.17 (Existence of Prime Factorization). Every positive integer either is equal to 1, or is a prime, or can be expressed as a product of primes.

This theorem is likely to be familiar to you. We can give an intuitive explanation as follows:

> If n is 1 or a prime, we're done. If not, then it can be factored as a product xy, with x and y integers greater than 1. If the factors x and y are primes, then we're done. If not, they can be factored as products. Keep on going until the factors can't be factored any more.

The phrase "keep on going" is a clue that some sort of induction argument is appropriate for a complete proof. Let $p(n)$ represent the open sentence "n is either 1 or a prime or can be expressed as a product of primes". The truth of $p(t + 1)$ is not directly connected to the truth of $p(t)$. For example, the factorization of 56 has no immediate relationship to the factorization of 55. However, it is related to the factorization of 4 and 14 (since $4 \times 14 = 56$).

Thus, we will have trouble proving the statement "$(\forall\, t)\,(p(t) \Rightarrow p(t + 1))$", i.e., that the truth set of $p(n)$ is inductive, but we will be able to prove that the truth set is weakly inductive.

Here is the proof:

Proof of Theorem 5.17: Let $p(n)$ be the open sentence "n is either 1 or a prime or can be expressed as a product of primes" (for the universe **N**), and let P be its truth set. We will show:

i) $1 \in P$;
ii) P is weakly inductive.

By Theorem 5.15, this will prove that $P = \mathbf{N}$, which is the assertion of this theorem.

The assertion "$1 \in P$" follows directly from the definition of the open sentence $p(n)$.

Now suppose that $\mathbf{N}_{[1,t]}$ is a subset of P. We need to prove $t + 1 \in P$, i.e., that $p(t + 1)$ is true.

Either $t + 1$ is a prime, or it isn't. If it is, then $p(t + 1)$ is true. So suppose $t + 1$ is not a prime.

Then there are integers x and y, with $1 < x < t + 1$ and $1 < y < t + 1$, such that $t + 1 = xy$. By our weak-induction hypothesis, that $\mathbf{N}_{[1,t]} \subseteq P$, we know that $p(x)$ and $p(y)$ are true. Therefore, each of x and y either is a prime or can be expressed as a product of primes. Replacing each of x and y (if necessary) by a product of primes in the equation $t + 1 = xy$ gives an expression for $t + 1$ as a product of primes. This completes the proof. ■

COMMENTS

1. We emphasize that the reason why the Second Principle of Induction is so useful in proving this theorem is that the truth of $p(t+1)$ depends on $p(x)$ and $p(y)$, rather than on $p(t)$.
2. We will prove in Section 5.3 that the expression of a composite positive integer as a product of primes is unique, except for the order of the factors.
3. Since the case $n = 1$ is almost irrelevant to the reasoning, this theorem would be a natural candidate for the use of the Generalized Second Principle of Induction, using $k = 2$.

The Well-Ordering Principle

The second alternative to the Principle of Induction is called the *Well-Ordering Principle*. It can be stated very simply:

THEOREM 5.18 (Well-Ordering Principle). Every nonempty subset of **N** has a least element.

Before proving this theorem, we want to clarify its meaning. The statement of the Well-Ordering Principle involves three hidden variables: a universally quantified variable for the subset of **N**, an existentially quantified variable for the least element, and then a universally quantified variable representing a general element of the subset, which is used to express the idea of "least". It may help to make these quantifiers and variables explicit with the following alternative statement of the Well-Ordering Principle:

THEOREM 5.18′ (Well-Ordering Principle).

$$(\forall\, S \subseteq \mathbf{N})\,[S \neq \varnothing \Rightarrow (\exists\, x \in S)\,(\forall\, y \in S)\,(x \leq y)]\,.$$

In this statement, the open sentence "$(\forall\, y \in S)\,(x \leq y)$", together with the requirement that x be an element of S, says that x is the least element of S.

Notice that the only object in Theorem 5.18′ which is not quantified is the set **N**. The Well-Ordering Principle is a statement about the set of natural numbers.

Here are some examples of subsets of **N** and their least elements:

Subset S	*Least element x*
$\{2, 4, 6, \ldots\}$	2
{odd primes}	3
$\{5, 6, 7, 8, \ldots\}$	5
$\{t \in \mathbf{N} : 5 \mid (3t+1)\}$	3 (Verify!)

The Well-Ordering Principle says that the only subset of **N** which does not have a least element is the empty set.

To continue to clarify the meaning of the Well-Ordering Principle, we emphasize that it does not apply to arbitrary sets, but specifically to subsets of **N**. (We will generalize the Well-Ordering Principle to subsets of any $\mathbf{N}_{[k,\infty)}$ in Theorem 5.19 below.)

For example, the interval $(3, 7)$ is a subset of **R** which does not have a least element. [It has lower bounds, and even a greatest lower bound, but none of those is a least element of $(3, 7)$, since they don't belong to the interval.]

The set **Z** does not have a least element, or even a lower bound.

Also, the Well-Ordering Principle is not simply saying that **N** itself has a least element (although the assertion that **N** has a least element is a special case of the Well-Ordering Principle). It is saying that *every subset* of **N**, except for the empty set, has a least element.

You may want to use the following idea for thinking about the Well-Ordering Principle:

Metaphor: Well-Ordering as "Least-Element Finder"

In Section 4.1, we discussed viewing statements of the form "$(\forall\ x)\,(\exists\ y)$" as "element finders". The Well-Ordering Principle has this form, and the elements it finds are least elements of certain sets.

Therefore, any time you have a nonempty subset of **N**, you can use the Well-Ordering Principle to locate its least element for you, and the Well-Ordering Principle will give you a certificate guaranteeing that this object is, indeed, the least element of the set.

Here is an informal argument showing that the Well-Ordering Principle is true:

> Suppose S is a nonempty subset of **N**. Since S is nonempty, there is some element in S, say n. Therefore, even if n itself is not the least element of S, that least element must be at most n, i.e., belong to $\{1, 2, \ldots, n\}$. So we look at the set $T = S \cap \{1, 2, \ldots, n\}$. T is a *finite,* nonempty set, and so it must have a least element. The least element of T is the least element of S.

COMMENT The part of this explanation that requires the Principle of Induction is the step that says "T is a *finite,* nonempty set, and so it must have a least element." Though this statement is true, it needs justification. Why is T finite, and why must a finite set have a least element? Such questions are tricky, because we don't have a formal definition of the word "finite". The concept of "finiteness" is actually a complex one, which we will discuss in Section 8.1.

We now give a formal proof of the Well-Ordering Principle. Our proof uses the Second Principle of Induction.

Proof of Theorem 5.18: In order to prove the Well-Ordering Principle, we assume that S is a nonempty subset of **N**, and prove that S has a least element. We will prove this by contradiction. We will assume that S does not have a least element, and get a contradiction by showing that S must be empty.

We will prove that S is empty by showing that $\mathbf{N} - S = \mathbf{N}$. For simplicity, let $T = \mathbf{N} - S$. We will prove two things:

a) $1 \in T$.
b) T is weakly inductive.

By the Second Principle of Induction, this will guarantee that $T = \mathbf{N}$, completing the proof of the theorem.

Since S has no least element, we know that $1 \notin S$. (If 1 were in S, it would be its least element.) Therefore $1 \in T$, proving a).

To prove that T is weakly inductive, suppose that $\mathbf{N}_{[1,t]} \subseteq T$. We need to show that $t + 1 \in T$. If not, then $t + 1 \in S$. But that would make $t + 1$ the least element of S, since all the natural numbers less than $t + 1$ are in T, and hence not in S. Therefore $t + 1$ can't be in S, so it is in T, completing the proof of b), and proving the theorem. ■

COMMENTS

1. This proof does not directly use the First Principle of Induction, but uses the Second Principle of Induction instead. We will make reference to this fact later when we discuss the interconnections among these principles.

2. The reasoning in this proof can be adapted to show that the Well-Ordering Principle and the Second Principle of Induction are like contrapositives of each other, as follows:

   If X is any subset of $\mathbf{N}$, it can be shown that the open sentence $p(X)$: "$1 \in X$ and X is weakly inductive" is equivalent to the open sentence $q(X)$: "$\mathbf{N} - X$ has no least element". (See Exercise 9.)

   Using this equivalence, the Second Principle of Induction can be rephrased as: "if $\mathbf{N} - X$ has no least element, then $X = \mathbf{N}$". Letting $Y = \mathbf{N} - X$, this is the same as saying "if Y has no least element, then $Y = \varnothing$", which is the contrapositive of the Well-Ordering Principle.

   This reasoning leads to a proof of the Second Principle of Induction based on the Well-Ordering Principle, without using the First Principle of Induction (see Exercise 10).

Generalized Well-Ordering

The Well-Ordering Principle, like both Principles of Induction, can be generalized to apply to a subset of any of the sets $\mathbf{N}_{[k,\infty)}$, rather than just to subsets of $\mathbf{N}$. We have the following theorem, whose proof is left as Exercise 11:

THEOREM 5.19 (Generalized Well-Ordering Principle). For any $k \in \mathbf{Z}$, if $S \subseteq \mathbf{N}_{[k,\infty)}$ and $S \neq \varnothing$, then S has a least element.

The ordinary Well-Ordering Principle is the case $k = 1$ of Theorem 5.19, since $\mathbf{N}_{[1,\infty)}$ is just $\mathbf{N}$. We will use the Generalized Well-Ordering Principle shortly with $k = 0$; i.e., we will use the fact that every nonempty subset of **W** has a least element. As with the different induction principles, we will not always distinguish between the ordinary and the Generalized Well-Ordering Principles.

(The proof of Theorem 5.19 is essentially identical to the proof of the ordinary Well-Ordering Principle, except that you need to use the *Generalized* Second Principle of Induction instead of the ordinary one.)

The following corollary is not much more than a restatement of Theorem 5.19. Its proof is left as Exercise 3.

COROLLARY 5.20. If S is a nonempty subset of **Z** which has a lower bound, then S has a least element.

The Division Algorithm: Using the Well-Ordering Principle

We illustrate the use of the Well-Ordering Principle by using it to prove a very important theorem about the integers, known as the *Division Algorithm*. We will state the theorem, discuss the idea it embodies, give a proof, and then comment on how the proof uses the Well-Ordering Principle.

THEOREM 5.21 (Division Algorithm).

a) (Existence) For any $a, b \in \mathbf{N}$, there are integers q and r such that:
  i) $a = bq + r$ and
  ii) $0 \leq r < b$.

b) (Uniqueness) For a given $a, b \in \mathbf{N}$, the integers q and r described in part a) are unique.

For now we discuss only the existence part of Theorem 5.21. We will deal with uniqueness later.

What is this theorem saying? You can explore the statement by taking particular values for the universally quantified variables a and b, and finding out what integers would work for q and r.

Try $a = 53$ and $b = 8$.

You should have found that the only possibility is $q = 6$ and $r = 5$. Where did these numbers come from? What's going on? What ordinary idea is involved in this theorem? (*Hint:* The theorem is called the *Division Algorithm*.)

? ? ?

6 is the *quotient* when 53 is divided by 8, and 5 is the *remainder*. (That's why we chose to label them q and r.) The Division Algorithm essentially tells us we can always divide and get a remainder that is smaller than the divisor. It says this without talking directly about division, but instead using equation i) which "checks" the division.

If condition i) represents "checking" the division, what is the role of condition ii)?

? ? ?

In anticipation of the way the proof of Theorem 5.21 is organized, try to find other pairs of integers for q and r, besides $q = 6$ and $r = 5$, that satisfy condition i) of the theorem, but not condition ii) (still using $a = 53$ and $b = 8$). In other words, find other integer solutions to the equation $53 = 8q + r$.

? ? ?

Here are some possibilities:

$q = 0$, $r = 53$. (This is the "easiest" solution.)
$q = 4$, $r = 21$.
$q = 9$, $r = -19$. (We didn't say r had to be positive.)
$q = -3$, $r = 77$. (We didn't say q had to be positive.)

You can think of the solution $q = 4$, $r = 21$, for example, as saying that 53 divided by 8 gives a quotient of 4 with remainder 21:

$$53 \div 8 = 4 \text{ R. } 21$$

Though this is not the standard way to divide 53 by 8, it does check: $53 = 4 \times 8 + 21$. Each of these solutions can be viewed as an alternative way to divide 53 by 8.

What distinguishes the "right" way to divide from these alternative ways? It is condition ii): $0 \leq r < b$. To divide "properly", we need to have a remainder that is nonnegative but less than the divisor. You should convince yourself experimentally that a solution for q and r satisfying both conditions i) and ii) will always exist, regardless of the choices for a and b. Notice that the "right" value for r (5, in our example) is the smallest nonnegative integer possible. That fact will be the key to our use of the Well-Ordering Principle.

Our proof of the Division Algorithm goes roughly as follows: among all "possible" ways to divide a by b, i.e., all solutions to condition i), choose the one that gives the smallest nonnegative remainder. Then prove that this remainder satisfies condition ii).

Here's the proof:

Proof of Theorem 5.21a) (Division Algorithm—Existence): Suppose $a, b \in \mathbf{N}$, and define a set R (the set of "possible remainders") as follows:

$$R = \{x \in \mathbf{W} : (\exists\, q \in \mathbf{Z})\,(a = bq + x)\}\,.$$

Then $R \subseteq \mathbf{W}$. Also $R \neq \varnothing$. (We have $a \in R$, since $a > 0$ and $a = b \cdot 0 + a$.) Therefore, by the Generalized Well-Ordering Principle, R has a least element, which we will call r.

The definition of the set R tells us that there is an integer q such that

$$a = bq + r \qquad (*)$$

What remains is to show that $0 \leq r < b$. We know, from the definition of R, that $R \subseteq \mathbf{W}$, so $r \geq 0$. So we need to show that $r < b$. We will prove this by contradiction.

Suppose, on the contrary, that $r \geq b$. Then $r - b \geq 0$. But we can also rewrite (*) as follows:

$$a = b(q + 1) + (r - b) \qquad (**)$$

This equation, combined with the fact that $r - b \geq 0$, tells us that $r - b$ fits the requirements to be an element of R.

But r is the least element of R, and $r - b < r$ (since $b > 0$). This is a contradiction, so the proof is complete. ■

COMMENTS

1. Notice how we used the Well-Ordering Principle as a "least-element finder". We created a nonempty subset of **W**, and were guaranteed, by the Well-Ordering Principle, that this set had a least element, which we named r. This is typical of the way in which the Well-Ordering Principle is used in proofs. We then showed that r had an additional property, condition ii), by contradiction, as follows: we proved that, if r did not have this property, there would be an element of R which was less than r, thus contradicting the defining property of r.

   A statement of the type "x is the least element of S", resulting from the use of the Well-Ordering Principle, is handled like any statement that results from application of a previously proved theorem. In the context of the given proof, it is considered to be a true statement. We can accomplish a proof by contradiction by finding an element of S which is less than x. This is a commonly used technique. The fact that we don't explicitly *describe* the element x, but instead used the Well-Ordering Principle to *find* it for us, does not in any way lessen the reliability of its being the least element. The Well-Ordering Principle provides a "guarantee certificate" for the fact that x is the least element of S.

2. The idea of working with the set R is a very creative element in this proof. As is often the case with existence theorems, we need to explore to get an intuitive idea of what the desired object is. We also need to come up with a way to describe this object that will work in every situation which fits the hypothesis of the theorem. Often, when a theorem asks you to prove the existence of an object satisfying more than one condition, it helps to look at the set of those objects which satisfy one of those conditions, and then develop a way to locate among them an object which satisfies the remaining conditions. Here we began with condition i): we looked at the set R of nonnegative integers which could play the role of r in that equation. From within this set of nonnegative integers, our exploration told us that the object we wanted was that set's smallest member. We then used the Well-Ordering Principle to select that least element for us.

3. Don't overlook the step of showing that the set R is nonempty. Though this is often the easiest part of a proof using the Well-Ordering Principle, it is still an essential part, just as proving "$1 \in P$" is an essential part of proof by induction. The next comment discusses generalizations of the Division Algorithm in which this step is a little more complicated.

4. We mention two ways in which the Division Algorithm can be generalized.

First, we can allow a to be an arbitrary integer, not necessarily positive. The only changed needed in the proof is in the step of showing that R is nonempty. [See Exercise 4a) and b).]

Second, we can let b be any element of $\mathbf{Z} - \{0\}$. (We can't let $b = 0$. Why not?) To make this generalization, we need to adjust condition ii) to say $0 \leq r < |b|$, replacing b by $|b|$. Here, in addition to adjusting the proof that R is nonempty, we also need to redo the end of the proof, working with $r - |b|$ instead of $r - b$. [See Exercise 4c) and d).]

5. The Division Algorithm is a very valuable tool in proving statements about divisibility. If, in the course of some proof, we have two integers a and b, with $b \neq 0$, and we wish to prove that $b \mid a$, we can invoke the Division Algorithm to tell as that there are integers q and r with $a = bq + r$ and $0 \leq r < b$. If we can show that $r = 0$, we can conclude that $b \mid a$.

This technique will be illustrated in the discussion of number theory in Section 5.3, particularly in Lemma 5.27.

Uniqueness in the Division Algorithm

Theorem 5.21b) states that the integers q and r are unique. We will set up the proof of this aspect of the theorem, and leave the details as an exercise.

Proof of Theorem 5.21b) (Division Algorithm—Uniqueness): Suppose $a, b \in \mathbf{N}$, and that there are two pairs of integers that "work". In other words, assume that q_1, r_1, q_2, and r_2 are integers such that we have both of the following pairs of conditions:

$$a = bq_1 + r_1 \quad \text{and} \quad 0 \leq r_1 < b$$

and

$$a = bq_2 + r_2 \quad \text{and} \quad 0 \leq r_2 < b\,.$$

We need to prove that $q_1 = q_2$, and $r_1 = r_2$.

We leave the details of the proof as Exercise 5. ■

Relationships among the Principles

We have stated three general principles about **N**:

i) the First Principle of Induction,
ii) the Second Principle of Induction,
iii) the Well-Ordering Principle.

As we indicated earlier, in some approaches, the second or third of these is used as the basic axiom in place of the First Principle of Induction. We have chosen the First Principle of Induction as our basic axiom in this text because we think it most directly captures the essential nature of the set of natural numbers.

We have already shown that the second of these principles follows from the first, and that the third follows from the second. (Recall that our proof of the Well-Ordering Principle was based on the Second Principle of Induction.) We restate these results here in a way that makes these connections explicit.

THEOREM 5.22. If the First Principle of Induction is true, then the Second Principle of Induction is true.

THEOREM 5.23. If the Second Principle of Induction is true, then the Well-Ordering Principle is true.

We can "complete the circle" by proving the following theorem:

THEOREM 5.24. If the Well-Ordering Principle is true, then the First Principle of Induction is true.

Proof: Assume that the Well-Ordering Principle is true. To prove the Principle of Induction, assume that S is a subset of $\mathbf{N}$ which contains 1 and is inductive. We need to prove that $S = \mathbf{N}$, which we do by contradiction.

Suppose $S \neq \mathbf{N}$. Then $\mathbf{N} - S$ is nonempty, so it has a least element, say w. This element w can't be 1, since $1 \in S$. Thus, $w > 1$.

Therefore, if we let $x = w - 1$, we have $x \in \mathbf{N}$. Since w is the least element of $\mathbf{N} - S$, we can't have $x \in \mathbf{N} - S$, so $x \in S$. But S is inductive, and so $x + 1 \in S$. But $x + 1 = w$, so $w \in S$, contradicting the fact that $w \in \mathbf{N} - S$.

This completes the proof. ∎

COMMENTS

1. By "completing the circle", we have shown that each of these three principles can be proved on the basis of either of the other two (see discussion following Example 8 in Section 4.2). In other words, the three principles are equivalent. As a consequence of this equivalence, it would make sense to use any one of the three principles as the axiom. Any theorem that can be proved using any one of these principles as the axiom could also be proved using either of the others.

2. The three connections among the principles that are not explicitly stated in these theorems can also be proved directly without using these theorems. Exercise 2 asks you to prove that the First Principle of Induction follows from the Second Principle of Induction. A proof that the Second Principle of Induction follows from the Well-Ordering Principle is sketched in Exercises 9 and 10. An outline of a proof that the Well-Ordering Principle follows directly from the First Principle of Induction is given in Exercise 14.

EXERCISES

1. Prove that any inductive subset of **N** is weakly inductive.
2. Use Exercise 1 to prove the First Principle of Induction using the Second Principle of Induction.
3. Prove the following:

   COROLLARY 5.20. If S is a nonempty subset of **Z** which has a lower bound, then S has a least element.

4. For $a, b \in \mathbf{Z}$, define the set R as in Theorem 5.21 (the Division Algorithm):
$$R = \{x \in \mathbf{W} : (\exists\, q \in \mathbf{Z})\,(a = bq + x)\}.$$
   Theorem 5.21 looked at the situation where $a, b \in \mathbf{N}$.
   a) Prove that, if a is an arbitrary integer and $b \in \mathbf{N}$, then R is nonempty. (*Hint:* Don't worry about finding the "best" x. Just explore and find some way to choose q, perhaps in terms of a and b, so that $a - bq \in \mathbf{W}$.
   b) Use part a) to give a proof of the Division Algorithm for the case where a is an arbitrary integer and $b \in \mathbf{N}$.
   c) Prove that, if a is an arbitrary integer and b is an arbitrary nonzero integer, then R is nonempty.
   d) Use part c) to give a proof of the Division Algorithm for the case where a is an arbitrary integer and b is an arbitrary nonzero integer. [You need to modify condition ii) of the theorem to say: $0 \leq r < |b|$.]
5. Complete the proof of the uniqueness part of Theorem 5.21 (Division Algorithm). [*Hint:* Write $bq_1 + r_1 = bq_2 + r_2$. Rearrange to get $b(q_1 - q_2) = r_2 - r_1$. Also, combine the inequalities for r_1 and r_2 to show $|r_2 - r_1| < b$.]
6. Prove that the number of subsets of $\{1, 2, \ldots, n\}$ which do not contain any pair of consecutive integers is equal to F_{n+2} (where F_i is the ith Fibonacci number). (*Hint:* Use the Second Principle of Induction. Split the set of subsets of $\{1, 2, \ldots, t + 1\}$ which do not contain any pair of consecutive integers into those that contain $t + 1$ and those that do not.)
7. (Exploration 1 in Section 1.1) In the game of SubDivvy, prove that, for $n \geq 2$:
   a) if n is even, then n is a winning position, and
   b) if n is odd, then n is a losing position.
   [*Hint:* Treat parts a) and b) as one theorem, and prove them simultaneously using the Second Principle of Induction. Start by proving both the case $n = 2$ and the case $n = 3$.]
8. In the game of Linear Nim (see Exercise 9 of Section 1.1), with N initial marks and a maximum number of t marks to be crossed out per turn, determine which player has a winning strategy, in terms of N and t, and prove your answer using the second Principle of Induction. (You'll need to do some exploring.)
9. Suppose $X \subseteq \mathbf{N}$. Prove that conditions a) and b) are equivalent, without using any of the three basic principles about **N**:
   a) $1 \in X$ and X is weakly inductive.
   b) $\mathbf{N} - X$ has no least element.
10. Use the result in Exercise 9 to prove, without reference to the First Principle of Induction, that if the Well-Ordering Principle is true, then the Second Principle of Induction is true.

11. Prove the Generalized Well-Ordering Principle, using the Generalized Second Principle of Induction.

12. **a)** Consider the following statement:

If n is an integer greater than 1, then n is divisible by some prime.

Without using Theorem 5.17, prove this statement in two ways:

i) using the Generalized Well-Ordering Principle (applied to the set $\{x \in \mathbf{N}_{[2,\infty)} : x \mid n\}$);

ii) using the Generalized Second Principle of Induction.

b) Which method of proof do you like better? Why?

13. Suppose that S is a nonempty subset of $\mathbf{Z}$ which has an upper bound. Prove that S has a greatest element. (You may use Exercise 3.)

14. This exercise outlines a proof of the Well-Ordering Principle directly from the First Principle of Induction. (Recall $\mathbf{N}_{[1,t]} = \{x \in \mathbf{N} : x \leq t\}$.)

a) Prove the following (without using the Well-Ordering Principle):

If $T \subseteq \mathbf{N}_{[1,t]}$ and $T \neq \varnothing$, then T has a least element.

(*Hint:* Prove this by induction on t.)

b) Use part a) to prove the Well-Ordering Principle. (*Hint:* use the idea of the informal argument for the Well-Ordering Principle stated in the text.)

15. According to a) of Exercise 14, the statement

"if $T \subseteq \{x \in S : x \leq n\}$ and $T \neq \varnothing$, then T has a least element"

is true if $S = \mathbf{N}$. Show, by giving specific sets T, that this statement is false if S is any of the sets $\mathbf{Z}$, $\mathbf{R}$, or $\mathbf{R}^+$.

16. Consider the set P of polynomials with rational coefficients. We will think of these as functions from $\mathbf{R}$ to $\mathbf{R}$, i.e., as functions of the form

$$f(x) = a_n x^n + a_{n-1} x^{n-1} + \cdots + a_0,$$

where each $a_i \in \mathbf{Q}$. If f is not the *zero polynomial* (i.e., the polynomial all of whose coefficients are zero), we define the *degree* of f [written $\deg(f)$] as the largest subscript i such that $a_i \neq 0$. (This is the standard definition used in high-school algebra. For example, the degree of "$5x^8 + 2x^4 - 7x^2 + 9$" is 8.)

We make two more definitions:

DEFINITION: A polynomial f is *irreducible* iff it cannot be expressed as the product of two nonconstant polynomials.

[For example, $x^2 + 1$ is irreducible, since it cannot be factored nontrivially. On the other hand, $x^2 - 1$ is not irreducible, since it can be factored as $(x + 1)(x - 1)$.]

DEFINITION: A real number r is *algebraic* iff there is a nonzero polynomial $f \in P$ of which r is a root [i.e., such that $f(r) = 0$].

[For example, $\sqrt{2}$ is algebraic, since it is a root of the polynomial $x^2 - 2$; i.e., $(\sqrt{2})^2 - 2 = 0$. There are real numbers which are not algebraic, such as π, but proof that specific numbers are not algebraic is very difficult.]

Prove the following, using these definitions and the Well-Ordering Principle:

If $r \in \mathbf{R}$, and r is algebraic, then there is an irreducible polynomial f such that $f(r) = 0$.

5.3 Basic Number Theory

The area of mathematics known as *number theory* generally deals with problems that specifically concern the integers. It is the branch of mathematics most accessible to the general public, and many people who were not trained as mathematicians have made valuable contributions over the centuries.

In this section, we will deal with two main questions. The first is the problem of Linear Diophantine Equations, first raised in Section 2.3. The second problem is the uniqueness of prime factorization. The proof of uniqueness depends on a result known as *Euclid's Lemma,* which in turn is proved using our solution of the Linear-Diophantine-Equation problem.

Linear Diophantine Equations

Recall the question raised in Exploration 3 in Section 2.3:

For what integers a, b, and c will the equation

$$ax + by = c$$

have integral solutions for x and y?

We gave a partial answer to this question in Corollary 2.14′: If the equation has a solution, then c must be divisible by $\text{GCD}(a, b)$. As we indicated in the comments in connection with that theorem, the converse is also true, and we turn now to examine that converse. [We are assuming, as we did in Corollary 2.14′, that a and b are not both zero; otherwise $\text{GCD}(a, b)$ is undefined.]

As with many problems in mathematics, we can get greater clarity in this problem by looking at it from a new perspective. Here is the same question asked in a slightly different way:

For particular integers a and b, what integers can be expressed in the form $ax + by$, where x and y are integers?

We can perhaps bring this question into clearer focus by forming a set out of the integers described here. For each pair of integers a and b, define a set $L_{a,b}$ as follows:

$$L_{a,b} = \{ax + by : x, y \in \mathbf{Z}\}.$$

The elements of $L_{a,b}$ are called *linear combinations of a and b.*

Let $d = \text{GCD}(a, b)$. Corollary 2.14′ says that $L_{a,b} \subseteq d\mathbf{Z}$, i.e., $L_{a,b}$ is a subset of the set of multiples of $\text{GCD}(a, b)$. The converse of Corollary 2.14′ says that the subset relationship holds in the reverse order, i.e., that $d\mathbf{Z} \subseteq L_{a,b}$. We will prove this converse. In other words, combining the two results, we will prove the following "solution" to the Linear-Diophantine-Equation problem:

THEOREM 5.25. Suppose $a, b \in \mathbf{Z}$, not both zero. Let $d = \text{GCD}(a, b)$, and suppose $c \in \mathbf{Z}$. Then the equation $ax + by = c$ has integral solutions for x and y if and only if $d \mid c$.

More simply stated in terms of $L_{a,b}$, we have:

THEOREM 5.25′. Suppose $a, b \in \mathbf{Z}$, not both zero, and let $d = \text{GCD}(a, b)$. Then $L_{a,b} = d\mathbf{Z}$.

We will refer to Theorem 5.25 (or Theorem 5.25′) as the *GCD–Linear-Combinations Theorem*.

We will prove this theorem, but first, we want to suggest some ways for you to think about it. As we pointed out above, Corollary 2.14′ already tells us that $L_{a,b} \subseteq d\mathbf{Z}$. Our proof that $d\mathbf{Z} \subseteq L_{a,b}$ is based on showing two things:

i) if $r \in L_{a,b}$, then $r\mathbf{Z} \subseteq L_{a,b}$;
ii) $d \in L_{a,b}$.

(Verify that the statement "$d\mathbf{Z} \subseteq L_{a,b}$" follows immediately from these two facts.)

Statement i) is easy to prove, and a little exploration should guide you to a proof: Suppose that $r \in L_{a,b}$, which means that there are integers x_0 and y_0 which solve the equation $ax + by = r$; in other words, we have $ax_0 + by_0 = r$. How does this show that $r\mathbf{Z} \subseteq L_{a,b}$? What would be a solution (for x and y) to the equation $ax + by = 2r$? to $ax + by = 5r$? to $ax + by = 23r$? to $ax + by = nr$?

? ? ?

The answers to these questions virtually constitute a proof of property i). We can formalize this as follows:

LEMMA 5.26. If $a, b \in \mathbf{Z}$ and $r \in L_{a,b}$, then $r\mathbf{Z} \subseteq L_{a,b}$.

Proof: Suppose $r \in L_{a,b}$, so there are integers, say m and n, such that

$$am + bn = r. \qquad (*)$$

Now suppose $u \in r\mathbf{Z}$, say $u = rt$. Multiplying $(*)$ by t, we get $(am)t + (bn)t = rt$. Thus, letting $x = mt$ and $y = nt$, we have $ax + by = u$, which proves that $u \in L_{a,b}$, as required. ■

We now turn to property ii), i.e., to showing that $d \in L_{a,b}$. The key to the proof of this is the following observation:

> If the statement we are trying to prove is true [i.e., if property ii) holds], then we have $L_{a,b} = d\mathbf{Z}$. If that is so, then d is the smallest positive element of $L_{a,b}$.

That conclusion suggests the following argument as a method of proof for ii): use the Well-Ordering Principle to select the smallest positive element of $L_{a,b}$, call it w, and then prove that d is equal to this element w.

We already know from Corollary 2.14′ that every element of $L_{a,b}$ (and in particular, w) is a multiple of d. Since $w > 0$, this tells us that $w \geq d$.

On the other hand, if we can show that w is a common divisor of a and b, then we will have $w \leq d$, since d is the greatest common divisor. Those two facts combine to guarantee that $w = d$.

Think about the plan just described. Before reading ahead, be sure you understand what we've just said.

To carry out this plan, we prove that the element w described above is a common divisor of a and b:

LEMMA 5.27. If $a, b \in \mathbf{Z}$, and w is the smallest positive element of $L_{a,b}$, then $w \mid a$ and $w \mid b$.

Comment: This proof uses Theorem 5.21—the Division Algorithm. You may wish to reread comment 5 following that result before reading this proof.

Proof: Assume that a, b, and w satisfy the hypothesis. By the Division Algorithm, there are integers q and r such that $a = wq + r$ and $0 \leq r < w$. If we can show that $r = 0$, we will have proved that $w \mid a$.

Comment: The way we will show $r = 0$ is to show that r can't be positive. We'll do that by showing that $r \in L_{a,b}$. Since w is the smallest positive element of $L_{a,b}$, and $r < w$, that will tell us that r is not positive.

We know that $w \in L_{a,b}$, so there are integers x and y with $w = ax + by$. Therefore

$$r = a - wq = a - (ax + by)q = a(1 - xq) + b(-yq)\,.$$

Thus $r \in L_{a,b}$. But w is the smallest positive element of $L_{a,b}$, and $r < w$, so r cannot be positive. (Otherwise it would be a positive element of $L_{a,b}$ smaller than w, which is impossible.)

Thus $r = 0$, so $w \mid a$. Similarly, $w \mid b$. ∎

That almost completes the argument. One missing detail is verifying that the Well-Ordering Principle is applicable, i.e., that the set of positive elements of $L_{a,b}$ is nonempty. We will include that in the following proof of the GCD–Linear-Combinations Theorem, which synthesizes our discussion. We restate the theorem here for convenience:

THEOREM 5.25′ (GCD–Linear-Combinations Theorem). Suppose $a, b \in \mathbf{Z}$, not both zero, and let $d = \mathrm{GCD}(a, b)$. Then $L_{a,b} = d\mathbf{Z}$.

Proof: By definition, $L_{a,b} = \{ax + by : x, y \in \mathbf{Z}\}$. Using $x = \pm 1$ and $y = 0$ as well as $y = \pm 1$ and $x = 0$, we get that a, $-a$, b, and $-b$ are all in $L_{a,b}$. Since a and b are not both zero, at least one of these four elements is positive, so $L_{a,b}$ contains at least one positive integer.

Therefore, by the Well-Ordering Principle, $L_{a,b} \cap \mathbf{N}$ has a smallest element. Call this element w.

By Lemma 5.27, $w|a$ and $w|b$, so $w \leq d$ (since d is the *greatest* common divisor). On the other hand, by Corollary 2.14′, $w \in d\mathbf{Z}$. Since $w > 0$, this guarantees that $w \geq d$. Thus $w = d$, so $d \in L_{a,b}$. Therefore, by Lemma 5.26, $d\mathbf{Z} \subseteq L_{a,b}$.
But $L_{a,b} \subseteq d\mathbf{Z}$, by Corollary 2.14′, so the proof is complete. ■

COMMENTS

1. This theorem, or more specifically Lemma 5.27, illustrates a common type of application of the Division Algorithm, specifically its use in combination with the Well-Ordering Principle. We showed that $r = 0$ by proving that, if r were positive, we would have a contradiction of the definition of w as the smallest positive element with a particular property, namely, belonging to $L_{a,b}$. (Carrying out that step required that we first prove that r itself actually belongs to $L_{a,b}$.)

2. If a and b are both 0, then $L_{a,b} = \{0\}$, but GCD(a, b) is undefined. Thus Theorem 5.25 does not apply in this case.

3. In *planning* the part of the proof that showed $d \in L_{a,b}$, we did something which you should never do as part of the proof itself: we assumed that the overall theorem ($L_{a,b} = d\mathbf{Z}$) was true. (See discussion immediately following Lemma 5.26.) We saw that, if the theorem were true, then d would be the smallest positive element of $L_{a,b}$, and so, in particular, we would have $d \in L_{a,b}$. That observation led to the idea of trying to *prove* that d is equal to the smallest positive element of $L_{a,b}$. Often, as in this case, statements that would be consequences of a theorem are helpful as hints about how to prove the theorem. But you should be very careful to distinguish between assuming the theorem as a way of exploring or getting insight, which is a legitimate process, and assuming the theorem as part of its proof, which is completely forbidden.

4. The proof of Theorem 5.25′ is rather complex. At times you may have felt that you "lost it"—that is, that you became bogged down in the many details, and lost the "big picture". This is a common phenomenon, even among professional mathematicians. (Even one of the authors felt that way about this proof, reading the manuscript in draft form. That's why this comment is here.) The appropriate response is usually to go back to the beginning of the discussion, and reread with the "big picture" in mind, ignoring the details. You may wish to write out an outline for your own clarification.

One case of Theorem 5.25 deserves special mention. We state it as a corollary:

COROLLARY 5.28. GCD$(a, b) = 1$ iff $1 \in L_{a,b}$.

Proof: If GCD$(a, b) = 1$, then $L_{a,b} = \mathbf{Z}$, so $1 \in L_{a,b}$.
On the other hand, if $1 \in L_{a,b}$, then GCD$(a, b)\,|\,1$, so GCD$(a, b) = 1$.

COMMENT We cannot replace 1 by any other value in Corollary 5.28. If some integer v is in $L_{a,b}$, that tells us that $d|v$, but not that $d = v$. Only for the special case $v = 1$ does knowing that $d|v$ guarantee us that $d = v$.

The important part of Corollary 5.28 is the direction that assumes $\text{GCD}(a, b) = 1$. This result is used in many applications, and is worth stating in its own right:

COROLLARY 5.29. If a and b are integers that are relatively prime, then there are integers x and y such that $ax + by = 1$.

We will look at some important applications of this corollary later in this section, but first we want to examine some aspects of the proof of Theorem 5.25 more carefully, specifically the proof of Lemma 5.27. The technique of this proof can be used to solve part of Exploration 2 in Section 1.1, which asked you to explore which subsets of **Z** are closed under addition or subtraction. You should review or redo the subtraction part of that exploration before continuing.

Any new ideas?

Notice that the set $L_{a,b}$ is closed under subtraction. (See Exercise 1.) In fact, Theorem 2.11 states that any set of the form $t\mathbf{Z}$, with $t \in \mathbf{Z}$, is closed under subtraction, and we just showed that $L_{a,b}$ has that form. Is every set that is closed under subtraction of the form $t\mathbf{Z}$?

Can you find any sets that are closed under subtraction that are not of the form $t\mathbf{Z}$?

As with the sets $L_{a,b}$, we need to follow the evidence of our exploration. You presumably did not find any nonempty sets that are closed under subtraction which are not of the form $t\mathbf{Z}$, since there are none. (The empty set *is* closed under subtraction, but that's a special case. Give yourself a pat on the back if you noticed this exception.) Of course, the fact that you didn't find any other nonempty sets closed under subtraction is not a guarantee that there aren't any, but it should suggest to you that you might try to prove that there aren't any.

Try it. Use the proof of Theorem 5.25 as inspiration for your method of proof.

Write down any ideas you have, even if they don't give you a complete proof.

Once again, the keys to the proof are the Well-Ordering Principle and the Division Algorithm: we consider the smallest positive element of the set, and show that the set consists of the multiples of this element. (Notice that any set of the form $t\mathbf{Z}$, with $t \neq 0$, *is* equal to the set of multiples of its smallest positive element.)

We state this result as a theorem, and outline the proof, leaving the details as exercises.

THEOREM 5.30. If S is a nonempty subset of **Z** which is closed under subtraction, then there is an integer $t \geq 0$ such that $S = t\mathbf{Z}$.

Outline of Proof: If $S = \{0\}$, then $S = 0\mathbf{Z}$, so the theorem is true in this case.

Suppose $S \neq \{0\}$. Since $S \neq \varnothing$, there is some nonzero integer, say k, in S. The proof then proceeds as outlined below:

Step 1. Show that $0 \in S$, and then show that $-k \in S$. Conclude that S must have some positive element.

Step 2. Let w be the smallest positive element of S (which exists by the Well-Ordering Principle). Prove, as in Lemma 5.27, that every element of S is divisible by w. [You need a generalization of the Division Algorithm—see parts a) and b) of Exercise 4 of Section 5.2.] Conclude that $S \subseteq w\mathbf{Z}$.

Step 3. By step 1, $0 \in S$. We can show (as in step 1) that $-w \in S$. Use these facts to prove by induction that $w\mathbf{N}$ and $-w\mathbf{N}$ are subsets of S. Conclude that $w\mathbf{Z} \subseteq S$.

Step 4. Combine the conclusions of steps 2 and 3. ■

The details are left as Exercises 2–4.

COMMENTS

1. If we had proved Theorem 5.30 first, we could have simplified the proof of Theorem 5.25 considerably, using the fact that $L_{a,b}$ is closed under subtraction.

2. It is often the case that we prove a theorem, and then notice later that the reasoning applies more generally than the situation described in the hypothesis of the theorem. Such observations lead to more general theorems, and making generalizations is the essence of doing abstract mathematics.

Uniqueness of Prime Factorization

We saw in Section 5.2 that every integer greater than 1 is either a prime or a product of primes. We will add to that result in this section by showing that a composite integer can only be written in one way as a product of primes (except for the order of the factors). To be more precise, we will prove the following theorem:

THEOREM 5.31 (Uniqueness of Prime Factorization). Suppose n is an integer greater than 1, and that we have both

$$n = lp_1 p_2 \cdots p_r$$

and

$$n = q_1 q_2 \cdots q_s$$

where $p_1, p_2, \ldots, p_r$ and $q_1, q_2, \ldots, q_s$ are primes, with $p_1 \le p_2 \le \cdots \le p_r$ and $q_1 \le q_2 \le \cdots \le q_s$. Then $r = s$, and for each i from 1 to r we have $p_i = q_i$.

This uniqueness result, together with the existence result, Theorem 5.17, makes up a theorem known as the *Fundamental Theorem of Arithmetic*. We will prove Theorem 5.31 after some preliminary work.

DISCUSSION If you have never seen this assertion before, you should take time to think about it. Start with an integer with many factors, like 720, and break it down by successive factoring: e.g., $720 = 12 \times 60$; then $12 = 4 \times 3$ and $60 = 6 \times 10$; then $4 = 2 \times 2$, $6 = 2 \times 3$, and $10 = 2 \times 5$. Thus, $720 = 2 \times 2 \times 3 \times 2 \times 3 \times 2 \times 5$. This process can be represented graphically as in Figure 5.1, which is known as a *factor tree*.

If you try another sequence of steps (for example, start by factoring 720 as 72×10), you will see that you end up with the same list of primes, no matter how you do it: four 2's, two 3's, and one 5.

Get Your Hands Dirty 1

Make three more factor trees for 720, different from the one in Figure 5.1. Compare your results. □

COMMENT Mathematicians specifically exclude the number 1 from the definition of the term "prime". Theorem 5.31 is one of the reasons for this exclusion. If we allowed 1 as a prime, the "factor tree" process would never end, and a natural number would have infinitely many distinct factorizations into primes.

If you are familiar with Theorem 5.31, it may strike you as "obvious"—too basic to be provable. But in fact it can be proved, and the proof is based on the GCD–Linear-Combinations Theorem, or more precisely, Corollary 5.29 for the relatively prime case. Recall that those results depended in turn on the Division Algorithm and the Well-Ordering Principle.

Euclid's Lemma

Corollary 5.29 is used in the proof of Theorem 5.31 through the following result, known as *Euclid's Lemma*. The general logic of Euclid's Lemma was discussed in Example 12 of Section 4.2. You may want to reread that discussion.

THEOREM 5.32 (Euclid's Lemma). If p is prime, and $a, b \in \mathbf{N}$, such that $p \mid ab$, then either $p \mid a$ or $p \mid b$.

Proof: We will assume $p \nmid a$, and prove that $p \mid b$.

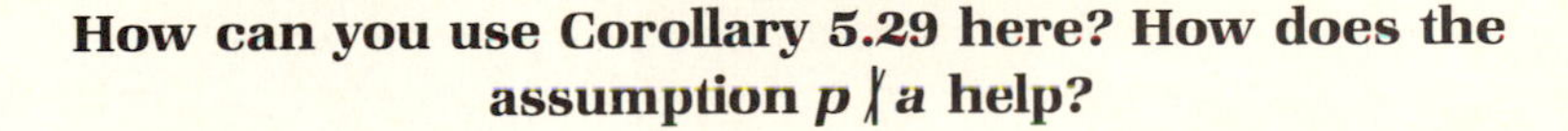
How can you use Corollary 5.29 here? How does the assumption $p \nmid a$ help?

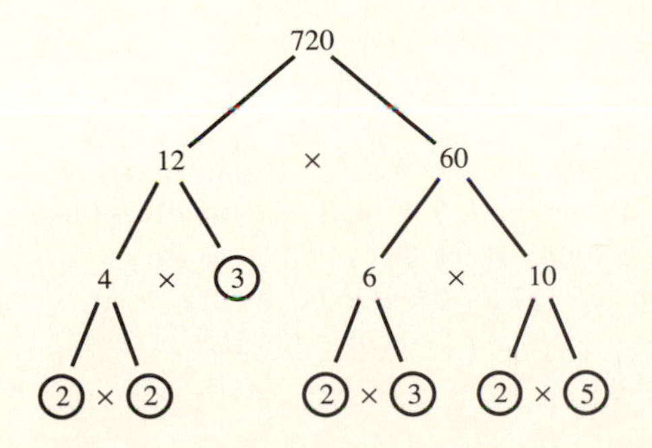

Figure 5.1. A factor tree for 720. Each term is repeatedly factored until only primes remain. These are circled, and their product is the original number. Thus $720 = 2 \times 2 \times 3 \times 2 \times 3 \times 2 \times 5$.

Since $p \nmid a$, and p is a prime, we have that $\mathrm{GCD}(a,p) = 1$. (Why?) By Corollary 5.29, we are guaranteed that there are integers x and y with $ax + py = 1$.

We can multiply both sides of this equation by b, giving us

$$(ab)x + p(by) = b\,. \qquad (*)$$

The hypothesis also says that $p \mid (ab)$, so there is also an integer w such that $ab = pw$. Substituting pw for ab in $(*)$, we get $(pw)x + p(by) = b$. But we can factor out p from this equation, giving $p(wx + by) = b$, which proves that $p \mid b$. ■

COMMENTS

1. The case $p = 2$ of Euclid's Lemma simply says that, if a product of integers is even, then one of the factors is even.
2. The algebra of this proof is very clever. The technique is a common one, however, and we will give you some other opportunities to use it—see Exercises 5 and 6.

We leave as Exercise 7 the proof of the following generalization of Theorem 5.32 to the situation where the prime p divides a product of more than two factors:

THEOREM 5.33. If p is a prime, and $a_1, a_2, \ldots, a_r \in \mathbf{Z}$ such that $p \mid a_1 a_2 \cdots a_r$, then p divides one of the factors a_i.

Proof of Uniqueness of Prime Factorization

Finally, we turn to the proof of Theorem 5.31 itself, which we restate here:

THEOREM 5.31 (Uniqueness of Prime Factorization). Suppose n is an integer greater than 1, and that we have both

$$n = p_1 p_2 \cdots p_r$$

and

$$n = q_1 q_2 \cdots q_s$$

where $p_1, p_2, \ldots, p_r$ and $q_1, q_2, \ldots, q_s$ are primes, with $p_1 \le p_2 \le \cdots \le p_r$ and $q_1 \le q_2 \le \cdots \le q_s$. Then $r = s$, and, for each i from 1 to r, $p_i = q_i$.

Proof: We prove this by induction on n, using the Second Principle of Induction. Since 2 is a prime, the only equation of the form $2 = p_1 p_2 \cdots p_r$, with the p_i all primes, is the case where $r = 1$ and $p_1 = 2$. Therefore the theorem is true for $n = 2$.

We now assume that it is true for $n \le t$, and we suppose that the integer $t + 1$ has two such factorizations into primes. Without loss of generality, we may assume $p_1 \le q_1$.

Since $p_1 \mid t + 1$, we know by Theorem 5.32 that p_1 must divide one of the factors q_i. But p_1 and q_i are primes, so this guarantees that $p_1 = q_i$. Since $q_1 \le q_i$, we get $p_1 \ge q_1$. On the other hand, we assumed $p_1 \le q_1$, so we must have $p_1 = q_1$.

We therefore have $p_2 \cdots p_r = q_2 \cdots q_s$, and this product is less than $t + 1$. We can thus apply our induction hypothesis to these factorizations, and conclude that $r - 1 =$

$s - 1$, and that, for each i from 2 to r, $p_i = q_i$.

This completes the proof. ■

Nonunique Factorization among Complex Numbers

We conclude this chapter with an example of a number system in which the principle of unique factorization fails to hold. The ideas discussed in this example are important in understanding the development of number theory. This example may seem a bit obscure. It is intended to give you a somewhat broader perspective on Theorem 5.31. Don't worry if you don't follow the details.

We have been discussing factorization within the set of integers. For various reasons, mathematicians are interested in generalizing this process to other, larger number systems. One such system arises within the set **C** of complex numbers as follows.

Consider some negative integer, $-n$ (with $n \in \mathbf{N}$), and its complex square root, $\sqrt{-n}$. Let S be the subset of **C** given by $S = \{a + b\sqrt{-n} : a, b \in \mathbf{Z}\}$. A routine computation shows that S is closed under addition, subtraction, and multiplication. We can ask how elements in S can be factored into products of other elements of S, and which elements cannot be factored *nontrivially*. [It turns out in this context, as in **Z**, that the only factorizations of a number x that are considered trivial are the factorizations $x = x \times 1$ and $x = (-x) \times (-1)$.]

It makes sense to use the term "prime" to mean a number other than ± 1 which has no nontrivial factorization. Some integers which are primes in the usual sense turn out not to be primes in this context. For example, if $n = 6$, the number 7 is no longer a prime: it can be factored as $(1 + \sqrt{-6})(1 - \sqrt{-6})$ (verify). However, it can be proved that 2, 5, $2 + \sqrt{-6}$, and $2 - \sqrt{-6}$, are all primes in this new system.

The interest in these specific values is as follows: within the set S, we can factor 10 in two different ways, namely, $10 = 2 \times 5$ and $10 = (2 + \sqrt{-6})(2 - \sqrt{-6})$ (verify). In other words, we have two distinct ways to factor 10 as a product of primes. The counterpart of Theorem 5.31 in this system is not true.

COMMENT It is well beyond the scope of this book to fully explain the reasons for mathematical interest in this question, but it may help to say that such factorization problems play a major role in efforts to understand Fermat's Last Theorem—(see Historical Note in Section 2.3).

Many efforts to prove this conjecture turned out to be incorrect, because they were based on the assumption that unique factorization carried over into these complex number systems. Eventually, the nineteenth-century mathematician E. E. Kummer, after falling into this trap himself, recognized the fallacy in his reasoning, and developed some important concepts of number theory to work around this difficulty. His pioneering work led to some partial results in connection with Fermat's Last Theorem.

The purpose of this discussion is to point out that what may seem "obvious" can be misleading. Even great mathematicians can be misled by their intuition. It is dangerous to accept anything in mathematics without proof.

EXERCISES

1. Prove directly (i.e., without using any theorems from this section) that, for any $a, b \in \mathbf{Z}$, the set $L_{a,b}$ is closed under subtraction.
2. (Theorem 5.30—step 1) Suppose S is a subset of $\mathbf{Z}$ which is closed under subtraction, and $k \in S$. Prove that $0 \in S$, and then use that to prove that $-k \in S$.
3. (Theorem 5.30—step 2) Suppose S is a subset of $\mathbf{Z}$ which is closed under subtraction, and that w is the smallest positive element of S. Prove that every element of S is divisible by w.
4. (Theorem 5.30—step 3) Suppose S is a subset of $\mathbf{Z}$ which is closed under subtraction, and that w, 0, and $-w$ are all in S.
   a) Prove, by induction on n, that $wn \in S$ for all $n \in \mathbf{N}$.
   b) Prove, by induction on n, that $-wn \in S$ for all $n \in \mathbf{N}$.
5. **a)** Prove the following generalization of Euclid's Lemma: If $a, b, r \in \mathbf{Z}$, with GCD $(a, r) = 1$, and $r \mid (ab)$, then $r \mid b$.
   b) Explain why the theorem in part a) is a generalization of Euclid's Lemma.
6. Prove the following:
   If $a, b, r \in \mathbf{Z}$, such that $a \mid r$, $b \mid r$, and GCD$(a, b) = 1$, then $ab \mid r$.
7. Prove the following:

   THEOREM 5.33. If p is a prime, and $a_1, a_2, \ldots, a_r \in \mathbf{Z}$ such that $p \mid a_1 a_2 \cdots a_r$, then p divides one of the factors a_i.

   (*Hint:* Prove this by induction on r, using Theorem 5.32 as the proof of the case $r = 2$.)
8. Suppose $n \in \mathbf{N}$, with $n = p_1^{a_1} p_2^{a_2} \cdots p_r^{a_r}$, where the p_i are distinct primes. Develop a formula, in terms of the exponents a_i, for the number of positive integers which are divisors of n.

6

Relations and Functions

Introduction

In Chapter 3, we introduced the formal definition of a *function,* which can be thought of as a way of relating the elements of one set with those of another (or the same) set. In this chapter, we will examine a generalization of the concept of function, which is called a *relation*.

In Section 6.1, we will define the general concept of relation, and look at some examples. We will then look at the special case of functions in more detail than we did in Chapter 3. In particular, we will provide some further discussion of the question raised in Exploration 3 in Section 3.2, concerning whether two functions being one-to-one or onto is connected to their composition being one-to-one or onto. (This question was partially answered in Section 4.1.) Section 6.1 concludes by defining some important general properties of relations, based on our examples.

In Section 6.2, we will be examining an important special type of relation, called an *equivalence relation,* and the closely related concept of a *partition*. In rough terms, equivalence relations are used to describe "similarities" among objects of a given set: the relations "is equal to", "is similar to" (for geometric figures), and "has the same parity as" (i.e., both are odd or both are even) are important examples of equivalence relations. The concept of an equivalence relation is formally defined in terms of the general properties of relations introduced in Section 6.1. The concept of a partition is a more intuitive way of expressing the same information about the objects in the set: the set is decomposed into subsets, and objects that are "similar" are placed in the same subset. We will examine the relationship between partitions and equivalence relations, and discuss the advantages and disadvantages of each.

Chapter 7 will develop the concepts of this chapter further, examining other special types of functions and relations.

6.1 Arbitrary Relations and Functions

Relations — Examples and Definition

We begin with two arbitrary sets, which we label A and B, and ask how the elements of these two sets might be "related" to each other.

This is an extremely vague question. There are many kinds of relationships that might exist. Here are some nonmathematical possibilities:

Example 1. Some Nonmathematical Relations

i) A = the set of students in a certain class; B = the set of grades $\{A, B, C, D, F\}$. (The grade B is not the same thing as the set B.) The relationship consists of a particular student having received a particular grade on one of the semester's exams. Each student may be "related" to several different grades. Very likely, each grade is related to several different students. We might use the statement "Student X is related to grade Z" to mean "Student X has received a grade of Z on at least one exam (in the given class)."

ii) A = the set of member countries of the United Nations; $B = A$. The relationship consists of the first country having an ambassador in the second country. In this case, every country is almost certainly related to at least one other country, and probably to many. The statement "country X is related to country Z" means "country X has an ambassador in country Z".

iii) A = the set of states in the United States; $B = A$. The relationship consists of the first state having a smaller population than the second (in the 1980 census). In this case, California is not related to any states, and Alaska is related to all other states. The statement "state X is related to state Z" means "X had a smaller population than Z in the 1980 census".

iv) A = the set of U.S. cities with population over 100,000 (in the 1980 census); $B = A$. The relationship consists of the first city being at most 500 miles from the second. One important feature of this example is that every element of A is related to itself. The statement "city X is related to city Z" means "The distance from X to Z is at most 500 miles". □

Each of these examples has some interesting properties which we will examine later in this section. At this point, we are primarily interested in formulating a precise definition of the general concept of relation.

Try to come up with a definition for the concept of "a relation from A to B" which includes all of the above examples:

Think about the definition of "function" (Definition 3.14 in Section 3.2).

In each case, certain elements of A are related to certain elements of B. There is a *statement* which expresses the relationship, and this statement is an open sentence with two variables, one from A and one from B. Not every choice of elements will make this statement true. Understanding the relationship involves being able to determine whether a particular pair of elements have this relationship to each other or not.

In other words, we want to be able to determine the truth set of the open sentence describing the relationship. Since the open sentence has two variables, its truth set is a set of ordered pairs, i.e., a subset of $A \times B$. This is the basis for the following definition:

DEFINITION 6.1: For any two sets A and B, a *relation from A to B* is a subset of $A \times B$. If B is the same set as A, we refer to the relation as a *relation on A*.

COMMENTS

1. Compare this definition with the definition of "function". A function from A to B is a subset of $A \times B$ which satisfies two particular conditions. A relation from A to B can be *any* subset of $A \times B$, including the empty set and $A \times B$ itself.

2. In the definition of relation, there is no reference to open sentences. Though most relations we are interested in can be described by means of an open sentence, there are some relations which cannot be expressed this way. We will clarify this assertion in Section 8.1 (see Comment 6 following GHYD 6 of that section).

3. When a particular relation is defined using an open sentence, we often think of it as if it *were* the open sentence, even though our formal definition is phrased in terms of the *truth set of* that open sentence. This is similar to the situation with a function, which we may think of intuitively in terms of the equation or rule which describes it. Thus, for example, in discussing functions, we may say "the function $y = x^2$" or "the squaring function", when formally we mean $\{(x, y) \in \mathbf{R} \times \mathbf{R} : y = x^2\}$. Similarly, we might speak about "the relation $x < y$" or "the relation 'is less than' " as a shorthand for $\{(x, y) \in \mathbf{R} \times \mathbf{R} : x < y\}$.

4. Another valuable way to think about a relation is in terms of a *graph:* if $A = B = \mathbf{R}$, then a set of ordered pairs can be interpreted as a set of points in terms of the usual coordinate system, with the first coordinate along the horizontal axis and the second coordinate along the vertical axis. We will refer to this idea occasionally as we describe different types of relations. Notice that the solution set to any equation in two variables will be a relation, even if the equation does not yield a function. For example, the graph of the equation "$x^2 - 3xy + 4y^2 + 5x = 12$" is an ellipse, and this equation does not represent a function. However, since this graph is certainly a subset of $\mathbf{R} \times \mathbf{R}$, the equation does represent a relation on $\mathbf{R}$.

5. The perspective of a relation as a graph makes the most sense when $A = B = \mathbf{R}$, but the idea is useful as a tool for building intuition even when the "vertical axis" and "horizontal axis" are arbitrary sets. You should just picture the elements of A "spread out" horizontally and those of B "spread out" vertically, as we suggested in our introductory discussion of Cartesian product in Section 2.1. The elements of the relation are "points" in this nonstandard coordinate system. A diagram such as Figure 6.1, in which the relation is represented by the shaded area S, suggests the general idea.

The relation "=", as a relation *on a set A,* is perhaps the most fundamental of all mathematical relations for any set. Its graph is just the "the main diagonal"—for $A = \mathbf{R}$, this is the line $y = x$. The next example gives some other standard mathematical relations.

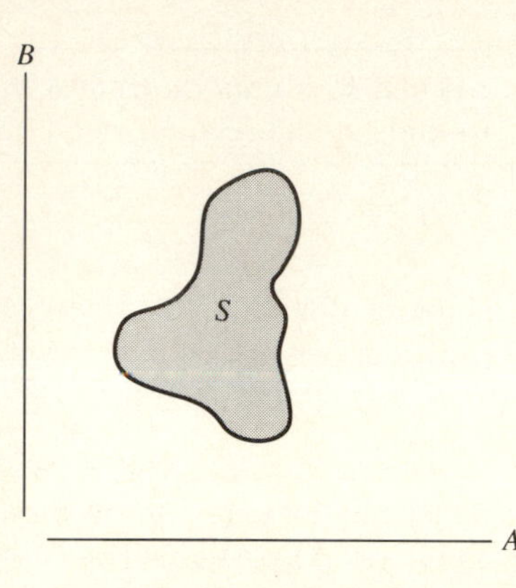

Figure 6.1. Here A and B are just sets, not necessarily $\mathbf{R}$. The shaded area S represents a relation from A to B. The ordered pairs of the relation are thought of as "points" in $A \times B$.

Example 2. **Some Mathematical Relations** There are many familiar examples of relations in mathematics. Here are some of them, expressed by symbolic open sentences [in v) and vi), S is any set]: □

	Symbolic Open Sentence	*Possible Choice for A × B*	*Verbal Description*
i)	$x < y$	$\mathbf{R} \times \mathbf{R}$	"is less than"
ii)	$x \neq y$	$\mathbf{R} \times \mathbf{R}$	"is not equal to"
iii)	$x \mid y$	$\mathbf{Z} \times \mathbf{Z}$	"divides"
iv)	$\lvert x \rvert = \lvert y \rvert$	$\mathbf{R} \times \mathbf{R}$	"has the same absolute value as"
v)	$x \subseteq y$	$\mathcal{P}(S) \times \mathcal{P}(S)$	"is a subset of"
vi)	$x \in y$	$S \times \mathcal{P}(S)$	"is an element of"

Since most familiar relations are expressed in terms of a relation symbol, it is sometimes suggestive to use a generic relation symbol for defining a new relation or speaking about relations in a general context. We will use the symbol "$\sim$" (read "squiggle") in this way. Thus, we will sometimes write "$x \sim y$" to mean "$(x, y) \in R$", and refer to "the relation $\sim$" instead of "the relation R". For example, we might define the relation in iii) of Example 1 by saying

"$X \sim Z$" means the population of state X is less than that of state Z.

The notation "$x \not\sim y$" means "$(x, y) \notin R$". [Some texts avoid introducing the "$\sim$" notation by writing "$x \; R \; y$" as an alternative to "$(x, y) \in R$".]

Some of the terminology used for functions can be adapted for use with respect to relations. We will use the following definitions:

DEFINITION 6.2: For sets A and B and a relation R from A to B:

The *domain* of R [written "Dom(R)"] is

$$\{u \in A : (\exists\ v \in B)((u, v) \in R)\}.$$

The *image* of R [written "Im(R)"] is

$$\{v \in B : (\exists\ u \in A)((u, v) \in R)\}.$$

In relation i) of Example 1, the domain consists of all those students who received any of the letter grades A, B, C, D, or F on a midterm. If a student dropped out before receiving any grade, or if the instructor used credit/no credit or some other grading system, this might not include all students. Also, the image of the relation could be any subset of {A, B, C, D, F}, although it is unlikely that it would be the empty set, unless letter grades weren't used.

Relation iii) is more straightforward: the domain consists of all states except California (which has the largest population), and the image consists of all states except Alaska (which has the smallest population).

Example 3. **Relations on Finite Sets** When the sets A and B are finite, it is often useful to describe a relation by simply listing the ordered pairs that belong to it. We emphasize that *any* set of ordered pairs gives a relation. Here are some examples, with $A = \{1, 2, 3\}$, $B = \{4, 5, 6\}$:

i) $R = \varnothing$: Although not a very interesting relation, the empty set is certainly a subset of $A \times B$. Its domain and image are both the empty set as well.

ii) $R = \{(1, 4), (2, 5), (3, 6)\}$: This example is actually a one-to-one, onto function from A to B. It can be described as $\{(x, y) \in A \times B : y = x + 3\}$. We have $\text{Dom}(R) = A$ and $\text{Im}(R) = B$.

iii) $\sim$ given by the condition "$x \sim y$ iff $|x - y| \leq 2$": In this case, $2 \sim 4$, $3 \sim 4$, and $3 \sim 5$, and no other elements are related to each other. In other words, R contains exactly three ordered pairs. The domain of this relation is $\{2, 3\}$, and its image is $\{4, 5\}$.

iv) $R = \{(1, 4), (2, 4), (2, 6)\}$: This set of ordered pairs was chosen without any particular condition or open sentence in mind. It is just some subset of $A \times B$, and is just as much a relation as the other examples. Its domain is $\{1, 2\}$ and its image is $\{4, 6\}$. □

Comparing Functions and Relations

Since a function from A to B is a particular type of subset of $A \times B$, we repeat the definition of function here, rephrased in terms of relations, so that we can make some interesting comparisons:

DEFINITION 3.14: For sets A and B, a *function from A to B* is a relation R from A to B which satisfies the following conditions:

a) $(\forall\ u \in A)(\exists\ v \in B)((u, v) \in R)$, and
b) $(\forall\ u \in A)(\forall\ v, w \in B)([(u, v) \in R \text{ and } (u, w) \in R] \Rightarrow v = w)$.

[Condition b) can be simplified, using "$f(x)$" notation, but in the context of relations, the notation above is more appropriate.]

As we pointed out in Section 3.2, if the roles of A and B in conditions a) and b) of this definition are reversed, we get the definitions of the terms "onto" and "one-to-one" for functions. Condition b) is often confused with the concept of "one-to-one". In the context of the more general idea of a relation, the distinction is clearer. We will find it convenient to have specific terminology for relations that satisfy condition a) or b) individually. We make the following definitions:

DEFINITION 6.3: For any sets A and B and any relation R from A to B:
"R is *domain-complete*" means

$$(\forall\ u \in A)(\exists\ v \in B)((u, v) \in R).$$

[This is condition a) of Definition 3.14.]
"R is *functionlike*" means

$$(\forall\ u \in A)(\forall\ v, w \in B)([(u, v) \in R \text{ and } (u, w) \in R] \Rightarrow v = w).$$

[This is condition b) of Definition 3.14.]

Note: This is not standard terminology. What we are calling "functionlike" is sometimes described by the phrase "single-valued", but, in our view, that terminology too closely resembles the phrase "one-to-one", which has a different meaning (see comment 1 below). There is no standard terminology for "domain-complete".

Intuitively, a relation R from A to B is domain-complete iff every element of A is used at least once as an ordered pair of R, and R is functionlike iff every element of A is used at most once as an ordered pair of R.

The following result is nothing more than a restatement of the definition of "function" using this new terminology:

THEOREM 6.1. For sets A and B and a relation R from A to B, R is a function from A to B iff R is domain-complete and functionlike.

COMMENTS

1. Notice the difference between "one-to-one" and "functionlike". "Functionlike" means that an element of A is not paired with two different elements from B. "One-

to-one" (applied to a function) means that an element of B is not paired with two different elements from A.

2. We can visualize the concepts of "domain-complete" and "functionlike" using the idea of a relation as a graph. We expand here on the idea of a vertical-line test, which was discussed when we first defined the concept of function. A relation is functionlike iff there is no vertical line which crosses the graph in more than one point. A relation is domain-complete iff every vertical line crosses the graph in at least one point. Exercise 2 asks about the possible significance of analogous "horizontal-line tests", when applied to functions.

3. The condition "R is domain-complete" can be written succinctly by the equation "$\mathrm{Dom}(R) = A$".

Get Your Hands Dirty 1

Which of the relations listed in Examples 1, 2, and 3 are domain-complete? Which are functionlike? □

We also make the following definition:

DEFINITION 6.4: For sets A and B and a relation R from A to B, the *inverse of* R (written R^{-1}, and read "R inverse") is the relation from B to A defined by

$$R^{-1} = \{(y, x) \in B \times A : (x, y) \in R\}.$$

We begin with some elementary observations about inverse relations:

THEOREM 6.2. For sets A and B and a relation R from A to B:

i) $\mathrm{Dom}(R^{-1}) = \mathrm{Im}(R)$;
ii) $\mathrm{Im}(R^{-1}) = \mathrm{Dom}(R)$;
iii) $(R^{-1})^{-1} = R$.

Proof of i): Both $\mathrm{Dom}(R^{-1})$ and $\mathrm{Im}(R)$ are subsets of B, and we need to show that they are equal. Suppose x is an element of B. Then we have the following:

$x \in \mathrm{Dom}(R^{-1})$ iff $(\exists\, y \in A)((x, y) \in R^{-1})$ (by definition of "domain")
iff $(\exists\, y \in A)((y, x) \in R)$ (by definition of "R^{-1}")
iff $x \in \mathrm{Im}(R)$ (by definition of image). ■

The proofs of ii) and iii) are equally straightforward, and are left as Exercise 4.

COMMENTS

1. Many common mathematical relations have inverses that are also used frequently. For example, the inverse of the relation "is less than" is the relation "is greater than".

2. Notice that a relation *on a set* A (i.e., in which $B = A$), can be equal to its own inverse. This is actually the case with iv), and probably ii), in Example 1, and with the mathematical relations "is not equal to" and "has the same absolute value as" from Example 2. Exercise 5 asks how to describe the fact that a relation is equal to its inverse in terms of the graph of the relation.

More about Functions

Since a function is a special type of relation, the concept of inverse can be applied to functions as well. Thus, if f is any function, there will be an *inverse relation* f^{-1}. But we must emphasize that this inverse relation will not necessarily be a function.

Get Your Hands Dirty 2

Make up a function whose inverse is not a function. □

To distinguish the case when the inverse of f *is* a function, we make the following definition:

DEFINITION 6.5: For a function $f: A \to B$, "f is *invertible*" means: f^{-1} is a function from B to A.

Get Your Hands Dirty 3

Let $f: \mathbf{R} \to \mathbf{R}$ be the function given by the equation "$f(x) = 3x + 4$".

i) Find five ordered pairs in f^{-1}.
ii) Show that f^{-1} is a function.
iii) Describe f^{-1} in terms of an equation. [Check that your ordered pairs from i) fit your equation.] □

The following is a natural question to ask, after Definition 6.5:

QUESTION 1 If f is a function from A to B, under what circumstances will f be invertible? Express your answer in terms of the following concepts from Section 3.2, whose definitions are repeated here:

DEFINITION 3.17: For any function $f: A \to B$, "f is *onto*" (or *surjective*) means:

$$(\forall\, v \in B)(\exists\, u \in A)(f(u) = v).$$

DEFINITION 3.18: For any function $f: A \to B$, "f is *one-to-one*" (or *injective*) means:

$$(\forall\, v \in B)(\forall\, u, w \in A)([f(u) = v \text{ and } f(w) = v] \Rightarrow u = w).$$

[As with condition b) of Definition 3.14 of "function", the condition in Definition 3.18 can be simplified, but we prefer to use this version here.]

? ? ?

The following theorem follows directly from the definition of inverse relation, using a comparison of the definitions of "domain-complete" and "functionlike" with those of "onto" and "one-to-one":

THEOREM 6.3. For any function $f: A \to B$:

i) f^{-1} is domain-complete (as a relation from B to A) iff f is onto.
ii) f^{-1} is functionlike (as a relation from B to A) iff f is one-to-one.

The next result, which answers Question 1, is an immediate consequence of Theorems 6.1 and 6.3:

COROLLARY 6.4. For any function $f: A \to B$, f is invertible iff f is one-to-one and onto.

Proof: The reasoning is a chain of equivalences:

f is invertible iff f^{-1} is a function (by Definition 6.5)
iff f^{-1} is domain-complete and functionlike (by Theorem 6.1)
iff f is onto and one-to-one (by Theorem 6.3). ∎

Thus, the terms "invertible" and "bijective" are equivalent conditions on functions.

COMMENT Intuitively, for f to be invertible, i.e., for f^{-1} to be a function, each element of B must be used as the first element of exactly one ordered pair of f^{-1}. This is the same as saying that each element of B is used as the second element in exactly one ordered pair of f, i.e., that f is one-to-one and onto.

The following corollary is equally important:

COROLLARY 6.5. For any function $f: A \to B$, if f is bijective, then f^{-1} is a bijective function.

Proof: Suppose that f is bijective. Corollary 6.4 tells us that f is invertible, so f^{-1} is a function. But f^{-1} is itself invertible, since its inverse is just f [by Theorem 6.2 iii)]. Since f^{-1} is invertible, it is bijective, by Corollary 6.4. ∎

We know from Corollary 4.7 that the composition of bijective functions is bijective. The following result describes a simple relationship between the inverse of this composition and the inverses of its components.

THEOREM 6.6. For bijective functions $f: A \to B$ and $g: B \to C$,

$$(g \circ f)^{-1} = f^{-1} \circ g^{-1}.$$

Proof: We know that $g \circ f$ is a function from A to C. Since f and g are bijective, so is $g \circ f$, so $g \circ f$ is invertible. Therefore, by Definition 6.5, $(g \circ f)^{-1}$ is a function from C

to A. Similarly, we know that f^{-1} is a function from B to A and g^{-1} is a function from C to B, so $f^{-1} \circ g^{-1}$ is also a function from C to A. We will prove that they are equal by showing that they have the same value for an arbitrary element z of their domain C.

So suppose $z \in C$. Since g is bijective, there is a unique element $y \in B$ such that $g(y) = z$. Similarly, there is a unique element $x \in A$ such that $f(x) = y$. Thus we have $(g \circ f)(x) = z$, and so $(g \circ f)^{-1}(z) = x$.

On the other hand, $g^{-1}(z) = y$, and $f^{-1}(y) = x$, so we also have $(f^{-1} \circ g^{-1})(z) = x$, which completes the proof. ■

Inverse Functions and Composition

There is another perspective on the concept of the inverse of a function which is important to point out. (This perspective will make more sense in the context of studying binary operations—see Section 7.1—but is sufficiently important to discuss here as well.)

Suppose f is a function from A to B, and that the inverse f^{-1} is also a function; i.e., f is one-to-one and onto. What can you say about the two compositions $f^{-1} \circ f$ and $f \circ f^{-1}$? We break this question down into two stages.

First: What are the domains and codomains of each of these functions?

? ? ?

By the definition of composition, $f^{-1} \circ f$ is a function from A to itself, and $f \circ f^{-1}$ is a function from B to itself.

Second: Precisely which functions are they? To understand this question, take two small sets—say $A = \{1, 2, 3\}$ and $B = \{4, 5, 6\}$—and a specific function $f: A \to B$ that is one-to-one and onto. Find f^{-1} and the two compositions explicitly.

? ? ?

What you should get for the two compositions, no matter what bijective function f you choose, are the functions $\{(1,1), (2,2), (3,3)\}$ and $\{(4,4), (5,5), (6,6)\}$. These functions are called the *identity functions* on A and B respectively. More generally, we have the following definition:

DEFINITION 6.6: For a set A, the *identity function* on A (written I_A) is the function from A to A defined by the equation $I_A(x) = x$ for all $x \in A$. (Formally, $I_A = \{(x, x) : x \in A\}$.)

What we are saying about f and its inverse is part a) of the following theorem:

THEOREM 6.7. For any sets A and B:

a) If $f: A \to B$ is invertible, then $f^{-1} \circ f = I_A$ and $f \circ f^{-1} = I_B$.

b) If $f: A \to B$ and $g: B \to A$ are two functions such that $g \circ f = I_A$ and $f \circ g = I_B$, then f is invertible and $g = f^{-1}$.

We will prove b), and leave the proof of a) to the reader as Exercise 7.

Proof of b): Suppose f and g are as described in b). By Corollary 6.4, in order to prove that f is invertible, we need to show that it is one-to-one and onto.

To show f is one-to-one, suppose $x, y \in A$, with $f(x) = f(y)$. Then $g(f(x)) = g(f(y))$. But $g \circ f = I_A$, so $g(f(x)) = x$ and $g(f(y)) = y$. Thus $x = y$, which proves that f is one-to-one.

To show f is onto, suppose $t \in B$. Since $f \circ g = I_B$, we have $f(g(t)) = t$. This shows that there is an element w in A, namely, $w = g(t)$, such that $f(w) = t$, which proves that f is onto.

So we have shown that f is invertible, i.e., f^{-1} is a function from B to A.

Finally, we need to show that $g = f^{-1}$. Both are functions from B to A. To show that they are equal, suppose that $t \in B$. We need to show that $g(t)$ and $f^{-1}(t)$ are equal.

Let $u = f^{-1}(t)$, so $(t, u) \in f^{-1}$. By definition of f^{-1}, we have $(u, t) \in f$, i.e., $f(u) = t$. Therefore $g(f(u)) = g(t)$. But $g(f(u)) = u$, since $g \circ f = I_A$. Thus, $g(t) = u = f^{-1}(t)$, which completes the proof. ∎

COMMENT Notice what we must do to prove that two functions are equal. Basically, we need to show that they consist of the same ordered pairs. We can do this by first showing that they have the same domain and then proving, for any element of that common domain, that the two functions have the same value.

In Section 7.1, we will use the terms "identity" and "inverse" in a different sense from that of Definitions 6.4 and 6.6, and we will see that the definitions here can be seen as a particular case of the later definitions.

Exploration 3 of Section 3.2 Revisited

In Exploration 3 of Section 3.2, we asked the following: Suppose f and g are functions, with $f: A \to B$ and $g: B \to C$. Is there any connection between whether either of the functions f or g is one-to-one or onto and whether the composition $g \circ f$ is one-to-one or onto?

This question was partially answered in Section 4.1, where we proved the following two theorems:

THEOREM 4.3. If $g: X \to Y$ and $h: Y \to Z$ are both injective functions, then $h \circ g: X \to Z$ is an injective function.

THEOREM 4.6. If $g: X \to Y$ and $h: Y \to Z$ are both surjective functions, then $h \circ g: X \to Z$ is a surjective function.

These theorems leave unanswered the question of their converses, which we pose directly here:

QUESTION 2 If the composition of two functions is injective (one-to-one) or surjective (onto), what does that tell us, if anything, about whether either or both of the two component functions is one-to-one or onto? (The answer may surprise you. Look at examples with small finite sets, and use arrow diagrams.)

? ? ?

It is certainly reasonable to guess that the converses of Theorems 4.3 and 4.5 might be true. In fact, they are not. It is instructive to see what might happen if you tried to prove one of these converses. Since the conclusion of each of the converses involves a conjunction, we sugggest here four "partial converses" for you to consider as conjectures.

In each of the following conjectures, X, Y, and Z are sets, g is a function from X to Y, and h is a function from Y to Z.

Conjecture 1

If $h \circ g : X \to Z$ is a surjective function, then $g : X \to Y$ is a surjective function.

Conjecture 2

If $h \circ g : X \to Z$ is a surjective function, then $h : Y \to Z$ is a surjective function.

Conjecture 3

If $h \circ g : X \to Z$ is an injective function, then $g : X \to Y$ is an injective function.

Conjecture 4

If $h \circ g : X \to Z$ is an injective function, then $h : Y \to Z$ is an injective function.

Which of these (if any) do you think are true?

Try to prove each of these conjectures. If you run into a stumbling block, try to find a counterexample instead.

We leave the resolution of this issue as Exercise 9.

General Properties of Relations

We now return from the topic of functions to a discussion of relations in general. We commented earlier that a relation on a set A could be equal to its inverse. This observation leads to the following definition:

DEFINITION 6.7: For a relation R on a set A, "R is *symmetric*" means:

$$R = R^{-1}.$$

We can see why the word "symmetric" is used for this concept if we express it using the "$\sim$" notation, as follows: A relation "$\sim$" on A is *symmetric* iff

$$(\forall\, x, y \in A)\,(x \sim y \Rightarrow y \sim x).$$

COMMENT You might expect a double implication ("$\Leftrightarrow$") in the quantifier version of this definition, since the set condition $R = R^{-1}$ requires showing both $R \subseteq R^{-1}$ and $R^{-1} \subseteq R$. In fact, it turns out that, if $\sim$ is symmetric, then the statement with a double implication holds as well; i.e., the single implication is equivalent to the double implication (see GYHD 4 below). Since the two statements are equivalent, we use as our definition the one that is easier to establish in proofs.

Get Your Hands Dirty 4

Prove that the following conditions on A and $\sim$ are equivalent [recall from Section 4.2 that this means you must show that i) implies ii) and ii) implies i).]:

i) $(\forall\, x, y \in A)\,(x \sim y \Rightarrow y \sim x)$,
ii) $(\forall\, x, y \in A)\,(x \sim y \Leftrightarrow y \sim x)$. □

For any set A, "is equal to" is a symmetric relation on A. Item iv) of Example 2, in which $x \sim y$ means $|x| = |y|$, is based on equality, and is also symmetric (since, if $x \sim y$, then $|x| = |y|$, so $|y| = |x|$, so $y \sim x$).

Get Your Hands Dirty 5

Which of the other relations in Example 2 are symmetric? □

For finite relations, it is easy to tell if a relation is symmetric just by looking at the set of ordered pairs. If the set is the same when all the pairs are reversed, then it is symmetric; otherwise it is not. For example, let $A = \{1, 2, 3\}$. The relation $\{(1, 2), (2, 1), (1, 3), (3, 1)\}$. is symmetric. The relation $\{(1, 2), (2, 3), (3,1)\}$ is not symmetric.

Among the relations discussed in Example 1, case iv) (in which $X \sim Y$ means "the distance from city X to city Y is at most 500 miles") is symmetric, and case ii) (in which $X \sim Y$ means "country X has an ambassador in country Y") is likely to be symmetric. Relation iii) (in which $X \sim Y$ means "state X has a smaller population than state Y") is not symmetric. [In i), we have $A \neq B$, so Definition 6.7 doesn't apply.]

There are important relations which are about as "unsymmetric" as possible; namely, the only way in which both "$x \sim y$" and "$y \sim x$" can be true is if $x = y$. The following definition encapsulates this idea:

DEFINITION 6.8: For a relation R on a set A, "R is *antisymmetric*" means:

$$(\forall\, x, y \in A)\,([(x, y) \in R \text{ and } (y, x) \in R] \Rightarrow x = y$$

[Equivalently, $\sim$ is antisymmetric iff $(\forall\, x, y \in A)\,([x \sim y \text{ and } y \sim x] \Rightarrow x = y)$.]

The relation "$\leq$" on $\mathbf{R}$ is an example of an antisymmetric relation (since, if $x \leq y$ and $y \leq x$, then $x = y$).

COMMENTS

1. The relation "$<$" is also antisymmetric. The condition defining "antisymmetric" is vacuously true for "is less than", because it is impossible to have numbers x and y for which both "$x < y$" and "$y < x$" are true statements.

2. The same basic relation can be antisymmetric on one set, but not antisymmetric on another. For example, the divisibility relation "$|$" is antisymmetric as a relation on $\mathbf{N}$, but not as a relation on $\mathbf{Z}$ (for example, $-2|2$ and $2|-2$, but $-2 \neq 2$).
3. "Antisymmetric" is not the same thing as "not symmetric". The relation "$|$" on $\mathbf{Z}$ is not symmetric, but it isn't antisymmetric either.

Get Your Hands Dirty 6

Which of the relations in Example 2 are antisymmetric? Which are neither symmetric nor antisymmetric? □

Get Your Hands Dirty 7

Do each of the following, or show that the given task is impossible:

a) Make up three relations on $\{1, 2, 3, 4, 5\}$ which are symmetric.
b) Make up three relations on $\{1, 2, 3, 4, 5\}$ which are antisymmetric.
c) Make up three relations on $\{1, 2, 3, 4, 5\}$ which are neither symmetric nor antisymmetric.
d) Make up three relations on $\{1, 2, 3, 4, 5\}$ which are both symmetric and antisymmetric. □

Another natural question concerns the relationship of an object to itself. In other words, we are interested in knowing whether $x \sim x$ (or, more precisely, for a given relation, we want to determine the truth set of the open sentence "$x \sim x$"). For some relations, this open sentence is true for all x in A. This is the idea behind the following definition:

DEFINITION 6.9: For a relation R on a set A, "R is *reflexive*" means:

$$(\forall\, x \in A)\,((x, x) \in R)$$

[Equivalently, $\sim$ is reflexive iff $(\forall\, x \in A)\,(x \sim x)$.]

The relation "is equal to" is a reflexive relation.

There are also relations for which "$x \sim x$" is not true for any values of x. These are described by the following definition:

DEFINITION 6.10: For a relation R on a set A, "R is *irreflexive*" means:

$$(\forall\, x \in A)\,((x, x) \notin R)$$

[Equivalently, $\sim$ is irreflexive iff $(\forall\, x \in A)\,(x \not\sim x)$.]

The relation "is a proper subset of" is an example of an irreflexive relation.

COMMENT As with symmetric and antisymmetric, a relation does not have to be either reflexive or irreflexive. Consider the following examples, for which S is some set and $A = \mathcal{P}(S)$:

i) Define $\sim$ by "$X \sim Y$ iff $X \subseteq Y$". This relation is reflexive, since "$T \sim T$" is true for all T; i.e., "$T \subseteq T$" is true for all T.
ii) Define $\sim$ by "$X \sim Y$ iff $X \not\subseteq Y$". This relation is irreflexive, since "$T \sim T$" is false for all T; i.e., "$T \not\subseteq T$" is false for all T.
iii) Define $\sim$ by "$X \sim Y$ iff $|X \cap Y| > 1$". This is not reflexive, since, for example, "$T \sim T$" is false when $T = \emptyset$; i.e., "$|\emptyset \cap \emptyset| > 1$" is false. On the other hand, if S has at least two elements, this relation is not irreflexive either, because "$T \sim T$" will be true when $T = S$; i.e., "$|S \cap S| > 1$" is true.

Get Your Hands Dirty 8

Which of the relations in Example 2 are reflexive? Which are irreflexive? Which are neither? □

Get Your Hands Dirty 9

Do each of the following, or show that the task is impossible:

a) Make up three relations on $\{1, 2, 3, 4, 5\}$ which are reflexive.
b) Make up three relations on $\{1, 2, 3, 4, 5\}$ which are irreflexive.
c) Make up three relations on $\{1, 2, 3, 4, 5\}$ which are neither reflexive nor irreflexive.
d) Make up three relations on $\{1, 2, 3, 4, 5\}$ which are both reflexive and irreflexive. □

The last of the general properties of relations we will introduce in this section is somewhat more complicated. We motivate it by mentioning two of our early theorems:

THEOREM 2.2. If $A \subseteq B$ and $B \subseteq C$, then $A \subseteq C$.

THEOREM 2.9. If $a, b, c \in \mathbf{Z}$, and $a \mid b$ and $b \mid c$, then $a \mid c$.

You can probably think of other relations for which a similar statement would be true. (Can you? What are some of them?)

The common property described by these two theorems is formalized in the following definition:

DEFINITION 6.11: For a relation R on a set A, "R is *transitive*" means:

$$(\forall\, x, y, z \in A)\,((x, y) \in R \text{ and } [(y, z) \in R] \Rightarrow (x, z) \in R)\,.$$

[Equivalently, $\sim$ is transitive iff $(\forall\, x, y, z \in A)\,[(x \sim y \text{ and } y \sim z) \Rightarrow (x \sim z)]$.]

Thus, Theorems 2.2 and 2.9 say that the relations "$\subseteq$" and "$|$" are both transitive. The fundamental relation "is equal to" has this property as well.

Get Your Hands Dirty 10

Which of the other relations in Example 2 are transitive? □

COMMENT If a relation $\sim$ is not transitive, and $a \sim b$ and $b \sim c$, that does not mean that the statement "$a \sim c$" is necessarily false. It may be false for some combinations of a, b, and c, and true for others. For example, consider the relation on **Z** defined by "$x \sim y$ iff $|x - y| < 5$". Taking $a = 2$, $b = 5$, and $c = 9$, we have $a \sim b$ and $b \sim c$, but $a \not\sim c$, so $\sim$ is not transitive. On the other hand, letting $d = 7$, we have $d \sim b$ and $b \sim c$, and we do have $d \sim c$.

Get Your Hands Dirty 11

Make up two relations on $\{1, 2, 3, 4, 5\}$ which are transitive. □

Exploration 1: Combinations of Properties of Relations

We have described many different types of relations in this section: domain-complete, functionlike, symmetric, antisymmetric, reflexive, irreflexive, and transitive. What combinations of these properties are possible? For example, is there a relation which is domain-complete, symmetric, and irreflexive, but not transitive? Does the answer depend on the choice of A and B? If so, in what way? (*Hint:* Work with small finite sets, as well as with familiar mathematical relations discussed earlier. Actually build relations with particular combinations of properties by listing specific ordered pairs to include. Try to come up with at least two combinations of properties that are impossible.)

• E X P L O R E •

We conclude this section with one more exploration:

Exploration 2: Composition of Relations

In Section 3.2, we defined an operation called "composition" for functions. Specifically, if A, B, and C are sets, and $f : A \to B$ and $g : B \to C$, then there is a function from A to

C, called the *composition of f and g*, written $g \circ f$, which is defined by the equation $(g \circ f)(a) = g(f(a))$ (see Definition 3.20).

Since a function is a special type of relation, it is reasonable to ask whether the concept of composition can be generalized to apply to arbitrary relations. In other words, if R is a relation from A to B, and S is a relation from B to C, is there some natural way to define the "composition" of R and S, which we would write as $S \circ R$, to get a relation from A to C? (Of course, we want this general definition to be the same as that for functions in the case where R and S are actually functions.)

If you come up with a definition that seems suitable, there are many avenues for further exploration. Here are some of them:

i) Examine how the composition of R^{-1} and S^{-1} is related to the composition of R and S.

ii) Explore whether the theorems about composition of functions generalize to theorems about composition of relations. For example, look at the extension of the concepts "injective" and "surjective" to relations (see Exercise 6). Examine whether Theorems 4.3 and 4.6 still hold true in this more general context.

iii) Examine whether there are composition theorems corresponding to Theorems 4.3 and 4.6 which deal with domain-complete and functionlike relations. (*Hint:* Use the definitions and theorems in Exercise 6.)

iv) Examine the possible converses of any theorems you develop in parts ii) and iii).

• E X P L O R E •

EXERCISES

1. **a)** Which of the relations in Examples 1, 2, and 3 are domain-complete? [For v) and vi) of Example 2, describe how the answer depends on the choice of the set S.]

b) Which of the relations in Examples 1, 2, and 3 are functionlike? [For v) and vi) of Example 2, describe how the answer depends on the choice of the set S.]

2. Suppose f is a function. Comment 2 following Theorem 6.1 discussed the significance of two vertical-line tests. Express each of these horizontal-line tests in terms of similar ideas about functions:

a) "There is no horizontal line which crosses the graph in more than one point."

b) "Every horizontal line crosses the graph in at least one point."

3. Suppose A and B are finite sets, with $|A| = s$ and $|B| = t$.

a) How many relations are there from A to B? (Compare this with Exercise 12 in Section 3.2.)

b) How many domain-complete relations are there from A to B? (Compare Exercise 13 in Section 3.2.) (*Hint:* Start with some simple cases, like $s = 2$, $t = 3$, and actually list all the relations which qualify. Then look at your results, both in terms of the numerical answers and in terms of how you got all the possible sets of pairs, and generalize the results.)

c) How many functionlike relations are there from A to B, in the case where $s = 1$? where $s = 2$? where $s = 3$? (*Warning:* The domain need not be all of A.) (Compare Exercise 13 in Section 3.2.)

4. Prove the following parts of Theorem 6.2:

THEOREM 6.2. For sets A and B and a relation A to B:

ii) $\text{Im}(R^{-1}) = \text{Dom}(R)$;
iii) $(R^{-1})^{-1} = R$.

5. Suppose R is a relation on $\mathbf{R}$. Give a geometric condition on the graph of R which is necessary and sufficient for R to be equal to its inverse.

6. Suppose we extend the definitions of surjective and injective to apply to relations in general, as follows:

DEFINITION. For any sets A and B and any relation R from A to B:
"R is *surjective*" means:

$$(\forall\, v \in B)(\exists\, u \in A)((u, v) \in R).$$

"R is *injective*" means:

$$(\forall\, v \in B)(\forall\, u, w \in A)([(u, v) \in R \text{ and } (w, v) \in R] \Rightarrow u = w).$$

(This terminology is not standard.)

a) Develop and prove a theorem which describes when a relation R is surjective in terms of some property of R^{-1}.
b) Develop and prove a theorem which describes when a relation R is injective in terms of some property of R^{-1}.

7. Prove the following:

THEOREM 6.7. a) For any sets A and B, if $f: A \to B$ is invertible, then $f^{-1} \circ f = I_A$ and $f \circ f^{-1} = I_B$.

8. Find three different functions from $\mathbf{R}$ to $\mathbf{R}$ which are equal to their own inverses.

9. For each of Conjectures 1–4 (following Question 2), provide either a proof or a counterexample. (Remember to use arrow diagrams to explore.)

10. Consider the following properties of relations: symmetric (S), antisymmetric (AS), reflexive (R), irreflexive (IR), and transitive (T). For each of the following relations, state which of the preceding properties it has, and justify your answer:

a) $\sim$ defined on $\mathbf{R}$ by "$x \sim y$ iff $x - y \in \mathbf{Z}$".
b) $\perp$ defined on the set of lines in the plane by "$L_1 \perp L_2$ iff L_1 is perpendicular to L_2".
c) $\sim$ defined on $\mathbf{R}^+$ by "$x \sim y$ iff $x/y \in \mathbf{Q}$".
d) $\sim$ defined on $\mathbf{R}$ by "$x \sim y$ iff $x - y \notin \mathbf{Q}$".
e) $\sim$ defined on $\mathcal{P}(Y)$, for some set Y, by "$M \sim N$ iff $M \cap N = \varnothing$".
f) $\sim$ defined on the set of continuous functions from $[0, 1]$ to $\mathbf{R}$ by "$f \sim g$ iff $\int_0^1 f(x)\,dx \leq \int_0^1 g(x)\,dx$".
g) $\sim$ defined on $\mathbf{N}$ by "$m \sim n$ iff $\{p : p \text{ is a prime and } p \mid m\} \subseteq \{p : p \text{ is a prime and } p \mid n\}$".

11. Suppose R is a relation on A. Let R' be the complementary relation; i.e., $(x, y) \in R'$ iff $(x, y) \notin R$.

a) For each of the properties S, AS, R, IR, and T (see Exercise 10), is it true that, if R has the property, then R' has the property?

b) For each of the properties S, AS, R, IR, and T (see Exercise 10), is it true that, if R has the property, then R' does not have the property?

12. Suppose R is a relation on A. For each of the properties S, AS, R, IR, and T (see Exercise 10), write the negation of the statement "R has the property" in terms of quantifiers and ordered-pair notation for relations. (Do not use the negation symbol "$\sim$". You may, however, use the symbol "$\notin$".)

13. Consider the following set of three properties: reflexive (R), symmetric (S), and transitive (T). For each subset of {R, S, T}, make up a relation on $\{1, 2, 3, 4\}$ which has the properties in the subset and does not have the others. (For example, for the subset {R, S} of {R, S, T}, make up a relation which is reflexive and symmetric, but not transitive. {R, S, T} has eight subsets, including the whole set and the empty set, so you must make up eight relations.)

14. Suppose R is a relation on A. For each of the following statements, either give a proof or make up a counterexample. (For a counterexample, state both the set A and the relation R.)

a) $R \cap R^{-1}$ is symmetric.
b) If R is reflexive, then $R \cap R^{-1} = I_A$.
c) If $R \cap R^{-1} = I_A$, then R is reflexive and antisymmetric.
d) If R is transitive, then $R \cup R^{-1}$ is transitive.

15. Suppose R and S are relations on A. For each of the following statements, either give a proof or make up a counterexample. (For a counterexample, state both the set A and the two relations R and S.)

a) If R and S are transitive, then $R \cup S$ is transitive.
b) If $R \cap S$ is symmetric, then either R or S is symmetric.
c) If $R \cap S$ is reflexive, then R and S are reflexive.

16. (GYHD 5 and 6) Which of the relations in Example 2 are symmetric? Which are antisymmetric? Which are neither? Justify your answers.

17. Suppose A is a finite set with $|A| = n$.

a) How many relations are there on A which are symmetric?
b) How many relations are there on A which are antisymmetric?

[*Hint for both* a) *and* b): Try a small value of n, such as $n = 4$; experiment until you get a feel for the problem.]

18. (GYHD 8) Which of the relations in Example 2 are reflexive? Which are irreflexive? Which are neither? Justify your answers.

19. Suppose A is a finite set with $|A| = n$.

a) How many relations are there on A which are reflexive?
b) How many relations are there on A which are irreflexive?
c) How many relations are there on A which are both reflexive and symmetric?

(See hint from Exercise 17.)

20. (GYHD 10) Which of the relations in Example 2 are transitive? Justify your answers.

21. Prove or disprove each of the following:

a) If R is a symmetric relation, then R^{-1} is also symmetric.
b) If R is an antisymmetric relation, then R^{-1} is also antisymmetric.
c) If R is a reflexive relation, then R^{-1} is also reflexive.
d) If R is an irreflexive relation, then R^{-1} is also irreflexive.
e) If R is a transitive relation, then R^{-1} is also transitive.

22. Suppose R is a relation from A to B, and S is a relation from C to D. Come up with a reasonable definition for the "product" of R and S, as a relation from $A \times C$ to $B \times D$.

23. Consider the following definition:

DEFINITION. For sets A, B, and C, and relations R from A to B and S from B to C, the *relative product of R and S* (written "$R|S$") is the relation from A to C defined as follows:

$$R|S = \{(x, z) \in A \times C : (\exists\, y \in B)\,[(x, y) \in R \text{ and } (y, z) \in S]\}.$$

a) Show that, if the relations R and S are actually functions, then $R|S$ is equal to the composition $S \circ R$.

b) Write the definition of relative product of relations in terms of the $\sim$ notation for relations. (Use $\approx$ for R and $\approxeq$ for S.)

c) Suppose that R is a relation on a set A. Prove that R is transitive iff $R|R \subseteq R$. (In thinking about this subset condition, remember that both R and $R|R$ are subsets of $A \times A$.)

d) What is the relationship between dom$(R|S)$ and dom(R)? (Both are subsets of A.) Under what circumstances are they equal?

e) What is the relationship between Im$(R|S)$ and Im(S)? (Both are subsets of B.) Under what circumstances are they equal?

f) How can the inverse of $R|S$ be described in a general way in terms of the inverses of R and S? Explain and justify your answer.

24. As in Exercise 23, A, B, and C are sets, and R and S are relations from A to B and from B to C respectively. (This exercise uses the extended versions of the definitions of surjective and injective, which apply to relations in general—see Exercise 6.)

a) Prove the following generalization of Theorem 6.3:

i) R^{-1} is domain-complete $\Leftrightarrow$ R is surjective;

ii) R^{-1} is functionlike $\Leftrightarrow$ R is injective.

b) Prove or disprove the following analogs to Theorems 4.3 and 4.6 (see Exercise 22 for the definition of $R|S$):

Conjecture 1:

If R and S are both surjective, then the relative product $R|S$ is surjective.

Conjecture 2:

If R and S are both injective, then the relative product $R|S$ is injective.

c) Explore possible converses to the conjectures in part b). (See "Exploration 3 of Section 3.2 Revisited" in this section).

25. Suppose f is a function from A to B. We can define two set-valued functions, both based on f, as follows: Define $f: \mathcal{P}(A) \to \mathcal{P}(B)$ by

$$f(S) = \{f(x) : x \in S\}.$$

This set $f(S)$ is called the *image of S under f*; notice that $f(A)$ is just Im(f). (*Apology:* It is somewhat sloppy, and perhaps confusing, to use the letter f both for the original function and this new function, but this notation and terminology are fairly standard.)

Define $f^{-1}: \mathcal{P}(B) \to \mathcal{P}(A)$ by

$$f^{-1}(T) = \{x \in A : f(x) \in T\}.$$

This set $f^{-1}(T)$ is called the *inverse image of T under f.* (Again, the notation and terminology are standard; do not confuse this *function* f^{-1} with the *relation* that is the inverse of the original function f.) These functions on the power sets are sometimes called *induced functions.* In terms of these definitions, discuss the following questions:

a) Are the new functions described above inverses of each other? If not in general, under what circumstances?

b) If $f: A \to B$ is surjective or injective, is the same true for the function $f: \mathcal{P}(A) \to \mathcal{P}(B)$? for the function $f^{-1}: \mathcal{P}(B) \to \mathcal{P}(A)$?

c) If S_1 and S_2 are subsets of A, how is $f(S_1 \cap S_2)$ related to $f(S_1)$ and $f(S_2)$? What about union? set complement?

26. Suppose f is a function from A to B, and C is a subset of A. We can define a function from C to B, called the *restriction of f to C* (denoted "$f\big|_C$") as follows:

$$f\big|_C = \{(x, y) \in f : x \in C\}.$$

In other words, $f\big|_C$ consists of those ordered pairs from f whose first element is in C, so for $x \in C, f\big|_C(x)$ and $f(x)$ are the same. In terms of this definition, discuss the following questions:

a) If f is surjective or injective, is the same true for the function $f\big|_C$?

b) If $f\big|_C$ is surjective or injective, is the same true for the function f?

6.2 Partitions and Equivalence Relations

Note: All relations in this section will be relations *on a set A*—that is, relations from A to itself.

Partitions

As suggested in the introduction to this chapter, equivalence relations are used to describe "similarities" between certain objects from a set. Such similarities are often used to decompose a set into several smaller subsets, by placing "similar" objects together in the same subset. This decomposition process, formally known as a *partition,* has a natural, intuitive flavor.

The concept of an *equivalence relation* is very closely related to the concept of a partition. An equivalence relation is defined to be a relation that is reflexive, symmetric, and transitive. To the student of mathematics, it may not be clear why this combination of properties should be important. We will spend some time examining the relationship between partitions and equivalence relations, before making the formal definition. In a sense, the concept of equivalence relation is a formal device designed to make working with partitions easier. Although partitions are perhaps easier to understand intuitively, equivalence relations are easier to work with in proofs, and so they are commonly used in contemporary mathematics.

The idea of a partition is to take a set and break it up into a bunch of smaller sets. A diagram such as Figure 6.2 suggests what a partition is all about.

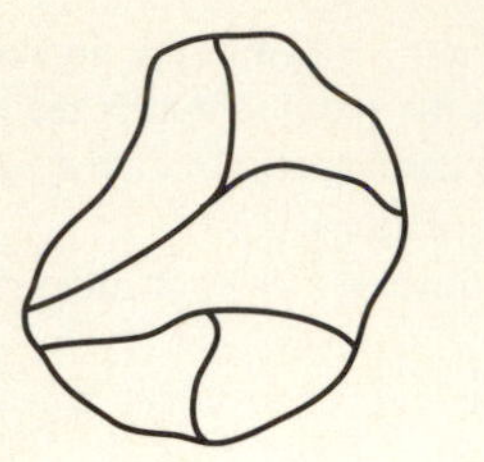

Figure 6.2. This set is partitioned. Each of the subsets is a *part* of the partition.

The formal definition of a partition is as follows:

> **DEFINITION 6.12:** For a set A, a *partition of* A is a set $\mathscr{C}$ of subsets of A (i.e., $\mathscr{C} \subseteq \mathscr{P}(A)$) such that:
>
> **a)** $(\forall\, x \in A)(\exists\, U \in \mathscr{C})(x \in U)$,
> **b)** $(\forall\, U \in \mathscr{C})(U \neq \varnothing)$,
> **c)** $(\forall\, U, V \in \mathscr{C})(U \cap V \neq \varnothing \Rightarrow U = V)$.
>
> We will refer to the elements of $\mathscr{C}$ as the *parts* of the partition. (This is not standard terminology. We use it to emphasize that the elements of $\mathscr{C}$ are sets.)

What does this mean? First of all, $\mathscr{C}$ is a subset of $\mathscr{P}(A)$. In other words, the elements of $\mathscr{C}$, i.e., the parts of the partition, represented by the variables U and V in the definition, are subsets of A.

In Figure 6.2, the overall figure is the set A, and the sections into which it is divided are the parts of the partition. Here is an example:

Example 1. A Sample Partition Let $A = \{1, 2, 3, 4, 5\}$ and let $\mathscr{C} = \{\{1, 2\}, \{3\}, \{4, 5\}\}$. The parts of this partition are the sets $\{1, 2\}$, $\{3\}$, and $\{4, 5\}$. □

We will look next at the three conditions in the definition of "partition", and see how they apply to Example 1:

Condition a) says that every element of A belongs to (at least) one of the elements of $\mathscr{C}$, i.e., to one of the parts. We can show this explicitly for Example 1: $1 \in \{1, 2\}$, $2 \in \{1, 2\}$, $3 \in \{3\}$, $4 \in \{4, 5\}$, and $5 \in \{4, 5\}$.

Condition b) says that none of the parts of the partition is empty. It is clear that this is true for Example 1 from the list of sets in $\mathscr{C}$.

Condition c) is the most complicated of the three conditions. It says that, if two of the parts are not disjoint, then they are equal; that is, if two parts have at least one element in common, then they are the same part. This may be clearer if the implication in c) is replaced by its contrapositive: $U \neq V \Rightarrow U \cap V = \varnothing$. This contrapositive says that, if you take two different parts of the partition, they will be disjoint, i.e., that different parts wiill not overlap. Such sets are called *pairwise disjoint*. Again, the partition $\mathscr{C}$ in Example 1 has this property.

COMMENTS

1. In working with a specific partition, such as Example 1, it is often easier to verify the contrapositive of c) than c) itself. However, as part of a proof about partitions in general, the original version of c) is usually easier to work with. (Isn't it nice that we know they mean the same thing?)

2. Conditions a) and c) combined say that each element of A belongs to exactly one of the parts of the partition. We can therefore define a function from A to $\mathscr{C}$, in which $f(x)$ is *the* element of $\mathscr{C}$ that contains x. This function is sometimes called the *quotient function* (or *canonical function*) *from A to* $\mathscr{C}$.

3. Condition b) is included in Definition 6.12 because having an "empty part" does not contribute anything to the notion and would lead to unnecessary complications.

You may find the following metaphor useful in thinking about partitions:

Metaphor: Partition as Countries; Elements as Citizens

Imagine the parts of a partition as representing different countries, and the elements of the underlying set as individual persons. For an element to belong to one of the parts of the partition is like an individual being a citizen of that country. The requirements for a partition tell us:

i) every individual is a citizen of some country,
ii) no country is citizen-less, and
iii) no individual has dual citizenship.

Get Your Hands Dirty 1

What is *not* a partition? For each of the following combinations of properties, find a subset $\mathscr{C}$ of $\mathscr{P}(\{1, 2, 3, 4, 5\})$ as described:

i) $\mathscr{C}$ satisfies conditions b) and c), but fails to satisfy condition a).
ii) $\mathscr{C}$ satisfies conditions a) and c), but fails to satisfy condition b).
iii) $\mathscr{C}$ satisfies conditions a) and b), but fails to satisfy condition c). □

Example 2. **Some Important Mathematical Partitions** The following are commonly used partitions of familiar mathematical sets:

i) $A = \mathbf{Z}: \mathscr{C} = \{\mathscr{E}, \mathscr{O}\}$, where $\mathscr{E}$ is the set of even integers and $\mathscr{O}$ is the set of odd integers: Two integers are in the same part of $\mathscr{C}$ iff they have the same parity, i.e., are both even or both odd.

ii) A = the set of all quadratic equations with real coefficients: $\mathscr{C}$ has three parts, which we label E_1, E_2, and E_3, and which are defined as follows:

E_1 = the set of all quadratic equations with no real roots,
E_2 = the set of all quadratic equations with two distinct real roots, and
E_3 = the set of all quadratic equations with one double root.

We can determine which part of the partition an equation "$ax^2 + bx + c = 0$" belongs to by evaluating its *discriminant* $d = b^2 - 4ac$. The quadratic formula shows that the equation belongs to E_1 if $d < 0$, to E_2 if $d > 0$, and to E_3 if $d = 0$.

iii) $A = \{X \in \mathcal{P}(\mathbf{Z}) : X \text{ is finite}\}$: Each part of $\mathscr{C}$ consists of all subsets of $\mathbf{Z}$ of a particular (finite) cardinality. For example, the set consisting of all one-element subsets of $\mathbf{Z}$ is one part of $\mathscr{C}$. Notice that $\{\varnothing\}$ is one of the parts of $\mathscr{C}$, since $\varnothing$ is the only subset of $\mathbf{Z}$ of cardinality 0. [This does not violate condition b): we are saying that $\{\varnothing\} \in \mathscr{C}$, not that $\varnothing \in \mathscr{C}$.] This partition has infinitely many parts, one for each nonnegative integer. □

Why Use Partitions?

Probably the primary reason for working with partitions of a set is to simplify a situation. If A is an infinite set, for example, there may be questions about A which can be simplified by being broken down into questions about finitely many separate cases. In order to use such a method, we need to be sure that every element of A falls under one of the cases.

The following example illustrates a situation where the informal use of a partition makes a problem much simpler:

Example 3. **Solving an Equation Using Parity** Suppose $A = \mathbf{Z}$, and consider the following question:

$$\text{Are there integers } x \text{ and } y \text{ such that } x^2 + y^2 = 74{,}203?$$

Certainly, this question could be answered by checking all the integers sufficiently small to be feasible. (x and y both must be less than or equal to $\sqrt{74{,}203}$.) But it is easier to examine this question by splitting it up into separate cases depending on the parity of x and y, as follows:

If x and y are both even, the sum of their squares is even, so there can't be a solution of this type. The same is true if both are odd.

So, if there is a solution, we can assume that one of x and y is even and the other is odd. Without loss of generality, we can assume that x is even and y is odd, so $x = 2n$ and $y = 2m + 1$, for some integers n and m.

Then $x^2 + y^2 = 4n^2 + 4m^2 + 4m + 1 = 4(n^2 + m^2 + m) + 1$. But if $x^2 + y^2 = 74{,}203$, then we have $4(n^2 + m^2 + m) + 1 = 74{,}203$, so that $4(n^2 + m^2 + m) = 74{,}202$. Since $4 \nmid 74{,}202$, this is impossible, so the original equation has no integer solution. □

Essentially, the purpose of using a partition is to simplify information, by replacing a set A by a smaller collection of its subsets. Any time that we summarize information, such as by means of a chart or graph, we "collapse" the information into categories in order to understand it better. The information is put into a more usable form, and we can focus more clearly on the relevant aspects of a situation.

For example, we might break down a census by age brackets, in order to understand population changes. In doing so, we are partitioning the set of possible ages into certain intervals that seem useful. Inevitably, however, when we do so, we also lose information, because we no longer indicate how many people there are of a particular age within an age bracket.

In Example 3, it turned out that the parity of x and y was the key to showing that the equation didn't have a solution. It didn't matter which even or odd integer x or y was, as long as we knew that one was even and one was odd.

Partition iii) in Example 2 is useful in doing proofs by induction. The proof of Theorem 5.9, which says that $|\mathcal{P}(A)| = 2^{|A|}$, is a good example of this.

The trick is to choose the partition in such a way that the information we lose is irrelevant to the problem. Demographers use their training to decide what age brackets will be useful. Mathematicians use their experience in solving mathematical problems to decide what partition is likely to be helpful.

COMMENT Remember always to distinguish between exploration and proof. In Example 3, a good student may ask why we chose to use the particular partition of odd and even. The student might completely understand the reasoning of the proof, and still have legitimate questions about the process by which we *found* the proof.

Knowing what partition will be appropriate to a given problem is not easy. This is a matter for exploration, trial and error, and patience. In the case above, experience with many similar questions would be helpful in suggesting that the parity partition is a good one to work with. There may be other partitions that would be equally helpful.

Making a Relation from a Partition

Intuitively, a partition is generally formed by grouping together the objects in A that share some common property. If $\mathscr{C}$ is a partition on the set A, then it makes sense to describe two elements of A as "similar" or "related" if they belong to the same part of the partition. We will use relations defined in terms of partitions as the prototype for what we will define as an equivalence relation.

We will actually define a relation starting with an arbitrary collection $\mathscr{C}$ of subsets of A, and then see what happens if $\mathscr{C}$ is a partition.

DEFINITION 6.13: For a set A, and for $\mathscr{C} \subseteq \mathcal{P}(A)$, the *relation associated with* $\mathscr{C}$ (denoted "$\sim_{\mathscr{C}}$" and read "squiggle $\mathscr{C}$") is the relation on A defined by

$$x \sim_{\mathscr{C}} y \quad \text{iff} \quad (\exists\, U \in \mathscr{C})\,(x \in U \text{ and } y \in U)\,.$$

In ordinary language, "$x \sim_{\mathscr{C}} y$" means that there is some element of $\mathscr{C}$ which contains both x and y.

Example 4. **The Relation Associated with a Collection of Subsets** Let $A = \{1, 2, 3, 4\}$ and $\mathscr{C} = \{\{1, 2\}, \{2, 3\}\}$. Then $\sim_{\mathscr{C}}$ is as follows:

$$\sim_{\mathscr{C}} = \{(1, 1), (2, 2), (3, 3), (1, 2), (2, 1), (2, 3), (3, 2)\}.$$

Notice that $\sim_{\mathscr{C}}$ includes pairs like $(1, 1)$: there is no reason why x and y have to be different. □

Get Your Hands Dirty 2

List the ordered pairs in $\sim_{\mathscr{C}}$ if $\mathscr{C} = \{\{1, 2\}, \{3\}, \{4, 5\}\}$. □

QUESTION 1 What properties does such a relation $\sim_{\mathscr{C}}$ have? What properties does it have if $\mathscr{C}$ is a partition?

? **?** **?**

No matter what $\mathscr{C}$ is, $\sim_{\mathscr{C}}$ will be symmetric, since, if an element U of $\mathscr{C}$ contains both x and y, it also contains both y and x.

Example 4 shows that $\sim_{\mathscr{C}}$ is not necessarily reflexive: in that example, $4 \not\sim_{\mathscr{C}} 4$, since 4 does not belong to any element of $\mathscr{C}$. But if every element of A does belong to an element of $\mathscr{C}$, then $\sim_{\mathscr{C}}$ is reflexive. In particular, if $\mathscr{C}$ is a partition, then $\sim_{\mathscr{C}}$ is reflexive.

What about transitivity?

? **?** **?**

Example 4 shows that, in general, $\sim_{\mathscr{C}}$ need not be transitive. In that example, we have $1 \sim_{\mathscr{C}} 2$ and $2 \sim_{\mathscr{C}} 3$, but $1 \not\sim_{\mathscr{C}} 3$. However, if $\mathscr{C}$ is a partition, then $\sim_{\mathscr{C}}$ is transitive. We will prove this as part of the following theorem, which states all three properties of $\sim_{\mathscr{C}}$ for partitions:

THEOREM 6.8. For a set A, and for $\mathscr{C} \subseteq \mathscr{P}(A)$, if $\mathscr{C}$ is a partition of A, then $\sim_{\mathscr{C}}$ is reflexive, symmetric, and transitive.

Proof: Suppose that $\mathscr{C}$ is a partition. We showed above that $\sim_{\mathscr{C}}$ is reflexive and symmetric.

In order to prove that $\sim_{\mathscr{C}}$ is transitive, suppose that $x, y, z \in A$, with $x \sim_{\mathscr{C}} y$ and $y \sim_{\mathscr{C}} z$. We need to show that $x \sim_{\mathscr{C}} z$.

By definition of $\sim_{\mathscr{C}}$, there exist parts $U, V \in \mathscr{C}$ such that $x, y \in U$ and $y, z \in V$. Since $y \in U \cap V$, we have $U \cap V \neq \varnothing$, and so $U = V$, by condition c) of the definition of partition. It follows from the definition of $\sim_{\mathscr{C}}$ that $x \sim_{\mathscr{C}} z$. ■

Equivalence Relations

In the above discussion, we began with an arbitrary subset $\mathscr{C}$ of $\mathscr{P}(A)$, and defined a relation $\sim_{\mathscr{C}}$ on A, the relation associated with $\mathscr{C}$. We saw that, if $\mathscr{C}$ is a partition, then $\sim_{\mathscr{C}}$ is reflexive, symmetric, and transitive.

These three properties are combined in the following definition:

DEFINITION 6.14: For a set A, an *equivalence relation on A* is a relation on A which is reflexive, symmetric, and transitive.

Our discussion in Section 6.1 shows that the fundamental relation "is equal to" is an equivalence relation on any set. Using Definition 6.14, we can restate Theorem 6.8 more concisely:

THEOREM 6.8′. If $\mathscr{C}$ is a partition, then $\sim_{\mathscr{C}}$ is an equivalence relation.

It is natural to ask the following question:

QUESTION 2 Can we reverse the process described in Definition 6.13 and in Theorem 6.8? That is, can we begin with an arbitrary relation $\sim$ on a set A, and define a subset $\mathscr{C}^{\sim}$ of $\mathscr{P}(A)$ to go with it? If so, can we draw any conclusions about the subset if the relation is an equivalence relation? Is the subset necessarily a partition?

? ? ?

We can answer all of these questions affirmatively. We begin by describing how $\mathscr{C}^{\sim}$ is formed. We begin with an arbitrary relation $\sim$ on a set A. Our intuitive picture is that "$x \sim y$" should mean that x and y are "similar" in some way. How might we go about creating a collection of subsets of A based on $\sim$?

? ? ?

The idea is to put objects together if they are "similar". To do this, we start with a general element of A, say t, and state which elements of A should go with t.

In other words, we form a set, which we will label C_t, consisting of those elements of A to which t is related. Formally,

$$C_t = \{z \in A : z \sim t\}.$$

This is certainly a subset of A, and we can create a subset $\mathscr{C}^{\sim}$ of $\mathscr{P}(A)$ by defining $\mathscr{C}^{\sim} = \{C_t : t \in A\}$.

We make this into a formal definition:

DEFINITION 6.15: For a set A and a relation $\sim$ on A, the *subset collection associated with* $\sim$ (denoted "$\mathscr{C}^{\sim}$" and read "$\mathscr{C}$ squiggle") is $\{C_t : t \in A\}$, where $C_t = \{z \in A : z \sim t\}$.

The following example gives a relation for which $\mathscr{C}^{\sim}$ is not a partition:

Example 5. **The Subset Collection Associated with a Relation** Let $\sim$ be the relation on $\{1, 2, 3, 4\}$ such that "$x \sim y$" means $|x - y| < 2$. (This relation is reflexive and symmetric, but not transitive.) Then $C_1 = \{1,2\}$, $C_2 = \{1,2,3\}$, $C_3 = \{2,3,4\}$, and $C_4 = \{3,4\}$. $\mathscr{C}^{\sim}$ is the collection $\{\{1,2\},\{1,2,3\},\{2,3,4\},\{3,4\}\}$, and is not a partition. □

Get Your Hands Dirty 3

Find each of the sets C_t for the relation "≤" on the set $A = \{1,2,3,4,5\}$. Is $\{C_t : t \in A\}$ a partition of A for this relation? □

With this discussion as background, we state the counterpart of Theorem 6.8.

THEOREM 6.9. For a set A and a relation $\sim$ on A, if $\sim$ is an equivalence relation, then $\mathscr{C}^{\sim}$ is a partition of A.

Proof: Suppose $\sim$ is an equivalence relation on A, and the sets C_t and the collection $\mathscr{C}^{\sim}$ are as given in Definition 6.15:

$$C_t = \{z \in A : z \sim t\} \quad \text{and} \quad \mathscr{C}^{\sim} = \{C_t : t \in A\}.$$

We need to show that the collection of sets $\mathscr{C}^{\sim}$ satisfies conditions a), b), and c) of the definition of partition. For convenience, we restate those conditions:

a) $(\forall\, x \in A)(\exists\, U \in \mathscr{C})(x \in U)$,
b) $(\forall\, U \in \mathscr{C})(U \neq \varnothing)$,
c) $(\forall\, U, V \in \mathscr{C})(U \cap V \neq \varnothing \Rightarrow U = V)$.

Because $\sim$ is reflexive, we have $x \sim x$, so $x \in C_x$ for any $x \in A$. This fact shows both that every x belongs to some element of $\mathscr{C}^{\sim}$, which is condition a), and that no element of $\mathscr{C}^{\sim}$ is empty, which is condition b).

So it remains to prove condition c). So suppose that we have two elements of the set $\mathscr{C}^{\sim}$, which we can label C_x and C_y, and suppose that $C_x \cap C_y \neq \varnothing$. Thus there is some element $z \in C_x \cap C_y$. We need to show that $C_x = C_y$. We will show $C_x \subseteq C_y$, and the proof of the reverse direction is similar. So suppose that $w \in C_x$. We need to show $w \in C_y$.

Comment: The rest of the proof is just applying the definition of the sets C_t again and again, and using the fact that $\sim$ is symmetric and transitive. (We have already used reflexivity.) Before reading on, you should try to put the pieces together on your own.

? ? ?

Since $w \in C_x$, we have $w \sim x$. Since $z \in C_x \cap C_y$, we have $z \sim x$ and $z \sim y$. We need to show $w \in C_y$, i.e., we need to show that $w \sim y$.

Comment: Again, if you didn't get it already, try to combine the information in the previous paragraph to get the desired result. It's not hard, and there is more than one way to do it. It just requires some trial and error. Look at the definitions of "symmetric" and "transitive".

Since $z \sim x$, and $\sim$ is symmetric, we have $x \sim z$. Since $x \sim z$, and $z \sim y$ (from above), and $\sim$ is transitive, we have $x \sim y$. Finally, since $x \sim y$, and $w \sim x$ (from above), and $\sim$ is transitive, we have $w \sim y$, as desired.

Thus we have shown that $C_x \subseteq C_y$. The proof of the reverse direction is identical, simply switching all the x's and y's. This completes the proof of the theorem. ■

Thus, Theorem 6.9 is essentially the converse of Theorem 6.8. The combination of results underscores the significance of the concept of equivalence relation. For an equivalence relation $\sim$, we have the following alternative standard terminology for the parts of the partition $\mathscr{C}^\sim$.

DEFINITION 6.16: For any equivalence relation $\sim$ on a set A and any $t \in A$, the *equivalence class of t under* $\sim$ (written $[t]$ or $[t]_\sim$) is $\{z \in A : z \sim t\}$.

Thus, the equivalence class of t is simply the set C_t, and the equivalence classes under $\sim$ are the same sets as the parts of $\mathscr{C}^\sim$. In this context, $\mathscr{C}^\sim$ is called the *partition associated with* $\sim$.

Note: It turns out that $\mathscr{C}^\sim$ can be a partition even if $\sim$ is not an equivalence relation—Exercise 7 asks you to create an example of this.

Equivalence Relations vs. Partitions

Theorems 6.8 and 6.9 essentially tell us that the two concepts of partition and equivalence relation are "equally good"—anything that can be expressed or explained in terms of one can also be expressed or explained in terms of the other. This fact will give us the freedom to use whichever one happens to be more convenient in a particular situation.

After all this discussion of the relationship between partitions and equivalence relations, you may wonder why we bother with both. After all, we just said that anything that can be expressed in terms of one can also be expressed in terms of the other. Why not just introduce one of these concepts and leave it at that?

Our answer reflects the fact that doing mathematics is a mixture of intuition and formalism. The concept of a partition is a simple, intuitive one. We can draw a picture to give a sense of what is going on. Unfortunately, writing proofs directly in terms of partitions is often clumsy, and even defining the parts of a partition can be cumbersome.

The concept of an equivalence relation, on the other hand, is comparatively easy to work with, and, in practice, mathematicians work much more with equivalence relations than they do with partitions. Defining a specific relation is usually straightforward, and proving that a particular relation is reflexive, symmetric, and transitive is generally a matter of several very simple steps. (See Definition 6.17 and Theorem 6.11 below, for example.)

The problem with the concept of an equivalence relation for the beginning mathematician is that it lacks the natural motivation inherent in partitions. As we mentioned before, it is not clear at first why one would want to single out this particular category of relations. It may help to think of an equivalence relation as a kind of "generalized equality." When an ordered pair (x, y) belongs to a given equivalence relation, we are saying that x and y are "the same in certain respects". The equivalence relation defines the attributes that we want to pay attention to, and two objects that are the same in these ways are considered equivalent. The familiar examples of congruence for triangles and equivalence for ordinary fractions are good illustrations of this. For example, though $\frac{1}{3}$ and $\frac{2}{6}$ are not the same fractions (they have different numerators and different denominators), they do represent the same numerical quantity, and so we call them equivalent. Within certain contexts, we can treat them as if they are interchangeable.

If we understand the ideas behind Theorems 6.8 and 6.9, then we can go back and forth comfortably between the two concepts, using partitions for our intuitive understanding, and working with equivalence relations when proving theorems.

Examples of Equivalence Relations

We begin with a specific relation, and look at its equivalence classes:

Example 6. Finding an Equivalence Class from an Equivalence Relation Consider the relation defined on $\mathbf{N}$ as follows:

$$m \sim n \quad \text{iff} \quad \{p : p \text{ is a prime and } p \mid m\} = \{p : p \text{ is a prime and } p \mid n\}.$$

(We will prove a theorem shortly that shows that this is an equivalence relation.)

This relation may be rather intimidating, but a little "getting our hands dirty" will get us through it. The two sets described depend, respectively, on m and n. Let's take a particular positive integer, say $m = 60$, and find its equivalence class. First of all, what is $\{p : p \text{ is a prime and } p \mid 60\}$?

? ? ?

We simply have to factor 60 into primes: $60 = 2^2 \times 3 \times 5$. Thus $\{p : p \text{ is a prime and } p \mid 60\} = \{2, 3, 5\}$.

Now, in order for some other positive integer to be "related" to 60, it must be divisible by 2, by 3, by 5 and by no other primes. For example, $30 \sim 60$; $300 \sim 60$; $960 \sim 60$. But $12 \not\sim 60$ (because 5 is not a divisor of 12), and $210 \not\sim 60$ (because another prime, 7, *is* a divisor of 210).

Thus, the equivalence class of 60 is $\{2^a \times 3^b \times 5^c : a, b, c \in \mathbf{N}\}$. □

Get Your Hands Dirty 4

For the relation of Example 6, find three elements (if possible) in each of the following equivalence classes:

a) [10],
b) [7],
c) [1].

Our next example puts Example 6 in the context of a more general process:

Example 7. **A "Generic" Equivalence Relation** One general way to define an equivalence relation on a set A is to start with a function f from A to some set B. We then define a relation on A, based on this function, as follows:

$$x \sim y \quad \text{iff} \quad f(x) = f(y)\,. \qquad \square$$

We can rephrase Example 6 in this model, by considering the function $f: \mathbf{N} \to \mathcal{P}(\mathbf{N})$ given by $f(t) = \{p : p \text{ is a prime and } p \mid t\}$. The relation is then defined by: $m \sim n$ iff $f(m) = f(n)$. The fact that this relation is an equivalence relation is a special case of the following theorem:

THEOREM 6.10. For any sets A and B and any function $f: A \to B$, the relation $\sim$ given by "$x \sim y$ iff $f(x) = f(y)$" is an equivalence relation.

Proof: We need to prove three things about $\sim$:

i) $\sim$ is reflexive: This says that, for any x, $f(x) = f(x)$.
ii) $\sim$ is symmetric: This says that, if $f(x) = f(y)$, then $f(y) = f(x)$.
iii) $\sim$ is transitive: This says that if $f(x) = f(y)$ and $f(y) = f(z)$, then $f(x) = f(z)$.

All three of these assertions follow from the fact that the relation "=" is an equivalence relation. ■

We will refer to the relation defined in Theorem 6.10 as the *equivalence relation induced on A by f*, and to the partition associated with this equivalence relation as the *partition induced on A by f*.

Congruence modulo *n*

We conclude this section with a very important family of equivalence relations on **Z** and their associated partitions. We begin with a formal definition:

DEFINITION 6.17: For $x, y \in \mathbf{Z}$ and $n \in \mathbf{N}$, "x is *congruent to y modulo n*" [written "$x \equiv y \pmod{n}$"] means: $n \mid (x - y)$.

For any particular $n \in \mathbf{N}$, the open sentence "$x \equiv y \pmod{n}$" describes a relation on **Z**, which we call *congruence modulo n* [usually shortened to "congruence mod n", and written "$\equiv \pmod{n}$"]. The integer n is sometimes called the *modulus*.

We will prove shortly that "$\equiv \pmod{n}$" is an equivalence relation. To get some insight into this relation, we look at a specific case, adopting the notation of equivalence classes in anticipation of our theorem:

Example 8. **Congruence mod 6** What are the sets C_t, i.e., the equivalence classes $[t]$, for this relation?

The simplest case is $t = 0$: applying the general definition of $\mathscr{C}^{\sim}$ to congruence mod 6, we have $[0] = \{x \in \mathbf{Z} : 6 \mid (x - 0)\}$. Thus, $[0]$ is just the set of multiples of 6; i.e., $C_0 = 6\mathbf{Z}$.

Similarly, for any t, $[t] = \{x \in \mathbf{Z} : 6 \mid (x - t)\}$. But if $6 \mid (x - t)$, then $x - t = 6n$ for some $n \in \mathbf{Z}$, so $x = t + 6n$. In other words, $[t] = \{t + 6n : n \in \mathbf{Z}\}$. For example,

$$[1] = \{\ldots, 1 - 12, 1 - 6, 1, 1 + 6, 1 + 12, \ldots\}.$$

Thus, we have the following:

$$[0] = \{\ldots, -12, -6, 0, 6, 12, 18, \ldots\};$$
$$[1] = \{\ldots, -11, -5, 1, 7, 13, 19, \ldots\};$$
$$[2] = \{\ldots, -10, -4, 2, 8, 14, 20, \ldots\};$$
$$[3] = \{\ldots, -9, -3, 3, 9, 15, 21, \ldots\};$$
$$[4] = \{\ldots, -8, -2, 4, 10, 16, 22, \ldots\};$$
$$[5] = \{\ldots, -7, -1, 5, 11, 17, 23, \ldots\}.$$

Notice that, if t is any multiple of 6, then $[t] = [0]$. More generally, for any integers m and n, if $m \in [n]$, then $[m] = [n]$ (see Exercise 8). □

Get Your Hands Dirty 5

Find 10 ordered pairs (x, y) such that $x \equiv y \pmod 4$. □

We now prove that these congruence relations are all equivalence relations:

THEOREM 6.11. For $n \in \mathbf{N}$, congruence modulo n is an equivalence relation on $\mathbf{Z}$.

Proof: Let n be an element of $\mathbf{N}$. We need to prove that the relation "$\equiv \pmod n$" is reflexive, symmetric, and transitive. So suppose $x, y, z \in \mathbf{Z}$.

First, we need to show that $x \equiv x \pmod n$. But we have $x - x = n \times 0$, and so $n \mid (x - x)$, as needed under the definition of "$\equiv \pmod n$". Thus, "$\equiv \pmod n$" is reflexive.

Second, suppose $x \equiv y \pmod n$, so that $n \mid (x - y)$. Thus, for some integer t, we have $x - y = n \cdot t$. But then $y - x = n \cdot (-t)$ (with $-t \in \mathbf{Z}$), and so $n \mid (y - x)$, i.e., $y \equiv x \pmod n$. Thus, "$\equiv \pmod n$" is symmetric.

Finally, suppose $x \equiv y \pmod n$ and $y \equiv z \pmod n$. Then there are integers s and t such that $x - y = n \cdot s$ and $y - z = n \cdot t$. Adding these two equations gives $(x - y) + (y - z) = n \cdot s + n \cdot t$, which simplifies to $x - z = n \cdot (s + t)$. Since $s + t \in \mathbf{Z}$, the last equation shows that $n \mid (x - z)$, i.e., $x \equiv z \pmod n$. Thus, "$\equiv \pmod n$" is transitive. ■

The equivalence relation "$\equiv$ (mod n)" is important for understanding many mathematical problems, as is the partition it induces on $\mathbf{Z}$. This partition is often labeled "$\mathbf{Z}_n$". For emphasis, we will make this a formal definition:

DEFINITION 6.18: For $n \in \mathbf{N}$, $\mathbf{Z}_n$ is the partition on $\mathbf{Z}$ induced by the equivalence relation "$\equiv$ (mod n)", defined for $x, y \in \mathbf{Z}$ by "$x \equiv y$ (mod n) iff $n \mid (x - y)$".

Thus, for example, $\mathbf{Z}_6$ is the set consisting of the six subsets of $\mathbf{Z}$, namely [0], [1], [2], [3], [4], and [5], described in Example 8. We will sometimes write "$[x]_n$" to represent the equivalence class of x under the equivalence relation "$\equiv$ (mod n)".

The partition $\mathbf{Z}_n$ is formally defined by the general results relating equivalence relations and partitions. The following results, which generalize Example 8, give a more concrete description of the partition and another way of describing the equivalence relation:

THEOREM 6.12. For $n \in \mathbf{N}$, every integer belongs to precisely one of the sets $[0]_n$, $[1]_n, [2]_n, \ldots, [n-1]_n$. Thus, $\mathbf{Z}_n = \{[0]_n, [1]_n, [2]_n, \ldots, [n-1]_n\}$.

THEOREM 6.13. For $x, y \in \mathbf{Z}$, $n \in \mathbf{N}$, $x \equiv y$ (mod n) iff x and y have the same remainder under division by n.

These theorems can be proved using the Division Algorithm; the proofs are left as Exercise 10.

We will return to the partition $\mathbf{Z}_n$ in Section 7.1 when we discuss binary operations.

EXERCISES

1. State whether each of the following collections of sets $\mathscr{C}$ is a partition of the indicated set A. If it is a partition, then describe the associated equivalence relation (either symbolically or in words). If it is not, state which of the three conditions in Definition 6.12 it fails to satisfy. (There may be more than one.) Justify your answers.
   a) $A = \mathbf{R}$, $\mathscr{C} = \{[n, n+1) : n \in \mathbf{Z}\}$, where each $[n, n+1)$ is an interval in $\mathbf{R}$.
   b) $A = \mathbf{Q}$, $\mathscr{C} = \{A_r : r \in \mathbf{N}\}$, where $A_r = \{t/r : t \in \mathbf{Z}\}$.
   c) $A = \mathbf{R} \times \mathbf{R}$, $\mathscr{C} = \{C_r : r \in \mathbf{R}\}$, where $C_r = \{(x, y) : y = x^2 + r\}$.
   d) $A = \mathbf{R} \times \mathbf{R}$, $\mathscr{C}$ = the set of lines with slope 1.
   e) $A = \mathbf{Z} \times \mathbf{Z}$, $\mathscr{C} = \{A_r : r \in \mathbf{Q}\}$, where
   $$A_r = \left\{(m, n) : m, n \in \mathbf{Z}, \frac{m}{n} = r\right\}.$$
   f) $A = \mathbf{N}$, $\mathscr{C} = \{A_r : r \in \mathbf{W}\}$, where $A_r = \{n : 2^r \mid n \text{ and } 2^{r+1} \nmid n\}$.
2. State whether each of the following relations is an equivalence relation on the indicated set. If it is not, state which of the three properties—reflexivity, symmetry, and transitivity—it fails to satisfy. (There may be more than one.) Justify your answers.

a) Define $\sim$ on $\mathbf{R}$ by "$x \sim y$ iff $x - y \notin \mathbf{Z}$".
b) Let A be a set and B a particular subset of A. Define $\sim$ on $\mathcal{P}(A)$ by "$X \sim Y$ iff $X \cap B = Y \cap B$".
c) Define $\sim$ on $\mathbf{R}^+$ by "$x \sim y$ iff $x/y \in \mathbf{Q}$".
d) Define $\sim$ on $\mathbf{N}$ by "$m \sim n$ iff $2 \mid mn$".
e) Define $\sim$ on $\mathbf{R} \times \mathbf{R}$ by "$(x, y) \sim (u, v)$ iff $2x - y = 2u - v$".
f) Let A be a set, and define $\sim$ on $\mathcal{P}(A)$ by

$$\text{“}X \sim Y \quad \text{iff} \quad X \cap Y \neq \varnothing\text{”}.$$

3. For each of the following equivalence relations, find two specific, distinct elements in the equivalence class of the given element (you do not need to prove that the relations defined here are equivalence relations):

a) Define $\sim$ on $\mathbf{N}$ by "$m \sim n$ iff

$$\{p : p \text{ is a prime and } p \mid m\} = \{p : p \text{ is a prime and } p \mid n\}\text{”}$$

Find $x, y \in [12]$.

b) Define $\sim$ on $\mathcal{P}(\{1, 2, 3, 4, 5\})$ by

$$\text{“}A \sim B \quad \text{iff} \quad A \cup \{1, 2\} = B \cup \{1, 2\}\text{”}.$$

Find $X, Y \in [\{2, 4, 5\}]$.

c) Define $\sim$ on the set of continuous functions from $[0, 1]$ to $\mathbf{R}$ by

$$\text{“}f \sim g \quad \text{iff} \quad \int_0^1 f(x)\, dx = \int_0^1 g(x)\, dx\text{”}.$$

Find $h_1, h_2 \in [h]$, where $h(x) = x^2$.

4. Suppose f is a function from A to B. In Exercise 25 of Section 6.1, we defined two "induced functions" $f : \mathcal{P}(A) \to \mathcal{P}(B)$ and $f^{-1} : \mathcal{P}(B) \to \mathcal{P}(A)$, based on f. We repeat those definitions here:

For $S \in \mathcal{P}(A)$: $\qquad f(S) = \{f(x) : x \in S\}$.
For $T \in \mathcal{P}(B)$: $\qquad f^{-1}(T) = \{x \in A : f(x) \in T\}$.

(This notation, which unfortunately uses the same symbol for the first induced function as for the original function, is standard.)

Based on these definitions, answer the following:

a) Suppose $\mathscr{C}$ is a partition on B. Under what condition(s) on f will $\{f^{-1}(X) : X \in \mathscr{C}\}$ be a partition of A? Justify your answer.
b) Suppose $\mathscr{C}^*$ is a partition on A. Under what condition(s) on f will $\{f(X) : X \in \mathscr{C}^*\}$ be a partition of B? Justify your answer.

5. Show that each of the following relations is an equivalence relation, and describe the equivalence classes in terms of more familiar sets:

a) The relation $\sim$ defined on $\mathbf{Z} \times (\mathbf{Z} - \{0\})$ by

$$(a, b) \sim (c, d) \quad \text{iff} \quad ad = bc.$$

b) The relation $\sim$ defined on $\mathbf{R} \times \mathbf{R}$ by

$$(a, b) \sim (c, d) \quad \text{iff} \quad a^2 + b^2 = c^2 + d^2.$$

6. Suppose A and B are sets, with partitions $\mathscr{C}_A$ on A and $\mathscr{C}_B$ on B. Is $\mathscr{C}_A \cup \mathscr{C}_B$ a partition on $A \cup B$? If so, prove it. If not, give a counterexample, and find some additional conditions which will make $\mathscr{C}_A \cup \mathscr{C}_B$ a partition on $A \cup B$.

7. **a)** We stated that it is possible for $\sim_{\mathscr{C}}$ to be an equivalence relation even if $\mathscr{C}$ is not a partition. Prove this by finding an example of a collection $\mathscr{C}$ of subsets of $\mathcal{P}(\{a, b, c\})$ which is not a partition, but for which $\sim_{\mathscr{C}}$ is an equivalence relation.

b) What about the reverse direction? Is it possible to find a relation $\sim$ which is not an equivalence relation, for which $\mathscr{C}^{\sim}$ is a partition? Justify your answer.

8. Prove the following: For any set A, any equivalence relation $\sim$ on A, and any $x, y \in A$,

$$\text{if} \quad x \in [y]_{\sim}, \quad \text{then} \quad [x]_{\sim} = [y]_{\sim} .$$

(*Hint:* $[x]_{\sim}$ and $[y]_{\sim}$ are both subsets of A. You must show they contain the same elements.)

9. Find a function from $\mathbf{Z}$ to $\{0, 1, 2, \ldots, n-1\}$ whose induced equivalence relation is congruence mod n.

10. Prove each of the following:

a) **THEOREM 6.12.** For $n \in \mathbf{N}$, every integer belongs to precisely one of the sets $[0]_n, [1]_n, [2]_n, \ldots, [n-1]_n$. Thus, $\mathbf{Z}_n = \{[0]_n, [1]_n, [2]_n, \ldots, [n-1]_n\}$.

b) **THEOREM 6.13.** For $x, y \in \mathbf{Z}$, $n \in \mathbf{N}$, $x \equiv y \pmod{n}$ iff x and y have the same remainder under division by n.

11. It is sometimes useful to combine relations on two individual sets A and B to create a relation on their product. We make the following definition:

DEFINITION: For sets A and B and relations $\sim$ and $\approx$ on A and B respectively, the *product* of $\sim$ and $\approx$ is the relation $\approxeq$ on $A \times B$ defined by

$$(x, y) \approxeq (u, v) \quad \text{iff} \quad x \sim u \text{ and } y \approx v .$$

a) Prove the following:

If $\sim$ and $\approx$ are equivalence relations on A and B respectively, then the product of $\sim$ and $\approx$ is an equivalence relation on $A \times B$.

b) Describe, as a set of subsets of $A \times B$, the partition associated with the product of $\sim$ and $\approx$ in terms of the partitions $\mathscr{C}^{\sim}$ associated with $\sim$ and $\mathscr{C}^{\approx}$ associated with $\approx$.

7

Further Ideas on Functions and Relations

Introduction

This chapter is a continuation of Chapter 6. In Section 7.1, we look at *binary operations,* which are formally defined as a special type of function. We will look at some familiar examples, and define and discuss some important concepts, such as identity element and inverse.

In Section 7.2, we examine *order relations,* which are a special type of relation. Order relations are used to set up "comparisons" between objects of a given set: "is less than", "divides", and "is a subset of" are three important examples of order relations.

7.1 Binary Operations

We have discussed many different operations, primarily operations on numbers and operations on sets. So far, our use of the word "operation" has been informal, to describe any method of combining objects to get another object. For example, the standard arithmetic operations are used to combine two numbers to get their sum, product, difference, or quotient. The basic operations on sets, such as union and intersection, are used similarly to create new sets from old ones.

In this section, we will give a precise definition for this term, look at some more examples, and study some of the general properties which operations have.

Definition of "Operation"

Formally, an operation is a type of function. What we will be working with more specifically are binary operations, which are functions of two variables, i.e., functions whose domains consist of ordered pairs. The definition is as follows:

DEFINITION 7.1: For a set A, a *binary operation on* A is a function from $A \times A$ to A.

There are also such things as *unary operations* (functions from A to A), *ternary operations* (functions from $A \times A \times A$ to A), etc. We will not be studying these in any general way, and, unless otherwise indicated, the word "operation" is used in this text specifically to mean "binary operation". We sometimes loosely refer to A as the *domain* of a binary operation on A, although, as a function, the domain of the operation is $A \times A$.

Thus, addition, subtraction, and multiplication are all considered binary operations on, say, **R**. Division, however, is not a binary operation on **R**: considered as a relation from $\mathbf{R} \times \mathbf{R}$ to **R**, division is not domain-complete, since division by zero is not defined. Also, subtraction is not a binary operation on **N**: in this case, subtraction is defined for all pairs of natural numbers, but **N** cannot be used as the codomain, since not all differences are positive.

Most of the familiar operations are written with an *operation symbol,* rather than using standard function notation. If we were to write addition to look like a function, we might express the sum of x and y as $+(x, y)$; instead we write "$x + y$". This operation-symbol notation is usually used for the general study of operations as well. We will use "$*$" as a generic symbol for a binary operation. Thus, if "$*$" represents a binary operation on A, we will write "$x * y$" (read "x star y") to mean the value of the operation $*$ at the ordered pair (x, y).

Example 1. **Common Mathematical Operations** The following are some of the most familiar operations in mathematics, with their operation symbols:

i) Addition ($+$), defined on any one of the following sets: **N**, **Z**, **Q**, **R**, **C**. (There are other sets on which addition could be considered an operation.)

ii) Subtraction ($-$): as noted above, this is not an operation on **N**, although it is an operation on the other sets in i). Henceforth, unless otherwise indicated, when we speak of the operation of subtraction, we will mean subtraction as defined on one of these other sets.

iii) Union ("$\cup$") and intersection ("$\cap$"), defined on $\mathcal{P}(S)$, for any set S.

iv) Composition ("$\circ$"), defined on the set of all functions from S to S, for any set S. We will use the notation $\mathcal{F}(S)$ for this set of functions. This set and operation have many applications in mathematics, and will be discussed further in this section.

v) Addition ("$+$"), defined on the set of all functions from S to **R**, for any set S. This operation is defined as follows: if f and g are any two functions from S

to **R**, then $f + g$ is the function from S to **R** whose value at an element x of S is given by

$$(f + g)(s) = f(s) + g(s) .$$

Other arithmetic operations on this set of functions are defined similarly (with the condition that, for division, the domain consists only of those functions which do not have 0 in their image).

vi) max: Short for "maximum", this is an operation on **R** which is written in function style, and is defined as follows:

For $x, y \in \mathbf{R}$,

$$\max(x, y) = \begin{cases} x & \text{if } x \geq y \\ y & \text{if } y > x \end{cases}$$

vii) Greatest common divisor: this is an operation on **N**, written in function style as GCD(x, y), for $x, y \in \mathbf{N}$. [Recall that by Definition 2.9 (Section 2.4), GCD(x, y) is the largest positive integer that divides both x and y.] □

COMMENT When the domain of a function f is a set of ordered pairs, it is standard practice to write "$f(x, y)$" instead of "$f((x, y))$". That is, we think of f as a function of two variables, rather than as a function with one ordered-pair variable. This shorthand is used in vi) and vii) above.

"Random" Operations

Not every operation is as interesting or as important as those listed in Example 1. As with functions, operations as such need only fit the formal definition. If A is a small, finite set, one common way to define an operation is by listing explicitly all the possible combinations of elements and the result of the operation on each pair. Here is an example:

EXAMPLE 2. **A "Random" Operation** Let $A = \{1, 2, 3\}$. Define an operation $*$ on A by the following conditions:

$$\begin{array}{lll} 1 * 1 = 2 & 2 * 1 = 2 & 3 * 1 = 1 \\ 1 * 2 = 1 & 2 * 2 = 3 & 3 * 2 = 3 \\ 1 * 3 = 2 & 2 * 3 = 1 & 3 * 3 = 1 \end{array}$$

You might be able to come up with an explanation of this operation, but none is intended. It is simply a function we picked at random whose domain is $A \times A$, and whose codomain is A. □

Operation Tables

The type of information displayed in Example 2 is often arranged in a table, known as an *operation table* or *Cayley table* (after a nineteenth century mathematician, Arthur Cayley), and which is essentially the same idea as an ordinary table of addition or multiplication facts. The elements of A are listed as headings across the top and along the left side of the table, and the result of the operation is listed in the body of the table. The operation symbol is shown in the upper left corner of the table. Where the row labeled by x meets the column headed by y, we place the value of $x * y$. The operation on $\{1, 2, 3\}$ described in Example 2 could be presented in the table shown in Figure 7.1, in which the 1 in a box indicates that $2 * 3 = 1$.

Get Your Hands Dirty 1

If A is a finite set with $|A| = n$, how many operations on A are there?

Closure

We mentioned the curious situation with subtraction on **N**, in which the operation is *defined* for all natural numbers, but it is not considered an operation *on* the set of natural numbers, because the result is not necessarily a natural number. This phenomenon is a common one. To help describe this type of situation, we have the following terminology:

> **DEFINITION 7.2:** For a set A, an operation $*$ on A, and a set $B \subseteq A$, "B is *closed under* $*$" means:
>
> $$(\forall x, y \in B)(x * y \in B).$$

For example, suppose we think of the basic arithmetic operations as operations on **R**. Then we can say that the set $\mathbf{R}^+$ is closed under multiplication, since the product of any two positive real numbers is a positive real number; but the set **N** is not closed under subtraction, since, for example, 3 and 8 are in **N** but $3 - 8 \notin \mathbf{N}$.

COMMENTS

1. Notice that $x * y$ is *defined* for all $x, y \in B$, since x and y are then also elements of A, and $*$ is an operation on A. The issue here is whether $x * y$ is necessarily an element of B or not.

*	1	2	3
1	2	1	2
2	2	3	[1]
3	1	3	1

Figure 7.1. An operation table for the operation $*$.

2. This is precisely how we used this terminology in Exploration 2 in Section 1.1, "Sets Closed under Addition or Subtraction", in which we asked which subsets of $\mathbf{Z}$ are closed under each of these operations.

Get Your Hands Dirty 2

Let $A = \{a, b, c\}$.

a) Give an example of a proper subset of $\mathcal{P}(A)$ which is closed under $\cup$ and a proper subset of $\mathcal{P}(A)$ which is not closed under $\cup$.

b) Make up an operation $*$ on A using a table, and give a proper subset of A which is closed under $*$ and a proper subset of A which is not closed under $*$. (*Caution:* Such sets do not exist for every operation—see Exercise 2.) □

Properties of Operations

In studying mathematical systems, it is often interesting to consider only operations with certain special properties. The next two definitions describe two of the most commonly studied conditions on operations:

DEFINITION 7.3: For a set A and an operation $*$ on A, "$*$ is *commutative*" means:

$$(\forall\, x, y \in A)\,(x * y = y * x)\,.$$

An operation which is not commutative is called *noncommutative*.

For example, $+$ is a commutative operation on $\mathbf{Z}$, since, for all $x, y \in \mathbf{Z}$, $x + y = y + x$. On the other hand, $\div$ is a noncommutative operation on $\mathbf{R}^+$, since, for example, $5 \div 7 \neq 7 \div 5$.

If two elements x and y satisfy the condition "$x * y = y * x$", we say that x and y *commute*.

COMMENT If an operation is noncommutative, that does not mean that there is no pair of elements x and y which commute; it just means that not every pair of elements commutes.

Get Your Hands Dirty 3

a) Name two other familiar operations which are commutative and two which are not.

b) Make up, using tables, two operations on $\{a, b, c\}$ which are commutative and two which are not. □

DEFINITION 7.4: For a set A and an operation $*$ on A, "$*$ is *associative*" means:

$$(\forall\, x, y, z \in A)[(x * y) * z = x * (y * z)].$$

An operation which is not associative is called *nonassociative*.

COMMENT If an operation is nonassociative, that does not mean that there are no elements x, y, and z satisfying the equation $(x * y) * z = x * (y * z)$; it just means that not every triple of elements satisfies the equation.

The operation $*$ here is not a ternary operation. In order to calculate $(x * y) * z$, we first find $x * y$, which is itself an element of A, and then combine this element $x * y$ with z. We do a similar process to find $x * (y * z)$. The issue here is whether these two processes necessarily give the same result.

For example, suppose $*$ is ordinary subtraction, considered as an operation on $\mathbf{Z}$, and $x = 9$, $y = 6$, and $z = 2$. Then

$$(x * y) * z = (9 - 6) - 2 = 3 - 2 = 1,$$

while

$$x * (y * z) = 9 - (6 - 2) = 9 - 4 = 5.$$

Thus, subtraction is not associative.

On the other hand, addition is an associative operation on $\mathbf{R}$, since $(x + y) + z = x + (y + z)$ for any real numbers x, y, and z.

Get Your Hands Dirty 4

a) Name two other familiar operations which are associative and two which are not.

b) Make up, using tables, two operations on $\{a, b\}$ which are associative and two which are not. (Convince yourself that your examples for associativity are really associative. Notice how many individual cases you have to check, even when the domain is so small.) □

Among the familiar arithmetic operations, addition and multiplication have both properties, i.e., each operation is both commutative and associative. On the other hand, subtraction and division have neither property, i.e., each operation is both noncommutative and nonassociative. These facts often create the incorrect impression among students that the two properties are interchangeable, or at least, that any operation that has one of these properties must have the other.

This is not the case. Here are examples to illustrate the independence of the two properties:

Example 3. A Commutative, Nonassociative Operation Consider a rectangle whose length and width are x and y, as in Figure 7.2:

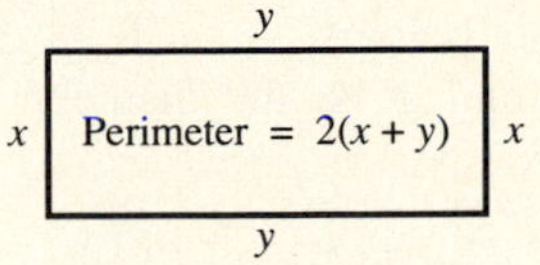

y

x Perimeter $= 2(x + y)$ x

y

Figure 7.2. The length and width of the rectangle are x and y. Its perimeter is $2(x + y)$.

We will create an operation on **R** whose definition is motivated by the formula for the perimeter of such a rectangle (just as the operation of multiplication can be interpreted in terms of area). The new operation, which we represent by the letter P because of its relationship to perimeter, is defined as follows:

$$x \; P \; y = 2(x + y) .$$

Notice that, if $x, y \in \mathbf{R}^+$, then $x \; P \; y$ is the perimeter of the rectangle in Figure 7.2, but the operation is defined for all $x, y \in \mathbf{R}$.

The operation P is commutative:

$$x \; P \; y = 2(x + y), \text{ while } y \; P \; x = 2(y + x) .$$

Since $2(x + y) = 2(y + x)$ (because of the commutativity of addition), we get $x \; P \; y = y \; P \; x$.

On the other hand, P is not associative. Suppose, for example, $x = 3$, $y = 5$, and $z = 2$. Then

$$(x \; P \; y) \; P \; z = (3 \; P \; 5) \; P \; 2 = 16 \; P \; 2 = 36$$

(using $3 \; P \; 5 = 16$), while

$$x \; P \; (y \; P \; z) = 3 \; P \; (5 \; P \; 2) = 3 \; P \; 14 = 34$$

(using $5 \; P \; 2 = 14$). Since the results are different, P is not associative.

COMMENT Notice that we showed that P is commutative by working with variables, since the definition of commutativity uses a universal quantifier. We showed that P is not associative using a specific substitution, since we were looking for a counterexample to a universally quantified statement.

Example 4. **An Associative, Noncommutative Operation** Let Y be any set, and consider the operation of composition on $\mathscr{F}(Y)$ (the set of all functions from Y to Y). This operation is associative, which we can show as follows.

Suppose $f, g, h \in \mathscr{F}(Y)$, and $t \in Y$. Then

$$[(f \circ g) \circ h](t) = (f \circ g)(h(t)) = f(g(h(t))) ,$$

while

$$[f \circ (g \circ h)](t) = f((g \circ h)(t)) = f(g(h(t))) .$$

Since the results are the same, the operation $\circ$ is associative.

However, for most choices of Y (see note below), $\circ$ will not be commutative. For example, suppose $Y = \{1, 2\}$, and let f and g be the elements of $\mathcal{F}(Y)$ defined as follows:

$$f(1) = 2 \qquad g(1) = 2$$
$$f(2) = 2 \qquad g(2) = 1$$

You should verify that the two compositions $f \circ g$ and $g \circ f$ are the following functions:

$$(f \circ g)(1) = 2 \qquad (g \circ f)(1) = 1$$
$$(f \circ g)(2) = 2 \qquad (g \circ f)(2) = 1$$

Since these functions are different, we have shown that composition is not commutative. □

Get Your Hands Dirty 5

Find two functions f and g from **R** to **R** such that $f \circ g \neq g \circ f$. (*Hint:* most functions will work.) □

Note: We leave as Exercise 5 for you to determine if there are any sets Y for which the operation of composition on $\mathcal{F}(Y)$ *is* commutative.

For future reference, we state part of Example 4 as a theorem:

THEOREM 7.1. For any set Y, the operation of composition on $\mathcal{F}(Y)$ is associative.

Identity Elements

Children learn early in school that 0 and 1 behave in a special way with regard to addition and multiplication respectively. Specifically, we know that, no matter what the number x is, we have

$$x + 0 = 0 + x = x \quad \text{and} \quad x \times 1 = 1 \times x = x\,.$$

A similar phenomenon occurs for many operations, and that leads to the following definition:

DEFINITION 7.5: For an operation $*$ on a set A, and $t \in A$, "t is an *identity element* for $*$" means:

$$(\forall\, x \in A)\,(x * t = x \text{ and } t * x = x)\,.$$

Thus 0 is an identity element for addition, and 1 is an identity element for multiplication. Another familiar example is $\varnothing$, which is an identity element for union [defined on $\mathcal{P}(Y)$, for any set Y].

COMMENT For various reasons, it turns out to be important to include both of the conditions "$x * t = x$" and "$t * x = x$" as part of the definition of an identity element.

Therefore the number 0 is not considered an identity element for subtraction, even though $x - 0 = x$ for all x, since $0 - x \neq x$. In fact, there is no identity element for subtraction. Similarly, there is no identity element for division.

Get Your Hands Dirty 6

We just saw that some operations do not have an identity element. Are there operations with more than one identity element? For the set $\{a, b, c\}$, either find an operation with more than onc identity element or show that no such operation exists. □

Since the definition of identity element includes both conditions "$x * t = x$" and "$t * x = x$", it might be supposed that only a commutative operation can have an identity element. After all, among the familiar arithmetic operations, addition and multiplication, which are both commutative, do have identity elements, while subtraction and division, which are both noncommutative, do not have identity elements.

But, in fact, this is not the case. For those familiar with matrix arithmetic, we point out that matrix multiplication, as defined on, say, 2×2 matrices with entries in $\mathbf{R}$, is a noncommutative operation with the identity element $\left(\begin{smallmatrix} 1 & 0 \\ 0 & 1 \end{smallmatrix}\right)$.

The following is another example, connected with our earlier work, of a noncommutative operation which has an identity element:

Example 5. **Identity Element for Composition** Suppose Y is any set. In Section 6.1, we defined something called the *identity function* on Y, written I_Y, as the element of $\mathcal{F}(Y)$ given by the equation $I_Y(t) = t$ for all $t \in Y$. The two uses of the word "identity"—identity function, as an element of $\mathcal{F}(Y)$, and identity element for an operation, as described by Definition 7.5—are closely related. Specifically, the identity *function*, I_Y, is the identity *element* for the operation of composition on $\mathcal{F}(Y)$. In other words, if g is any element of $\mathcal{F}(Y)$, then $g \circ I_Y = g$ and $I_Y \circ g = g$. (Verify—apply each composition to an arbitrary element of Y.)

As discussed in Example 4, composition on $\mathcal{F}(Y)$ can be a noncommutative operation, and so there is indeed a noncommutative operation which has an identity element. □

For future reference, we state part of Example 5 as a theorem:

THEOREM 7.2. For any set Y, the identity function I_Y is the identity element for the operation of composition on $\mathcal{F}(Y)$.

Notice that we are gradually building up information about $\mathcal{F}(Y)$.

A companion question is the following:

Get Your Hands Dirty 7

Is there a commutative operation which does not have an identity element? □

As we indicated earlier, an operation on a set need not have any identity element. As your work on GYHD 6 may have suggested, an operation *cannot* have more than one identity element. This assertion is contained in the following theorem:

THEOREM 7.3. For a set A and an operation $*$ on A, there is at most one identity element for $*$.

For the purpose of organizing the proof, we can rephrase the assertion in the following typical manner for uniqueness proofs: if there are two objects, s and t, which are both identity elements, then $s = t$.

Before proceeding, try to complete the proof with this introduction. Examine your work from GYHD 6.

Proof: Suppose that s and t are both identity elements for $*$. Then, by Definition 7.5, we have both of the following conditions:

$$(\forall\, x \in A)\,(x * s = x \text{ and } s * x = x)\,;$$

$$(\forall\, x \in A)\,(x * t = x \text{ and } t * x = x)\,.$$

Substituting $x = t$ in the first equation of the first condition gives $t * s = t$. Similarly, substituting $x = s$ in the second equation of the second condition gives $t * s = s$.

These two equations together tell us that $s = t$, as desired. ■

Inverse Elements

The last of our general concepts for operations is only applicable to operations which have an identity element. For this discussion, $*$ is an operation on a set A which has an identity element. We will follow fairly standard practice and use the letter $\mathbf{e}$ for this identity element. (By Theorem 7.3, it makes sense to refer to $\mathbf{e}$ as *the* identity element, since there cannot be more than one.)

The concept we are about to define is a generalization of the familiar ideas of the negative and the reciprocal of a number. For every real number x, there is a real number u such that the sum of x and u is 0, which is the identity for addition. Similarly, except when $x = 0$, there is a number v such that the product of x and v is 1, which is the identity for multiplication.

These examples motivate the following definition:

DEFINITION 7.6: For an operation $*$ on A with identity $\mathbf{e}$, and for $x, y \in A$:

"y is an *inverse* of x for $*$" means:

$$y * x = x * y = \mathbf{e}\,.$$

"x is *invertible under* $*$" means: There is an element which is an inverse of x for $*$.

For example, for the operation of addition on $\mathbf{R}$, and $x \in \mathbf{R}$, $-x$ is the inverse of x, and all real numbers are invertible under this operation. For the operation of multiplication on $\mathbf{R}$, for $x \in \mathbf{R} - \{0\}$, $1/x$ is the inverse of x, and all nonzero real numbers are invertible under this operation.

COMMENTS

1. As with identity elements, we require two-sidedness for inverses: both $y * x = \mathbf{e}$ and $x * y = \mathbf{e}$. And, as with identity elements, existence of inverses is not necessarily related to the commutativity or noncommutativity of the operation.

2. For a particular operation with an identity, there may be some elements that have an inverse and others that do not. We will show in Theorem 7.7, however, that, for associative operations, a particular element cannot have more than one inverse.

3. Because of Theorem 7.7 below, it makes sense (at least for associative operations) to refer to *the* inverse of an element. For a generic associative operation $*$, the standard notation for the inverse of x is x^{-1}, even if the operation is not related to ordinary multiplication.

Before looking at examples, we state two elementary theorems about inverses that we will use in Chapter 10. The first is a further development from comment 1 above:

THEOREM 7.4. For an operation $*$ with identity $\mathbf{e}$ on a set A, and element $s \in A$, if s is invertible, then s^{-1} is also invertible. Specifically, the inverse of s^{-1} is s.

Proof: Assume that s is invertible, with inverse s^{-1}. By Definition 7.6, we have

$$s^{-1} * s = s * s^{-1} = \mathbf{e}.$$

These conditions (written in a different order) are precisely what we must show in order to conclude that s^{-1} is the inverse of s. ■

We can express this theorem by the formula $(s^{-1})^{-1} = s$.

Our second elementary result is the following:

THEOREM 7.5. For an associative operation $*$ with identity $\mathbf{e}$ on a set A, and elements s and $t \in A$, if s and t are invertible, then $s * t$ is invertible. Specifically, the inverse of $s * t$ is $t^{-1} * s^{-1}$.

Proof: Assume that s and t are invertible, with respective inverses s^{-1} and t^{-t}. Thus, by Definition 7.6, we have

$$s^{-1} * s = s * s^{-1} = \mathbf{e} \quad \text{and} \quad t^{-1} * t = t * t^{-1} = \mathbf{e}.$$

Therefore,

$$\begin{aligned}
(t^{-1} * s^{-1}) * (s * t) &= t^{-1} * [s^{-1} * (s * t)] && \text{(by associativity)} \\
&= t^{-1} * [(s^{-1} * s) * t] && \text{(by associativity)} \\
&= t^{-1} * (\mathbf{e} * t) && (\text{since } s^{-1} * s = \mathbf{e}) \\
&= t^{-1} * t && \text{(by definition of identity element)} \\
&= \mathbf{e} && (\text{since } t^{-1} * t = \mathbf{e}).
\end{aligned}$$

We leave it to you to verify, similarly, that

$$(s * t) * (t^{-1} * s^{-1}) = \mathbf{e}.$$

These two conditions prove the theorem. ■

We can express this theorem by the formula

$$(s * t)^{-1} = t^{-1} * s^{-1}.$$

COMMENTS

1. Notice that the inverse of $s * t$ is the product of their individual inverses *in reverse order*. The inverse of $s * t$ is $t^{-1} * s^{-1}$, not $s^{-1} * t^{-1}$. You may find it helpful to think of this in terms of the following metaphor: In "dressing" your feet, you first put on your socks and then put on your shoes. To "undress" your feet, you reverse the order: first you take off your shoes, and then you take off your socks.
2. Theorems 7.4 and 7.5 can be summed up by the statements "the inverse of an invertible element is invertible" and "the product of invertible elements is invertible". Notice that the second of these results requires that the operation be associative.

Get Your Hands Dirty 8

Prove (as needed in Theorem 7.5) that $(s * t) * (t^{-1} * s^{-1}) = \mathbf{e}$. □

Example 6. **Inverse under Composition** We already have seen that the identity *function* I_Y turned out to be the identity *element* for the operation of composition on $\mathcal{F}(Y)$. A similar situation occurs for inverse: for any relation f, we defined its inverse *relation* f^{-1} by reversing all the ordered pairs (Definition 6.4) and defined f to be *invertible as a function* if this relation f^{-1} is a function (Definition 6.5). We have now defined another kind of inverse, namely, inverse under the operation of composition. The following theorem asserts that these two types of inverse and invertibility coincide. Specifically:

THEOREM 7.6. For $f \in \mathcal{F}(Y)$, if f is invertible either as a function or under composition, then f is invertible in the other sense also, and the inverses are the same.

Proof: This is precisely what parts a) and b) of Theorem 6.7 say, in the case where $A = B = Y$ [using the fact that I_Y is the identity element for composition of $\mathcal{F}(Y)$]. ■

COMMENT This theorem justifies our use of the words "inverse" and "invertible" and the "exponent -1" notation both for "inverse of a relation" (Definition 6.4) and for "inverse under an operation" (Definition 7.6).

The following result is a companion to Theorem 7.3:

THEOREM 7.7. If $*$ is an associative operation on a set A with identity element $\mathbf{e}$, and t is an element of A, then t has at most one inverse for $*$.

Proof: Suppose that the hypothesis holds, and that u and v are both inverses of t for $*$. We need to show that $u = v$. By the definition of inverse, we have

$$t * u = u * t = \mathbf{e} \quad \text{and} \quad t * v = v * t = \mathbf{e}.$$

We look at $u * (t * v)$ and $(u * t) * v$:

$$u * (t * v) = u * \mathbf{e} = u$$

(since $t * v = \mathbf{e}$, and $\mathbf{e}$ is the identity element for $*$), and similarly

$$(u * t) * v = \mathbf{e} * v = v.$$

But we know that $u * (t * v) = (u * t) * v$, by the associativity of $*$. Therefore $u = v$. ■

We leave as Exercise 9 to show that Theorem 7.7 is false without the assumption of associativity.

Arithmetic in $\mathbf{Z}_n$

We defined "congruence mod n" (for $n \in \mathbf{N}$) by the condition

$$x \equiv y \pmod{n} \quad \text{iff} \quad n \mid x - y.$$

We showed that this is an equivalence relation on $\mathbf{Z}$; the associated partition is called $\mathbf{Z}_n$. We would now like to define operations of "addition" and "multiplication" on $\mathbf{Z}_n$, based on the corresponding operations on $\mathbf{Z}$.

The outline of the method is as follows: We start with two elements of $\mathbf{Z}_n$, say $[x]$ and $[y]$. (Recall that the elements of $\mathbf{Z}_n$ are equivalence classes.) To add $[x]$ and $[y]$, we would like to simply add the *integers* x and y, and then define the sum of $[x]$ and $[y]$ as the equivalence class that contains the integer $x + y$. We would like to define multiplication similarly.

Stop! Think about this. There is something subtle going on here that could lead to problems.

The potential problem is this: $[x]$ is only one of the "names" for a particular equivalence class. If $x \equiv u \pmod{n}$, then $[x]$ and $[u]$ are the same thing, the same element of $\mathbf{Z}_n$. A similar statement holds if $y \equiv v \pmod{n}$. If we label the equivalence class of x simply as "A", and the equivalence class of y as "B", then the "sum" of A and B is both the "sum" of $[x]$ and $[y]$ and the "sum" of $[u]$ and $[v]$. So do we define "$A + B$" as $[x + y]$ or as $[u + v]$?

? ? ?

The wonderful thing about this particular situation is that it doesn't matter; the two equivalence classes $[x + y]$ and $[u + v]$ are the same, as are $[x \cdot y]$ and $[u \cdot v]$.

Do the following to help understand what's going on:

Get Your Hands Dirty 9

a) Choose two specific elements (i.e., equivalence classes) A and B in $\mathbf{Z}_6$.
 i) Find $a + b$ for different combinations of elements with $a \in A$ and $b \in B$. Are the resulting sums equivalent?
 ii) Find $a \cdot b$ for different combinations of elements with $a \in A$ and $b \in B$. Are the resulting products equivalent?

b) Let $\sim$ be the equivalence relation on $\mathbf{Z}$ given by "$x \sim y$ iff $|x| = |y|$". As in a), choose two specific equivalence classes A and B for this relation.
 i) Find $a + b$ for different combinations of elements with $a \in A$ and $b \in B$. Are the resulting sums equivalent?
 ii) Find $a \cdot b$ for different combinations of elements with $a \in A$ and $b \in B$. Are the resulting products equivalent? □

Compare the various parts of this GYHD, and try to understand what's going on.

We will state as a formal theorem what happens with $\mathbf{Z}_n$:

THEOREM 7.8. If $n \in \mathbf{N}$, and $x, y, u, v \in \mathbf{Z}$, with $x \equiv u \pmod{n}$ and $y \equiv v \pmod{n}$, then:

i) $x + y \equiv u + v \pmod{n}$, and
ii) $x \cdot y \equiv u \cdot v \pmod{n}$

The proof is straightforward, and you should try to do it before reading on.

Proof: Assume that n, x, y, u, and v are as described. Since $x \equiv u \pmod{n}$, we have $n \mid (x - u)$, and similarly, $n \mid (y - v)$. Thus, there are integers s and t such that $x - u = ns$ and $y - v = nt$. We need to find integers z and w such that $[(x + y) - (u + v)] = nz$ and $[(x \cdot y) - (u \cdot v)] = nw$.

If you weren't sure how to start the proof, try it again now with this beginning.

It helps to rewrite the equations about s and t as

$$x = u + ns \quad \text{and} \quad y = v + nt\,.$$

Substituting these expressions for x and y gives

$$\begin{aligned}[(x + y) - (u + v)] &= [(u + ns) + (v + nt)] - (u + v) \\ &= ns + nt \\ &= n(s + t)\,.\end{aligned}$$

Therefore the integer $z = s + t$ fulfills the required condition on z.

The case of multiplication is similar:

$$\begin{aligned}[(x \cdot y) - (u \cdot v)] &= (u + ns)(v + nt) - uv \\ &= uv + nsv + unt + nsnt - uv \\ &= n(sv + ut + nst)\,.\end{aligned}$$

Therefore the integer $w = sv + ut + nst$ fulfills the required condition on w. ■

Using this theorem, we can make the following definitions:

DEFINITION 7.7: For $n \in \mathbf{N}$ and $S, T \in \mathbf{Z}_n$:

The *sum* of S and T (written $S +_n T$) is the equivalence class containing $x + y$, where x is any element of S and y is any element of T. This operation is called *addition mod n*.

The *product* of S and T (written $S \cdot_n T$) is the equivalence class containing $x \cdot y$, where x is any element of S and y is any element of T. This operation is called *multiplication mod n*.

Theorem 7.8 makes this definition legitimate, by proving that the results do not depend on which elements x and y are chosen from the equivalence classes S and T. This point is discussed more fully below.

Get Your Hands Dirty 10

a) Make operation tables for addition mod 4 and multiplication mod 4.

b) Determine which elements of $\mathbf{Z}_4$ have inverses under $\cdot_4$.

c) Show that the equation $[x]^2 +_4 [y]^2 = [3]$ in $\mathbf{Z}_4$ has no solution, i.e., there are no integers x and y that make it true. (If necessary, try all four possibilities for each of $[x]$ and $[y]$—16 combinations altogether. This result will be used in Exercise 10.) □

Well-Defined Functions—a Digression

Our discussion of the process of defining the arithmetic operations in $\mathbf{Z}_n$ reflects a more general problem when making definitions that involve equivalence relations and equivalence classes—namely, we must be very careful that the definitions truly reflect the equivalence classes, and are not based on properties of individual elements of those classes. That is, we must show that the definition does not depend on the *choice of representative* for the equivalence class.

To illustrate this more clearly, consider the following problem: We begin with the partition $\mathbf{Z}_{12}$ of the integers, and use $[x]$ to represent the equivalence class of x in this partition. For each positive integer n, we would like to define a function f_n from $\mathbf{Z}_{12}$ to $\mathbf{Z}$

such that $f_n([x])$ is the remainder obtained when x is divided by n (as described in the Division Algorithm). For example, if 9 is divided by 5, the remainder is 4 (i.e., $9 = 5 \times 1 + 4$), so we would like $f_5([9])$ to be 4.

But there is a problem: notice that [9] and [21] are the same equivalence class, since $21 \equiv 9 \pmod{12}$. However, when 21 is divided by 5, the remainder is 1, and so our definition would require $f_5([21]) = 1$. But we saw above that we need $f_5([9])$ to be 4. Since [9] and [21] are equal, we can't have $f_5([9]) = 4$ and $f_5([21]) = 1$. In other words, there is no function satisfying the requirements for f_5.

We can get a better understanding of the difficulty by starting with a *relation*, rather than a function. Whether or not a function f_n exists as described above, we can define a *relation* f_n from $\mathbf{Z}_{12}$ to $\mathbf{Z}$ that is similar to what we want, namely,

$$f_n = \{([x], r) \in \mathbf{Z}_{12} \times \mathbf{Z} : x \in \mathbf{Z} \text{ and } r \text{ is the remainder when } x \text{ is divided by } n\}.$$

In this relation, each equivalence class $[x]$ is paired with *all* the remainders obtained when elements of this equivalence class are divided by n. For example, both ([9], 4) and ([21], 1) belong to the relation f_5.

In order for this relation to be a function from $\mathbf{Z}_{12}$ to $\mathbf{Z}$, every equivalence class in $\mathbf{Z}_{12}$ must be paired with exactly one element of $\mathbf{Z}$; that is, every element of a given equivalence class must have the same remainder. In other words, we are saying that we need the relation f_n to be *functionlike* (see Definition 6.3 in Section 6.1).

Since $[9] = [21]$ (for $n = 5$), the relation f_5 contains both ([9], 4) and ([9], 1), and so is not functionlike.

In mathematical writing, the process of defining a function such as f_n is often handled by beginning as if there were such a function, saying something like: "if $[x] \in \mathbf{Z}_{12}$, define $f_n([x])$ as the remainder when x is divided by n." The mathematician then acknowledges the difficulty by saying: "we need to show that *the function f_n is well defined*" (our emphasis).

The language in that last phrase is potentially confusing because it seems to be assuming that there already is a function f_n and then proving something about this function. By the phrase "the function f_n is well defined", mathematicians mean what we would express instead by saying that we need to show that the *relation f_n* (which definitely does exist) is *functionlike*. (Indeed, it is largely for the purpose of dealing with this issue that we created the term "functionlike".)

Thus, the following three statements all assert the same thing:

i) there is a function $f_n : \mathbf{Z}_{12} \to \mathbf{Z}$ such that $f_n([x])$ is the remainder when x is divided by n;

ii) if f_n is defined by

$$f_n([x]) = \text{the remainder when } x \text{ is divided by } n,$$

then f_n is a well-defined function;

iii) the following relation f_n is functionlike:

$$f_n = \{([x], r) \in \mathbf{Z}_{12} \times \mathbf{Z}:$$
$$x \in \mathbf{Z} \text{ and } r \text{ is the remainder when } x \text{ is divided by } n\}.$$

In Exercise 13, we ask you to determine for which values of n the relation f_n as defined above is functionlike, i.e., for which n the relation f_n is a function.

Properties of Arithmetic Operations in $\mathbf{Z}_n$

Many properties of $+_n$ and $\cdot_n$ follow from the corresponding properties of ordinary addition and multiplication. Review the general concepts for operations introduced already in this section, and decide how they apply to these new operations.

? ? ?

The following theorem summarizes the basic properties. We will prove a) as a sample, and leave the rest as Exercise 11.

THEOREM 7.9. For $n \in \mathbf{N}, x, y, z \in \mathbf{Z}$:

a) $+_n$ is commutative.
b) $\cdot_n$ is commutative.
c) $+_n$ is associative.
d) $\cdot_n$ is associative.
e) $[0]$ is the identity element for $+_n$.
f) $[1]$ is the identity element for $\cdot_n$.
g) $[-x]$ is the inverse of $[x]$ for $+_n$.
h) $[x] \cdot_n ([y] +_n [z]) = ([x] \cdot_n [y]) +_n ([x] \cdot_n [z])$.

Proof of a): Suppose $[x], [y] \in \mathbf{Z}_n$. Then

$$[x] +_n [y] = [x + y]_n = [y + x]_n = [y] +_n [x],$$

which proves a). ■

COMMENT Property h) illustrates another general concept about operations, involving the relationship between *two* operations. Known loosely as the "distributive property", this concept is defined as follows:

DEFINITION 7.8: For operations $*$ and $\bullet$ on a set A: "$*$ is *distributive* over $\bullet$" means: For all $x, y, z \in A$,

i) $x * (y \bullet z) = (x * y) \bullet (x * z)$ and
ii) $(y \bullet z) * x = (y * x) \bullet (y * x)$.

Thus, for example, as operations on **R**, ordinary multiplication is distributive over ordinary addition, since both "$x(y + z) = xy + xz$" and "$(y + z)x = yx + zx$" are true for all real numbers x, y, and z. Condition h) of Theorem 7.9 can be combined with b) to assert that $\cdot_n$ is distributive over $+_n$. Exercise 12 examines this concept further.

The item conspicuously absent from the list in Theorem 7.9 is any discussion of inverses for $\cdot_n$. This is a more complicated question. As you may have observed in GYHD 10, not every element of $\mathbf{Z}_n$ has a multiplicative inverse.

It shouldn't be surprising that [0] has no inverse for $\cdot_n$, since the integer 0 has no inverse for multiplication. But you may have been surprised to see that $[2]_4$ has no inverse for $\cdot_4$, and that it is possible to multiply two elements of $\mathbf{Z}_n$, neither of which is [0], and get [0] as the product. For example, in $\mathbf{Z}_4$, $[2] \cdot_4 [2] = [0]$.

We leave this phenomenon as a topic for you to explore:

Exploration 1: Multiplicative Inverses in $\mathbf{Z}_n$

Which elements in $\mathbf{Z}_n$ have inverses under $\cdot_n$? For what values of n does every element of $\mathbf{Z}_n$ except [0] have a multiplicative inverse?

Experiment. Try several values of n, and test every element of $\mathbf{Z}_n$.

• E X P L O R E •

EXERCISES

1. Suppose A is a finite set with $|A| = n$.

  a) (GYHD 1) How many operations are there on A?

  b) How many commutative operations are there on A? (*Note:* There is no simple formula for the number of associative operations.)

2.

  a) Make up an operation $*$ on $A = \{a, b, c\}$, using a table, in which every subset of A is closed under $*$.

  b) Make up an operation $*$ on $A = \{a, b, c\}$, using a table, in which no nonempty proper subset of A is under $*$.

  c) Suppose A is an arbitrary set. Is it always possible to find an operation $*$ in which every subset of A is closed under $*$?

3. Which of the following operations (on the sets indicated) are commutative? Which are associative? Justify your answers.

  a) Exponentiation ($x * y = x^y$), on $\mathbf{R}^+$.

  b) GCD, on **N**.

  c) max, on **R**.

  d) Symmetric difference $[A * B = (A - B) \cup (B - A)]$, on $\mathcal{P}(\mathbf{Z})$. (Use a Venn diagram for the justification of your answer.)

  e) Addition, on the set of functions from $\{1, 2, 3\}$ to **R** (see v) of Example 1).

4. Suppose you have the table for an operation on a finite set. How can you tell from a glance at the table:
   a) whether or not the operation is commutative?
   b) whether or not the operation has an identity element?

5. Is there a finite set Y such that the operation of composition, defined on the set $\mathcal{F}(Y)$ of all functions from Y to Y, is a commutative operation? If so, what are all the possibilities for Y?

6. Which of the following operations have identity elements? If the operation has an identity element, which elements have inverses? Justify your answers.
   a) Exponentiation ($x * y = x^y$), on $\mathbf{R}^+$.
   b) GCD, on $\mathbf{N}$.
   c) LCM, on $\mathbf{N}$.
   d) max, on $\mathbf{R}$.
   e) Symmetric difference $[A * B = (A - B) \cup (B - A)]$, on $\mathcal{P}(\mathbf{Z})$.
   f) Addition, on the set of functions from $\{1, 2, 3\}$ to $\mathbf{R}$.
   g) Intersection, on $\mathcal{P}(\mathbf{R})$.
   h) P [perimeter from Example 3: $x P y = 2(x + y)$], on $\mathbf{R}$.

7. If A is a finite set with $|A| = n$, and $x \in A$, how many operations are there on A with x as an identity element?

8. Make up a commutative operation on $\{1, 2, 3\}$, using a table, which does not have an identity element.

9. a) Make up an operation on $\{1, 2, 3\}$, using a table, in which 1 is the identity element and in which both 2 and 3 are inverses of 2.
   b) According to Theorem 7.7, your answer to part a) will have to be a nonassociative operation. Show by explicit example that your operation is nonassociative.

10. a) Show, using Definition 7.7, that if x and y are integers with $x^2 + y^2 = 74{,}203$, then $[x]^2 +_4 [y]^2 = [3]$ in $\mathbf{Z}_4$. (*Note:* $[t]^2$ means $[t] \cdot_4 [t]$.)
    b) Combine part a) with GYHD 10c) to show that the equation $x^2 + y^2 = 74{,}203$ has no integer solutions.
    c) We proved the statement of part b) in Example 3 of Section 6.2. Compare the two approaches.

11. Prove parts b)–h) of Theorem 7.9.

12. For each of the following pairs of operations, state whether the first is distributive over the second, and justify your answers:
    a) multiplication over subtraction (on $\mathbf{R}$);
    b) addition over multiplication (on $\mathbf{R}$);
    c) $\cup$ over $\cap$ [on $\mathcal{P}(A)$, for some set A];
    d) $\cap$ over $\cup$ [on $\mathcal{P}(A)$, for some set A].

13. Define a relation f_n from $\mathbf{Z}_{12}$ to $\mathbf{Z}$, as in the text, by the following:
    $$f_n = \{([x], r) \in \mathbf{Z}_{12} \times \mathbf{Z} : x \in \mathbf{Z} \text{ and } r \text{ is the remainder when } x \text{ is divided by } n\}.$$
    For what positive integers n is f_n a function (i.e., for what n is the relation functionlike)?

14. In each of the following examples, $\sim$ is an equivalence relation on some set, and $\mathcal{C}$ is the partition consisting of the equivalence classes for that equivalence relation. In each case, state whether or not the given function f is well defined. Justify each answer by a proof or coun-

terexample. (You do not need to prove that $\sim$ is an equivalence relation.)

a) $\sim$ is defined on $\mathbf{R}$ by "$x \sim y$ iff $|x| = |y|$". Define $f: \mathscr{C} \to \mathbf{R}$ by "$f([x]) = x^3$".

b) $\sim$ is defined on $\mathbf{R}^+$ by "$x \sim y$ iff $x/y \in \mathbf{Q}$". Define $f: \mathscr{C} \to \mathbf{R}^+$ by "$f([x]) = x$".

c) $\sim$ is defined on $\mathbf{Z}$ by "$x \sim y$ iff $x \equiv y \pmod 6$". Define $f: \mathscr{C} \to \mathbf{Z}$ by $f([x]) = x^3$.

d) $\sim$ as in c). Define $f: \mathscr{C} \to \mathbf{Z}$ by "$f([x]) = |x|$".

e) Let S be some finite set; $\sim$ is defined on $\mathscr{P}(S)$ by "$X \sim Y$ iff $|X| = |Y|$". Define $f: \mathscr{C} \to \mathscr{P}(S)$ by "$f([X]) = X'$" (complement of X).

f) $\sim$ as in e). Define $f: \mathscr{C} \to \mathbf{W}$ by "$f([X]) = |X'|$".

g) Let $A = \mathscr{P}(\{1,2,3,4,5\})$; $\sim$ is defined on A by "$X \sim Y$ iff $X \cup \{1,2\} = Y \cup \{1,2\}$". Define $f: \mathscr{C} \to \mathbf{W}$ by "$f([X]) = |X|$".

h) $\sim$ is defined on $\{1,2,3\}$ by

$$\sim\ = \{(1,1),(2,2),(3,3)\}.$$

Define $f: \mathscr{C} \to \{1,2,3\}$ by "$f([x]) = x$".

i) $\sim$ is defined on $\{1,2,3\}$ by

$$\sim\ = \{(1,1),(1,2),(2,1),(2,2),(3,3)\}.$$

Define $f: \mathscr{C} \to \{1,2,3\}$ by "$f([x]) = x$".

15. In each of the following examples, $\sim$ is an equivalence relation on some set, and $\mathscr{C}$ is the partition consisting of the equivalence classes for that equivalence relation. In each case, state whether or not the given operation $*$ described on $\mathscr{C}$ is well defined. Justify each answer by a proof or counterexample. (You do not need to prove that $\sim$ is an equivalence relation.)

a) $\sim$ is defined on $\mathbf{Q}$ by "$x \sim y$ iff $x - y \in \mathbf{Z}$". Define $*$ on $\mathscr{C}$ by

$$[x] * [y] = [x + y]$$

b) $\sim$ as in a). Define $*$ on $\mathscr{C}$ by

$$[x] * [y] = [x \cdot y]$$

c) $\sim$ is defined on $\mathbf{N}$ by "$m \sim n$ iff the largest power of 2 that divides m is the same as the largest power of 2 that divides n". (For example, $12 \sim 20$, since both are divisible by 4 but not by 8, while $24 \not\sim 84$, since 24 is divisible by 8, but 84 is not.) Define $*$ on $\mathscr{C}$ by "$[x] * [y] = [x + y]$".

d) $\sim$ as in c). Define $*$ on $\mathscr{C}$ by "$[x] * [y] = [x \cdot y]$".

7.2 Order Relations

As indicated in the introduction to this chapter, order relations are relations used to describe "comparisons" between the elements of a set. We will define several different types of order relations, focusing primarily on the concept of a *partial ordering*. Before formally defining these concepts, we will look at a very important example.

The Subset Relation

Suppose Y is some set, with $U = \mathscr{P}(Y)$, and R is the usual subset relation on U: i.e., for $S, T \in U$, we have $(S,T) \in R$ iff $S \subseteq T$.

Get Your Hands Dirty 1

Use $Y = \{1, 2, 3\}$, and list the ordered pairs from $U \times U$ that are in the subset relation. □

A Lattice Diagram for Subsets

A diagram is often a helpful tool for picturing the relationship among different subsets of a larger set. Suppose that A and B are two subsets of Y. We can represent the relationship among the sets Y, A, B, $A \cup B$, $A \cap B$, and $\varnothing$ by Figure 7.3, which is known as a *lattice diagram* for subsets. In such a diagram, if a set S is connected by a line segment to a set T "above" it, then S is a subset of T. Figure 7.3, for example, shows that $A \cap B$ is a subset both of A and of B, and that A and B are both subsets of $A \cup B$.

Partial Ordering

We would like to use the subset relation on $\mathcal{P}(Y)$ as something of a prototype of what a partial ordering on any set A should be like. Roughly speaking, we want a partial ordering to describe a relationship between elements x and y in which x is either equal to y or "is part of", "comes earlier than", "is smaller than", or otherwise "precedes" y.

Which of the general properties of relations described in Section 6.1 should such a relation have?

? **?** **?**

We noted, when we defined "transitive", that the subset relation is transitive. (In fact, Theorem 2.2, which asserts the transitivity of "$\subseteq$", helped motivate that definition.) You should verify that "$\subseteq$" is also reflexive and antisymmetric.

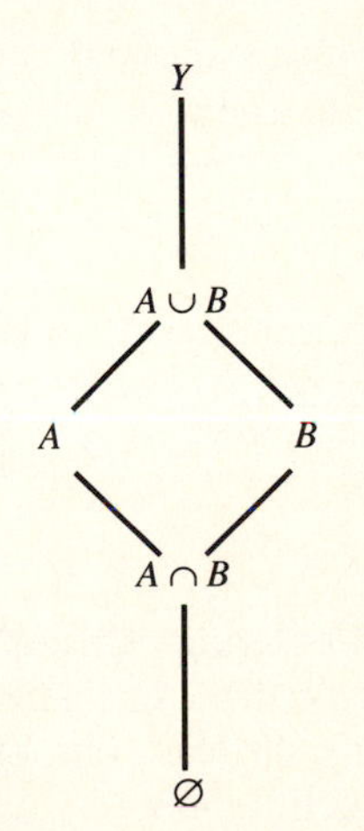

Figure 7.3. This lattice diagram shows that certain sets are subsets of certain others.

Based on these properties of "$\subseteq$", we make the following definition:

DEFINITION 7.9: A *partial ordering* on a set A is a relation on A which is reflexive, antisymmetric, and transitive.
A set together with a partial ordering on that set is called a *partially ordered set*, sometimes abbreviated as *poset*.

We have set up the definition so that "$\subseteq$" is a partial ordering on $\mathcal{P}(Y)$ for any set Y. We will use the suggestive symbol "$\precsim$" (read "precedes") to represent a generic partial ordering, and occasionally use the symbol "$\succsim$" (read "follows") to represent the inverse of "$\precsim$". Also, "$x \prec y$" means "$x \precsim y$ and $x \neq y$," and similarly, "$x \succ y$" means "$x \succsim y$ and $x \neq y$."

Another familiar partial ordering is the divisibility relation on $\mathbf{N}$, in which $(x, y) \in R$ iff $x|y$. The following proof of this assertion illustrates how to prove that a particular relation is a partial ordering:

THEOREM 7.10. The divisibility relation is a partial ordering on $\mathbf{N}$.

Proof: If $x \in \mathbf{N}$, then $x = x \times 1$, so $x|x$, and so "$|$" is reflexive.
If $x|y$ and $y|x$, then there are positive integers m and n such that $y = mx$ and $x = ny$. Therefore $y = m(ny)$, and so $mn = 1$. Since m and n are both positive integers, this means $m = n = 1$, and so $x = y$. This shows that "$|$" is antisymmetric.
Finally, Theorem 2.9 states that "$|$" is transitive. ■

Get Your Hands Dirty 2

Prove that each of the following relations is a partial ordering:

i) The relation "$\leq$" on $\mathbf{R}$.
ii) The relation on the set of English words in which one word is related to another iff the first is alphabetically prior to, or the same as, the second.
iii) The relation on $\mathbf{R} \times \mathbf{R}$ defined by

$$(x, y) \sim (u, v) \quad \text{iff} \quad x \leq u \text{ and } y \leq v\,.$$

iv) The relation "$\precsim$" defined on the set of all functions from $\mathbf{R}$ to $\mathbf{R}$ by

$$f \precsim g \quad \text{iff} \quad (\forall\, x \in \mathbf{R})\,(f(x) \leq g(x))\,.$$ □

Total Ordering

As GYHD 2 states, the relation "$\leq$" on $\mathbf{R}$ is a partial ordering. This relation has an additional property which you may think of as part of "order", but which is not part of Definition 7.9. Given any two real numbers x and y, at least one of the statements "$x \leq y$" and "$y \leq x$" will be true. (They are both true iff $x = y$.)

If a partial ordering is supposed to describe "comparison", why isn't a condition like this included in the definition?

? ? ?

Our reason is that we want our definition to be broad enough to include the subset relation, which does not generally have this property. For example, on the set $\mathcal{P}(\{1,2\})$, if $S = \{1\}$ and $T = \{2\}$, then $S \not\subseteq T$ and $T \not\subseteq S$; i.e., neither "$S \subseteq T$" nor "$T \subseteq S$" is true.

Definition 7.9 allows for a relation $\precsim$ such that, for some choices of x and y, neither $x \precsim y$ or $y \precsim x$ will hold. That is the reason that this type of relation is called a *partial* ordering. If we want every pair of objects in A to be "comparable", we need to add an additional condition on the relation. This suggests the following terminology:

DEFINITION 7.10: For a partial ordering $\precsim$ on a set A, and $x, y \in A$, "x and y are *comparable*" means:

$$x \precsim y \quad \text{or} \quad y \precsim x\,.$$

DEFINITION 7.11: A *total ordering* on a set A is a partial ordering on A such that

$$(\forall\, x, y \in A)\,(x \text{ and } y \text{ are comparable})\,.$$

A total ordering is also known as a *linear ordering*, a *simple ordering*, or a *complete ordering*. Note that a total ordering is a special type of partial ordering.

Get Your Hands Dirty 3

Which of the partial orderings in GYHD 2 are total orderings? □

COMMENT It is not hard to show that, if R is a partial ordering on some set A, then R^{-1} is also a partial ordering on A [see Exercise 2a)]. For example, "$\supseteq$", which is the inverse of "$\subseteq$", is also a partial ordering on $\mathcal{P}(Y)$ for any set Y. Also, if R is total, then so is R^{-1} [see Exercise 2b)].

There is not really anything that can be learned from studying "$\supseteq$" that can't be learned instead by studying "$\subseteq$". For the sake of being able to use consistent intuitive language, mathematicians generally choose arbitrarily to think in terms of the "is less than or equal" type of relation when discussing partial orderings, rather than the "is greater than or equal" type of relation.

Partial Orderings for Finite Sets

For small, finite partially ordered sets A, it is often helpful to visualize the partial ordering by means of a lattice diagram (see discussion and Figure 7.3 at the beginning of this section). In such a diagram, each element of A is shown, with certain objects connected by line segments so that, for $a \neq b$, there is an "upward path" from a to b iff $a \lesssim b$. It is easy to tell from such a diagram if the ordering is total, because the lattice for a total ordering will not have any "branches". It will be essentially a single line.

Example 1. A Lattice Diagram Suppose $A = \{1, 2, 3, 4, 5\}$, and the partial ordering R is the following set of pairs:

$$\{(1,1),(1,2),(1,4),(1,5),(2,2),(2,4),(2,5),(3,3),(3,5),(4,4),(5,5)\}.$$

(Verify that R is reflexive, antisymmetric, and transitive.)

The lattice diagram for R would look something like Figure 7.4. □

COMMENT To avoid complicating the figure unnecessarily, we do not draw a line directly from 1 to 4 in Figure 7.4. Since R is transitive, the lines from 1 to 2 and from 2 to 4 are sufficient to tell us that the pair $(1, 4)$ is also in the relation. As noted above, all we need is an "upward path".

Get Your Hands Dirty 4

Find some other partial orderings, as follows: Begin with a small, finite set (e.g., five elements), and find at least four different partial orderings. Choose them so as to make their lattice diagrams as varied as you can. (You decide exactly what that means.) You may wish to start with a lattice diagram and derive the partial ordering from that. □

It may be also helpful to see some examples of relations which resemble partial orderings, but in fact are not.

Example 2. Some Relations Which Are Not Partial Orderings Each of the following relations is not a partial ordering, for the reasons stated:

i) The relation "$<$" on **R**: This relation is antisymmetric and transitive, but not reflexive. (In fact, "$<$" is irreflexive.)

ii) The relation on **Q** (the rational numbers) defined by

$$x \sim y \quad \text{iff} \quad |x| \leq |y|.$$

This relation is reflexive and transitive, but not antisymmetric. For example, $-2 \sim 2$ and $2 \sim -2$, but $-2 \neq 2$.

iii) The relation on the points of the unit circle in the plane (i.e., $\{(x, y) \in \mathbf{R} \times \mathbf{R} : x^2 + y^2 = 1\}$) given by "$P \sim Q$ iff either $P = Q$ or the shorter arc from P

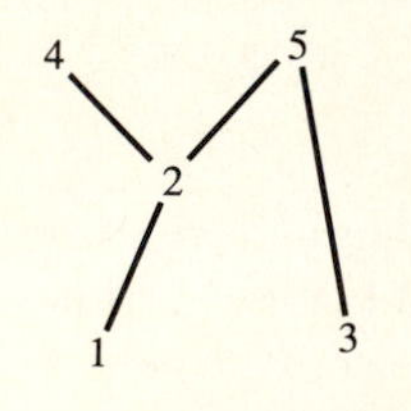

Figure 7.4. A lattice diagram for the relation R of Example 1.

to Q is clockwise". (If P and Q are opposite points on the circle, then $P \not\sim Q$.) This relation is reflexive and antisymmetric, but not transitive (see Figure 7.5). □

Generalizing Upper Bounds, etc.

In Section 3.2, we introduced the concepts of upper and lower bounds, sets bounded above and below, etc., and in Section 4.1, we proved some simple theorems about those concepts. Those concepts were built around the relation "$\leq$" on **R**. The concept of partial ordering allows us to pursue those ideas in a more general context. (You may wish to reread our earlier development before continuing.)

For this discussion, A represents some set, "$\lesssim$" represents a partial ordering on A, and "$\gtrsim$" is its inverse relation.

We begin with the generalization of the concept of "upper bound" (Definition 3.11 in Section 3.2):

DEFINITION 7.12: For $x \in A$ and $S \subseteq A$, "x is an *upper bound* for S" means:

$$(\forall\, y \in S)(x \gtrsim y)).$$

The next definition generalizes terminology introduced in Exercise 3 of Section 4.1:

DEFINITION 7.13: For $x \in A$ and $S \subseteq A$, "x is the *maximum element* of S" means:

$$x \in S \quad \text{and} \quad (\forall\, y \in S)(x \gtrsim y).$$

The use of the word "the" in the previous definition is justified by the following theorem, whose proof is part of Exercise 8:

THEOREM 7.11. S cannot have more than one maximum element.

A set need not have any maximum element. You should examine the concept of "maximum" with the following:

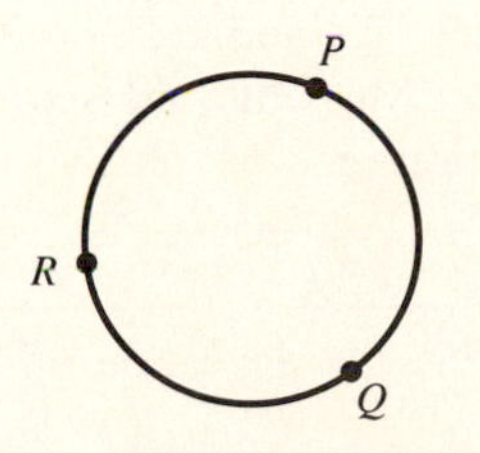

Figure 7.5. Using relation iii) of Example 2, we have $P \sim Q$ and $Q \sim R$, but $P \not\sim R$.

Get Your Hands Dirty 5

a) Give an example of a subset of **R** which has no maximum element.
b) Which subsets of $\mathcal{P}(\{1,2\})$ have maximum elements under "$\subseteq$"? [List them explicitly. $\mathcal{P}(\{1,2\})$ has 4 elements and 16 subsets.] □

The following definition should be compared carefully with Definition 7.13. It reflects the complications which arise from the "nontotality" of some partial orderings:

DEFINITION 7.14: For $x \in A$ and $S \subseteq A$, "x is a *maximal element* of S" means:

$$x \in S \quad \text{and} \quad \sim(\exists\, y \in S)(y > x).$$

COMMENTS

1. We repeat: be sure to distinguish carefully between *a maximal* element and *the maximum* element. A *maximal* element x need not be "greater than or equal to" every element of S; we only require that "$y > x$" not be true for any y. There may be some values of y for which neither "$x \gtrsim y$" not "$y > x$" is true—the negation of "$x \gtrsim y$" is not "$y > x$". For example, suppose $A = \mathcal{P}(\{1,2,3\})$, and $S = \{X \in A : |X| \leq 2\}$ (using "$\subseteq$" as the partial ordering on A). Then $\{1,2\}$ is a maximal element of S, since no element of S properly contains $\{1,2\}$; but $\{1,2\}$ is not the maximum element of S, since $\{3\} \in S$, and $\{1,2\} \not\supseteq \{3\}$. We can describe this distinction in terms of lattice diagrams: a maximal element is at the top of *some* branch of the lattice, while a maximum element is at the "top" of *all* branches—that is, all branches lead up to it.

2. A set can have more than one maximal element (but not more than one maximum element). In the example described in comment 1, the sets $\{1,2\}$, $\{1,3\}$, and $\{2,3\}$ are all maximal elements of S.

The terms "bounded above" and "least upper bound" have similar generalizations, as do the theorems on upper bounds and sets bounded above in Section 4.1 [see Exercise 8b) and c)].

Strict Orderings

We conclude this section with an irreflexive variation on the concept of partial ordering. We use as our prototype such pairs of relations as "$\subseteq$" and "$\subset$" on $\mathcal{P}(Y)$ or "$\leq$" and "$<$" on **R**. In each of these pairs, the first is reflexive, while the second is irreflexive. This distinction is formalized in the following definition:

DEFINITION 7.15: A *strict ordering* on a set A is a relation on A which is irreflexive and transitive.

Note: A relation which is irreflexive and transitive must be antisymmetric (see GYHD 6 below), so the condition of being antisymmetric is not specifically included in the definition of a strict ordering.

Unless $A = \varnothing$, a relation on A cannot be both reflexive and irreflexive, so a strict ordering on a nonempty set is not a partial ordering. We can describe the connection between these two types of relations by the following theorem:

THEOREM 7.12.

a) If "$\lesssim$" is a partial ordering on a set A, then the relation "$<$" defined by

$$x < y \quad \text{iff} \quad x \lesssim y \text{ and } x \neq y$$

is a strict ordering on A.

b) If "$<$" is a strict ordering on a set A, then the relation "$\lesssim$" defined by

$$x \lesssim y \quad \text{iff} \quad x < y \text{ or } x = y$$

is a partial ordering on A.

We will prove a) and leave the proof of b) as Exercise 10.

Proof of a) of Theorem 7.12: We have to show that $<$ is irreflexive and transitive.

To show that $<$ is irreflexive, recall Definition 6.10: a relation $\sim$ is called irreflexive iff $(\forall x)(x \not\sim x)$. The condition "$x < x$" cannot hold, since the definition of "$x < y$" includes the requirement that $x \neq y$.

To prove transitivity, suppose $x < y$ and $y < z$. Then $x \lesssim y$ and $y \lesssim z$, so $x \lesssim z$, by the transitivity of "$\lesssim$". We need only show that $x \neq z$, which we will do by contradiction. So suppose that $x = z$. Since we already knew that $y \lesssim z$, and we now have $x = z$, it follows that $y \lesssim x$. Thus, $x \lesssim y$ and $y \lesssim x$, so by antisymmetry of "$\lesssim$", we have $x = y$. But this is impossible, since the hypothesis "$x < y$" includes the requirement that $x \neq y$.

That concludes the proof. ■

Get Your Hands Dirty 6

Prove that every strict ordering is vacuously antisymmetric. □

EXERCISES

1. For what sets Y is "$\subseteq$" a complete ordering? In other words, for what sets Y is the following statement true:

$$(\forall S, T \in \mathcal{P}(Y))(S \subseteq T \text{ or } T \subseteq S)? \qquad (*)$$

To answer this question and justify that answer, you need to give a description in elementary terms of the sets Y that have the property $(*)$, and then prove two things:

a) Any set that fits your description has the property $(*)$.

b) Any set with the property $(*)$ fits your description.

2. Let R be a relation on a set A. Prove the following:
   a) If R is a partial ordering, then R^{-1} is also a partial ordering.
   b) If R is a total ordering, then R^{-1} is also a total ordering.

3. Which of the following relations are partial orderings? For each of those that is not, state which properties it fails to have. Justify your answers.
   a) A = the set of polynomials with coefficients in $\mathbf{Q}$ (see Exercise 16 of Section 5.2). Define "$\lesssim$" on A by
   $$f \lesssim g \quad \text{iff} \quad \deg(f) \leq \deg(g) .$$
   b) A = the set of continuous functions from $[0, 1]$ to $\mathbf{R}$. Define $\lesssim$ by
   $$f \lesssim g \quad \text{iff} \quad \int_0^1 f(x)\,dx \leq \int_0^1 g(x)\,dx .$$
   c) A as in ii). Define $\lesssim$ by
   $$f \lesssim g \quad \text{iff} \quad (\forall\, x \in [0, 1])\,(f(x) \leq g(x)) .$$
   d) $A = \mathbf{N}$. Define $\lesssim$ by
   $$m \lesssim n \quad \text{iff} \quad \{p : p \text{ is a prime and } p \mid m\} \subseteq \{p : p \text{ is a prime and } p \mid n\} .$$

4. Let A be a finite set with $|A| = n$. How many total orderings on A are there? Explain your answer.

5. a) Explicitly list, as sets of ordered pairs, all the partial orderings on $\{1, 2, 3\}$. (There are 512 relations on this set, but the number of these which are partial orderings is manageable.)
   b) Partition the set of partial orderings by putting together those partial orderings whose lattices are "the same shape". (You decide and explain exactly what that means.)
   c) How many "lattice types" are there for partial orderings on $\{1, 2, 3, 4\}$? Explain your answer.

6. Suppose $\lesssim$ is a partial ordering on a set A, and x any object not in A. Let $B = A \cup \{x\}$. Define, in precise language, a partial ordering on B in which x is the maximum element, while the ordering on elements of A is the same as that given by $\lesssim$.

7. Suppose $\lesssim_1$ is a partial ordering on a set A, and $\lesssim_2$ is a partial ordering on a set B.
   a) Is the product of $\lesssim_1$ and $\lesssim_2$ (see Exercise 11 of Section 6.2) a partial ordering on $A \times B$? Justify your answer.
   b) Describe a partial ordering on $A \times B$, based on $\lesssim_1$ and $\lesssim_2$, which resembles the way in which two-letter words are alphabetized using the standard ordering on the alphabet.

8. Suppose that A is a set partially ordered by $\lesssim$, S and T are subsets of A, and x and z are elements of A. Prove each of the following:
   a) **THEOREM 7.11.** S cannot have more than one maximum element.
   b) (Generalization of Theorem 4.2) If x is an upper bound for S, and $z \gtrsim x$, then z is an upper bound for S.
   c) (Generalization of Theorem 4.5) If S is bounded above, and $T \subseteq S$, then T is bounded above.

9. Suppose that $\lesssim$ is a partial ordering on a set A for which every two-element subset has a least upper bound. Prove that any finite, nonempty subset of S has a lub. (*Hint:* Use induction.)

10. Prove the following:

THEOREM 7.12b). If "$\prec$" is a strict ordering on a set A, then the relation "$\precsim$" defined by

$$x \precsim y \quad \text{iff} \quad x \prec y \text{ or } x = y$$

is a partial ordering on A.

11. Prove that a strict ordering $\prec$ has the following property:

$$(\forall\, x, y \in A)\,(x \prec y \Rightarrow y \nprec x)\,.$$

(A relation with this property is sometimes called *asymmetric*.)

PART III

Topics in Mathematics

8

Cardinality

Introduction

In this chapter, we will look at the idea of *counting,* which is probably the most fundamental notion in all of mathematics. The idea of counting may seem too elementary to be discussed in a book like this, but we will see that this notion contains deep and complex principles. We will formally define *cardinality,* which will give us a way to talk about the "number of elements" in a set, even when the set is infinite. Strange as it may seem, we will see that there are "just as many" natural numbers as there are rational numbers. Given that fact, you may then be even more surprised to find that some infinite sets *are* "bigger" than others. In particular, we will show that there are "more" real numbers than there are natural numbers.

There is one key theorem — the *Cantor–Schroeder–Bernstein Theorem* — which is part of the development of Section 8.1, and whose proof is rather technical. We have placed this proof in a separate section, Section 8.2, in order to avoid disrupting the general development.

8.1 Cardinality

Formalizing Cardinality

Here are two sets: $A = \{3, 11, 17, -2\}$, $B = \{9, 2, -5, 41\}$. They have the same number of elements. We can count the objects in each set, and see that each set has four members.

Here's another situation: a crowd of people enters an auditorium and they all sit down in the seats, each person in a separate seat. If there are no vacant seats, we know immediately that the auditorium has the same number of seats as there are people in the audience.

What does this second example tell us about the ideas of "how many" and "the same number"? How does this example compare with the first situation?

? ? ?

The most important principle we want you to appreciate from this is that we don't have to count to compare the sizes of two sets. The first main goal of this section is to define a relation on "the set of all sets" (see comment below) which:

a) will express our intuitive idea of one set being "the same size as" another, and
b) can be expressed precisely in terms of formal mathematical ideas that have already been discussed.

We will then modify that relation to get one which describes what it means for one set to be "smaller than" another.

COMMENT The idea of "the set of all sets" turns out to lead to contradictory statements, as we will discuss at the end of this section. We will avoid such contradictions by letting A be some set and letting $\mathbf{U} = \mathcal{P}(A)$ be our universe of discourse. Our relation will be defined on $\mathbf{U}$.

We pose this task for you to think about:

QUESTION 1 We want to define a relation R on $\mathbf{U} = \mathcal{P}(A)$, for which "$(S,T) \in R$" has the intuitive meaning "the number of elements in S is equal to the number of elements in T"

How can we define such a relation without directly talking about "number of elements"?

***Hint:* How did you know that there were the same number of seats as people?**

Set Equivalence

Back in Section 3.2, we commented on the intuitive connection between bijective functions and cardinality. We said that, if there was a bijective (one-to-one and onto) function from a finite set S to a finite set T, then S and T would appear to have the same cardinality. We will make the following definition, based on this idea:

DEFINITION 8.1: For sets S and $T \in \mathbf{U}$, "S is *equivalent* to T" (written $S \approx T$) means: There exists a bijective function from S to T. (As usual, "$S \not\approx T$" means "S is not equivalent to T".)

Note: This relation has many names in mathematics. Some of the other terms for sets having this relationship are "equipotent", "equinumerous", and "equipollent".

This definition describes a relation on $\mathbf{U} = \mathcal{P}(A)$, which is called *set equivalence*. The use of this terminology is justified by following theorem:

THEOREM 8.1. "$\approx$" is an equivalence relation.

Proof: "$\approx$" is reflexive because, for any S, the identity function I_S [defined by $I_S(x) = x$] is a bijective function from S to itself.

"$\approx$" is symmetric because, by Corollary 6.5, if f is a bijective function from S to T, then f^{-1} is a bijective function from T to S.

Finally, "$\approx$" is transitive because, by Corollary 4.7, if $f: S \to T$ and $g: T \to U$ are both bijective functions, then $g \circ f$ is a bijective function from S to U. ■

We will use "$\overline{\overline{S}}$" to represent the equivalence class of S under the relation "$\approx$", and "$\mathscr{C}$" to represent the partition associated with this equivalence relation. (You may recognize the "$\overline{\overline{S}}$" from Exercise 9 of Section 4.1. You will see shortly how our use of the notation here relates to its use there.)

Thus, the statement "$\overline{\overline{S}} = \overline{\overline{T}}$" is synonymous with "$S \approx T$". Which one we use will depend on whether we want to think about individual sets or about their equivalence classes. Intuitively, either of these statements means that S and T have the same number of elements. In other words, "$\overline{\overline{S}} = \overline{\overline{T}}$" seems to mean the same thing as "$|S| = |T|$".

! ! !

We seem to have come up with something that is just as good as cardinality. In fact, it's better, because it makes perfectly good sense even if we are talking about infinite sets.

The Meaning of Counting

It's time for a brief digression on the meaning of counting and cardinality from the perspective of the young child.

The first thing a child usually learns to do with numbers is recite the number words by rote: "one, two, three, four, . . ." After a while, the child learns to use this recitation to count the number of things in a set, by pointing to the objects, one at a time, while saying the number words. If this is done properly (i.e., one word for each object, with no repetitions), then the last number word recited tells "how many".

This process of matching the objects with the number words is equivalent to defining a specific one-to-one, onto function from the set of objects to a set of the type $\{1, 2, \ldots, n\}$—the set we called $\mathbf{N}_{[1,n]}$ at the start of Section 5.2. (Recall that we defined $\mathbf{N}_{[1,0]} = \varnothing$.)

So $\mathbf{N}_{[1,0]}$ is our prototype of an n-element set. Intuitively, when we say that a set has n elements, we mean that there is a one-to-one, onto function from that set to $\mathbf{N}_{[1,n]}$. (This works even for $n = 0$.) When we count the objects in the set, reciting the numbers from 1 to n, we are defining that function.

Essentially we are saying that, for finite sets at least, the idea of "how many" is pretty well described by the relation of set equivalence. Two sets seem to have the same "how-many-ness", i.e., the same cardinality, if and only if they belong to the same equivalence class under this relation. We say "seem to" because we have not formally defined "cardinality" yet. So far we have just worked with this concept on an intuitive level.

We can establish this principle, and at the same time start to talk about infinite sets, by using the idea of equivalence classes to formally define cardinality for arbitrary sets.

The definition is as follows:

DEFINITION 8.2: For any set $S \in \mathbf{U}$, the *cardinality* of S (or the *cardinal number* of S) is the equivalence class $\overline{\overline{S}}$ of S under the relation of set equivalence.

Thus, for example, the cardinality of $\{1, 5\}$ is the set consisting of all 2-element sets (within the universe **U**).

Now you might say that we are overly complicating things, and that the cardinality of $\{1, 5\}$ should just be the number 2. But what is "2"? Our discussion above should suggest that there is an idea of "twoness", and all two-element sets represent this idea. Essentially, what we are doing is defining the cardinality of $\{1, 5\}$ to be this "two-ness idea", as represented by the collection of all two-element sets.

It follows immediately from this definition that two sets have the same cardinality iff they are in the same equivalence class. After all, the cardinality *is* the equivalence class.

COMMENTS

1. This may seem like cheating. We have simply defined cardinality in a way that makes the desired principle true. We could make any statement true if we just defined the terms in the right way. The justification for this process is that it fits our intuitive concept for finite sets. In fact, the discussion above points out that, when we say "$|S| = n$", what we mean is that S is equivalent to $\mathbf{N}_{[1,n]}$. When forced to explain thoroughly our intuitive idea of cardinality, we actually end up talking about set equivalence.

2. *Reminder:* As mentioned earlier, we must always keep the whole discussion within the framework of some universe **U**. For technical reasons, we can't talk about "all sets" at once. But we can keep mentally adjusting the universe to include any specific sets we have in mind.

3. The partition $\mathscr{C}$ is thus the set of all cardinalities of sets that are elements of the universe **U**.

A Partial Ordering for Cardinalities

Now that we know what it means to say that two sets have *the same* cardinality, it would be nice to have a way of *comparing* cardinalities. This suggests that we need to develop a partial ordering. We will begin by looking for such a relation on **U**, our set of sets, and then see that we need to switch to a relation on $\mathscr{C}$, the set of cardinalities, to get what we want.

Historical Note: It wasn't until the late nineteenth century that the idea of cardinality got serious attention from mathematicians. Georg Cantor (1845–1918) laid the foundations for this subject, introducing most of the central ideas of this section. He showed that many sets that seem "bigger" than **N** actually have the same cardinality as **N**, and proved that there are infinite sets which have different cardinalities. Though his work was received with considerable skepticism by his contemporaries, it is now universally accepted as a major contribution to mathematical thinking.

We ask the following:

QUESTION 2 How can we define a relation "$\lesssim$" on **U** for which "$S \lesssim T$" has the intuitive meaning: "the number of elements in S is less than or equal to the number of elements in T"?

Hint: Keep using functions.

We can answer Question 2 with the following definition:

DEFINITION 8.3: For sets S and T, "$S \lesssim T$" means: There is an injective function from S to T.

Think about this definition. In the exercises in Section 3.2, we asked about the number of functions of various types from S to T, in terms of the number of elements of the finite sets S and T. You may have noticed that, if $|S| > |T|$, there are no injective functions from S to T. Intuitively, that is because there are not enough places for elements of S to go. Conversely, if $|S| \leq |T|$, then there are enough places, and such functions should exist. It is precisely that intuition which motivates Definition 8.3.

Also note that this definition of "$\lesssim$" is consistent with our definition of "$\approx$"; i.e., if $S \approx T$, then we also have $S \lesssim T$: the first condition requires a bijective function, and the second just requires an injective function.

Example 1 **Working with "$\lesssim$"** We can prove that two sets have this relation to each other by giving an explicit one-to-one function from one to the other. For instance:

i) Let $S = \{a, b, c\}$ and $T = \{1, 2, 3, 4, 5\}$, and let f be the function from S to T consisting of the following set of ordered pairs: $\{(a, 1), (b, 4), (c, 2)\}$. Then f is injective, and so $\{a, b, c\} \lesssim \{1, 2, 3, 4, 5\}$. (Many other functions would be equally satisfactory for proving this.)

ii) Let $S = \mathbf{Z}$, $T = \mathbf{Q}^+$ (positive rational numbers). Then $S \lesssim T$, because the function $g: S \to T$ given by $g(x) = 2^x$ is injective. (Again, there are many other functions that would work.) ■

Get Your Hands Dirty 1

Prove that if $A \subseteq B$, then $A \lesssim B$. □

We said preceding Question 2 that we were looking for a partial ordering. It is easy to see that "$\lesssim$" is reflexive (since the identity function from a set to itself is injective) and transitive (since composition of injective functions is injective—Theorem 4.3).

But "$\lesssim$" is not a partial ordering, because it is not antisymmetric. If $S \lesssim T$ and $T \lesssim S$, we can't conclude that $S = T$: it could be that S and T are just different sets of the same size.

For this reason, it is better for us to work with equivalence classes, i.e., cardinalities, so we want to adapt "$\precsim$" to define a relation on $\mathscr{C}$. In order to talk about elements of $\mathscr{C}$ without tying them down to specific sets, we will use the Greek letters κ ("kappa") and λ ("lambda") to represent elements of $\mathscr{C}$. Keep in mind that these objects are each sets of sets.

We make the following definition:

DEFINITION 8.4: For cardinalities κ and λ in $\mathscr{C}$, "$\kappa \leq \lambda$" means: For some $S \in \kappa$ and $T \in \lambda$, $S \precsim T$.

We would like to be sure that this definition does not depend on the particular elements of κ and λ selected. The following result says exactly that:

LEMMA 8.2. For $\kappa, \lambda \in \mathscr{C}$, if $\kappa \leq \lambda$, and $U \in \kappa$ and $V \in \lambda$, then $U \precsim V$.

It may help to rephrase this directly in terms of functions and sets. The condition "$\kappa \leq \lambda$" tells us that there are sets $S \in \kappa$ and $T \in \lambda$ with $S \precsim T$. The following version of Lemma 8.2 refers to all four sets S, T, U, and V:

LEMMA 8.2′. For $S, T, U, V \in \mathbf{U}$, if $S \approx U$, $T \approx V$, and there is an injective function from S to T, then there is an injective function from U to V.

The proof is left to the reader as Exercise 1.

We will go back and forth between using $\precsim$ as a relation on $\mathbf{U}$ and using $\leq$ as a relation on $\mathscr{C}$.

The relation "$\leq$" on $\mathscr{C}$ seems to answer our needs. For finite sets, at least, the condition "$\overline{\overline{S}} \leq \overline{\overline{T}}$" fits our intuitive understanding of the statement "$|S| \leq |T|$", and it seems to be a reasonable way to get a handle on infinite sets. Furthermore, as with the relations "$\precsim$" and "$\approx$" on $\mathbf{U}$, the definition of "$\leq$" on $\mathscr{C}$ is consistent with our definition of equality of cardinalities: if $\overline{\overline{S}} = \overline{\overline{T}}$, then $\overline{\overline{S}} \leq \overline{\overline{T}}$.

As usual with orderings, "$\overline{\overline{S}} < \overline{\overline{T}}$" means "$\overline{\overline{S}} \leq \overline{\overline{T}}$ but $\overline{\overline{S}} \neq \overline{\overline{T}}$". Similarly, "$S \prec T$" means "$S \precsim T$ but $S \not\approx T$".

But we still have an unresolved issue:

QUESTION 3 Is $\leq$ a partial ordering on $\mathscr{C}$?

Is it reflexive? Antisymmetric? Transitive?

Two of these requirements are easy. To show that "$\leq$" is reflexive, let κ be any element of $\mathscr{C}$, and let S be any set in κ. We already know that "$\precsim$" is reflexive, so $S \precsim S$. The definition of "$\leq$" (Definition 8.4) then tells us $\kappa \leq \kappa$. The transitivity of "$\leq$" follows similarly from that of "$\precsim$", with the help of Lemma 8.2 (see Exercise 2).

But what about antisymmetry? Suppose that $\overline{\overline{S}} \leq \overline{\overline{T}}$ and $\overline{\overline{T}} \leq \overline{\overline{S}}$, so there are injective functions f from S to T, and g from T to S. Does that mean that $\overline{\overline{S}} = \overline{\overline{T}}$? Are f and g necessarily bijective?

Caution: The last two questions are not the same.

The answer to the second question is no. For example, suppose $S = T = \mathbf{Z}$, and $f:\mathbf{Z} \to \mathbf{Z}$ and $g:\mathbf{Z} \to \mathbf{Z}$ are given by the equations $f(x) = x^3$ and $g(x) = x^3$. Then both f and g are injective, but neither is surjective. In this case, however, even though neither f nor g is bijective, it is easy to show that $\overline{\overline{S}} = \overline{\overline{T}}$, since we can use the identity function on $\mathbf{Z}$ to prove $S \approx T$.

Now suppose that $S = \mathbf{R}$ and $T = \mathbf{R}^{\geq 0}$ (the set of nonnegative real numbers). Let $f:\mathbf{R} \to \mathbf{R}^{\geq 0}$ be the function given by $f(x) = 2^x$, and let $g:\mathbf{R}^{\geq 0} \to \mathbf{R}$ be the function given by $g(x) = x$. Again, both functions are injective, so $\overline{\overline{S}} \leq \overline{\overline{T}}$ and $\overline{\overline{T}} \leq \overline{\overline{S}}$. And again, neither function is bijective, so we don't yet have $\overline{\overline{S}} = \overline{\overline{T}}$.

Get Your Hands Dirty 2

Why isn't the function f above (given by $f(x) = 2^x$) a bijection from $\mathbf{R}$ to $\mathbf{R}^{\geq 0}$ [with inverse $h(x) = \log_2 x$]? □

Are $\boldsymbol{R}$ and $\boldsymbol{R}^{\geq 0}$ "the same size"? Is there a bijective function from one to the other? (Try to find one.)

The Cantor–Schroeder–Bernstein Theorem

The answer to these questions is yes. But the construction of such a function is not easy. Strange as it may seem, it is quite difficult to show that $\leq$ is antisymmetric. This assertion is known as the *Cantor–Schroeder–Bernstein Theorem*. We state it here, and postpone the proof to Section 8.2:

THEOREM 8.3: (Cantor–Schroeder–Bernstein Theorem). If S and T are sets such that $S \lesssim T$ and $T \lesssim S$, then $S \approx T$.

We can rephrase this in terms of cardinalities as follows:

THEOREM 8.3′: (Cantor–Schroeder–Bernstein Theorem). If κ and λ are cardinalities such that $\kappa \leq \lambda$ and $\lambda \leq \kappa$, then $\kappa = \lambda$.

COMMENT The proof of this theorem requires constructing a bijective function from S to T, based on two injective functions, one from S to T and the other from T to S. The standard proof is quite complex and very interesting. It works by partitioning the sets S and T, and piecing together "parts" of each of the two given functions, using a rather clever and intricate method, to build the desired function. The proof is rather technical, and certainly not one you should expect to be able to come up with on your own.

The following is an immediate consequence of Theorem 8.3′ and our earlier discussion:

COROLLARY 8.4. "$\leq$" is a partial ordering on $\mathscr{C}$.

Note: You might ask: "Isn't '$\leq$' actually a *total* ordering?" After all, given any two sets, it would seem that there should be an injective function from the "smaller" one to the "larger" one. Unfortunately, it's not so simple, as explained later in comment 8 following GYHD 6 (p. 309).

Theory of Cardinality for Finite Sets

To a large extent, this abstract discussion of cardinality is motivated by the complexities of dealing with infinite sets. We will look at infinite sets shortly.

But first, we want to illustrate what this theory can tell us about finite sets. One of the basic assumptions we make about counting is that a finite set has only one "size". Another way to put this is as follows: if two people count the number of objects in a set and neither makes a mistake, then they must come up with the same answer. This principle allows us to talk about *the* number of elements in a set.

We can express this idea more formally using the language of set equivalence. Recall that, in the beginning of Section 5.2, we introduced the notation $\mathbf{N}_{[1,n]}$ to represent the set $\{1, 2, \ldots, n\}$ (with $\mathbf{N}_{[1,0]}$ representing the empty set). We think of $\mathbf{N}_{[1,n]}$ as our prototype of an n-element set. The assertion that a finite set cannot have two different sizes is expressed by the following:

THEOREM 8.5. For $m, n \in \mathbf{W}$,

$$\text{if} \quad m \neq n, \quad \text{then} \quad \mathbf{N}_{[1,m]} \not\approx \mathbf{N}_{[1,n]} \text{ and } \overline{\overline{\mathbf{N}_{[1,m]}}} \neq \overline{\overline{\mathbf{N}_{[1,n]}}}.$$

The two conclusions are equivalent. The proof is by induction, and is outlined in Exercise 3.

COMMENT As our discussion shows, Theorem 8.5 is not at all surprising. Indeed, it would be rather strange if this assertion were false. But you may be surprised that something so basic is actually provable. One of the benefits of giving formal definitions to intuitive ideas is that it allows us to prove very fundamental statements about those ideas.

If $m < n$, the function $f(x) = x$ from $\mathbf{N}_{[1,m]}$ to $\mathbf{N}_{[1,n]}$ is injective, so $\mathbf{N}_{[1,m]} \leq \mathbf{N}_{[1,n]}$. By Theorem 8.5, we have $\overline{\overline{\mathbf{N}_{[1,m]}}} \neq \overline{\overline{\mathbf{N}_{[1,n]}}}$. Therefore we have the following:

COROLLARY 8.6. For m, $n \in \mathbf{W}$,

$$\text{if} \quad m < n, \quad \text{then} \quad \overline{\overline{\mathbf{N}_{[1,m]}}} < \overline{\overline{\mathbf{N}_{[1,n]}}}.$$

We can use these ideas to formally define "finite" and "infinite," and to formally define $|S|$ for finite sets.

DEFINITION 8.5: For a set S,

"S is *finite*" means: $(\exists\, t \in \mathbf{W})(\overline{\overline{S}} = \overline{\overline{\mathbf{N}_{[1,t]}}})$.
"S is *infinite*" means: S is not finite.

DEFINITION 8.6: For a finite set S, $|S|$ is the (unique) whole number t such that $\overline{\overline{S}} = \overline{\overline{\mathbf{N}_{[1,t]}}}$.

(The uniqueness follows from Theorem 8.5.)

Infinite Sets

Perhaps the most surprising thing in the entire subject of cardinality is the following fact, which we prove in this section:

Some infinite sets are "bigger" than others.

That is, there are infinite sets S and T such that there is no bijective function from one to the other. We shall see that **N** and **R** are such sets.

If you don't find that surprising, then perhaps you will be surprised by the following fact, which we also prove:

N and **Q** are equivalent.

This assertion is an example of a more general phenomenon: an infinite set can be equivalent to a proper subset of itself.

The study of cardinality of infinite sets is a challenge to the intuition. You may need to rethink your concept of "how-many-ness" in order to accommodate the ideas presented here.

Countable Sets

Our first surprise is a simple one:

Example 2 **Equivalence of N and W** Let f be the function from **N** to **W** given by $f(x) = x - 1$. We leave it to you to verify that this function is bijective. Thus **N** and **W** are equivalent. □

Perhaps more surprising is that **Z** and 2**Z** (the even integers) are equivalent, since only "half" of the integers are even. We ask you to prove this:

Historical Note: The great Italian physicist and astronomer Galileo (1564–1642) was among those who noted that an infinite set can be equivalent to a proper subset of itself. Much later, in 1888, the German mathematician Richard Dedekind (1831–1916) used this property as a way of distinguishing infinite sets from finite ones. A set which is equivalent to a proper subset of itself is sometimes called *Dedekind-infinite*.

Get Your Hands Dirty 3

Prove that $\mathbf{Z}$ and $2\mathbf{Z}$ are equivalent. (You just need to find a bijective function from $\mathbf{Z}$ to $2\mathbf{Z}$.) □

We now move on to a harder example, the equivalence of $\mathbf{N}$ and $\mathbf{Q}$. There would seem to be "infinitely many times as many" rational numbers as natural numbers, and yet, once again, the cardinalities turn out to be equal.

Creating an appropriate bijective function is more difficult in this case, and we will prove the equivalence in two stages, first showing $\mathbf{N} \approx \mathbf{Z}$ and then showing $\mathbf{Z} \approx \mathbf{Q}$. The second proof uses the Cantor–Schroeder–Bernstein Theorem.

THEOREM 8.7. $\mathbf{N} \approx \mathbf{Z}$.

Proof: Define $g: \mathbf{N} \to \mathbf{Z}$ as follows:

$$g(x) = \begin{cases} \dfrac{x-1}{2} & \text{if } x \text{ is odd.} \\ \dfrac{-x}{2} & \text{if } x \text{ is even.} \end{cases}$$

Since distinct odd positive integers get mapped to distinct nonnegative integers, and distinct even positive integers get mapped to distinct negative integers, g is injective. We leave it to you to prove (GYHD 4) that g is surjective.

Thus g is bijective, so the two sets are equivalent. ■

Get Your Hands Dirty 4

Prove that the function g in the above proof is surjective. □

COMMENT For the function g in this proof, we have $g(1) = 0$, $g(2) = -1$, $g(3) = 1$, $g(4) = -2$, $g(5) = 2$, etc. Thus, the function creates a "list" of the integers: $0, -1, 1, -2, 2, \ldots$. A common way to demonstrate that a set is equivalent to $\mathbf{N}$ is to "list" its elements in this fashion. It is important to verify that every object from the set is listed, and that no object is listed twice.

THEOREM 8.8. $\mathbf{Z} \approx \mathbf{Q}$.

Proof: The function $f: \mathbf{Z} \to \mathbf{Q}$ given by $f(x) = x$ is injective, so $\mathbf{Z} \precsim \mathbf{Q}$. We will define an injective function h from $\mathbf{Q}$ to $\mathbf{Z}$, showing $\mathbf{Q} \precsim \mathbf{Z}$. The theorem will then follow from the Cantor–Schroeder–Bernstein Theorem.

In order to define h, we first define a function $g: \mathbf{Q}^+ \to \mathbf{N}$, as follows:

If $x \in \mathbf{Q}^+$, write $x = p/q$, with $p, q \in \mathbf{N}$ and $\text{GCD}(p, q) = 1$. Define $g(x) = 2^p 3^q$ (see comment 1 below). To prove that g is injective, suppose $x = p/q$ and $y = r/s$, with

$p, q, r, s \in \mathbf{N}$, $\text{GCD}(p, q) = 1$, and $\text{GCD}(r, s) = 1$. If $g(x) = g(y)$, then $2^p 3^q = 2^r 3^s$. It follows from the Unique-Factorization Theorem (Theorem 5.31) that $p = r$ and $q = s$, so $x = y$, as needed.

Now define $h: \mathbf{Q} \to \mathbf{Z}$ by

$$h(x) = \begin{cases} g(x) & \text{if } x > 0 \\ 0 & \text{if } x = 0 \\ -g(-x) & \text{if } x < 0\,. \end{cases} \qquad (*)$$

It remains only to show that h is injective. So suppose that $h(u) = h(v)$. Notice first that $h(x)$ always has the same sign as x (or both are 0). Therefore, u and v must have the same sign (or both be 0), and so $h(u)$ and $h(v)$ are defined by the same line in $(*)$. It follows that we have one of three possibilities:

i) $g(u) = g(v)$,
ii) $h(u) = h(v) = 0$,
iii) $-g(-u) = -g(-v)$.

But g is injective, so in the first case, we get $u = v$; in the second case, we get $u = v = 0$; and in the third case, we get $g(-u) = g(-v)$, so $-u = -v$ (since g is injective), and again $u = v$.

This completes the proof. ■

COMMENTS

1. We should be careful here, to be sure that g is well defined. It can be proved using a generalization of Euclid's Lemma that a positive rational number can only be expressed in one way as a quotient of two relatively prime positive integers (see Exercise 4). Therefore, there is no ambiguity in the definition of g.

2. The use of 2 and 3 in the definition of g was arbitrary—any two distinct primes would work.

Since $\mathbf{N} \approx \mathbf{Z}$ (Theorem 8.7) and $\mathbf{Z} \approx \mathbf{Q}$ (Theorem 8.8), we have the following (since $\approx$ is an equivalence relation):

COROLLARY 8.9. $\mathbf{N} \approx \mathbf{Q}$.

COMMENTS

1. Perhaps the most peculiar aspect of this sequence of results is the idea that a set can be equivalent to a proper subset of itself. It can be proved, using Theorem 8.5, that this cannot happen with finite sets (see Exercise 5). But we have just seen several instances where this does happen with infinite sets. (In some texts, this phenomenon is used as a definition of "infinite".)

2. An alternative approach to proving Corollary 8.9 is to create a "list" of the rational numbers, as described in the comment following Theorem 8.18. The following *diagonal method* is often used to show $\mathbf{N} \approx \mathbf{Q}^+$ (and we can use that to show $\mathbf{N} \approx \mathbf{Q}$). We arrange the elements of $\mathbf{Q}^+$ in an infinite array, as in Figure 8.1, in which each row contains the fractions with a given denominator. To form a list of positive rational numbers, we follow the arrows as indicated: we begin at $\frac{1}{1}$, then list the elements of the diagonal containing $\frac{2}{1}$ and $\frac{1}{2}$, then the diagonal containing $\frac{3}{1}$, $\frac{2}{2}$, and $\frac{1}{3}$, etc.

This "list" defines a function f from $\mathbf{N}$ to $\mathbf{Q}^+$, in which $f(n)$ is simply the nth fraction in the list. Unfortunately, f isn't injective, because the list contains duplications (e.g., it contains both $\frac{2}{1}$ and $\frac{4}{2}$). However, we can modify f by having it skip any repetitions, and that will accomplish our goal of a bijection from $\mathbf{N}$ to $\mathbf{Q}^+$.

Get Your Hands Dirty 5

Suppose f is a bijective function from $\mathbf{N}$ to $\mathbf{Q}^+$. Describe how to use f to build a bijective function g from $\mathbf{Z}$ to $\mathbf{Q}$. □

Sets that are equivalent to $\mathbf{N}$, such as $\mathbf{Z}$ and $\mathbf{Q}$, are of particular interest, and are described by the following terminology:

DEFINITION 8.7: For a set S, "S is *countably infinite*" (or *denumerable*) means:

$$S \approx \mathbf{N}.$$

The cardinality of $\mathbf{N}$ is generally represented by the symbol "$\aleph_0$". [This is read "aleph-nought"; "$\aleph$" (aleph) is the first letter in the Hebrew alphabet.]

COMMENT Cantor himself chose this notation, presumably because of his Jewish ancestry (although he was raised as a Christian). When you discover or invent an important idea in mathematics, you will get to name it whatever you want.

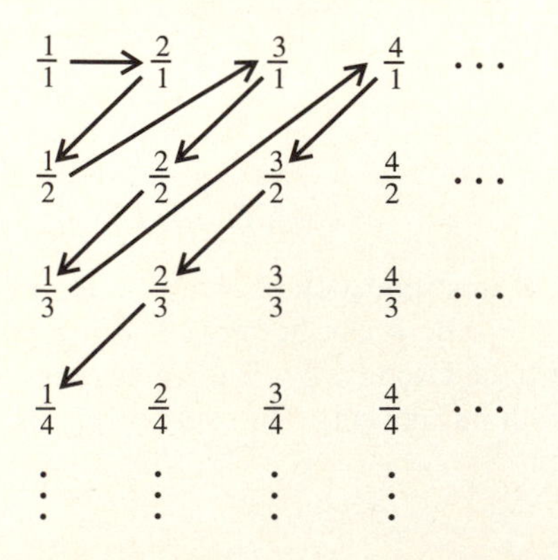

Figure 8.1 Following the arrows gives something like a list of the positive rational numbers.

The following definition combines the countably infinite sets with the finite sets:

> **DEFINITION 8.8:** For a set S, "S is *countable*" means: S is finite or countably infinite.
> A set which is not countable is called *uncountable*.

Uncountability of R

You may be thinking that the term "uncountable" is unnecessary. After the last several theorems, it begins to seem as if all infinite sets have the same cardinality. That, after all, is not so unreasonable. "Infinite is infinite", you might say, "how can you distinguish different types of infinity?"

The answer to that question is part of Cantor's achievement. We proceed to prove the existence of a specific uncountable set, namely, the set of real numbers. The basic idea of the proof is due to Cantor.

We first introduce some notation: every real number has an *integer part* and a *decimal part*. For example, the integer part of 17.3892 is 17, and its decimal part is .3892.

We will use the notation $t|_n$ to represent the nth digit in the decimal part of the number t. For example, if $t = 17.3892$, then $t|_1 = 3$, $t|_2 = 8$, $t|_3 = 9$, $t|_4 = 2$, and $t|_n = 0$ for $n > 4$. (There is an ambiguity with infinite decimals like .999 . . . , since this also equals 1.000 . . . , but we will take care that this ambiguity does not affect our proof—see comment 1 below.)

Now for the proof:

THEOREM 8.10. **R** is uncountable.

Proof: Suppose, on the contrary, that **R** is countable. Since **R** is infinite, this must mean that **R** is equivalent to **N**. So suppose there is an bijective function $f : \mathbf{N} \to \mathbf{R}$. We will derive a contradiction by demonstrating that there is a real number x which is not in $\text{Im}(f)$.

DISCUSSION Before continuing with the formal proof, we give an intuitive description of how it works:

We will define a real number x that differs from the real number $f(1)$ in its first decimal place, from $f(2)$ in its second decimal place, and so on. In this way, x will differ from every $f(n)$ in at least one decimal place, and so x cannot equal $f(n)$ for any n.

Try to define x in precise language before reading on.

The formalization of the above process is as follows:

Define a real number $x \in [0, 1]$ as follows:

$$x|_n = \begin{cases} 2 & \text{if } f(n)|_n \neq 2 \\ 3 & \text{if } f(n)|_n = 2 \end{cases}$$

(see comment 1 below). We will prove that $x \notin \text{Im}(f)$. For suppose $m \in \mathbf{N}$, with $f(m) = x$. If $f(m)|_m = 2$, then $x|_m = 3$. If $f(m)|_m \neq 2$, then $x|_m = 2$. In either case, $f(m)|_m \neq x|_m$, and so $f(m) \neq x$.

Thus $x \notin \text{Im}(f)$, which contradicts the assumption that f was bijective.

This completes the proof of the theorem. ■

COMMENTS

1. The use of the particular digits 2 and 3 in the construction of x was somewhat arbitrary. We just needed to avoid the digits 0 and 9 which are used in representing certain decimals in more than one way; e.g., $1 = .9999\ldots = 1.0000\ldots$.

2. The method of constructing x is known as the *diagonal argument*. The reason for this terminology is as follows: Suppose we write the real numbers $f(1)$, $f(2)$, etc., one above the other, with decimal points aligned, something like this:

$$f(1) = a_0.a_1a_2a_3\ldots$$
$$f(2) = b_0.b_1b_2b_3\ldots$$
$$f(3) = c_0.c_1c_2c_3\ldots$$
$$\vdots$$

   where $a_0, b_0, c_0, \ldots$ are integers, and the other letters represent digits in the decimal parts of the numbers. We can form a "diagonal" decimal by using a_1 in the first decimal place, b_2 in the second, c_3 in the third, and so on. Thus, the decimal $.a_1b_2c_3\ldots$ is defined to have the same "nth place" as the real number $f(n)$. We have defined x as a kind of "antidiagonal" element: that is, we have defined it so that it *differs* from this diagonal decimal in every decimal place. Therefore x differs from every number of the form $f(n)$.

Cardinality and Power Sets

Theorem 8.10 shows that not all infinite sets have the same cardinality. We turn now to showing that there is an endless list of cardinalities of infinite sets. More specifically, no matter how "big" a set is, there is always one that is "bigger".

The proof of this assertion uses the idea of power sets. We begin by recalling a result about finite sets:

THEOREM 5.9. For a finite set A,

$$|\mathcal{P}(A)| = 2^{|A|}.$$

We also proved:

THEOREM 5.2. For $n \in \mathbf{N}$, $n < 2^n$.

This inequality is also true for $n = 0$, so we can combine these results with Corollary 8.6 (and the definition of finite) to get the following:

COROLLARY 8.11. If S is finite, then $\overline{\overline{S}} < \overline{\overline{\mathcal{P}(S)}}$.

Our next theorem states that the "inequality" in this corollary holds even for infinite sets. The result is rather remarkable, and the proof is ingenious:

THEOREM 8.12. If S is any set, then $\overline{\overline{S}} < \overline{\overline{\mathcal{P}(S)}}$.

Proof: We need to show $\overline{\overline{S}} \leq \overline{\overline{\mathcal{P}(S)}}$ and $\overline{\overline{S}} \neq \overline{\overline{\mathcal{P}(S)}}$. We prove the first statement by finding an injective function from S to $\mathcal{P}(S)$. Define $f: S \to \mathcal{P}(S)$ by $f(x) = \{x\}$. If $x \neq y$, then $\{x\} \neq \{y\}$, so this function is injective. Therefore we have $\overline{\overline{S}} \leq \overline{\overline{\mathcal{P}(S)}}$.

Now we need to show $\overline{\overline{S}} \neq \overline{\overline{\mathcal{P}(S)}}$. We prove this by contradiction. Suppose that $\overline{\overline{S}} = \overline{\overline{\mathcal{P}(S)}}$, and that g is a bijective (and hence surjective) function from S to $\mathcal{P}(S)$. We need to get a contradiction.

For each $x \in S$, $g(x)$ is a subset of S. This subset may or may not contain x as one of its elements. Define a subset T of S as follows:

$$T = \{x \in S : x \notin g(x)\}.$$

It may clarify our discussion to consider the open sentence "$x \notin g(x)$". Thus, T is defined as the truth set of this open sentence. Since T is a subset of S, i.e., an element of $\mathcal{P}(S)$, and we have assumed that g is surjective, there must be some element y of S such that $g(y) = T$.

Now we ask: is $y \in T$? We have two cases:

Case 1: $y \in T$.

In this case, since T is assumed to be equal to $g(y)$, we have $y \in g(y)$. Thus the statement "$y \notin g(y)$" is false, i.e., y is not in the truth set of the open sentence "$x \notin g(x)$". Since T is the truth set of this open sentence, it follows that $y \notin T$. Thus, we have a contradiction for case 1.

Case 2: $y \notin T$.

In this case [again, because T is equal to $g(y)$], we have $y \notin g(y)$; i.e., the open sentence "$x \notin g(x)$" is true for $x = y$. Since T is the truth set for this open sentence, it follows that $y \in T$, which is a contradiction for case 2.

Thus we get a contradiction in either case. Therefore our assumption that $\overline{\overline{S}} = \overline{\overline{\mathcal{P}(S)}}$ must be false. This completes the proof of the theorem. ■

The reasoning that led to the final contradiction may be difficult to follow, particularly if you have never seen anything like it before. The same idea is used in a variety of mathematical brainteasers, of which the following is perhaps the best known:

In a certain town, there is one man who is the barber. Some men in the town shave themselves, and some are shaved by the barber. The barber shaves all those men, and only those men, who do not shave themselves.

Who shaves the barber?

? **?** **?**

To match this problem up with the reasoning in Theorem 8.12, let **U** be the universe of men in the town, and define T as follows:

$$T = \{x \in \mathbf{U} : x \text{ does not shave himself}\}.$$

The problem states that the barber shaves all members of T, and no one else.

Does the barber shave himself? The barber shaves himself if, and only if, he is a member of T. But he is a member of T if, and only if, he does not shave himself. Thus he shaves himself if and only if he does not shave himself.

This is impossible. (It doesn't help to suggest that perhaps the barber doesn't get shaved at all. If that were true, then it would be the case that he doesn't shave himself, and so he would be a member of T. But he shaves all the members of T.)

As in Theorem 8.12, we have a contradiction. The "solution" to the brainteaser is that such a town cannot exist. The conclusion in our theorem is that such a function g cannot exist.

COMMENT The method of proof of Theorem 8.12 is similar to the diagonal argument by which we proved **R** uncountable. We created a set T which differed from every $f(x)$, by putting x in T if and only if x was not in $f(x)$.

The following result follows immediately from Theorem 8.12:

COROLLARY 8.13. $\mathcal{P}(\mathbf{N})$ is uncountable.

We now have two uncountable sets, **R** and $\mathcal{P}(\mathbf{N})$. Our last theorem of this section states that they are equivalent. We will sketch the proof, providing the necessary functions and leaving verification of their properties as exercises:

THEOREM 8.14. $\mathbf{R} \approx \mathcal{P}(\mathbf{N})$.

Proof: We will prove the following sequence of "inequalities":

$$\mathbf{R} \precsim \mathcal{P}(\mathbf{Q}) \precsim \mathcal{P}(\mathbf{N}) \precsim \mathbf{R}.$$

By the Cantor–Schroeder–Bernstein theorem (and transitivity of $\precsim$), this will prove the theorem.

Step 1: $\mathbf{R} \precsim \mathcal{P}(\mathbf{Q})$. Define $g : \mathbf{R} \to \mathcal{P}(\mathbf{Q})$ by $g(r) = \{x \in \mathbf{Q} : x \leq r\}$.

We leave it as Exercise 6 to show that g is injective, which proves step 1.

Step 2: $\mathcal{P}(\mathbf{Q}) \precsim \mathcal{P}(\mathbf{N})$. By Corollary 8.9, $\mathbf{N} \approx \mathbf{Q}$, so there is an bijective function $f : \mathbf{Q} \to \mathbf{N}$. Let $\bar{f}$ be the function from $\mathcal{P}(\mathbf{Q})$ to $\mathcal{P}(\mathbf{N})$ defined by $\bar{f}(S) = \{f(x) : x \in S\}$. We leave it as Exercise 7a) to show that $\bar{f}$ is injective, which proves step 2. [$\bar{f}$ is actually bijective—see Exercise 7c).]

Step 3: $\mathcal{P}(\mathbf{N}) \lesssim \mathbf{R}$. Define $h : \mathcal{P}(\mathbf{N}) \to \mathbf{R}$ as follows: for $S \subseteq \mathbf{N}$, let $h(S)$ be the real number $x \in [0, 1]$ whose nth decimal digit is given by

$$x|_n = \begin{cases} 2 & \text{if} \quad n \in S \\ 3 & \text{if} \quad n \notin S. \end{cases}$$

We leave it as Exercise 8 to show that h is injective, which proves step 3. ∎

Get Your Hands Dirty 6

Find $h(2\mathbf{N})$, where h is the function defined in step 3 above. □

COMMENTS We conclude this topic with several observations about countable and uncountable sets. Some of these will be discussed in the exercises; others will be left without proof.

1. $\aleph_0$ is the smallest infinite cardinality. In other words, if S is an infinite set, then $\mathbf{N} \lesssim S$. (We will not prove this.)
2. If S is an infinite set and T is a finite set, then $S \cup T \approx S$. (We will not prove this.)
3. The following sets are also countably infinite:

   a) $2\mathbf{Z}$ [Exercise 9a)],
   b) $\mathbf{Z} \times \mathbf{Z}$ [Exercise 9c)],
   c) $S \times S$ for any countably infinite set S [Exercise 9c)],
   d) the set of polynomials with rational coefficients,
   e) the set of algebraic numbers.

   (See Exercise 16 of Section 5.2 for definitions for the last two examples; we will not give proofs for these.)
4. The interval $[0, 1]$ is uncountable: this is actually demonstrated by the proof given of Theorem 8.10 (Exercise 10).
5. Any interval in $\mathbf{R}$ other than a single point or the empty set is equivalent to $\mathbf{R}$ (Exercise 10).
6. We noted in Section 6.1, following the definition of "relation" (Definition 6.1), that not every relation can be expressed by an open sentence. This can be proved using the concepts of this chapter. Let's take the case of relations on $\mathbf{R}$. The argument is roughly as follows: suppose the "language" for open sentences includes all real numbers, and at most countably many other symbols. Let $\mathbf{S}$ be the set of all open sentences that can be written in this language. It can then be shown that $\mathbf{S} \approx \mathbf{R}$. On the other hand, the set of all relations on $\mathbf{R}$ is $\mathcal{P}(\mathbf{R} \times \mathbf{R})$. But by Theorem 8.12, $\overline{\overline{\mathbf{R}}} < \overline{\overline{\mathcal{P}(\mathbf{R})}}$. Thus, $\overline{\overline{\mathbf{S}}} = \overline{\overline{\mathbf{R}}} < \overline{\overline{\mathcal{P}(\mathbf{R})}} \leq \overline{\overline{\mathcal{P}(\mathbf{R} \times \mathbf{R})}}$. In other words, there are really "more" relations than open sentences, so there must be relations that cannot be expressed as open sentences.

7. You may wonder if there are any cardinalities "between" **Q** and **R**. In other words, is there a set T such that $\mathbf{Q} < T$ and $T < \mathbf{R}$? It has been proved that, with the standard axioms about sets, this question is "unanswerable". In other words, consider the following two statements:

   a) $(\exists\, T \subseteq \mathbf{R})(\mathbf{Q} < T \text{ and } T < \mathbf{R})$.
   b) $\sim(\exists\, T \subseteq \mathbf{R})(\mathbf{Q} < T \text{ and } T < \mathbf{R})$.

   Since ii) is the negation of i), it would seem that one of these must be true. Yet it has been proved that neither of these two statements can be proved from the standard set theory axioms. Either one of these statements (but not both) can be added as an additional axiom without creating a contradiction.

   Ponder that statement for awhile.

   Mathematicians generally agree that their intuitive picture of **R** suggests that there should not be such a set T; i.e., they "believe" that statement ii) is true. This statement is known as the *Continuum Hypothesis*.

8. Finally, we mentioned after Corollary 8.4 the question of whether or not $\leq$ is a total ordering. This question is in a similar category to that of the previous one: the answer depends on the system of axioms for sets. The axiom involved here is called the *Axiom of Choice*. Like the Continuum Hypothesis, either it or its negation can be added to the other standard axioms without creating a contradiction. However, there are many mathematical problems in which this axiom is needed, and most mathematicians accept the Axiom of Choice as true. Many books treat it as a standard axiom.

"The Set of All Sets"—an Impossibility

We noted at the beginning of the discussion of cardinality that we could not work with "the set of all sets." The problem with this idea is similar to that of the problem of the barber. Here's why.

Suppose there were a "universal set" **U**, which contained every possible set as an element. Consider the open sentence "$S \notin S$", and let T be the truth set of this open sentence, i.e.,

$$T = \{S \in \mathbf{U} : S \notin S\}.$$

We ask the crucial question:

Does T itself belong to this set?

If so, then it fits the defining condition for membership in the set, namely, it must be in the truth set of the open sentence "$S \notin S$". In other words, if $T \in T$, then "$T \notin T$" must be true.

On the other hand, if T does not belong to T, i.e., if "$T \notin T$" is true, then T is in the truth set of the open sentence "$S \notin S$", but, by definition, this truth set is T itself. In other words, if "$T \notin T$" is true, then $T \in T$.

Thus we have the following awful situation:

$$T \in T \quad \text{if and only if} \quad T \notin T.$$

This violates the fundamental principles of logic—a statement can't be both true and false.

The conclusion from all of this is like the conclusion about the barber. The "set" S does not exist.

COMMENT It's not exactly that S doesn't exist. Perhaps it does, and perhaps it doesn't, but, in any case, S is not a set. Some mathematicians who study the principles of set theory would refer to S as a "class".

EXERCISES

1. Prove the following:

LEMMA 8.2.′ If $S \approx U$, $T \approx V$, and there is an injective function from S to T, then there is an injective function from U to V.

2. Prove that "$\leq$" is a transitive relation on $\mathscr{C}$. (*Hint:* You need to use Lemma 8.2.)

3. Use steps a)–c) to prove the following:

THEOREM 8.5. For $m, n \in \mathbf{W}$, if $m \neq n$, then $\mathbf{N}_{[1,m]} \not\approx \mathbf{N}_{[1,n]}$.

a) Prove that if $n \in \mathbf{N}$, then $\mathbf{N}_{[1,n]} \not\approx \mathbf{N}_{[1,0]}$. (*Hint:* $\mathbf{N}_{[1,0]} = \varnothing$.)

Historical note: The realization that not every description of a collection of objects gives rise to a set sent shock waves through the mathematical world. The problem of the "set of all sets" is known as *Russell's paradox,* after the great British philospher and mathematician Bertrand Russell (1872–1970). Russell did pioneering work with one approach toward setting the foundations of mathematicians on solid ground. He constructed a *theory of types,* which distinguished among elements, sets of elements, sets of sets of elements, etc. Today, the axiom system for sets developed by Ernest Zermelo in 1908 and improved by Abraham Fraenkel and Thoralf Skolem in 1922–23 is generally considered the standard approach. This system is usually referred to as "ZF" (which unfortunately deprives Skolem of proper credit). More information about different ways of dealing with the problem of Russell's Paradox can be found in *Foundations of Set Theory,* by A. Fraenkel and Y. Bar-Hillel, North-Holland Publishing Company, 1958.

b) Let $S = \{m \in \mathbf{W} : (\forall\, n \in \mathbf{W})\ (n \neq m \Rightarrow \mathbf{N}_{[1,n]} \not\approx \mathbf{N}_{[1,m]})\}$. [Recall that $\mathbf{W}$ is the set of nonnegative integers. Part a) says "$0 \in S$".] Prove that S is inductive.
c) Combine a) and b) to prove Theorem 8.5.

4. Prove that a positive rational number can only be expressed in lowest terms in one way. That is, suppose p, q, r, and s are positive integers, with $p/q = r/s$, such that $\mathrm{GCD}(p, q) = 1$ and $\mathrm{GCD}(r, s) = 1$. Prove that $p = r$ and $q = s$. [*Hint:* You may use the result of Exercise 5a) in Section 5.3.]

5. Prove that if S is finite and $T \subset S$, then $S \not\approx T$. (*Hint:* Use Theorem 8.5.)

6. Define $g : \mathbf{R} \to \mathcal{P}(\mathbf{Q})$ by $g(r) = \{x \in \mathbf{Q} : x \leq r\}$. Prove that g is injective. (You may use the fact that $\mathbf{Q}$ is *dense* in $\mathbf{R}$, i.e., that between any two distinct real numbers, there is a rational number.)

7. Suppose f is a function from S to T, and let $\overline{f}$ be the function from $\mathcal{P}(S)$ to $\mathcal{P}(T)$ defined by
$$\overline{f}(X) = \{f(t) : t \in X\}.$$
Prove the following:
a) If f is injective, then $\overline{f}$ is injective.
b) If f is surjective, then $\overline{f}$ is surjective.
c) If f is bijective, then $\overline{f}$ is bijective.

8. Let h be the function from $\mathcal{P}(\mathbf{N})$ to $\mathbf{R}$ defined in Theorem 8.14 as follows: for $S \subseteq \mathbf{N}$, $h(S)$ is the real number $x \in [0, 1]$ whose nth decimal digit is given by
$$x|_n = \begin{cases} 2 & \text{if } n \in S \\ 3 & \text{if } n \notin S. \end{cases}$$
Prove that h is injective.

9. Prove that each of the following sets is countably infinite:
a) $n\mathbf{Z}$, for any $n \in \mathbf{N}$,
b) $\{x \in \mathbf{Q} : 0 < x < 1\}$,
c) $S \times S$, for any countably infinite set S.

10. Prove that, if $a, b \in \mathbf{R}$ with $a < b$, then each of the following is equivalent to $\mathbf{R}$:
a) $[a, b]$,
b) $[a, b)$,
c) $(a, b]$,
d) (a, b),
e) $(-\infty, b)$,
f) $(-\infty, b]$,
g) (a, ∞),
h) $[a, \infty)$.

11. (Hard, but fun.) Without using the Cantor–Schroeder–Bernstein Theorem, prove that the following pairs of sets are equivalent (in other words, construct an explicit bijective function from one to the other):
a) $(0, 1)$ and $[0, 1)$,
b) $(0, 1)$ and $(0, 1) \times (0, 1)$.

8.2 Proof of the Cantor–Schroeder–Bernstein Theorem

This section is devoted to a proof of the following important result:

THEOREM 8.3: (Cantor–Schroeder–Bernstein Theorem). If S and T are sets such that $S \lesssim T$ and $T \lesssim S$, then $S \approx T$.

We begin the proof by expressing the relations in the hypothesis in terms of their definitions.

Proof: Suppose $S \lesssim T$ and $T \lesssim S$, so there are injective functions $f: S \to T$ and $g: T \to S$.

To complete the proof, we will need to create a bijective function from S to T. In order to do so, we need to develop some general concepts.

Restriction of a Function

As indicated in Section 8.1, the key to this proof is to partition the sets S and T, and to piece together certain "parts" of each of the functions f and g. In order to describe these "parts" of functions, we make the following definition:

DEFINITION 8.9: For a function $h: A \to B$, and a set $C \subseteq A$, the *restriction of h to C* (written "$h|_C$") is the function from C to B such that, for $x \in C$, $h|_C(x) = h(x)$.

Thus, as a set of ordered pairs, $h|_C$ is a subset of h, given by

$$h|_C = \{(x, y) \in A \times B : x \in C \text{ and } y = h(x)\}.$$

COMMENT Since $\text{Im}(h|_C)$ may be smaller than $\text{Im}(h)$, we often think of $h|_C$ as a function from C to this smaller set, rather than to B.

We need two basic facts about restrictions:

LEMMA 8.15. If $h: A \to B$ is an injective function, and $C \subseteq A$, then $h|_C$ is injective.

Proof: Suppose, by way of contradiction, that $h|_C$ is not injective. Then there are elements $u, v \in C$, with $u \neq v$, such that $h|_C(u) = h|_C(v)$. But this means that $h(u) = h(v)$, and u and v are distinct elements of A, and this contradicts the assumption that h is injective. ■

In the second result, we think of functions as sets of ordered pairs. If several functions have disjoint domains, then their union is a function whose domain is the union of the individual domains. The following result tells us that, if the individual functions are bijective, and if their codomains are also disjoint, then the union is also a bijective function.

LEMMA 8.16. Suppose that $A_1, A_2, \ldots, A_n$ are pairwise disjoint sets, and that $B_1, B_2, \ldots, B_n$ are pairwise disjoint sets. Suppose further that, for each $i = 1, 2, \ldots, n$, we have a bijective function $h_i : A_i \to B_i$.

Then $h_1 \cup h_2 \cup \cdots \cup h_n$ is a bijective function from $A_1 \cup A_2 \cup \cdots \cup A_n$ to $B_1 \cup B_2 \cup \cdots \cup B_n$.

Proof: Because the domains A_i are disjoint, every element of $A_1 \cup A_2 \cup \cdots \cup A_n$ appears as the first element of exactly one pair of one of the functions h_i, and so $h_1 \cup h_2 \cup \cdots \cup h_n$ is a function with domain $A_1 \cup A_2 \cup \cdots \cup A_n$.

Similarly, since the codomains B_i are disjoint, and each of the functions h_i is bijective, each element of $B_1 \cup B_2 \cup \cdots \cup B_n$ appears as the second element of exactly one pair of one of the functions h_i, and so $h_1 \cup h_2 \cup \cdots \cup h_n$ is bijective. ■

Ancestry of an Element

Let $f : S \to T$ and $g : T \to S$ be injective functions. The key to the proof of the Cantor–Schroeder–Bernstein Theorem is the partition of S and T, which will allow us to create appropriate restrictions of f and g. We will then combine these using Lemma 8.16.

The definition of this partition is itself dependent on both of the given functions, and proceeds as follows.

Let x be an element of S. If x is in $\text{Im}(g)$, then there is a unique element y in T such that $g(y) = x$. (The uniqueness is a result of the assumption that g is injective.) We will call that element the *predecessor of* x. Notice that, unless g is actually bijective, not every element of S will have a predecessor.

Similarly, if w is an element of T, and $w \in \text{Im}(f)$, then there is a unique element v in S such that $f(v) = w$. Again, we call that element the predecessor of w; and again, unless f is actually bijective, there will be elements of T without predecessors. Figure 8.2 shows how this relationship of elements and their predecessors works.

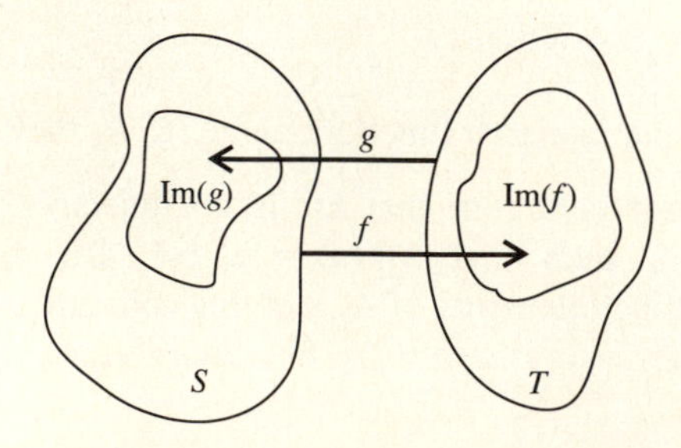

Figure 8.2. Elements of S which belong to $\text{Im}(g)$ have predecessors in T; similarly, elements of T which belong to $\text{Im}(f)$ have predecessors in S.

Now here's the tricky part. We define the *ancestry* of an element, whether in S or in T, to consist of its predecessor (if it has one), the predecessor of its predecessor (if there is one), the predecessor of that (if there is one), and so on. Any element obtained in this process is called an *ancestor* of the element with which we began. For some elements, this tracing of its ancestry may go on forever. In that case, we say that the element has *infinite ancestry*. For others, there may at some point be an ancestor which has no predecessor. This is called an *ultimate ancestor*. (*Comment:* Technically, we are using definition by recursion here.)

Get Your Hands Dirty 1

Let $S = T = \mathbf{Z}$. Define $f: S \to T$ by $f(x) = 2x$, and define $g: T \to S$ by $g(x) = 3x$. Trace the ancestry of each of the following elements:

a) the integers 6, 12, -18, and 0, considered as elements of S,
b) the integers 6, 12, -18, and 0, considered as elements of T. □

Partitioning the Sets

We therefore divide the elements of both S and T into three types:

i) elements with infinite ancestry,
ii) elements with an ultimate ancestor in S,
iii) elements with an ultimate ancestor in T.

Notice that an element in S or T can have an ultimate ancestor in either set (or have infinite ancestry).

We designate the resulting subsets of S and T as follows:

$$S_\infty = \text{the set of elements of } S \text{ with infinite ancestry},$$
$$S_S = \text{the set of elements of } S \text{ with an ultimate ancestor in } S,$$
$$S_T = \text{the set of elements of } S \text{ with an ultimate ancestor in } T.$$

and

$$T_\infty = \text{the set of elements of } T \text{ with infinite ancestry},$$
$$T_S = \text{the set of elements of } T \text{ with an ultimate ancestor in } S,$$
$$T_T = \text{the set of elements of } T \text{ with an ultimate ancestor in } T.$$

Our discussion shows that $S = S_\infty \cup S_S \cup S_T$, and that these three subsets of S are pairwise disjoint. (In other words, except possibly for the requirement that they be nonempty, these sets form a partition of S.) Similarly, $T = T_\infty \cup T_S \cup T_T$, and these three subsets of T are pairwise disjoint. These conditions on the sets will allow us to apply Lemma 8.16.

Get Your Hands Dirty 2

As in GYHD 1 let $S = T = \mathbf{Z}$, define $f: S \to T$ by $f(x) = 2x$, and define $g: T \to S$ by $g(x) = 3x$. Find each of the sets S_∞, S_S, S_T, T_∞, T_S, and T_T. □

Creating the Bijection

With this terminology in mind, we now resume the proof of the Cantor–Schroeder–Bernstein Theorem, beginning with the following lemma:

LEMMA 8.17. With the definitions described above:

i) $f|_{S_\infty}$ is a bijective function from S_∞ to T_∞;
ii) $f|_{S_S}$ is a bijective function from S_S to T_S;
iii) $g|_{T_T}$ is a bijective function from T_T to S_T.

Before proving this lemma, we show how the theorem follows from it:

Proof of Theorem 8.3 (using Lemma 8.17): Since $g|_{T_T}$ is a bijective function from T_T to S_T, its inverse $[g|_{T_T}]^{-1}$ is a bijective function from S_T to T_T. Thus, we have bijective functions from each of the parts of the partition of S to the corresponding part of the partition of T.

It follows from Lemma 8.16 that $h = f|_{S_\infty} \cup f|_{S_S} \cup [g|_{T_T}]^{-1}$ is a bijective function from S to T. (Figure 8.3 shows the partitions of S and T, together with the restriction functions $f|_{S_\infty}$, $f|_{S_S}$, and $[g|_{T_T}]^{-1}$.)

We conclude the proof of the Cantor–Schroeder–Bernstein Theorem by proving Lemma 8.17:

Proof of Lemma 8.17: We first need to justify describing $f|_{S_\infty}$, $f|_{S_S}$, and $g|_{T_T}$ as functions to the specified codomains; namely, we have to show:

i) if $x \in S_\infty$, then $f|_{S_\infty}(x) \in T_\infty$;
ii) if $x \in S_S$, then $f|_{S_S}(x) \in T_S$;
iii) if $x \in T_T$, then $g|_{T_T}(x) \in S_T$.

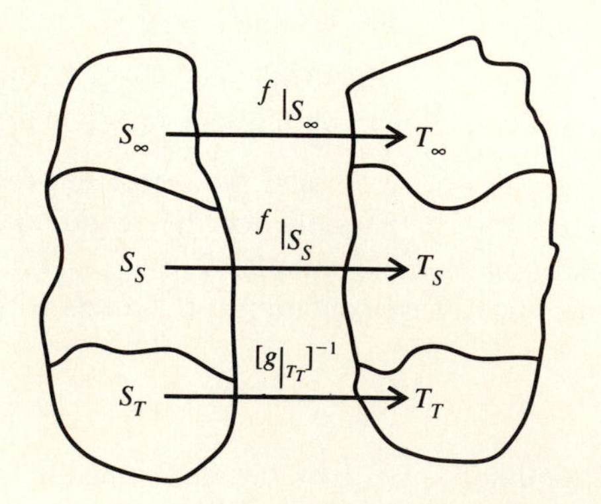

Figure 8.3 Each of the functions, $f|_{S_\infty}$, $f|_{S_S}$, and $[g|_{T_T}]^{-1}$, is bijective for the domain and codomain shown. Together, they make up a bijection from S to T.

To prove i), suppose $x \in S_\infty$. By definition, $f|_{S_\infty}(x) = f(x)$, and the ancestry of $f(x)$ consists precisely of the element x together with its ancestors. Therefore, if x has infinite ancestry, then so does $f|_{S_\infty}(x)$, and so $f|_{S_\infty}(x) \in T_\infty$. The proofs of ii) and iii) are similar.

It remains to show that the three restrictions are bijective. Since f and g are injective, it follows from Lemma 8.15 that their restrictions are injective. We need only show that they are surjective (for the given codomains).

To prove that $f|_{S_\infty} : S_\infty \to T_\infty$ is surjective, suppose $w \in T_\infty$. Then w has a predecessor, say x. (Otherwise, w would be its own ultimate ancestor, and w would belong to T_T.) This predecessor x is an element of S such that $f(x) = w$.

But then the ancestry of x is the same as the ancestry of w (except that x is an ancestor of w). Therefore, x also has infinite ancestry, i.e., $x \in S_\infty$. It follows that $(x, w) \in f|_{S_\infty}$, and so $w \in \text{Im}(f|_{S_\infty})$, which proves that $f|_{S_\infty} : S_\infty \to T_\infty$ is surjective.

Similarly, if $w \in T_S$, it has a predecessor x which is in S_S, and so $(x, w) \in f|_{S_S}$. Thus, $w \in \text{Im}(f|_{S_S})$, which proves that $f|_{S_S} : S_S \to T_S$ is surjective.

Finally, if $x \in S_T$, it has a predecessor y which is in T_T, and so $(y, x) \in g|_{T_T}$. Thus, $x \in \text{Im}(g|_{T_T})$ which proves that $g|_{T_T} : T_T \to S_T$ is surjective.

That completes the proof of the Cantor–Schroeder–Bernstein Theorem. ■

EXERCISES

1. (GYHD 1) Let $S = T = \mathbf{Z}$. Define $f : S \to T$ by $f(x) = 2x$ and define $g : T \to S$ by $g(x) = 3x$. Trace the ancestry of each of the following elements:
   a) the integers 6, 12, -18, and 0, considered as elements of S;
   b) the integers 6, 12, -18, and 0, considered as elements of T.
2. (GYHD 2) Let S, T, f, and g be as in Exercise 1. Find each of the sets S_∞, S_S, S_T, T_∞, T_S, and T_T.

9

Limits and Continuity

Introduction

In our final two chapters, we will take an introductory look at two major areas of modern mathematics: real analysis and abstract algebra. Both topics arose out of classical mathematical problems and concerns. Real analysis developed from the study of functions from **R** to **R** and the field of calculus. Abstract algebra has its origins in the study of the solution of polynomial equations.

Our discussion of these two topics will only scratch the surface. It is intended to suggest the flavor of these subject areas, and introduce you to some of the basic concepts in a way that is both intuitive and rigorous. We will be building on our use of quantifiers, our study of functions and operations, and our exploratory approach to introducing new ideas.

Some of the ideas in this chapter may be familiar to you from studying calculus. They may also seem new, because they are presented in a different way. You should try to build on the intuition you have developed in your past study, and look for the ways in which the formal definitions we give here relate to that intuition.

One of the central concepts in the study of real analysis is the idea of a *limit*. We will examine two related but distinct versions of this concept: *limit of a sequence* and *limit of a function*. We will begin our study of limits in Section 9.1 with a thorough exploration of what is meant by the limit of a sequence. Based on study of specific examples, we gradually build the complex three-quantifier definition. We then proceed in Section 9.2 to illustrate the use of this definition. We prove statements both about specific examples and about general situations, including examples of how to prove that something is *not* the limit of a sequence. Section 9.2 culminates with the proof that a sequence cannot have more than one limit.

In Section 9.3, we begin by defining the limit concept for functions, applying the insights gained in working with sequences, and prove analogs of some of the theorems of Section 9.2.

We then go on to the related idea of *continuity*. The usual intuitive description of a continuous function is that it is a function whose graph "has no breaks"; that is, the graph can be drawn "without lifting your pencil". We will prove a result about continuity, called the *Intermediate Value Theorem,* which actually confirms this intuitive picture.

In the final section of this chapter, we look at more advanced developments of the concepts of continuity and limit of a sequence. We introduce the ideas of *uniform continuity* and *Cauchy sequence,* and prove an important theorem about each. Both of these proofs make sophisticated use of the completeness property of the real numbers.

9.1 Limit of a Sequence — Developing a Definition

Defining "Sequence"

Before we can talk about the limit of a sequence, we have to answer a more fundamental question. What is a sequence? You probably have seen problems that ask: "What is the next term in this sequence?" We need to know exactly what this word means.

? ? ?

To get us started, here are some sample sequences, some of which may be familiar to you. It will be important in what follows that we have many sequences to use as examples.

Example 1. Some Familiar Sequences

i) $1, 2, 3, 4, 5, \ldots.$

ii) $1, 2, 4, 8, 16, \ldots.$

iii) $2, 4, 6, 8, 10, \ldots.$

iv) $1, 4, 9, 16, 25, \ldots.$

v) $\frac{1}{2}, \frac{1}{4}, \frac{1}{8}, \frac{1}{16}, \frac{1}{32}, \ldots.$

vi) $1, -1, 1, -1, 1, \ldots.$

vii) $\frac{1}{2}, \frac{2}{3}, \frac{3}{4}, \frac{4}{5}, \frac{5}{6}, \ldots.$

viii) $\frac{1}{2}, -\frac{1}{4}, \frac{1}{8}, -\frac{1}{16}, \frac{1}{32}, \ldots.$

ix) $\frac{1}{2}, \frac{1}{3}, \frac{1}{4}, \frac{1}{9}, \frac{1}{8}, \frac{1}{27}, \frac{1}{16}, \ldots.$

x) $\cos(1°), \cos(2°), \cos(3°), \cos(4°), \cos(5°), \ldots.$

xi) $\frac{1}{1}, -\frac{2}{3}, \frac{3}{5}, -\frac{4}{7}, \frac{5}{9}, -\frac{6}{11}, \ldots.$ □

A sequence seems to be simply a bunch of numbers strung out one after the other. The numbers in a sequence generally follow some sort of pattern or rule.

It may help to point out a few things that are not sequences. The set **Z** of integers is not a sequence. The interval [0, 1] of real numbers is not a sequence. The set **Q** of rational numbers is not a sequence. And even the set **N** of positive integers is not a sequence.

You may be somewhat surprised by the last statement. Isn't the first sequence in Example 1 the set **N**?

? ? ?

No, it isn't. There is a difference between a *set* and a *sequence*. In all of the sequences of Example 1, the numbers involved have been placed in a particular order. There is a first term, a second term, a third term, etc. The same set can be used to create many different sequences. For example, "2, 1, 4, 3, 6, 5, . . ." uses the same set of terms as "1, 2, 3, 4, 5, 6, . . .", but these are different sequences.

One other example of a nonsequence may be helpful. The following is not a sequence:

$$1, 2, 3, \ldots, \frac{1}{2}, \frac{2}{2}, \frac{3}{2}, \ldots, \frac{1}{3}, \frac{2}{3}, \frac{3}{3}, \ldots.$$

In this nonsequence, we first list all the positive integers, then the positive fractions with denominator 2, then the positive fractions with denominator 3, and so on.

The reason we don't consider this a sequence is that the fraction terms don't have a "numerical position." You can't say that $\frac{1}{2}$ is the 5th or 100th or nth term. You might be tempted to describe it as something like "the infinity-plus-first term", but that is not permitted in a sequence.

So what is a sequence?

Formulate a formal definition. Be precise. (*Hint:* Use the language of functions.)

The key idea is that there is one term in the sequence for each positive integer—a 1st term, a 2nd term, a 3rd term, and so on. So we make the following definition:

> **DEFINITION 9.1:** A *sequence* is a function whose domain is the set **N**. The *nth term* of the sequence is the value of the function at n.

Notation

We often represent sequences simply by listing the first few terms, as in Example 1. Another common notation is "$\langle f(n) \rangle$", where $f(n)$ is the nth term of the sequence. For example, sequence ii) in Example 1 is the function $f: \mathbf{N} \to \mathbf{N}$ defined by $f(x) = 2^{x-1}$, and the sequence can be written as $\langle 2^{n-1} \rangle$.

Some sequences may be harder than ii) to describe by a formula. For example, the function for sequence ix) of Example 1 requires a two-part definition, such as the following:

$$f(n) = \begin{cases} 2^{-(n+1)/2} & \text{if } n \text{ is odd} \\ 3^{-n/2} & \text{if } n \text{ is even.} \end{cases}$$

(Verify that this describes the sequence correctly.)

When we want to discuss a generic sequence, we often use subscripts to indicate the terms. For example, the nth term of a sequence might be represented as x_n, and the sequence written as "$x_1, x_2, x_3, x_4, \ldots$" or as "$\langle x_n \rangle$". You can picture a sequence as an "infinitely long n-tuple": $(x_1, x_2, x_3, x_4, \ldots, x_n, \ldots)$.

Whatever informal notation is used, keep in mind that a sequence *is a function*. Sometimes a formal proof will require us to go back to this aspect of the definition.

COMMENTS

1. There is no restriction on the codomain for the function. The sequences in Example 1 can all be thought of as sequences of real numbers—i.e., as functions from **N** to **R**—but, in general, the terms of a sequence can be elements of any set. For example, real analysis often studies sequences whose individual terms are themselves functions.
2. The function that defines a sequence does not have to be one-to-one. For instance, in vi) of Example 1, the number 1 is the 1st term, but it is also the 3rd term, the 5th term, the 7th term, etc.
3. Although a sequence is a particular type of function, the concept of "limit of a sequence" is not just a special case of "limit of a function". The latter phrase refers to functions whose domain is an interval of **R**. Although the two types of limits are closely related, they are distinct definitions.

Get Your Hands Dirty 1

Give an equation for the function for each of the sequences in Example 1 [except for ii) and ix), which are done above]. [*Hint* on vi), viii), and xi): use $(-1)^{n+1}$ as part of the formula.] □

Limit of a Sequence—Tentative Definitions

Note: For the rest of this section, **R** will be our universe of discourse. All of our sequences will be functions with codomain **R**.

You may have heard or read statements like "the limit of $\langle x_n \rangle$ is L" or "the sequence $\langle x_n \rangle$ approaches L", where L is a real number. (We will use the second phrase as a colloquial synonym of the first.) What does this mean?

Don't try to answer this fully yet. Just begin thinking about it.

This turns out to be a very complicated idea. The limit concept did not emerge full blown from the brain of some individual mathematician, but was developed gradually as its uses and applications became clearer. The first efforts at defining "limit" turned out to be inadequate for later developments.

You may already have learned the formal definition of the concept of limit of a sequence. If so, you should read the following discussion with the goal of understanding that definition better.

If you know the "official" definition, write it down, using quantifiers as appropriate, and think about what it means. If you don't know the "official definition", think about what the phrase "$\langle x_n \rangle$ approaches L" ought to mean. Really take some time to do this.

In what follows, we will guide you toward the definition that is now standard. Before we start that process, we want you to develop your own intuition about working with sequences. So do the following:

Get Your Hands Dirty 2

For each of the sequences in Example 1, decide if you *think* the sequence approaches some limit, and, if so, what that limit is. When appropriate, use a calculator to express the terms as decimals, to get a better sense of their values. □

In working toward the "official" definition of "limit", we will suggest several possible ways of defining "limit" that are reasonable. There are two methods we want you to use to evaluate each tentative definition:

i) Does it fit your intuitive idea of what "$\langle x_n \rangle$ approaches L" ought to mean? How does it compare with your work in GYHD 2?

ii) Does it fit the "official" answers? We will tell you below what mathematicians have decided, after much experience, about each of the sequences in Example 1.

It may turn out that these two criteria conflict. If that happens, you should try to figure out why mathematicians adopted the definition they did, and how it differs from your intuitive picture.

It doesn't mean that your intuition is wrong, but rather that, for some reason, mathematicians chose to use the word "limit" in a different way from the way you would use it. You should work on adjusting your picture to fit the "official definition", because that is the one that is used in all the books, so you have to get used to it.

A good definition of the limit concept should make true those statements about limits that you want to be true, and make false those statements that you want to be false. So the quality of a formal definition depends on whether it correctly embodies the answers that you want to get.

Limits of Specific Sequences

As we indicated above, the concept of limit evolved over the years. After experimenting with various possible definitions, mathematicians arrived at one which really seemed to work—that is, a definition that always gave the answers that made the most sense in terms of how the concept was to be used.

According to the "official definition", the results for the sequences of Example A are as follows:

i)–iv): These sequences do not have a limit.
v): This sequence approaches 0.
vi): This sequence does not have a limit.
vii): This sequence approaches 1.
viii)–ix): These sequences approach 0.
x)–xi): These sequences do not have a limit.

Note: we sometimes say that sequences i)–iv) "approach infinity". This concept is considered in Exercise 17 of Section 9.2, but is distinct from the concept being developed here. In any case, it does not apply to vi), x), and xi).

Check your definition and your work in GYHD 2 against the preceding "official answers". Do they agree?

We will proceed to work our way toward the "official definition", by looking at specific examples. Consider sequence v) from Example 1:

$$\frac{1}{2}, \frac{1}{4}, \frac{1}{8}, \frac{1}{16}, \frac{1}{32}, \ldots$$

We will represent this sequence as $\langle a_n \rangle$, where $a_n = 2^{-n}$. As you probably decided for yourself in GYHD 2, and as mathematicians have agreed, this sequence approaches zero. What exactly does that mean?

? ? ?

In this example, we have each term closer to zero than the term before it. By the "distance" between two numbers, we mean the absolute value of their difference, so the assertion that each term is closer to zero than the previous term can be expressed as follows:

$$(\forall\, n \in \mathbf{N})(|a_{n+1} - 0| < |a_n - 0|). \qquad (*)$$

Verify that $(*)$ is a true statement for our sequence $\langle a_n \rangle$.

Tentative Definition 1

So let's take the following as a tentative definition for the idea of limit:

TENTATIVE DEFINITION 1: For a sequence $\langle x_n \rangle$ and a real number L, "the *limit* of $\langle x_n \rangle$ is L" means:

$$(\forall\, n \in \mathbf{N})(|x_{n+1} - L| < |x_n - L|).$$

It is likely that something like this was the way mathematicians described limits at first. Think about this definition, and review your work on GYHD 2. Does this tentative definition describe your intuitive feeling about limits? Will it give the "official" answers described above? Does this definition fit all the situations you would like to describe by the statement "the sequence approaches L", and does it rule out all the situations you would not describe that way?

Get Your Hands Dirty 3

According to Tentative Definition 1, which of the following statements are true?

a) Sequence iii) of Example 1 approaches 100.
b) Sequence v) of Example 1 approaches -1.
c) Sequence ix) of Example 1 approaches 0.

(Be prepared for surprises.) □

As you probably noticed, Tentative Definition 1 does not always agree with the "official" answers. This definition has at least one major flaw. The statement

$$\text{"}(\forall\, n \in \mathbf{N})(|a_{n+1} - 0| < |a_n - 0|)\text{"} \qquad (*)$$

would still be true if the number 0 were replaced by any negative number, such as -1. (Recall: $a_n = 2^{-n}$). For example, the following statement is also true:

$$(\forall\, n \in \mathbf{N})(|a_{n+1} - (-1)| < |a_n - (-1)|)\,. \qquad (**)$$

(Verify.) Thus, according to Tentative Definition 1, every negative number is also the limit of the sequence. Do you want to say that the sequence $\langle 2^{-n} \rangle$ approaches -1?

? ? ?

You probably would not. In any case, as mathematicians worked with and developed the concept of limit, they decided that a particular sequence should not have more than one limit. And so, they rejected the above tentative definition.

What caused this difficulty? In what way did Tentative Definition 1 fail to capture the intuitive picture?

? ? ?

Although this definition guarantees that the terms get closer to the limit, it doesn't say how close they will get. Thus, although the terms of the sequence $\langle 2^{-n} \rangle$ are getting closer to -1, mathematicians don't think they are getting close enough. For example, they never even get within $\frac{1}{2}$ of -1.

How can we modify Tentative Definition 1 so that the sequence $\langle 2^{-n} \rangle$ approaches 0, but doesn't approach any other number?

Work on this. Write an alternative tentative definition.

Tentative Definition 2

Here is one possible response (though it's probably not the one you came up with): Since the terms are decreasing toward 0, the number 0 is a lower bound for the set $\{2^{-n}\}$. Negative numbers are also lower bounds, but we can distinguish 0 as the *greatest* lower bound. So consider the following:

> **TENTATIVE DEFINITION 2:** For a sequence $\langle x_n \rangle$ and a real number L, "the *limit* of $\langle x_n \rangle$ is L" means: L is the greatest lower bound for the set $\{x_n : n \in \mathbf{N}\}$.

Under this definition, the sequence $\langle 2^{-n} \rangle$ has a unique limit, namely 0. But that doesn't mean that we have the right definition. Again, we ask you to test the definition against the "official" answers:

Get Your Hands Dirty 4

According to Tentative Definition 2, which of the following statements are true?

a) Sequence i) of Example 1 approaches 1.
b) Sequence v) of Example 1 approaches 0.
c) Sequence vii) of Example 1 approaches 1. □

Again, we have problems. One difficulty is that many sequences will *increase*, rather than decrease, toward their limit. For example, $\frac{1}{2}, \frac{2}{3}, \frac{3}{4}, \frac{4}{5}, \ldots$ "officially" approaches 1, but does not approach 1 according to Tentative Definition 2. We could try some sort of mixture of least upper bound and greatest lower bound, but that would be very messy.

Let's go back to our first effort, and see if we can fix it up in a more constructive way. What's missing in Tentative Definition 1 is some way of saying that the terms get arbitrarily close to the limit. We want to say, more or less, that no matter how close you want the terms to be to the limit, they eventually get that close.

How can you say that more precisely?

We need some quantifiers. The phrase "no matter how close" suggests a universal quantifier—"for any closeness." We're talking about a distance, so we want the variable to represent a nonnegative real number. And we don't want to insist that the terms be *equal* to the limit, so we actually want the distance to be positive.

The Greek letter "ε"—epsilon—has become the standard symbol for the variable representing this "closeness". So we want a definition that begins: "For every positive number ε" [in symbols: "$(\forall \varepsilon > 0)(\cdots)$"—recall that our universe is $\mathbf{R}$, so ε is a real number].

What next? We want to say that "eventually" the terms get to within ε of the limit. The word "eventually" means "if you go far enough along the sequence". If we rephrase that as "there is some point beyond which . . .", it suggests an existential quantifier. The quantified variable will refer to the subscript of some term of the sequence, so this variable represents a positive integer. We'll use the letter N for this variable.

So now our new version begins: "For every positive number ε, we can find a positive integer N, such that . . ." [in symbols: "$(\forall \varepsilon > 0)(\exists N \in \mathbf{N})(\cdots)$"].

Such that what? What do we want to say about the *Nth* term of the sequence?

Tentative Definition 3

Perhaps we want to say that it is within ε of the limit L, which we can do using absolute value—we simply write: "$|x_N - L| < \varepsilon$".

So here is our new version:

> **TENTATIVE DEFINITION 3:** For a sequence $\langle x_n \rangle$ and a real number L, "the *limit* of $\langle x_n \rangle$ is L" means:
>
> $$(\forall \varepsilon > 0)(\exists\, N \in \mathbf{N})(|x_N - L| < \varepsilon).$$

We can prove that the sequence $\langle 2^{-n} \rangle$ approaches 0, *according to Tentative Definition 3*, as follows: Suppose ε is any positive real number. Let N be any positive integer satisfying the inequality

$$N > \log_2\left(\frac{1}{\varepsilon}\right). \qquad (*)$$

We just need to show that this N satisfies the condition required by Tentative Definition 3, namely, $|2^{-N} - 0| < \varepsilon$.

Since N satisfies $(*)$, we have $2^N > 2^{\log_2(1/\varepsilon)}$. But $2^{\log_2(1/\varepsilon)} = 1/\varepsilon$. Thus $2^N > 1/\varepsilon$, and so $2^{-N} < \varepsilon$. But this is the same as the condition $|2^{-N} - 0| < \varepsilon$, so we're done.

COMMENT For future use, we note the fact that the inequality $|2^{-m} - 0| < \varepsilon$ holds for *any* integer m greater than $\log_2(1/\varepsilon)$.

With this example in mind, we ask you to test Tentative Definition 3 using some other sequences:

Get Your Hands Dirty 5

According to Tentative Definition 3, which of the following statements are true?

a) Sequence ii) of Example 1 approaches 1.
b) Sequence vi) of Example 1 approaches -1.
c) Sequence viii) of Example 1 approaches $\frac{1}{2}$. □

This doesn't work either. Tentative Definition 3 just says that *one* of the terms is within ε of L. That isn't good enough. Under that definition, the sequence $\langle 2^{-n} \rangle$ would approach $\frac{1}{16}$ (as well as approach 0), because, no matter what the value of ε, we could just choose $N = 4$.

So now what?

Tentative Definition 4

One way to fix this flaw is to combine our latest version with our first try. If one particular term is within ε of L, and each term is closer to L than the one before, then all the terms from that particular term on will be within ε of L. So here's another try:

TENTATIVE DEFINITION 4: For a sequence $\langle x_n \rangle$ and a real number L, "the *limit* of $\langle x_n \rangle$ is L" means the following two conditions hold:

i) $(\forall \varepsilon > 0)(\exists N \in \mathbf{N})(|x_N - L| < \varepsilon)$;
ii) $(\forall n \in \mathbf{N})(|x_{n+1} - L| < |x_n - L|)$.

Condition i) says that, no matter how close you want it, there is at least one term "that close" to the limit. Condition ii) says that each term is even closer than the previous one. That looks good.

In fact, Tentative Definition 4 is very good. It may, in fact, express your original intuition perfectly, although it is not the "official" definition. The trouble with our earlier versions was that they allowed certain limit statements to be considered true which we didn't want to be true. For example, they allowed $\langle 2^{-n} \rangle$ to have more than one limit.

Tentative Definition 4 does not have this flaw. We have already seen that, for the sequence $\langle 2^{-n} \rangle$, both i) and ii) are true with $L = 0$. We now ask you to show for this same sequence that, if $L \neq 0$, then the statements i) and ii) of Tentative Definition 4 cannot both be true.

Get Your Hands Dirty 6

For the sequence with $x_n = 2^{-n}$, and for each nonzero real number L, figure out which of conditions i) and ii) is false. (Your answer will depend on L. If both conditions are false, say so.) Justify your answers by finding an ε that "doesn't work", as required for a counterexample to i), or an n that "doesn't work", as required for a counterexample to ii). □

The fact that we now have a unique limit for the sequence $\langle 2^{-n} \rangle$ suggests that we are getting closer to the "true" definition. Test it with the following:

Get Your Hands Dirty 7

According to Tentative Definition 4, is the following statement true?

Sequence ix) of Example 1 approaches 0. □

As we suggested earlier, Tentative Definition 4 does not allow any false statements to sneak through as true. Unfortunately, it has the opposite flaw; that is, there are some statements about limits which mathematicians want to consider true which Tentative Definition 4 would exclude.

The question in GYHD 7 illustrates this problem. Sequence ix), which we will label $\langle b_n \rangle$, is the following:

$$\frac{1}{2}, \frac{1}{3}, \frac{1}{4}, \frac{1}{9}, \frac{1}{8}, \frac{1}{27}, \frac{1}{16}, \ldots.$$

We pointed out earlier that we have

$$b_n = \begin{cases} 2^{-(n+1)/2} & \text{if } n \text{ is odd} \\ 3^{-n/2} & \text{if } n \text{ is even.} \end{cases}$$

The "official" answer says that this sequence approaches 0, although it seems to do so somewhat unevenly. Condition i) of Tentative Definition 4 is true, but, as you probably discovered, condition ii) is false. Writing the terms as decimals, we see that there is a general progress toward zero, but that there is a forward-a-little, back-a-little feeling about it:

$b_1 = .5$ $\quad b_5 = .125$ $\quad b_9 = .03125$

$b_2 = .3333\ldots$ $\quad b_6 = .0370\ldots$ $\quad b_{10} = .004115\ldots$

$b_3 = .25$ $\quad b_7 = .0625$ $\quad b_{11} = .015625$

$b_4 = .1111\ldots$ $\quad b_8 = .01234\ldots$ $\quad b_{12} = .001371\ldots$

Think about this example. Either the sequence $\langle b_n \rangle$ does not approach 0, or we need to improve our definition once more.

For various reasons having to do with later theorems, intuition, and experience, it turns out to be more useful to have a definition that makes the statement "$\langle b_n \rangle$ approaches 0" true than a definition that makes this statement false.

The sequence $\langle b_n \rangle$ is like two sequences, $\langle 2^{-n} \rangle$ and $\langle 3^{-n} \rangle$, woven together. Each is approaching 0, but at different rates. No matter how close we want the terms to get to 0, "eventually" all the terms, from both strands, are "that close".

Mathematicians have decided to adopt a definition that allows this forward-a-little, back-a-little kind of approach to the limit. In the case of $\langle b_n \rangle$, if we go far enough out in the sequence, both strands will be within ε of the limit. There may be some earlier terms that happen to be close as well, but that is not a problem. No matter how small ε is, there will be some point in the sequence beyond which all the terms will be within ε of the limit. Generally, the smaller ε is, the further along the sequence we would expect to have to go in order to achieve this.

How can you say this using quantifiers? (*Hint:* Use a modified version of condition i), in which you add one more quantified variable, referring to "all terms".)

The "Official" Definition

Here is the long-awaited "official" definition of the limit concept for sequences:

DEFINITION 9.2: For a sequence $\langle x_n \rangle$ and a real number L, "the *limit* of $\langle x_n \rangle$ is L" means:

$$(\forall\, \varepsilon > 0)(\exists\, N \in \mathbf{N})(\forall\, m \in \mathbf{N})(m > N \Rightarrow |x_m - L| < \varepsilon)\,.$$

In words, this definition says:

For every positive number ε, we can find a positive integer N such that every term of the sequence after the Nth term is within ε of L.

The condition "the limit of $\langle x_n \rangle$ is L" is usually expressed by the notation

$$\text{“}\lim_{n\to\infty} x_n = L\text{”}$$

which is read:

"the limit of x_n, as n approaches infinity, equals L".

COMMENTS

1. We have taken a lot of time to develop this definition carefully, and show why the definition is what it is. As we said at the outset, the limit concept is a complex one. Excellent mathematicians have been fooled by their intuition about limits, and so it is important that the concept be absolutely precise.

2. Working with definitions like this, with three quantified variables, can be quite difficult and confusing. In the next section, we will do proofs, using specific sequences, of theorems of the form "$\lim_{n\to\infty} x_n = L$". Like proofs with other definitions that start "$(\forall \ldots)(\exists \ldots)$", proofs about limits can be thought of as a "challenge" process (see comment following Theorem 4.6 in Section 4.1): "someone" gives you an ε, and you have to find an appropriate N. There will be more than one possibility for a successful N, and you are allowed to express N in terms of ε, but you then have to prove that your N works. With proofs about limits, the step of proving that N works involves a third variable, labeled "m" in Definition 9.2.

3. Perhaps just as important as proving statements of the form "$\lim_{n\to\infty} x_n = L$" is proving statements of the form "$\lim_{n\to\infty} x_n \neq L$". Just writing down the statement "$\langle x_n \rangle$ does not approach L" in terms of quantified variables is a significant task. (We'll ask you to do that in the next section.) Writing a proof of such a statement is even harder. (We'll do one of those for you.) There are many sequences that do not have any limit, as our "official answers" indicated, and we'll prove a statement like that as well. Our work with "nonlimits" will be used for the culminating theorem of the next section, which says that a sequence cannot have more than one limit.

EXERCISES

1. (GYHD 1) Give an equation for the function for each of the sequences in Example 1 [except for ii) and ix), which were done in the text]. [*Hint* on vi), viii), and xi): Use $(-1)^{n+1}$ as part of the formula.]
2. (GYHD 6) For the sequence with $x_n = 2^{-n}$, and for each nonzero real number L, figure out which of the following conditions [conditions i) and ii) of Tentative Definition 4] is false:
   i) $(\forall\, \varepsilon > 0)(\exists\, N \in \mathbf{N})(|x_N - L| < \varepsilon)$,
   ii) $(\forall\, n \in \mathbf{N})(|x_{n+1} - L| < |x_n - L|)$.

Your answer will depend on L. If conditions i) and ii) are both false, say so. Justify your answers by finding an ε that "doesn't work", as required for a counterexample to i), or an n that "doesn't work", as required for a counterexample to ii).

3. Suppose $\langle x_n \rangle$ is a sequence and L is a real number. Consider the following five ideas about what "$\langle x_n \rangle$ approaches L" might mean [statements i)–iv) are Tentative Definitions 1–4, and v) is the "official" definition (Definition 9.2)]:

i) $(\forall\, n \in \mathbf{N})(|x_{n+1} - L| < |x_n - L|)$.

ii) L is the greatest lower bound for the set $\{x_n : n \in \mathbf{N}\}$.

iii) $(\forall\, \varepsilon > 0)(\exists\, N \in \mathbf{N})(|x_N - L| < \varepsilon)$.

iv) Both of the following hold:

$$(\forall\, \varepsilon > 0)(\exists\, N \in \mathbf{N})(|x_N - L| < \varepsilon)$$

and

$$(\forall\, n \in \mathbf{N})(|x_{n+1} - L| < |x_n - L|)\,.$$

v) $(\forall\, \varepsilon > 0)(\exists\, N \in \mathbf{N})(\forall\, m \in \mathbf{N})(m > N \Rightarrow |x_m - L| < \varepsilon)$.

a) Compare each of the tentative definitions with the "official" definition, by determining which of the following "if . . . , then . . ." statements are true. If you think the statement is true, indicate why. If you think it is false, give a counterexample; i.e., give a specific sequence $\langle a_n \rangle$ and a real number L which make the hypothesis true and the conclusion false.

A) If i) is true, then v) is true.
B) If v) is true, then i) is true.
C) If ii) is true, then v) is true.
D) If v) is true, then ii) is true.
E) If iii) is true, then v) is true.
F) If v) is true, then iii) is true.
G) If iv) is true, then v) is true.
H) If v) is true, then iv) is true.

b) Compare the tentative definitions with each other, by determining which of the following "if . . . , then . . ." statements are true. If you think the statement is true, indicate why. If you think it is false, give a counterexample; i.e., give a specific sequence $\langle a_n \rangle$ and a real number L which make the hypothesis true and the conclusion false.

A) If i) is true, then ii) is true.
B) If ii) is true, then i) is true.
C) If i) is true, then iii) is true.
D) If iii) is true, then i) is true.
E) If i) is true, then iv) is true.
F) If iv) is true, then i) is true.
G) If ii) is true, then iii) is true.
H) If iii) is true, then ii) is true.
I) If ii) is true, then iv) is true.
J) If iv) is true, then ii) is true.
K) If iii) is true, then iv) is true.
L) If iv) is true, then iii) is true.

4. Write the following familiar theorem from calculus in completely symbolic form, using appropriate quantifiers:

Every continuous function from a finite closed interval to $\mathbf{R}$ has a maximum value.

(Use $S_{a,b}$ to represent the set of all continuous functions from a finite closed interval $[a, b]$ to **R**. The word "continuous" will be defined in Section 9.3, but you don't need to know its meaning to do this exercise.)

5. Express each of the following completely symbolic statements in ordinary mathematical language (these are true statements):

a) $(\forall\ u, v \in \mathbf{N})(\exists\ w \in \mathbf{N})[w \mid u \text{ and } w \mid v \text{ and } (\forall\ s \in \mathbf{N})(s \mid u \text{ and } s \mid v \Rightarrow s \leq w)]$. (*Hint:* First figure out and express in ordinary mathematical language what the statement in brackets says about the natural numbers u, v, and w.)

b) $(\forall\ \langle a_n \rangle \in T)[(\forall\ s, t \in \mathbf{N})(s > t \Rightarrow a_s \leq a_t) \Rightarrow (\exists\ c \in \mathbf{R}^{\geq 0})(\lim_{n\to\infty} a_n = c)]$, where T is the set of all sequences with codomain $\mathbf{R}^{\geq 0}$ (the set of nonnegative real numbers), and a generic element of T is represented as $\langle a_n \rangle$. [*Hint:* First figure out what the condition "$(\forall\ s, t \in \mathbf{N})(s > t \Rightarrow a_s \leq a_t)$" says about the sequence $\langle a_n \rangle$. Then express this condition in intuitive mathematical language. (We have not formally defined the terminology needed here, but it is a natural analog of terminology we have used.) Similarly, decide what the condition "$(\exists\ c \in \mathbf{R}^{\geq 0})(\lim_{n\to\infty} a_n = c)$" says. The overall statement above says that any sequence satisfying the first condition must satisfy the second condition.]

9.2 Limit of a Sequence—Using the Definition

In the last section, we went through considerable effort to develop a definition for "limit of a sequence" which would give the answers that mathematicians have decided are the right ones.

For convenience, we repeat that crucial definition here:

DEFINITION 9.2: For a sequence $\langle x_n \rangle$ and a real number L, "the *limit* of $\langle x_n \rangle$ is L" means:

$$(\forall\ \varepsilon > 0)(\exists\ N \in \mathbf{N})(\forall\ m \in \mathbf{N})(m > N \Rightarrow |x_m - L| < \varepsilon)\,.$$

Some Specific Examples

We now get down to the business of applying the work of Section 9.1. We start with some specific examples of how to prove that a particular sequence approaches a given limit.

A Simple Case

We will begin with the sequence $\langle n/(2n + 1) \rangle$, and we will show (as you might expect) that

$$\lim_{n\to\infty} \frac{n}{2n + 1} = \frac{1}{2}\,.$$

To prove this statement, we need to show the following:

$$(\forall\, \varepsilon > 0)(\exists\, N \in \mathbf{N})(\forall\, m \in \mathbf{N})\left(m > N \Rightarrow \left|\frac{m}{2m+1} - \frac{1}{2}\right| < \varepsilon\right).$$

Given a value of $\varepsilon > 0$, we need to find a value of N (based on ε) that works. As is often true for proofs of existence statements, we need to do some exploration first.

So suppose ε is some positive real number we've been given. We ultimately want $|m/(2m+1) - \frac{1}{2}| < \varepsilon$, which we can rewrite, using a common denominator for the two fractions, as

$$\left|\frac{2m}{2(2m+1)} - \frac{2m+1}{2(2m+1)}\right| < \varepsilon; \qquad \text{i.e.,} \quad \frac{1}{2(2m+1)} < \varepsilon\,.$$

This inequality can be rewritten as $2(2m+1) > 1/\varepsilon$, and then solved for m to give:

$$m > \frac{1}{2}\left(\frac{1}{2\varepsilon} - 1\right).$$

If we just choose the variable N to be any integer greater than $\frac{1}{2}(1/2\varepsilon - 1)$, then the last inequality will hold for all $m > N$.

We now write the proof.

THEOREM 9.1.

$$\lim_{n\to\infty} \frac{n}{2n+1} = \frac{1}{2}.$$

Proof: Suppose $\varepsilon > 0$. Choose N to be some positive integer with $N > \frac{1}{2}(1/2\varepsilon - 1)$. Now, if $m > N$, then $m > \frac{1}{2}((1/2\varepsilon) - 1)$, so $2m + 1 > 1/2\varepsilon$. Therefore, we have $1/(2m+1) < 2\varepsilon$, and so $1/[2(2m+1)] < \varepsilon$.

But a straight computation gives $|m/(2m+1) - \frac{1}{2}| = 1/[2(2m+1)]$, and so we get $|m/(2m+1) - \frac{1}{2}| < \varepsilon$, which completes the proof. ■

Simplifying the Algebra

Our next example shows a typical trick in finding a successful value for N. The trick is based on the fact that we don't have to find the *best* N—just one that works.

Consider the sequence $c_n = n/(n^2 + 1)$. Your intuition should suggest to you that, as n gets bigger, the denominator will grow much faster than the numerator, and so this sequence should approach zero. We will show this, but first, as before, we'll explore:

As required by Definition 9.2, for a given ε, we need to find a value of N such that the inequality

$$\frac{m}{m^2 + 1} < \varepsilon$$

holds for all $m > N$.

Solving this inequality is rather clumsy. But we don't need *all* the solutions—we just want to find a value of N such that the inequality is true for all $m > N$.

The trick is to recognize that $m/(m^2 + 1) < m/m^2$. If we just make the second fraction less than ε, then the first will certainly be less than ε. In order to get $m/m^2 < \varepsilon$ (i.e., $1/m < \varepsilon$) for all $m > N$, we will choose $N > 1/\varepsilon$.

So now we're ready for the next stage. Here's a formal proof:

THEOREM 9.2.

$$\lim_{n\to\infty} \frac{n}{n^2 + 1} = 0$$

Proof: Suppose $\varepsilon > 0$. Let N be some positive integer greater than $1/\varepsilon$, so $N \in \mathbf{N}$ and $1/N < \varepsilon$. Now suppose $m > N$. We must show $|m/(m^2 + 1) - 0| < \varepsilon$.

But

$$|m/(m^2 + 1) - 0| = m/(m^2 + 1) < \frac{1}{m} < \frac{1}{N} < \varepsilon,$$

as desired. ∎

COMMENTS

1. Notice the difference (once again) between exploration and proof. Exploration is informal and intuitive. We often start by looking at our goal, and think backwards to figure out what will get us there. Proof is formal and logical. It follows the structure imposed by the theorem.
2. The value of N chosen in this proof is not necessarily the best, in the sense that there may be smaller positive integers that would work. But there is nothing that requires us to find the smallest possible N. In fact, this value may be the best in the sense that it works and it is easy to work with.

Combining Good *N*'s

We give one more technique for the use of Definition 9.2 to prove that a sequence has a specific limit. It is a technique that we will repeat in the first of our general theorems on limits.

Consider the following sequence:

$$\frac{1}{2}, \frac{1}{9}, \frac{1}{8}, \frac{1}{81}, \frac{1}{32}, \frac{1}{729}, \ldots$$

whose terms can be described by the following formula:

$$d_n = \begin{cases} 2^{-n} & \text{if } n \text{ is odd} \\ 3^{-n} & \text{if } n \text{ is even.} \end{cases}$$

[This is a variation on ix) from Example 1 of Section 9.1. We are interweaving the two sequences $\langle 2^{-n} \rangle$ and $\langle 3^{-n} \rangle$ in a different way.]

Each of the sequences $\langle 2^{-n} \rangle$ and $\langle 3^{-n} \rangle$ individually approaches 0. For a given $\varepsilon > 0$, our earlier work (following Tentative Definition 3 in Section 9.1—see comment) shows that any integer N_1 which is greater than $\log_2(1/\varepsilon)$ works for the first sequence; i.e., if $N_1 > \log_2(1/\varepsilon)$ and $m > N_1$, then $|2^{-m} - 0| < \varepsilon$. Similarly, any integer $N_2 > \log_3(1/\varepsilon)$ works for the sequence $\langle 3^{-n} \rangle$; i.e., if $m > N_2$, then $|3^{-m} - 0| < \varepsilon$.

The trick for finding a successful N for $\langle d_n \rangle$ is to choose the larger of the two values N_1 and N_2. That way, when we have "$m > N$", we will have both "$m > N_1$" and "$m > N_2$". Therefore, no matter whether m is odd or even, we can apply the known result about the appropriate N_i, and get the conclusion $|d_m - 0| < \varepsilon$. We leave the details for you.

Get Your Hands Dirty 1

Adapt the method just described in order to prove that the sequence $\langle b_n \rangle$ given by

$$b_n = \begin{cases} 2^{-(n+1)/2} & \text{if } n \text{ is odd} \\ 3^{-n/2} & \text{if } n \text{ is even} \end{cases}$$

satisfies $\lim_{n\to\infty} b_n = 0$.

[This is sequence ix) from Example 1 of Section 9.1.] □

Sums of Sequences

Like any functions with codomain **R**, sequences can be added, subtracted, multiplied, and, except for division by zero, divided. Here are the formal definitions:

DEFINITION 9.3: For sequences $\langle a_n \rangle$ and $\langle b_n \rangle$, and $c \in \mathbf{R}$:

the *sum* of $\langle a_n \rangle$ and $\langle b_n \rangle$ (written $\langle a_n \rangle + \langle b_n \rangle$) is the sequence whose nth term is the sum $a_n + b_n$.

the *difference, product,* and *quotient* of $\langle a_n \rangle$ and $\langle b_n \rangle$ are, respectively, the sequences whose nth terms are the difference $a_n - b_n$, the product $a_n \cdot b_n$, and the quotient $a_n \div b_n$, with the condition that the quotient of the sequences is only defined if all terms of $\langle b_n \rangle$ are nonzero.

the *product* of c and $\langle a_n \rangle$ is the sequence whose nth term is $c \cdot a_n$.

(These operations will be defined for functions in general in Section 9.3.)

The Triangle Inequality—a Digression

Before proving a theorem concerning Definition 9.3, we make a brief interruption to discuss an important principle for working with absolute value inequalities—the *Tri-*

angle Inequality. This result says that, for any two real numbers x and y, we have $|x + y| \leq |x| + |y|$.

If x and y are both positive, we can just ignore the absolute-value signs, and we actually have equality. But for other sign combinations, we have to be more careful.

Get Your Hands Dirty 2

Verify the Triangle Inequality for the following pairs of values, each of which gives a different sign combination for x and y:

a) $x = 4$, $y = -2$.
b) $x = -7$, $y = 3$.
c) $x = -5$, $y = -8$. □

In Exercise 4, we ask you to prove the Triangle Inequality case by case, using the definition of absolute value.

COMMENT The name "Triangle Inequality" comes from the principle in geometry that the length of any side of a triangle is less than (or equal to, for a degenerate triangle) the sum of the other two sides—see Figure 9.1. We can relate our algebraic version to this geometric principle by interpreting x and y as two-dimensional vectors, and absolute value as length.

Limit of the Sum of Two Sequences

The following theorem, concerning arithmetic of sequences as defined in Definition 9.3, is an illustration of how the definition of limit is used in a general proof:

THEOREM 9.3. For sequences $\langle a_n \rangle$ and $\langle b_n \rangle$, if

$$\lim_{n\to\infty} a_n = L_1 \quad \text{and} \quad \lim_{n\to\infty} b_n = L_2\,,$$

then

$$\lim_{n\to\infty} (a_n + b_n) = L_1 + L_2\,.$$

Proof: Suppose $\lim_{n\to\infty} a_n = L_1$ and $\lim_{n\to\infty} b_n = L_2$. What we need to prove is the following:

$$(\forall\, \varepsilon > 0)(\exists\, N \in \mathbf{N})(\forall\, m \in \mathbf{N})\,(m > N \Rightarrow |(a_m + b_m) - (L_1 + L_2)| < \varepsilon)\,.$$

So suppose $\varepsilon > 0$. We need to "find" an appropriate value of N.

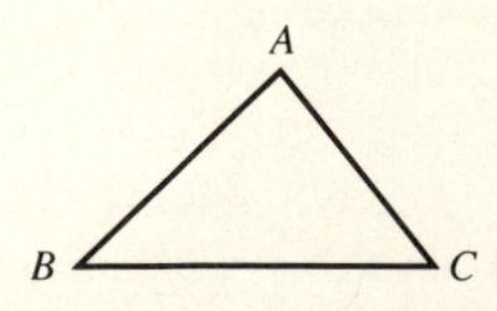

Figure 9.1. The length of BC is less than the sum of the lengths of AB and AC.

DISCUSSION How do we find N? We know that there is an integer N_1 beyond which the terms of $\langle a_n \rangle$ are close to L_1, and, similarly, an integer N_2 for the sequence $\langle b_n \rangle$ and its limit L_2. How close do we need the terms of these sequences to be to their limits in order to guarantee that $a_n + b_n$ is within ε of $L_1 + L_2$?

? ? ?

The key to this analysis is to notice that we can write the difference $(a_m + b_m) - (L_1 + L_2)$ as $(a_m - L_1) + (b_m - L_2)$. Therefore, we can make $(a_m + b_m) - (L_1 + L_2)$ less than ε by making each of the differences $a_m - L_1$ and $b_m - L_2$ less than $\varepsilon/2$. This idea is combined with the Triangle Inequality to prove the theorem.

We now resume the proof:

Since $\lim_{n\to\infty} a_n = L_1$, there is an integer N_1 such that

$$(\forall\, m \in \mathbf{N})\left(m > N_1 \Rightarrow |a_m - L_1| < \frac{\varepsilon}{2}\right).$$

Similarly, there is an integer N_2 such that

$$(\forall\, m \in \mathbf{N})\left(m > N_2 \Rightarrow |b_m - L_2| < \frac{\varepsilon}{2}\right).$$

How do we use these two integers?

Let $N = \max(N_1, N_2)$ (i.e., the larger of the two; if they are equal, use that common value for N). We will show that, if $m > N$, then $|a_m + b_m - (L_1 + L_2)| < \varepsilon$.

By the choice of N, if $m > N$, then $m > N_1$ and $m > N_2$. Therefore, $|a_m - L_1| < \varepsilon/2$ and $|b_m - L_2| < \varepsilon/2$. The desired inequality is obtained as follows:

$$\begin{aligned} |a_m + b_m - (L_1 + L_2)| &= |(a_m - L_1) + (b_m - L_2)| \\ &\le |a_m - L_1| + |b_m - L_2| \qquad \text{(by the Triangle Inequality)} \\ &< \frac{\varepsilon}{2} + \frac{\varepsilon}{2} \\ &= \varepsilon\,. \end{aligned}$$

■

COMMENTS

1. The hypotheses that $\lim_{n\to\infty} a_n = L_1$ and $\lim_{n\to\infty} b_n = L_2$ can be thought of as "good-N finders"—that is, they guarantee the existence of positive integers with a certain desirable property. (See the discussion in Section 4.1 about the technique we call "name it—then use it", used when the hypothesis of a theorem contains existence statements.)
2. This proof illustrates a typical use of the Triangle Inequality. Mathematicians often use this principle freely, without explicit identification of it.

The next theorem, whose proof is left for exercises, gives similar results for the other arithmetic operations:

THEOREM 9.4. For sequences $\langle a_n \rangle$ and $\langle b_n \rangle$, and $c \in \mathbf{R}$, suppose

$$\lim_{n\to\infty} a_n = L_1 \quad \text{and} \quad \lim_{n\to\infty} b_n = L_2 .$$

Then:

i) $\lim\limits_{n\to\infty}(a_n - b_n) = L_1 - L_2$,

ii) $\lim\limits_{n\to\infty}(a_n \cdot b_n) = L_1 \cdot L_2$,

iii) $\lim\limits_{n\to\infty}(c \cdot a_n) = c \cdot L_1$.

If, furthermore, $\langle b_n \rangle$ has no terms equal to zero and $L_2 \neq 0$, then:

iv) $\lim\limits_{n\to\infty}(a_n/b_n) = L_1/L_2$. ■

COMMENT The proofs of i) and iii) are very similar to the proof of Theorem 9.3, and are left as Exercise 5. Parts ii) and iv) are trickier, because the algebra is more complicated. In Section 9.3, we will prove a result for limits of functions which is analogous to ii), and ask you to prove ii) of Theorem 9.4 as an exercise. The proof of both iv) and its function analog are also exercises in Section 9.3.

Uniqueness of Limits

One of the principles that motivated our definition of the limit of a sequence was the desire that a sequence not have more than one limit. We now prove that Definition 9.2 achieves this goal:

THEOREM 9.5. A sequence cannot have two distinct limits.

This is not an easy theorem, largely because of the complexity of Definition 9.2. We will go through the proof very carefully, because it illustrates both the use of the quantifiers and the importance of exploration.

We will prove this theorem somewhat differently from most uniqueness theorems. Rather than assume that L_1 and L_2 are both limits, and prove them equal, we will assume that L_1 is a limit and that L_2 is any real number different from L_1, and prove that L_2 is not a limit. Here are the first two sentences of our proof:

Suppose that $\langle x_n \rangle$ is a sequence with limit L_1, and suppose that $L_2 \neq L_1$. We need to show that $\lim_{n\to\infty} x_n \neq L_2$.

The intuitive idea of the proof is that, if the terms of the sequence are "sufficiently close" to L_1, then they can't also be "arbitrarily close" to L_2. The hard work of the proof is to use the definition of limit exactly as it is stated, in order to make these "closenesses" precise.

Here is an outline of how we will proceed:

a) Since we need to show that something is *not* a limit, we will begin with a discussion of the negation of the statement "$\lim_{n\to\infty} x_n = L$".

b) We will then give an intuitive application of the discussion in a) by showing that $\frac{1}{3}$ is not the limit of $\langle 2^{-n} \rangle$.
c) We next will explore how to generalize the reasoning of the specific example in b).
d) We will then go on to complete the proof, with comments interspersed.
e) We will conclude with a restatement of the proof, without interruptions.

Negation of the Statement "$\lim_{n\to\infty} x_n = L$"

Since we are going to prove that something is not the limit of a sequence, we need to know what that means.

Write down, using appropriate quantifiers and variables, the statement "$\lim_{n\to\infty} x_n \neq L$".

We have to apply the principles of negating quantified statements to the three-quantifier condition in Definition 9.2. If we apply these principles formally, the result is the following:

$$(\exists\ \varepsilon > 0)\,(\forall\ N \in \mathbf{N})\,(\exists\ m \in \mathbf{N})\,(m > N \text{ and } |x_m - L| \geq \varepsilon)\,. \qquad (*)$$

(After its quantifiers, Definition 9.2 has the form $p \Rightarrow q$, an implication. When this implication is negated, we get the statement that we have a counterexample: p and $\sim q$.)

It is also important to understand this negation on an intuitive level. The statement that L *is* the limit of $\langle x_n \rangle$ can be paraphrased as follows:

> For any "closeness", there is a place in the sequence beyond which all terms are "that close" to L.

Thus, to say that L is *not* the limit is to say the following:

> There is a "closeness" such that, no matter how far out you go, there are terms which are beyond that point and which are not "that close" to L.

That is precisely what $(*)$ says. Here is the phrase-by-phrase translation:

$(\exists\ \varepsilon > 0)$	There is a "closeness" such that,
$(\forall\ N \in \mathbf{N})$	no matter how far out you go,
$(\exists\ m \in \mathbf{N})$	there are terms
$(m > N$	which are beyond that point
and $\|x_m - L\| \geq \varepsilon)$	and which are not "that close" to L.

Working with a Specific Example

We turn now to see how this negation is applied to a specific situation. Suppose we want to prove that the sequence $\langle 2^{-n} \rangle$ has no limit other than 0. To be more specific, we will prove the following, after appropriate investigation:

THEOREM 9.6.

$$\lim_{n\to\infty} 2^{-n} \neq \frac{1}{3}.$$

According to our discussion above, the statement $\lim_{n\to\infty} 2^{-n} \neq \frac{1}{3}$ is an *existence statement*—it says that we need to find an ε which satisfies a certain condition. As we have emphasized repeatedly, proving existence statements often requires exploration. We have to understand the situation well enough to know how to choose a successful ε.

Find an ε that works. Draw a picture.

First, an unsuccessful ε: we can't use $\varepsilon = 1$, because all of the terms of our sequence *are* within 1 of $\frac{1}{3}$. In fact, any value of ε greater than $\frac{1}{3}$ will be unsuccessful. (Why?)

So let's see, for example, what happens if we try using $\varepsilon = \frac{1}{4}$. We must show that, no matter how far out you go, there are terms which are beyond that and which are at least $\frac{1}{4}$ from $\frac{1}{3}$.

How do we get terms which are at least $\frac{1}{4}$ from $\frac{1}{3}$? Examine Figure 9.2—we want the terms to be outside the shaded interval around $\frac{1}{3}$. We should be able to keep our terms outside this interval by making them sufficiently close to their actual limit, which we (intuitively) know to be 0.

How close do the terms have to be to zero, in order for us to be guaranteed that they are at least $\frac{1}{4}$ from $\frac{1}{3}$? How small does the interval around 0 have to be in order not to intersect the shaded interval?

? ? ?

That's a matter of straight arithmetic: the left-hand end of the shaded interval is $\frac{1}{3} - \frac{1}{4} = \frac{1}{12}$, so any term that is within $\frac{1}{12}$ of 0 will be outside the shaded interval. Therefore, any term from $\frac{1}{16}$ (the 4th term of the sequence) and beyond will work.

Go over this carefully until it is clear to you.

With this analysis in mind, we can formally prove that $\lim_{n\to\infty} 2^{-n} \neq \frac{1}{3}$ as follows:

Proof of Theorem 9.6: Choose $\varepsilon = \frac{1}{4}$, and let N be any positive integer. We need to find an integer $m > N$ such that $|2^{-m} - \frac{1}{3}| \geq \frac{1}{4}$.

Choose $m = N + 3$, so that $m > N$ and $m \geq 4$. Then $2^{-m} \leq \frac{1}{16}$, so 2^{-m} is further from $\frac{1}{3}$ than $\frac{1}{16}$ is. Thus, $|2^{-m} - \frac{1}{3}| \geq |\frac{1}{16} - \frac{1}{3}| > \frac{1}{4}$, proving the result. ■

COMMENTS

1. Notice how the actual proof follows the pattern of the quantifiers in (*): $(\exists\ \varepsilon)(\forall\ N)(\exists\ m)$. We chose ε, we were "given" N, and then we chose m in terms of N. We will see this again when we prove Theorem 9.5. The specific choices for ε and m were based on our earlier analysis, or exploration, of the situation.

2. Any positive real number less than $\frac{1}{3}$ would have worked for ε (where, in this context, "works" means "is a counterexample to the condition on ε in the statement

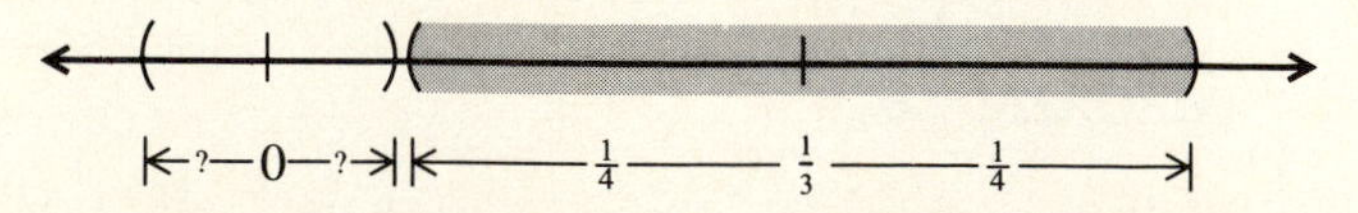

Figure 9.2. We want an interval around 0 that "misses" the shaded interval around $\frac{1}{3}$.

$\lim_{n\to\infty} 2^{-n} = \frac{1}{3}$"). But all we must do is find one value for ε that works, so there are many proofs for a limit not being a certain number. In general, a larger choice of ε will require a larger choice of m.

Though the analysis involved in preparation for Theorem 9.6 may seem intimidating, you really must do that kind of dirty work in order to fully understand the definition of "limit". Therefore, we give you the following:

Get Your Hands Dirty 3

Prove $\lim_{n\to\infty} n/(3n + 1) \neq 1/2$. □

Generalizing the Specific Example

Before turning back to the proof of Theorem 9.5, let's look briefly at how to generalize the above analysis of Theorem 9.6.

Suppose we know that $\lim_{n\to\infty}\langle x_n\rangle = L_1$, and that we want to show that $\lim_{n\to\infty}\langle x_n\rangle \neq L_2$. As in Theorem 9.6, we need to find, or choose, a successful value for ε. Since the terms are actually close to L_1, it should work to choose ε to be any value less than $|L_2 - L_1|$. Figure 9.3 is a generalization of Figure 9.2. How close must the terms be to L_1 to guarantee that they are at least ε from L_2, i.e., outside the shaded interval around L_2?

? ? ?

The diagram shows that, if the terms are within $|L_2 - L_1| - \varepsilon$ of L_1, then they are a distance at least ε from L_2.

Since ε can be any value less than $|L_2 - L_1|$, we will make a choice that simplifies the algebra, namely $|L_2 - L_1|/2$, i.e., half the distance between L_1 and L_2. With that choice for ε, we know that, if a term is less than ε from L_1, it must be more than ε from L_2 (see Figure 9.4).

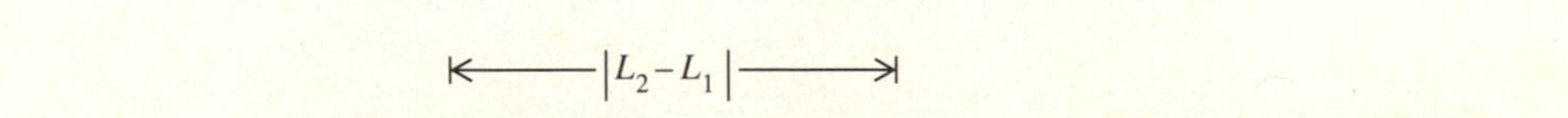

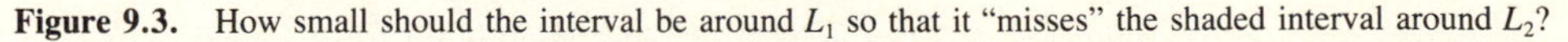

Figure 9.3. How small should the interval be around L_1 so that it "misses" the shaded interval around L_2?

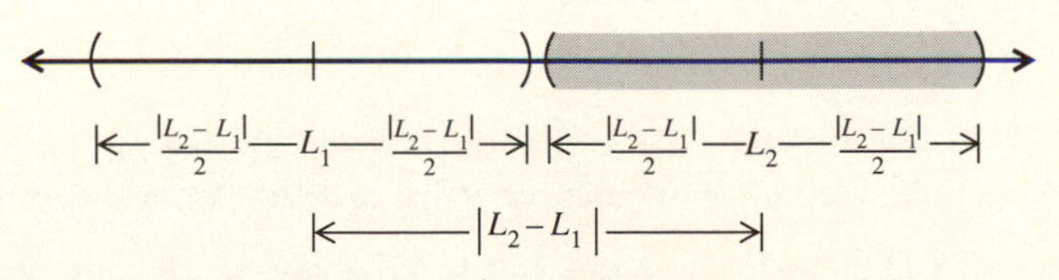

Figure 9.4. If a term is inside the interval of size $|L_2 - L_1|/2$ on either side of L_1, then it is outside the similar interval around L_2.

For use in the proof of Theorem 9.5, we summarize this inequality in the following:

LEMMA 9.7. For $x, L_1, L_2 \in \mathbf{R}$, and $\varepsilon_0 = |L_2 - L_1|/2$,

$$\text{if } |x - L_1| < \varepsilon_0, \text{ then } |x - L_2| > \varepsilon_0.$$

The proof involves the Triangle Inequality—see Exercise 7.

Proof of Uniqueness of Limits, with Commentary

We now return to our proof of Theorem 9.5. We will start the formal proof from scratch, interspersing it with comments referring to our discussions above.

THEOREM 9.5. A sequence cannot have two distinct limits.

Proof: Suppose that $\langle x_n \rangle$ is a sequence with limit L_1, and suppose that $L_2 \neq L_1$. We need to show that L_2 is not the limit of $\langle x_n \rangle$. Let $\varepsilon_0 = |L_2 - L_1|/2$.

Comment: We've chosen a specific value for ε which, according to our exploration, ought to work. With this specific value, we can use Lemma 9.7.

To show $\lim_{n\to\infty} x_n \neq L_2$, we will show that, for all $N \in \mathbf{N}$, there exists $m > N$ such that $|x_m - L_2| \geq \varepsilon_0$.

Comment: We're claiming that ε_0 "works", i.e., gives us a counterexample to the statement that L_2 is the limit.

So suppose $N \in \mathbf{N}$. To complete the proof, we need find m as described above.

Comment: In order to find the right m, we use the insight gained in our exploration of Example 4. We will find a term in the sequence which is "sufficiently close" to L_1, using Lemma 9.7.

Since $\lim_{n\to\infty} x_n = L_1$, there is a positive integer M such that

$$(\forall\, t \in \mathbf{N})\,(t > M \Rightarrow |x_t - L_1| < \varepsilon_0)\,. \qquad (**)$$

Comment: We are specializing the assumption that $\lim_{n\to\infty} x_n = L_1$. We've used a specific, carefully chosen value of the universally quantified variable ε which appears in the definition of "limit".

That assumption guarantees the existence of a positive integer beyond which all terms are close to L_1. By Lemma 9.7, we know that terms within ε_0 of L_1 will be more than ε_0 from L_2.

Let $m = M + N + 1$.

Comment: We're choosing a specific value for m which is bigger than both M and N. We will show that our value of m works. Many other choices would also work.

Since $m > N$, the proof will be complete if we show that $|x_m - L_2| \geq \varepsilon_0$. But $m > M$, so we have $|x_m - L_1| < \varepsilon_0$, by (**).

Comment: We're specializing the statement (**), using the fact that $m > M$.

The desired conclusion $|x_m - L_2| \geq \varepsilon_0$ now follows from Lemma 9.7. ■

The Complete Proof (without Interruption)

Here is the complete proof (slightly reworded) without interruption:

THEOREM 9.5. A sequence cannot have two distinct limits.

Proof: Suppose that $\langle x_n \rangle$ is a sequence with limit L_1, and suppose that $L_2 \neq L_1$. We need to show that L_2 is not the limit of $\langle x_n \rangle$, so we need to find an appropriate ε.

Let $\varepsilon_0 = |L_2 - L_1|/2$. To show $\lim_{n\to\infty} x_n \neq L_2$, we will show that, for all $N \in \mathbf{N}$, there exists $m > N$ such that $|x_m - L_2| \geq \varepsilon_0$. So suppose $N \in \mathbf{N}$.

Since $\lim_{n\to\infty} x_n = L_1$, there is a positive integer M such that

$$(\forall\, t \in \mathbf{N})\,(t > M \Rightarrow |x_t - L_1| < \varepsilon_0)\,. \qquad (**)$$

Let $m = M + N + 1$. Since $m > N$, the proof will be complete if we show that $|x_m - L_2| \geq \varepsilon_0$.

But $m > M$, so we have $|x_m - L_1| < \varepsilon_0$, by (**). By Lemma 9.7, this implies $|x_m - L_2| > \varepsilon_0$, which is our desired conclusion. ■

COMMENTS

1. We repeat our earlier comment that this is a complicated proof. Do not be discouraged if you find it difficult.

2. One important factor in that difficulty is the presence, *in the hypothesis,* of a condition using quantifiers in a complex form—namely, the assumption that $\lim_{n\to\infty} x_n = L_1$: We specialized that assumption in two ways. To help review this aspect of the proof, here is that assumption in quantifier form:

$$(\forall\, \varepsilon > 0)\,(\exists\, N \in \mathbf{N})\,(\forall\, m \in \mathbf{N})\,(m > N \Rightarrow |x_m - L_1| < \varepsilon)\,.$$

   We first specialized the variable ε, using $\varepsilon = \varepsilon_0 = |L_2 - L_1|/2$. This guaranteed the existence of an integer M such that

$$(\forall\, t \in \mathbf{N})\,(t > M \Rightarrow |x_t - L_1| < \varepsilon_0)\,. \qquad (**)$$

   (Notice the relabeling of variables.) We then specialized (**) as well, substituting $t = m = M + N + 1$, which told us that $|x_m - L_1| < \varepsilon_0$.

3. The last step in the proof was anticipated by our exploration in Theorem 9.6, and prepared for in Lemma 9.7. It often makes the reading of a proof clearer if some mechanical details like this are discussed before the proof itself begins.

4. The written proof itself does not explain how we chose a particular value for ε—it simply uses the selection successfully. This is typical of mathematical writing.

Sequences with No Limits

We have just seen that a sequence cannot have two distinct limits. As we noted in the "official answers" (in Section 9.1), some sequences do not have any limit. We conclude by giving an example of how to prove such an assertion about a specific sequence.

THEOREM 9.8. Let $\langle e_n \rangle$ be the sequence with $e_n = n$. Then there is no real number L such that $\lim_{n\to\infty} e_n = L$.

Proof: Suppose $L \in \mathbf{R}$. To prove that $\lim_{n\to\infty} e_n \neq L$, let $\varepsilon = 1$, and suppose that N is any positive integer. We need to find an integer $m > N$ such that $|e_m - L| \geq 1$.

Choose N_1 to be any integer greater than L, and let $m = \text{Max}(N + 1, N_1 + 1)$. Then $m > N$, and $m \geq N_1 + 1 > L + 1$. Since $e_m = m$, this gives $|e_m - L| = |m - L| > |(L + 1) - L| = 1$, as desired. ■

COMMENT In Exercise 17, we will give meaning to the statement "the limit of $\langle x_n \rangle$ is $+\infty$", and ask you to show that this applies to the sequence of the above theorem. However, that phrase does not apply to the sequence in the following GYHD:

Get Your Hands Dirty 4

Prove that the sequence $\langle (-1)^{n+1} \rangle$ [sequence vi) from Example 1 in Section 9.1] does not have a limit. □

Since some sequences have limits, and others do not, mathematicians have introduced a term to describe this distinction:

DEFINITION 9.4: For a sequence $\langle a_n \rangle$, "$\langle a_n \rangle$ is *convergent*" means:

$$(\exists\, L \in \mathbf{R})\left(\lim_{n\to\infty} a_n = L\right).$$

In simple language: a sequence is *convergent* iff it has a limit.

EXERCISES

1. (GYHD 1) Define the sequence $\langle b_n \rangle$ by

$$b_n = \begin{cases} 2^{-(n+1)/2} & \text{if } n \text{ is odd} \\ 3^{-n/2} & \text{if } n \text{ is even.} \end{cases}$$

Prove: $\lim_{n\to\infty} b_n = 0$.

2. (GYHD 3) Prove:

$$\lim_{n\to\infty} \frac{n}{3n + 1} \neq \frac{1}{2}.$$

3. (GYHD 4) Prove that the sequence $\langle(-1)^{n+1}\rangle$ does not have a limit.
4. The Triangle Inequality says that, for any real numbers x and y, $|x + y| \leq |x| + |y|$. Consider each of the following combinations of signs for x and y:
   a) $x \geq 0,\ y \geq 0$;
   b) $x \geq 0,\ y < 0$;
   c) $x < 0,\ y \geq 0$;
   d) $x < 0,\ y < 0$.

   Prove that the Triangle Inequality holds in each case. Use the following definition of absolute value:

$$|x| = \begin{cases} x & \text{if } x \geq 0 \\ -x & \text{if } x < 0. \end{cases}$$

5. Prove the following parts of Theorem 9.4:

   THEOREM 9.4. For sequences $\langle a_n\rangle$ and $\langle b_n\rangle$ and $c \in \mathbf{R}$, suppose
$$\lim_{n\to\infty} a_n = L_1 \text{ and } \lim_{n\to\infty} b_n = L_2 .$$
   Then:
   i) $\lim_{n\to\infty}(a_n - b_n) = L_1 - L_2$,
   iii) $\lim_{n\to\infty}(c \cdot a_n) = c \cdot L_1$.

6. Determine the limit of each of the sequences whose nth term is given below. (If there is no limit, say so.) Prove your answers.
   a) $x_n = \dfrac{2n}{3n + 1}$,
   b) $x_n = \dfrac{n^2}{2n^2 + 1}$,
   c) $x_n = (-1)^n 2^{-n}$,
   d) $x_n = \dfrac{\sin(n)}{n}$,
   e) $x_n = \dfrac{n!}{2^n}$.
7. Prove the following, using the Triangle Inequality:

   LEMMA 9.7. For $x, L_1, L_2 \in \mathbf{R}$, and $\varepsilon_0 = |L_2 - L_1|/2$,
$$\text{if } |x - L_1| < \varepsilon_0, \text{ then } |x - L_2| > \varepsilon_0 .$$
8. Suppose $\lim_{n\to\infty} a_n = L$, and $K < L < M$. Prove that $\langle a_n\rangle$ has only a finite number of terms less than K and only a finite number of terms greater than M. (*Hint:* Prove each assertion separately by choosing a suitable ε.)
9. Suppose $\lim_{n\to\infty} a_n = L$, where $L > 0$. Prove:
$$(\exists\, N \in \mathbf{N})\,(\forall\, m \in \mathbf{N})\,(m > N \Rightarrow a_m > L/2) .$$
   (*Hint:* Use Exercise 8 with $K = L/2$.)
10. Suppose $\lim_{n\to\infty} a_n = L$, where $L > 0$. Prove: $\lim_{n\to\infty}(1/a_n) = 1/L$. (*Hint:* First apply Exercise 9 to find a term of $\langle a_n\rangle$ beyond which all terms are at least $L/2$. Then work from there.)
11. A sequence $\langle x_n\rangle$ is called *increasing* iff
$$(\forall\, n \in \mathbf{N})\,(x_{n+1} \geq x_n) .$$

A sequence is called *bounded above* iff the set of all its terms, $\{a_n : n \in \mathbf{N}\}$, is bounded above. Suppose $\langle a_n \rangle$ is an increasing sequence which is bounded above. By the completeness property of $\mathbf{R}$ (which you may assume), $\{a_n : n \in \mathbf{N}\}$ has a least upper bound, say L. Prove:

$$\lim_{n\to\infty} a_n = L\,.$$

Hints:

a) Draw a picture that illustrates what's going on.

b) Prove the assertion by contradiction, by showing that if L is not the limit, then either L is not an upper bound for $\{a_n : n \in \mathbf{N}\}$ or there is a number less than L which is an upper bound for $\{a_n : n \in \mathbf{N}\}$. In either case, you would conclude that L is not the lub for $\{a_n : n \in \mathbf{N}\}$, which is a contradiction.

12. **a)** Make up a sequence $\langle a_n \rangle$ whose terms are all in $\mathbf{Q}$, but whose limit is in $\mathbf{R} - \mathbf{Q}$.

b) Make up a sequence $\langle b_n \rangle$ whose terms are all in $\mathbf{R} - \mathbf{Q}$, but whose limit is in $\mathbf{Q}$.

13. Suppose that $\lim_{n\to\infty} a_n = L$, that k is a specific positive integer, and that a sequence $\langle b_n \rangle$ is defined by the condition $b_n = a_{n+k}$. Prove:

$$\lim b_n = L\,.$$

14. Suppose $\langle a_n \rangle$ and $\langle b_n \rangle$ are two sequences, with $\lim_{n\to\infty} a_n = 0$, and such that, for some positive real number M, we have $|b_n| \le M$ for all $n \in \mathbf{N}$. Prove:

$$\lim_{n\to\infty}(a_n b_n) = 0\,.$$

15. Suppose $\lim_{n\to\infty} a_n = L$, and that all terms of $\langle a_n \rangle$ beyond some point are nonnegative. Prove:

$$L \ge 0\,.$$

16. Prove each of the following, without citing Theorem 9.5:

a) $\lim_{n\to\infty} \dfrac{n}{n+1} \ne 2,$

b) $\lim_{n\to\infty} \sin\left(\dfrac{n\pi}{2}\right) \ne 0.$

17. Consider the following:

DEFINITION. For a sequence $\langle x_n \rangle$, "the limit of $\langle x_n \rangle$ is $+\infty$" (written $\lim_{n\to\infty} x_n = +\infty$) means:

$$(\forall\, L \in \mathbf{R})\,(\exists\, N \in \mathbf{N})\,(\forall\, m \in \mathbf{N})\,(m > N \Rightarrow x_m > L)\,.$$

a) Prove: If $\langle a_n \rangle$ is increasing (see Exercise 11) and $\{a_n : n \in \mathbf{N}\}$ is not bounded above, then $\lim_{n\to\infty} a_n = +\infty$.

b) Prove that, for the sequence $\langle e_n \rangle$ with $e_n = n$ (see Theorem 9.8), we have $\lim_{n\to\infty} e_n = +\infty$

c) Make up a sequence $\langle a_n \rangle$ for which $\{a_n\}$ is not bounded above but whose limit is not $+\infty$.

9.3 Limits of Functions and Continuity

In this section, we will develop another aspect of the limit concept, and look at a very important property of most familiar functions—continuity.

Throughout this section, $[a, b]$ is some interval in $\mathbf{R}$, with $a < b$, and c will be an element of $[a, b]$. (We allow the possibility that c is one of the endpoints a or b.) We set

$A = (a, b) - \{c\}$. All functions under discussion will have codomain **R**, and, except for the general definition of arithmetic of functions in Definition 9.6 below, all functions have a domain that includes A.

The ideas in this section center on the following type of question: what happens to the value of $f(x)$ as x gets closer to c? Here is an example:

EXAMPLE 1 **Behavior of x^2 Near $x = 3$** Suppose f is the function from **R** to **R** defined by the equation $f(t) = t^2$, and x is a real number near 3. What can be said about $f(x)$?

? ? ?

It seems reasonable to expect that $f(x)$ should be near $f(3)$, i.e., near 9. This is, in fact, the case. The distance between $f(x)$ and $f(3)$ is $|x^2 - 3^2|$. We can write this as $|x + 3| \cdot |x - 3|$. The first factor is near 6, and the second is near 0, so the product is also near 0.

We can make this more precise as follows:

> No matter how close we want $f(x)$ to be to $f(3)$, we can achieve that closeness by making x sufficiently close to 3.

The desired closeness of $f(x)$ to $f(3)$ is generally denoted by ε (similar to the use of ε in the last section to represent the desired closeness of a_n to L). The Greek letter "δ" (delta) is usually used to indicate how close x must be to 3 to achieve the "desired closeness" of $f(x)$ to $f(3)$. In quantifier notation, our statement becomes the following:

$$(\forall\, \varepsilon > 0)\,(\exists\, \delta > 0)\,(\forall\, x)\,(|x - 3| < \delta \Rightarrow |x^2 - 3^2| < \varepsilon)\,.$$

We have not proved this statement, but our intuitive discussion earlier, based on factoring $x^2 - 3^2$, should convince you that it is at least reasonable. □

Limit of a Function

We will use the discussion in Example 1 as the basis for the central definition of this section. One important adjustment is that, in general, the function under consideration must be defined *near* c, but is not necessarily defined *at* c. We will use "L" in the role played above by $f(3)$, i.e., to denote the real number that $f(x)$ is supposed to get close to. Recall $A = (a, b) - \{c\}$.

DEFINITION 9.5: For a set B such that $A \subseteq B$, a function f defined on B, and $L \in$ **R**, "the *limit* of $f(x)$, as x approaches c, is L", written

$$\text{“}\lim_{x \to c} f(x) = L\text{”}$$

means:

$$(\forall\, \varepsilon > 0)\,(\exists\, \delta > 0)\,(\forall\, x \in B)\,(0 < |x - c| < \delta \Rightarrow |f(x) - L| < \varepsilon)$$

We sometimes write "$f(x) \to L$ as $x \to c$" [read "$f(x)$ approaches L as x approaches c"] as an alternative way of expressing this condition.

COMMENTS

1. Notice that the hypothesis of the implication is "$0 < |x - c| < \delta$" rather than "$|x - c| < \delta$". Definition 9.5 intentionally says nothing at all about $f(c)$ [i.e., the value of $f(x)$ when $|x - c| = 0$]. This is important, because c may not even be in the domain of f. Perhaps the most important situation where this happens is in the definition of derivative, which involves limits of expressions of the form $[f(x) - f(c)]/(x - c)$, which are undefined at $x = c$.

   A simpler example is $\lim_{x\to 0}(\sin x)/x$. In Exercise 4, we ask you to show that this limit is equal to 1.

2. The variable δ plays a role here analogous to that of N in the definition of the limit of a sequence in Section 9.1: N indicates how far along in the sequence one needs to go in order to guarantee that the terms are within ε of L; δ indicates how close to c x needs to get in order to guarantee that $f(x)$ is within ε of L. The implication in Definition 9.5 says, in essence, "if x is close enough to c, then $f(x)$ will be the desired closeness to L".

3. In terms of this definition, Example 1 states that $\lim_{x\to 3} x^2 = 9$. (We'll give a formal proof of this in a moment.)

4. In Exercise 10 we will define what it means for $f(x)$ to approach infinity as x approaches c.

Figure 9.5 illustrates the basic idea behind Definition 9.5. The number ε represents the width on either side of L of the horizontal band in Figure 9.5. Definition 9.5 says:

> No matter how narrow that band around L, there will be some vertical band around c, of width δ in each direction, such that all points on the graph within the vertical band, with the possible exception of a point with $x = c$, also lie within the horizontal band.

That is, all points $(x, f(x))$ on the graph with $0 < |x - c| < \delta$ must satisfy the condition $|f(x) - L| < \varepsilon$.

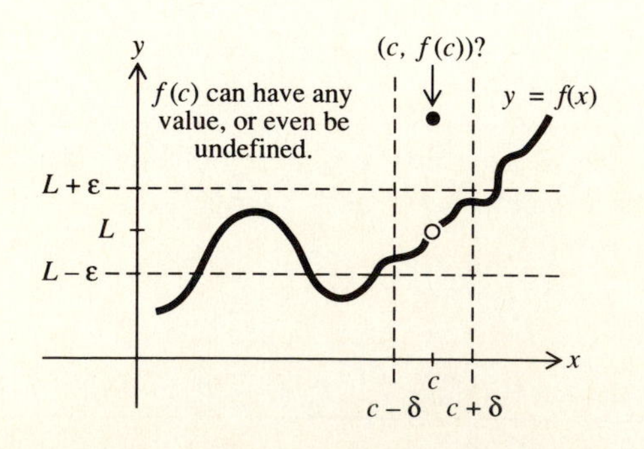

Figure 9.5. All points on the graph within the vertical band around c, except possibly $(c, f(c))$, stay within the horizontal band around L. [There is no restriction on $f(c)$; it can even be undefined.] If $\lim_{x\to c} f(x) = L$, then such a vertical band can be found for every horizontal band.

The width of the vertical band will, in general, depend on the width of the horizontal band. If the horizontal band were made narrower, we would have to choose a narrower vertical band to accomplish the desired condition.

A Sample Proof

We will illustrate the use of Definition 9.5 of the limit of a function by formally proving what we described intuitively in Example 1:

THEOREM 9.9.

$$\lim_{x \to 3} x^2 = 9.$$

According to Definition 9.5, given a positive value for ε, we need to find a positive δ such that, if $0 < |x - 3| < \delta$, then $|x^2 - 9| < \varepsilon$.

The discussion in Example 1 shows that $x^2 - 9$ is *approximately* $6(x - 3)$. If we make $|x - 3|$ less than $\varepsilon/6$, then $x^2 - 9$ will be *approximately* less than ε. Unfortunately, that isn't quite good enough.

The difficulty is that $x + 3$ is not equal to 6, but only close to 6. How can we control the value of $x + 3$?

? ? ?

The key is to realize that we are being asked to choose an *interval* of values for x. Though $x + 3$ may get bigger than 6, we can keep it from getting too much bigger, by keeping x close to 3. Once we control $x + 3$, we can adjust our restriction on $x - 3$ accordingly. For example, it will suit our needs perfectly well to make sure that $x + 3$ gets no bigger than 7, and we can accomplish that by requiring x to be within 1 of the value $x = 3$. If we then also make $|x - 3|$ less than $\varepsilon/7$, we'll have what we want.

Here's the proof:

Proof of Theorem 9.9: Suppose $\varepsilon > 0$. Let $\delta = \text{Min}(\varepsilon/7, 1)$ (i.e., the smaller of the two values), so $\delta > 0$, and suppose $0 < |x - 3| < \delta$. We need only prove that $|x^2 - 9| < \varepsilon$.

But $\delta \le 1$, so $2 \le x \le 4$, and so $|x + 3| \le 7$. Also, $\delta \le \varepsilon/7$, so $|x - 3| < \varepsilon/7$.

Therefore, $|x^2 - 9| = |x + 3| \cdot |x - 3| < 7 \cdot \varepsilon/7 = \varepsilon$, completing the proof. ■

COMMENTS

1. Once again, to prove that something exists (namely, a value of δ, in terms of ε, that works), we first explore to find a right value for δ. (There are many "right" values.) Only the results of that exploration are visible in the actual proof.

2. The device of choosing δ to be the smaller of two values is a common method for achieving two conditions at once. Here we combined the need to have $x + 3$ close to 6 with the need for $x - 3$ to be small. (This is similar to our technique of choosing the larger of two N's in working with sequences.)

Get Your Hands Dirty 1

Suppose f is the function from $\mathbf{R}$ to $\mathbf{R}$ defined by $f(x) = x^2 + 5x - 3$. Find $\lim_{x \to 2} f(x)$, and prove your answer. □

Uniqueness of Limits

Many theorems about limits of sequences have analogs for limits of functions. The following uniqueness theorem is the "function version" of Theorem 9.5:

THEOREM 9.10. For a function f defined on $[a, b] - \{c\}$, $\lim_{x \to c} f(x)$ cannot be equal to two different values.

The proof is very similar to that of Theorem 9.5 and is left as Exercise 7.

Arithmetic of Limits

Our next theorem is the analog of Theorems 9.3 and 9.4. First, we need to define formally the arithmetic operations on functions. In the following definition, the domain D can be any set. (It doesn't even have to be a subset of $\mathbf{R}$.)

DEFINITION 9.6: For functions f and g defined on a domain D with codomain $\mathbf{R}$:

the *sum* of f and g (written $f + g$) is the function from D to $\mathbf{R}$ defined by the equation $(f + g)(x) = f(x) + g(x)$.

the *difference, product,* and *quotient* of f and g (written $f - g$, $f \cdot g$, and f/g respectively) are the functions defined, respectively, by the equations

$$(f - g)(x) = f(x) - g(x),$$

$$(f \cdot g)(x) = f(x) \cdot g(x),$$

$$\left(\frac{f}{g}\right)(x) = \frac{f(x)}{g(x)}.$$

Both $f - g$ and $f \cdot g$ have domain D. The domain of f/g consists of those elements x in D for which $g(x) \neq 0$.

Also, for $w \in \mathbf{R}$, the function $w \cdot f$ is defined by the equation $(w \cdot f)(x) = w \cdot f(x)$, and has domain D. The function $w \cdot f$ is called a *scalar multiple* of f.

We can now state the following theorem on the arithmetic of limits:

THEOREM 9.11. Suppose that $w \in \mathbf{R}$ and that f and g are functions defined on $(a, b) - \{c\}$, such that $\lim_{x \to c} f(x) = L_1$ and $\lim_{x \to c} g(x) = L_2$. Then:

i) $\lim\limits_{x \to c}(f + g)(x) = L_1 + L_2,$

ii) $\lim_{x \to c}(f - g)(x) = L_1 - L_2$,

iii) $\lim_{x \to c}(f \cdot g)(x) = L_1 \cdot L_2$,

iv) $\lim_{x \to c}(w \cdot f)(x) = w \cdot L_1$.

Further, if $L_2 \neq 0$, then

v) $\lim_{x \to c}\left(\frac{f}{g}\right)(x) = \frac{L_1}{L_2}$.

The proofs of i), ii), and iv) are very similar to proofs of the analogous results for sequences [Theorem 9.3 and Theorem 9.4, parts i) and iii)], and are left as parts of Exercise 5. We will prove iii); the proof of v) is similar, and is also part of Exercise 5.

Proof of iii). To prove that $\lim_{x \to c}(f \cdot g)(x) = L_1 \cdot L_2$, suppose $\varepsilon > 0$. We want to make the difference $|(f \cdot g)(x) - L_1 \cdot L_2|$ less than ε, by requiring x to be sufficiently close to c.

There is an algebraic trick (see comment 1 below) involved in this proof, namely, we rewrite $f(x) \cdot g(x) - L_1 \cdot L_2$ as

$$f(x) \cdot g(x) - f(x) \cdot L_2 + f(x) \cdot L_2 - L_1 \cdot L_2$$

and express this in turn as

$$f(x)[g(x) - L_2] + L_2[f(x) - L_1]. \qquad (*)$$

Our goal will be to make each of the products in (*) less than $\varepsilon/2$ in absolute value. Both expressions in brackets can be made arbitrarily small by requiring x to be sufficiently close to c. If we also control the first factor $f(x)$ by making it "reasonably close" to L_1 (as we made $x + 3$ "reasonably close" to 6 in Theorem 9.9), we can make the entire expression less than ε.

Here are the details:

If $L_2 \neq 0$, choose δ_1 such that, if $0 < |x - c| < \delta_1$, then $|f(x) - L_1| < \varepsilon/2|L_2|$. (This is possible, since $\lim_{x \to c} f(x) = L_1$, and $\varepsilon/2|L_2|$ is a positive number.) If $L_2 = 0$, let δ_1 be any positive number, say $\delta_1 = 1$. Thus, in either case, if $0 < |x - c| < \delta_1$, then $|L_2| \cdot |f(x) - L_1| < \varepsilon/2$.

Similarly, since $\lim_{x \to c} g(x) = L_2$, we can choose δ_2 such that, if $0 < |x - c| < \delta_2$, then $|g(x) - L_2| < \varepsilon/[2(|L_1| + 1)]$.

Finally, choose δ_3 such that, if $0 < |x - c| < \delta_3$, then $|f(x) - L_1| < 1$. The latter inequality gives $|f(x)| < |L_1| + 1$. Let $\delta = \text{Min}(\delta_1, \delta_2, \delta_3)$, i.e., the smallest of the three values. We will complete the proof by showing that, if $0 < |x - c| < \delta$, then $|(f \cdot g)(x) - L_1 \cdot L_2| < \varepsilon$.

So suppose $0 < |x - c| < \delta$. Because of the way δ was chosen, we have all three of the following inequalities:

$$|L_2| \cdot |f(x) - L_1| < \frac{\varepsilon}{2}$$

$$|g(x) - L_2| < \frac{\varepsilon}{2(|L_1| + 1)}$$

$$|f(x)| < |L_1| + 1$$

Therefore (using the Triangle Inequality to get the second line below),

$$\begin{aligned}
|(f \cdot g)(x) - L_1 \cdot L_2| &= |f(x)[g(x) - L_2] + L_2[f(x) - L_1]| \\
&\leq |f(x)| \cdot |g(x) - L_2| + |L_2| \cdot |f(x) - L_1| \\
&< (|L_1| + 1) \cdot \frac{\varepsilon}{2(|L_1| + 1)} + \frac{\varepsilon}{2} \\
&= \frac{\varepsilon}{2} + \frac{\varepsilon}{2} \\
&= \varepsilon
\end{aligned}$$

as desired. ■

COMMENTS

1. Students often wonder where tricks like the one in this theorem come from. Basically, they come from exploration—that is, from lots of playing around and experimentation, using pages and pages of scratch paper.

2. Notice that in part v) of this theorem, we require $L_2 \neq 0$, but don't require that $g(x)$ be nonzero. We can do this because, if $\lim_{x \to c} g(x) \neq 0$, then we are guaranteed that $g(x) \neq 0$ for values of x close enough to c (see Exercise 8).

The following theorem, whose proof we leave as Exercise 9, connects the concept of "sequential limit" directly with the concept of "function limit":

THEOREM 9.12. For a function f defined on $B = [a, b] - \{c\}$, a sequence $\langle a_n \rangle$ with terms in B, and a real number L,

$$\text{if} \quad \lim_{n \to \infty} a_n = c \text{ and } \lim_{x \to c} f(x) = L, \quad \text{then} \quad \lim_{n \to \infty} f(a_n) = L.$$

The intuitive explanation of this theorem is as follows: we are given that, as n gets bigger, the terms a_n get closer to c. We are also told that, as the terms a_n get closer to c, the values $f(a_n)$ get closer to L. Thus, as n gets bigger, $f(a_n)$ gets closer to L.

Continuity

One of the questions often asked by students when they first encounter the limit concept is: why not just "plug in" $x = c$ to find out what happens when x is close to c? Examples 1 and 2 suggest that we could do just that.

We give two answers to this reasonable question. First of all, there are situations where "plugging in" is simply impossible, because the expression under consideration is not defined for $x = c$. As mentioned earlier, the concept of derivative is an important example of this phenomenon.

Second, even if c is in the domain of the function, there is no guarantee that the value of $f(x)$ will be close to $f(c)$, no matter how close x itself is to c. For example, Figure 9.6 shows the graph of the following function:

$$f(x) = \begin{cases} x + 3 & \text{if } x \neq 2 \\ 1 & \text{if } x = 2 . \end{cases}$$

For this function, as x gets closer to 2, the value $f(x)$ gets close to 5, rather than to $f(2)$, which is 1. Although this example may feel contrived, it is nevertheless a legitimate function, and so we have to be careful about jumping to conclusions.

To distinguish "normal" functions from "abnormal" situations like that of Figure 9.6, we make the following definition:

DEFINITION 9.7: For a function f defined on an interval $D \subseteq \mathbf{R}$, an element $t \in D$, and a set $S \subseteq D$:

"f is *continuous at* t" means:

$$\lim_{x \to t} f(x) = f(t);$$

"f is *continuous on* S" means: f is continuous at t for every $t \in S$.

COMMENTS

1. For a function f to be continuous at t, we really require three things:

   a) $\lim_{x \to t} f(x)$ must exist,
   b) $f(t)$ must be defined, and
   c) the values of $\lim_{x \to t} f(x)$ and $f(t)$ must be equal.

   The definition above puts these three conditions into one equation.

2. The statement "$\lim_{x \to t} f(x) = f(t)$" can be expressed in terms of the quantifier definition of limit, as follows:

$$(\forall\, \varepsilon > 0)(\exists\, \delta > 0)(\forall\, x \in D)(0 < |x - t| < \delta \Rightarrow |f(x) - f(t)| < \varepsilon). \quad (*)$$

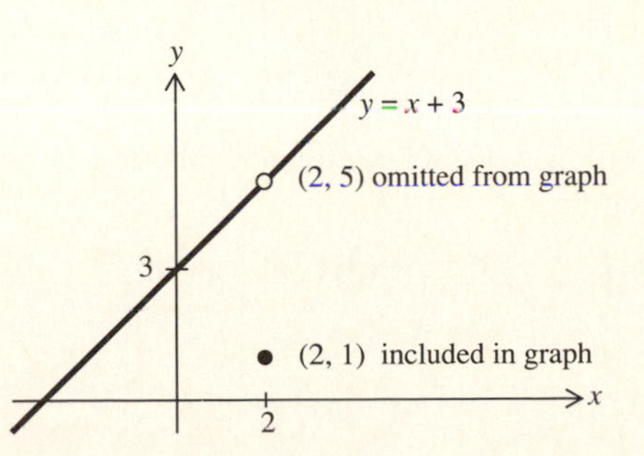

Figure 9.6 One point of the graph of f is not on the line $y = x + 3$.

In the general definition of the expression "$\lim_{x \to t} f(x)$", we don't want to say anything about $f(t)$. Therefore, the hypothesis in $(*)$ says "$0 < |x - t| < \delta$", and not just "$|x - t| < \delta$" (see comment 1 following Definition 9.5 of a function limit, and the example just preceding Definition 9.7 of continuity.)

However, when we are discussing limits in the context of continuity, we don't have that problem. We have a function which is defined at $x = t$, and the conclusion "$|f(x) - f(t)| < \varepsilon$" is always true when $x = t$, since ε is positive. Therefore, we can simplify $(*)$ slightly by dropping the condition "$0 < |x - t|$" from the hypothesis of the implication.

In other words, the condition $(*)$ is equivalent to the following:

$$(\forall\, \varepsilon > 0)(\exists\, \delta > 0)(\forall\, x \in D)(|x - t| < \delta \Rightarrow |f(x) - f(t)| < \varepsilon). \qquad (**)$$

We will use $(**)$ in our discussions below.

The following result is an immediate consequence of this definition, Definition 9.6 ("arithmetic of functions"), and Theorem 9.11 ("arithmetic of limits"):

COROLLARY 9.13. For functions f and g, and $w \in \mathbf{R}$, suppose that f and g are continuous at c. Then $f + g$, $f - g$, $f \cdot g$, and $w \cdot f$ are also continuous at c. If, further, $g(c) \neq 0$, then f/g is continuous at c.

By choosing functions for f and g that we know are continuous, we can use this result to prove that many other functions are continuous. In order to get this process started, we need to take care of the simplest examples first. The following lemma states that constant functions and the identity function are continuous:

LEMMA 9.14. For $w \in \mathbf{R}$, let f and g be the functions from $\mathbf{R}$ to $\mathbf{R}$ defined as follows:

$$f(x) = w \qquad \text{for all} \quad x \in \mathbf{R},$$

$$g(x) = x \qquad \text{for all} \quad x \in \mathbf{R}.$$

Then both f and g are continuous on $\mathbf{R}$.

Proof: Suppose $c \in \mathbf{R}$. We will show that both f and g are continuous at c. We have $f(c) = w$ and $g(c) = c$, so we need to show

$$\lim_{x \to c} f(x) = w \quad \text{and} \quad \lim_{x \to c} g(x) = c.$$

So suppose $\varepsilon > 0$. By Definition 9.5 (and comment 2 above), we have to find positive real numbers δ_1 and δ_2 such that:

$$(\forall\, x \in \mathbf{R})(|x - c| < \delta_1 \Rightarrow |f(x) - w| < \varepsilon) \qquad \text{(i)}$$

and

$$(\forall\, x \in \mathbf{R})(|x - c| < \delta_2 \Rightarrow |g(x) - c| < \varepsilon). \qquad \text{(ii)}$$

The condition (i) is true for any value of δ_1, since its conclusion is true for all x. [Since $f(x) = w$, we have $|f(x) - w| = 0$, and so $|f(x) - w| < \varepsilon$.] To be specific, we will arbitrarily choose $\delta_1 = 17.3$ (just for fun).

For (ii), we only need to be a little more choosy. Using $g(x) = x$, the conclusion "$|g(x) - c| < \varepsilon$" becomes simply "$|x - c| < \varepsilon$". Therefore, we can choose δ_2 to be equal to ε. With that choice, the condition (ii) now says:

$$\text{"if } |x - c| < \varepsilon, \text{ then } |x - c| < \varepsilon\text{"},$$

which is certainly true.

That concludes the proof. ■

We can "build" all *polynomial functions*—that is, all functions of the form $f(x) = a_n x^n + a_{n-1}x^{n-1} + \cdots + a_0$—by applying the arithmetic operations of addition, multiplication, and scalar multiplication (Definition 9.6) repeatedly to the two functions in this lemma. By using division, we can then also get all *rational functions*—i.e., all functions of the form $f(x)/g(x)$, where f and g are polynomials.

Thus the following result is an immediate consequence of Corollary 9.13 and Lemma 9.14:

COROLLARY 9.15. All polynomial functions and all rational functions are continuous at every element of their domain.

Now that we have proved that a large class of familiar functions are continuous, we turn to an important property of continuous functions, which helps to justify their name.

The Intermediate Value Theorem

One of the intuitive descriptions of a continuous function is "a function without any breaks"; i.e., you can draw its graph without "picking up your pencil." Our next theorem establishes the legitimacy of this intuitive picture (see Figure 9.7).

THEOREM 9.16. For a function f defined and continuous on $[a, b]$, suppose that $f(a)$ and $f(b)$ have opposite signs (i.e., one is positive and the other is negative). Then there is a real number t in (a, b) such that $f(t) = 0$.

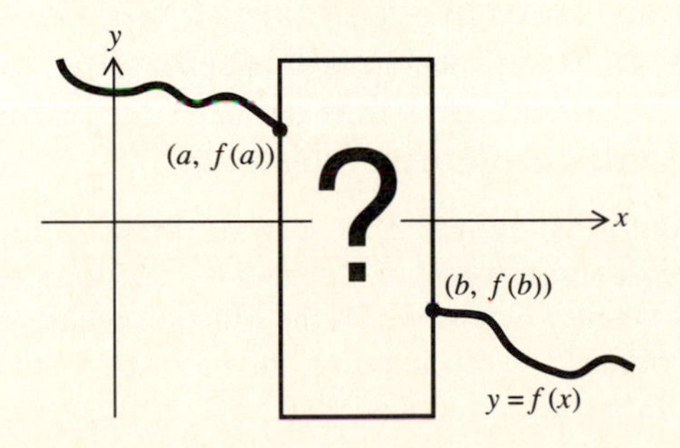

Figure 9.7. If the function f is continuous on $[a, b]$, then it intersects the x-axis somewhere in the missing part of the graph.

Before discussing or proving this result, we state and prove an important consequence, a generalization called the *Intermediate Value Theorem:*

COROLLARY 9.17. For a function g defined and continuous on $[a, b]$, suppose that k is a real number between $g(a)$ and $g(b)$ [i.e., either $g(a) < k < g(b)$ or $g(a) > k > g(b)$]. Then there is a real number t in (a, b) such that $g(t) = k$.

Proof: We prove this by applying Theorem 9.16 to the function f on $[a, b]$ given by the equation $f(x) = g(x) - k$. Thus, $f(a) = g(a) - k$ and $f(b) = g(b) - k$.

If $g(a) < k < g(b)$, then $f(a) < 0$ and $f(b) > 0$. On the other hand, if $g(a) > k > g(b)$, then $f(a) > 0$ and $f(b) < 0$. In either case, $f(a)$ and $f(b)$ have opposite signs.

Thus, we can apply Theorem 9.16 to f. (The fact that f is continuous follows from Lemma 9.14 and Corollary 9.13, and the fact that g is continuous.)

We therefore get a real number t in (a, b) with $f(t) = 0$. Since $g(x) = f(x) + k$, it follows immediately that $g(t) = k$. ■

Discussion—Theorem 9.16 The proof of this result depends on the completeness of **R**—the fact that every subset of **R** which is bounded above has a least upper bound. As discussed in Chapter 4, we are assuming this property of the real numbers, since its proof would require a formal definition of **R** that would take us far afield.

Theorem 9.16 is an existence theorem. It asserts the existence of a real number, t, with a certain property—namely, $f(t) = 0$. In order to prove this theorem, we need to be able to locate, or identify, the right value for t. As usual, this requires intuition and exploration.

In order to be specific, we will assume that $f(a)$ is negative and $f(b)$ is positive. The proof of the opposite case is virtually identical [or can be done by considering the function $(-1) \cdot f$].

Before reading any further, think about the following question:

How can we "identify" or "point to" a real number *t* where the function *f* will be zero? (*Hint:* Describe *t* as the lub of some set.)

Theorem 9.16 is one of the most advanced results in this book, and its proof is not simple. If you haven't already come up with an answer to the last question, here is one more hint:

The function *f* may change sign many times between *x* = *a* and *x* = *b*. How can you describe the value of *x* where it becomes positive for the last time? (Drawing a picture may help.)

The essential idea of the proof is as follows: if we let S be the set of elements in $[a, b]$ where f is negative, then the least upper bound of S should be a "border" between values where f is negative and values where f is positive. If the function really "has no breaks", then f should be zero at this "border". We'll interrupt the proof of Theorem 9.26 as we go along, to ask you to think through the individual steps.

Before beginning, we take care of one detail on inequalities (see Figure 9.8), so that we don't have to interrupt the proof.

LEMMA 9.18. For $a, b \in \mathbf{R}$, $b \neq 0$, if $|a - b| < |b|/2$, then a has the same sign as b.

Proof: Assume $|a - b| < |b|/2$, so $-|b|/2 < a - b < |b|/2$. If $b > 0$, we have $-b/2 < a - b$, so $a > b + (-b/2) = b/2 > 0$, as desired. The case where $b < 0$ is similar. ■

We now begin the formal proof of Theorem 9.16:

Proof of Theorem 9.16: Assume that $f(a) < 0$ and $f(b) > 0$. [If the opposite case is true, then apply this proof to the function $(-1)f$, and get the desired conclusion.]

Define the set S as follows:

$$S = \{x \in [a, b]: f(x) \leq 0\}.$$

Since $S \subseteq [a, b]$, b is an upper bound for S. Therefore S has a least upper bound, which we will call t.

We will prove, by contradiction, that $f(t) = 0$. More precisely, we will show that, if $f(t) \neq 0$, then t is not the least upper bound for S.

We have identified what should be a successful value for t. We now need to prove that $f(t) = 0$.

Why is this true?

What happens if $f(t) \neq 0$? Consider the two cases: $f(t) > 0$ and $f(t) < 0$. Why can't either of these be true? How would that contradict the definition of t?

? ? ?

We will show that, if $f(t) < 0$, then t is not an upper bound for S, and if $f(t) > 0$, then t is not the *least* upper bound for S.

So suppose $f(t) \neq 0$. There are two cases: either $f(t) < 0$ or $f(t) > 0$. We consider them separately:

Case 1: $f(t) < 0$.

How can you show that this is impossible? How can you use continuity to find a number w greater than t for which $f(w) < 0$? (Examine Figure 9.9.)

Let $\varepsilon = |f(t)|/2$. Since f is continuous, there is a positive number δ such that, if $|x - t| < \delta$, then $|f(x) - f(t)| < \varepsilon$. Since $f(t) < 0$, we know $t \neq b$. Therefore we

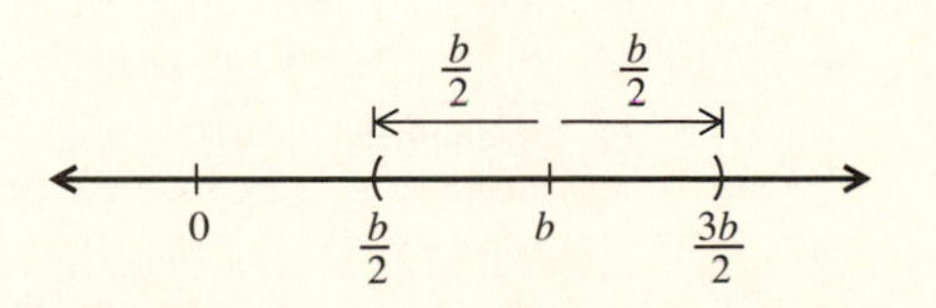

Figure 9.8. The case "$b > 0$" of Lemma 9.18. If $|a - b| < |b|/2$, then a must be between $b - b/2$ and $b + b/2$, i.e., between $b/2$ and $3b/2$.

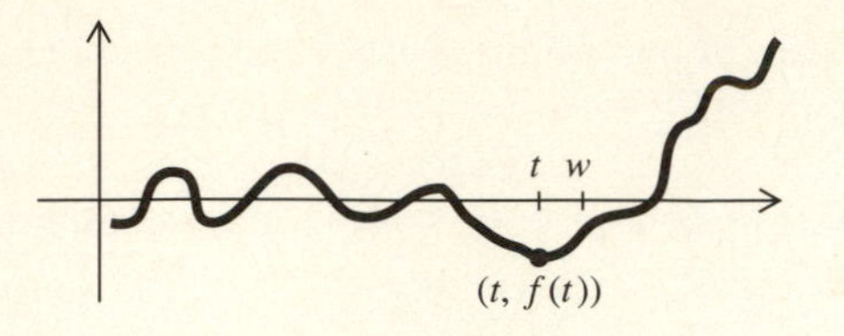

Figure 9.9. If $f(t) < 0$, and f is continuous, there should be a number w greater than t for which $f(w) < 0$, i.e., with $w \in S$.

can find a value w in $[a, b]$ with $t < w < t + \delta$, and so $|w - t| < \delta$. Thus $|f(w) - f(t)| < |f(t)|/2$. It follows from Lemma 9.18 that $f(w)$ has the same sign as $f(t)$, i.e., $f(w) < 0$, and so $w \in S$. This contradicts the assumption that t is an upper bound for S. Therefore case 1 is impossible.

Case 2: $f(t) > 0$.

How can you show that this is impossible?
How can you use continuity to find a number u less than t which is also an upper bound for S? Examine Figure 9.10.

Let $\varepsilon = |f(t)|/2$. Since f is continuous, there is a positive number δ such that, if $|x - t| < \delta$, then $|f(x) - f(t)| < \varepsilon$. Since $f(t) > 0$, we know $t \neq a$. Therefore we can find a value u in $[a, b]$ with $t - \delta < u < t$.

(Notice how similar this is to case 1.)

Then, for any $x \in [u, t]$, we have $|x - t| < \delta$, so $|f(x) - f(t)| < |f(t)|/2$. It follows from Lemma 9.18 that, for all $x \in [u, t]$, $f(x)$ has the same sign as $f(t)$, i.e., $f(x) > 0$. In other words, we have $S \cap [u, t] = \varnothing$. Since t is an upper bound for S, and since we have just shown that there are no elements of S between u and t, we thus also have that u is an upper bound for S. Since $u < t$, this contradicts the assumption that t is the *least* upper bound so case 2 is also impossible.

Thus, we must have $f(t) = 0$, and the theorem is proved. ■

EXERCISES

1. Prove $\lim_{x \to 4} (3x^2 - 7x - 5) = 15$ directly—i.e., find an appropriate δ in terms of ε, as was done in Example 1. Do not use the theorems of this section in your proof.

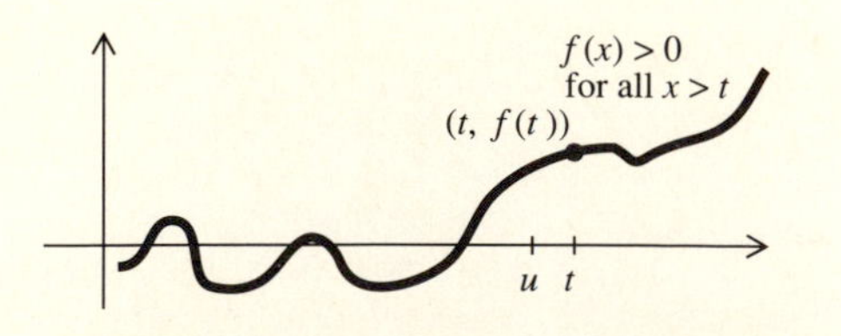

Figure 9.10. If $f(t) > 0$, and f is continuous, there should be a number u less than t which is an upper bound for S.

2. Define the function f as follows:

$$f(x) = \begin{cases} \sin x & \text{if } x \text{ is rational} \\ 0 & \text{if } x \text{ is irrational}. \end{cases}$$

Find $\lim_{x\to 0} f(x)$ and $\lim_{x\to\pi/2} f(x)$, and justify your answers. [If either limit does not exist, say so. You may assume that the function $g(x) = \sin x$ is continuous.]

3. Find each of the following limits (if the limit does not exist, say so, and explain why not):

   a) $\lim\limits_{x\to 0} \sin\left(\dfrac{1}{x}\right)$,

   b) $\lim\limits_{x\to 0} x^2 \sin\left(\dfrac{1}{x}\right)$,

   c) $\lim\limits_{x\to 1} \dfrac{\sqrt{x} - 1}{x - 1}$.

4. Consider the circle shown in Figure 9.11, in which radius OA has length 1.

   By comparing areas of triangle OAD, sector OAD, and triangle OAB, we have the following compound inequality:

$$\frac{\sin x}{2} < \frac{x}{2} < \frac{\tan x}{2}. \qquad (*)$$

   [Triangle OAD has base OA of length 1, and altitude CD of length $\sin x$, so its area is $(1 \cdot \sin x)/2$. The sector represents the fraction $x/2\pi$ of the area of the circle, so its area is $(x/2\pi)(\pi \cdot 1^2)$. Triangle OAB has base OA and altitude AB, which has length $\tan x$, so its area is $(1 \cdot \tan x)/2$.]

   a) Use $(*)$ to show that $\cos x < (\sin x)/x < 1$.

   b) Use part a) to prove that $\lim_{x\to 0} (\sin x)/x = 1$. (You may assume that $\lim_{x\to 0} \cos x = 1$. Use that assumption as a guarantee of "good δ's".)

5. Prove the following parts of Theorem 9.11 [for part v), use a trick similar to that shown in the text for part iii)]:

THEOREM 9.11. Suppose $w \in \mathbf{R}$ and that f and g are functions defined on $(a, b) - \{c\}$, such that $\lim_{x\to c} f(x) = L_1$ and $\lim_{x\to c} g(x) = L_2$.
Then:

i) $\lim\limits_{x\to c} (f + g)(x) = L_1 + L_2$,

ii) $\lim\limits_{x\to c} (f - g)(x) = L_1 - L_2$,

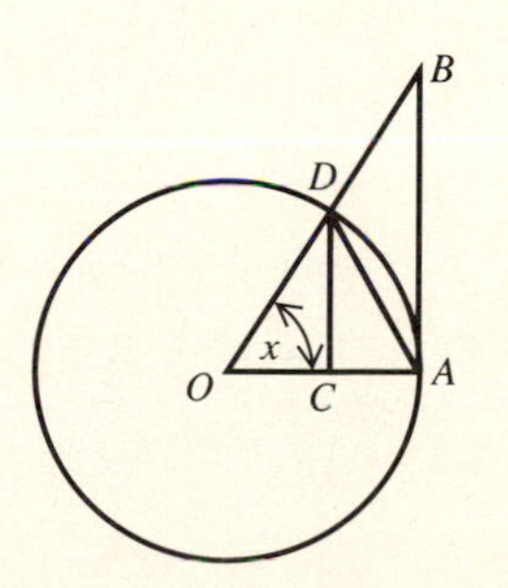

Figure 9.11. The circle has radius $OA = 1$, and central angle x as shown.

iv) $\lim_{x \to c} (wf)(x) = wL_1$.

Further, if $L_2 \neq 0$, then

v) $\lim_{x \to c} \left(\frac{f}{g}\right)(x) = \frac{L_1}{L_2}$.

6. Prove the following parts of Theorem 9.4 (from Section 9.2) [for ii), use the proof of part iii) of Theorem 9.11 as a model. For iv), use a similar algebraic trick]:

THEOREM 9.4. For sequences $\langle a_n \rangle$ and $\langle b_n \rangle$, suppose

$$\lim_{n \to \infty} a_n = L_1 \quad \text{and} \quad \lim_{n \to \infty} b_n = L_2$$

Then

ii) $\lim_{n \to \infty} (a_n \cdot b_n) = L_1 \cdot L_2$.

If, additionally, $\langle b_n \rangle$ has no terms equal to zero and $L_2 \neq 0$, then:

iv) $\lim_{n \to \infty} \left(\frac{a_n}{b_n}\right) = \frac{L_1}{L_2}$.

7. Prove:

THEOREM 9.10. For a function f defined on $[a, b] - \{c\}$, $\lim_{x \to c} f(x)$ cannot be equal to two distinct values.

8. Suppose $\lim_{x \to c} g(x) = L$, where $L \neq 0$. Prove:

$$(\exists\, \delta > 0)(\forall x \in A)(0 < |x - c| < \delta \Rightarrow g(x) \neq 0)$$

(*Hint:* Consider $L > 0$ and $L < 0$ as separate cases. This result is analogous to Exercise 9 of Section 9.2.)

9. Prove the following:

THEOREM 9.12. For a function f defined on $B = [a, b] - \{c\}$, a sequence $\langle a_n \rangle$ with terms in B, and a real number L,

$$\text{if} \quad \lim_{n \to \infty} a_n = c \text{ and } \lim_{x \to c} f(x) = L, \quad \text{then} \quad \lim_{n \to \infty} f(a_n) = L.$$

10. The following definition is analogous to that given in Exercise 17 of Section 9.2:

DEFINITION. For a set B such that $(a, b) - \{c\} \subseteq B$, a function f defined on B, and $L \in \mathbf{R}$, "the limit of $f(x)$, as x approaches c, is $+\infty$", written $\lim_{x \to c} f(x) = +\infty$", means:

$$(\forall N > 0)(\exists\, \delta > 0)(\forall x \in B)(0 < |x - c| < \delta \Rightarrow f(x) > N).$$

a) Prove: $\lim_{x \to 0} \frac{1}{x^2} = +\infty$.

b) Explain why $\lim_{x \to 0} \frac{1}{x} \neq +\infty$.

11. The following definition is similar to the definition of limit of a sequence:

DEFINITION. For a function f defined on $S = (a, \infty)$, and $L \in \mathbf{R}$, "the limit of $f(x)$, as x approaches ∞, is L", written "$\lim_{x \to \infty} f(x) = L$", means:

$$(\forall \varepsilon > 0)(\exists N > 0)(\forall x \in S)(x > N \Rightarrow |f(x) - L| < \varepsilon).$$

a) Prove that $\lim_{x\to\infty} x/(2x+1) = \frac{1}{2}$.
b) Prove that, if f is defined on $(0,\infty)$ and $\lim_{x\to\infty} f(x) = L$, then the sequence $\langle f(n)\rangle$ also approaches L.
c) Give an example of a function f defined on $(0,\infty)$ such that the sequence $\langle f(n)\rangle$ approaches some limit L, but the limit of $f(x)$, as x approaches ∞, is not L. [Part b) shows that $\lim_{x\to\infty} f(x)$ cannot be anything other than L, so, in your example, $\lim_{x\to\infty} f(x)$ will not exist.]

12. Prove that if f is continuous at c, then $|f|$ is continuous at c. [$|f|$ is the function defined by "$|f|(x) = |f(x)|$".]

13. Suppose that f is a continuous function defined on $[a,b]$. Prove that $\mathrm{Im}(f)$ is an interval. (Use the formal definition of "interval" given in Exercise 15 of Section 3.2.)

14. Suppose that f is a continuous, injective function defined on $[a,b]$. Prove that f is either strictly increasing or strictly decreasing. [A function f on a set D is strictly increasing iff $(\forall x, y \in D)(x < y \Rightarrow f(x) < f(y))$. For the definition of strictly decreasing, the last inequality is reversed.] (*Hint:* Use the Intermediate Value Theorem.)

9.4 Uniform Continuity and Cauchy Sequences

We conclude this chapter with a discussion of two related but independent topics. The first applies to functions, and is a variation of the concept of continuity, called *uniform continuity*. "Ordinary" continuity is a concept defined individually at each value t in the domain. A function f is continuous at t if numbers "close enough" to t have images that are "close to $f(t)$". Uniform continuity requires that the "close enough" be independent of t. Our main theorem is that any function which is continuous at each point in a closed interval must be uniformly continuous.

There is a somewhat similar concept for sequences, called *Cauchy sequences*. In Section 9.2, we defined a sequence to be *convergent* if the terms "get close to" some specific limit. In a Cauchy sequence, we simply require the terms to "get close to *each other*". We will prove that every Cauchy sequence is actually convergent.

With both topics, our discussion uses the *completeness property* of the real numbers—the property that every nonempty subset of **R** with an upper bound must have a least upper bound. This property has been one of our basic assumptions about **R**. Its proof requires a formal definition of the set of real numbers, which we have not given in this book.

Topic 1: Uniform Continuity

Recall that, if a function is defined on an interval D, then f is *continuous on D* iff f is continuous at t for every $t \in D$; i.e., if $\lim_{x\to t} f(x) = f(t)$ for every $t \in D$. If we express this in terms of the quantifier definition of limit, we get the following: f is continous on D iff

$$(\forall\, t \in D)(\forall\, \varepsilon > 0)(\exists\, \delta > 0)(\forall\, x \in D)(|x - t| < \delta \Rightarrow |f(x) - f(t)| < \varepsilon)\,. \quad (*)$$

(See comment 2 following Definition 9.7 of continuity in Section 9.3.)

The statement $(*)$ is rather complex, involving four quantifiers. We are going to introduce a new concept, called *uniform continuity,* by simply changing the order of the quantifier expressions in $(*)$. Specifically, we will move the expression ("$\forall\, t \in D$)" so that it comes *after* the expression "($\exists\, \delta > 0$)". This will mean that δ cannot be chosen in terms of t.

Here is the definition:

DEFINITION 9.8: For a function f defined on an interval D, "f is *uniformly continuous* on D" means:

$$(\forall\, \varepsilon > 0)(\exists\, \delta > 0)(\forall\, t \in D)(\forall\, x \in D)(|x - t| < \delta \Rightarrow |f(x) - f(t)| < \varepsilon)\,.$$

The condition of uniform continuity says, intuitively, no matter how small ε is, there is a "closeness" $\delta > 0$ such that any two elements of D that are within δ of each other have images under f which are within ε of each other.

The following theorem asserts that uniform continuity is, indeed, a special kind of continuity.

THEOREM 9.19. For a function f defined on an interval D, if f is uniformly continuous on D, then f is continuous on D.

The proof, which we leave as Exercise 1, follows from general principles about rearranging quantifiers. Essentially, the proof consists of pointing out that, if we can choose δ without knowing t, we can certainly choose δ if we do know t.

However, the converse of Theorem 9.19 is not true. A function can be continuous on an interval without being uniformly continuous. Example 1 below illustrates this phenomenon. However, before reading this example, do the following:

Get Your Hands Dirty 1

Write the statement "f is not uniformly continuous on D" in terms of quantifiers. □

Example 1 **A Continuous Function Which Is Not Uniformly Continuous** Consider the function f defined on the open interval $(0, 1)$ by the equation $f(x) = 1/x$.

Since f is a rational function, we know (Corollary 9.15) that f is continuous on $(0, 1)$. We will, however, prove the following:

THEOREM 9.20. The function f defined on $(0, 1)$ by the equation $f(x) = 1/x$ is not uniformly continuous on $(0, 1)$.

Proof: For f to *not* be uniformly continuous on D means the following:

$$(\exists\ \varepsilon > 0)(\forall\ \delta > 0)(\exists\ t \in D)(\exists\ x \in D)$$
$$(|x - t| < \delta \text{ and } |f(x) - f(t)| \geq \varepsilon)\,. \qquad (**)$$

(We have formally negated the definition of uniform continuity by changing all the quantifiers and negating the predicate. The negation of an implication is the statement that its hypothesis is true and its conclusion is false.)

We claim that $\varepsilon = 1$ "works". (Actually, any positive number will work, with appropriate changes needed in the proof.)

To prove this, we have to show something about "every positive δ", so suppose $\delta > 0$. We need to find real numbers t and x such that $|x - t| < \delta$ and $|f(x) - f(t)| \geq 1$.

To take care of certain inequalities later on, we set $\delta^* = \text{Min}(\delta, 1)$, so $\delta^* \leq \delta$ and $0 < \delta^* \leq 1$. We therefore know that, if $|x - t| < \delta^*$, then $|x - t| < \delta$. We will find t and x so that $|x - t| < \delta^*$ and $|f(x) - f(t)| \geq 1$.

Let $x = \delta^*/2$ and $t = \delta^*/4$, so $|x - t| = \delta^*/4 < \delta^*$. To complete the proof, we only need to show that $|f(x) - f(t)| \geq 1$.

But

$$|f(x) - f(t)| = \left|\frac{1}{x} - \frac{1}{t}\right| = \left|\frac{t - x}{xt}\right| = \frac{\delta^*/4}{[\delta^*]^2/8} = \frac{2}{\delta^*} \geq 2 > 1$$

as desired. ∎

COMMENTS

1. You may wonder where we came up with the choices for x and t. Of course, we found them by exploration. We began by playing with the conclusion, or more specifically, the expression $|f(x) - f(t)|$. It seemed like a good guess to try "fractions of δ" for x and t, and a little algebra helped us find denominators which would make $|f(x) - f(t)| \geq 1$.

2. Intuitively, the reason why f is not uniformly continuous on $(0, 1)$ is that its graph gets steeper and steeper, as x approaches 0, so that the value $f(x)$ changes very rapidly for very small changes in x, when x is close to 0. We will see in Example 2 that if we look at an interval "away from 0", the function is uniformly continuous.

Although Example 1 showed a function which was not uniformly continuous, the property of uniform continuity is not impossible to fulfill. We will show that, if we restrict the set D to be a (finite) *closed* interval, then the converse of Theorem 9.19 is true. That is, we will prove the following:

THEOREM 9.21. For a function f defined on a closed interval $D = [a, b]$, if f is continuous on D, then f is uniformly continuous on D.

The proof of Theorem 9.21 depends on the *completeness* of **R**—the property which says that every nonempty subset of **R** that is bounded above has a least upper bound. As earlier in the text, we will assume this property. Its proof requires a formal definition of the real numbers, which is beyond the scope of this book.

Before looking at the proof of Theorem 9.21, we will give an example of how to prove that a specific function *is* uniformly continuous.

Example 2 **The Function $f(x) = 1/x$ Is Uniformly Continuous on $[1,2]$** We will prove the following theorem:

THEOREM 9.22. The function f defined on $[1, 2]$ by the equation $f(x) = 1/x$ is uniformly continuous on $[1, 2]$.

Proof: Suppose $\varepsilon > 0$. Let $\delta = \varepsilon$. We claim that this choice of δ works, i.e., that

$$(\forall\, t \in [1, 2])\,(\forall\, x \in [1, 2])\,(|x - t| < \delta \Rightarrow |f(x) - f(t)| < \varepsilon)\,.$$

Suppose x and t satisfy the hypothesis, i.e., x and t are elements of $[1, 2]$, and $|x - t| < \delta$. Then (as in Theorem 9.20), $|f(x) - f(t)| = |1/x - 1/t| = |(t - x)/xt|$. But x and t are both greater than or equal to 1, so $|(t - x)/xt| \le |x - t|$. Since we know that $|x - t| < \delta$, and $\delta = \varepsilon$, we can conclude that $|(t - x)/xt| < \varepsilon$, as desired. ■

COMMENT The key to the algebra of this proof is keeping x and t away from 0. The function $f(x) = 1/x$ is not defined at 0 (i.e., 0 is not in its domain), and the value $f(x)$ gets arbitrarily large as x gets close to 0. That is the cause of the difficulty when we consider the interval $(0, 1)$ instead of $[1, 2]$.

Get Your Hands Dirty 2

Prove that the function given by the equation $f(x) = 1/x$ is uniformly continuous on each of the following intervals:

a) $\left[\frac{1}{2}, 2\right]$,

b) $\left(\frac{1}{8}, 2\right)$,

c) $\left[\frac{1}{10}, 100\right]$. □

Uniform Continuity on a Closed Interval—Discussion

We now turn to Theorem 9.21. This is a very deep theorem, and its proof involves the introduction of an important concept about the real numbers. We would like to motivate this concept by a discussion of the theorem.

We begin the proof as follows:

Proof: Suppose f is continuous on D, and $\varepsilon > 0$.

The theorem then requires us to find a positive real number for δ such that, whenever two numbers x and t in D are within δ, their images under f, $f(x)$ and $f(t)$, are within ε.

Intuitively, we want values in the domain that are "close" to give images that are "close".

The fact that f is "plain" continuous on D tells us that, if we fix t, we can be assured that values of x close to t have images close to $f(t)$. More specifically, there is a positive real number δ_t such that, if $|x - t| < \delta_t$ (and $x \in D$—see caution below), then $|f(x) - f(t)| < \varepsilon$. In other words, for each $t \in D$, there is an open interval $(t - \delta_t, t + \delta_t)$ such that the condition "$|f(x) - f(t)| < \varepsilon$" holds for all $x \in D$ that belong to the open interval.

The main problem is that there are infinitely many of these intervals, each with a "radius" δ_t that depends on t. If there were only a finite number of different values for δ_t under consideration, we could just take the smallest of them, and use that for δ.

But there are infinitely many values of t under consideration, so there are also infinitely many values of δ_t. Although these δ_t's are all positive, there may not be a positive number δ which is less than all of them.

The main work of the proof is to show that we can restrict our consideration to only finitely many intervals.

CAUTION We have to exercise some caution that we don't spill over the boundaries of the interval D in which f is defined. Keep in mind, therefore, that whenever we talk about $f(x)$ or $f(t)$, the elements x and t are assumed to be in D.

The proof of Theorem 9.21 has two basic parts: first, we show that we can always replace the set of infinitely many open intervals by finitely many; second, we show that the ability to use finitely many open intervals allows us to prove uniform continuity.

The Heine–Borel Theorem

The first part of the proof is a classic theorem of real analysis, called the *Heine–Borel Theorem:*

THEOREM 9.23: (Heine–Borel Theorem). Suppose D is a closed interval $[a, b]$, and that S is a set of open intervals, such that every element of D belongs to some element of S.

Then there is a finite subset T of S such that every element of D belongs to some element of T.

Note that S is a set whose elements are sets—that is, the elements of S are certain subsets of $\mathbf{R}$. The set of intervals described in our discussion above has precisely the property given for S in the hypothesis of this theorem. This condition on S is expressed by the following terminology:

DEFINITION 9.9: For a set $D \subseteq \mathbf{R}$ and a subset S of $\mathcal{P}(\mathbf{R})$:

"S is a *covering* of D" means: every element of D belongs to some element of S.

"S is an *open covering* of D" means: S is a covering of D whose elements are open intervals.

(**Note:** The term "open covering" usually is used to include "open sets" more general than intervals. We do not need the more general concept here.)

To describe the role of T, we have the following terminology:

DEFINITION 9.10: For a set D, a covering S of D, and another subset T of $\mathcal{P}(D)$, "T is a *subcovering* of S (for D)" means:

a) T is a subset of S, and
b) T is itself a covering of D.

Condition a) here simply says that T is obtained by using *some* of the sets that belong to S. In particular, if the elements of S are all open intervals, then so are the elements of T; i.e., a subcovering of an open covering is itself an open covering.

We can express Theorem 9.23 more succinctly using this terminology;

THEOREM 9.23′: (Heine–Borel Theorem). For a closed interval $D = [a, b]$, any open covering of D has a finite subcovering.

Our earlier discussion essentially showed the existence of certain kinds of open coverings of D. Before proving the Heine–Borel Theorem, we show how it can be used to prove Theorem 9.21:

Proof of Theorem 9.21: Suppose f is continuous on $D = [a, b]$, and that $\varepsilon > 0$. Then $\varepsilon/2$ is also positive, and so, for each $t \in D$, there is a positive number δ_t such that, if $|x - t| < \delta_t$ (and $x \in D$), then $|f(x) - f(t)| < \varepsilon/2$. (We will see below why we switch to $\varepsilon/2$.)

Let S_t be the open interval defined as follows:

$$S_t = \left(t - \frac{\delta_t}{2}, t + \frac{\delta_t}{2}\right)$$

(Again, we will see below why we switch.)

Let $S = \{S_t : t \in D\}$. For each $t \in D$, there is at least one element of S to which t belongs, namely S_t, so we know that S is an open covering of D. Therefore, by the Heine–Borel Theorem, there is a finite subcovering T of S. In other words, there is a finite set $\{t_1, \ldots, t_n\}$, with corresponding "radii" $\delta_{t_1}/2, \delta_{t_2}/2, \ldots, \delta_{t_n}/2$, such that every element of D belongs to one of the sets $S_{t_1}, S_{t_2}, \ldots, S_{t_n}$.

Let $\delta = \text{Min}(\delta_{t_1}/2, \delta_{t_2}/2, \ldots, \delta_{t_n}/2)$. We claim that this works to show uniform continuity.

For suppose $x, y \in D$, with $|x - y| < \delta$. We need to show that $|f(x) - f(y)| < \varepsilon$.

Since $y \in D$, we know that y belongs to at least one of the sets $S_{t_1}, S_{t_2}, \ldots, S_{t_n}$, say S_{t_i}. Therefore $|y - t_i| < \delta_{t_i}/2$. Also, $|x - y| < \delta$, and $\delta \leq \delta_{t_i}/2$, so (by the Triangle Inequality), $|x - t_i| < \delta_{t_i}$. In other words, both x and y are within δ_{t_i} of t_i. (This is why we used $\delta_{t_i}/2$ instead of δ_{t_i} in defining δ.)

Therefore, because of the way each δ_t was defined in the opening paragraph, we have that $f(x)$ and $f(y)$ are both within $\varepsilon/2$ of $f(t)$. Hence, (again using the Triangle Inequality), $f(x)$ and $f(y)$ are within ε of each other.

That completes the proof. ■

COMMENT Continuity only tells us about images under f for numbers close to one of the t_i. For uniform continuity, we need to know about numbers close to each other. By halving the width of the intervals, we were able to show that, if x and y are "close enough to each other" and one of them is "close enough to t_i", then they are both "close enough to t_i". We could then conclude that $f(x)$ and $f(y)$ were both "close to $f(t)$". By starting out using $\varepsilon/2$, we could then conclude that $f(x)$ and $f(y)$ were within ε of each other. All of this careful detail is, of course, the result of exploration.

We now turn to the proof of the Heine–Borel Theorem, where we will use the completeness of **R**:

Proof of Theorem 9.23 (the Heine-Borel Theorem): Suppose D is the closed interval $[a, b]$, and that S is an open covering of D. We need to show that S has a finite subcovering.

We begin by defining a set $D^* \subseteq D$, as follows:

$$D^* = \{t \in D\text{: there is a finite subcovering of } S \text{ for the interval } [a, t]\}.$$

Note: If $v \in D$, and there is some element $w \in D^*$ that is greater than v, then we also have $v \in D^*$ (since the finite subcovering for $[a, w]$ is also a finite subcovering for $[a, v]$). Thus, D^* is like an "initial segment" of D.

We can express the conclusion of the theorem succinctly in terms of D^*: it says simply "$b \in D^*$". We will prove this by showing that $D^* = D$.

Comment: To show $D^* = D$, we will use a method similar to induction, in which we first show $a \in D^*$, and then, as our "induction step", show we can always "enlarge" D^*. A slightly different version of this "Principle of Continuous Induction" is described in Exercise 4.

We begin by showing $a \in D^*$. Since $a \in D$, there is an open interval $X \in S$ such that $a \in X$. Therefore, $\{X\}$ is a finite subcovering of S for $[a, a]$. In fact, if $X = (r, s)$, then $\{X\}$ is a finite subcovering of S for any interval of the form $[a, w]$ such that $a < w < s$.

Thus $D^* \neq \varnothing$. Since $D^* \subseteq D$, we know that D^* is bounded above (b is an upper bound), and so, by the completeness of **R**, we know that D^* has a least upper bound. Set $c = \text{lub}(D^*)$, so $c \leq b$.

We will first show that $[a, c) \subseteq D^*$, then show that c itself is in D^*, and finally show $c = b$, completing the proof.

To show that $[a, c) \subseteq D^*$, suppose $d \in [a, c)$, so that $d < c$. Then d is not an upper bound for D^* (since c is the *least* upper bound of D^*). Therefore, there is some element of D^* which is greater than d, and so d itself must also be in D^* (see note above). Thus, we have $[a, c) \subseteq D^*$.

We next show that c itself belongs to D^*.

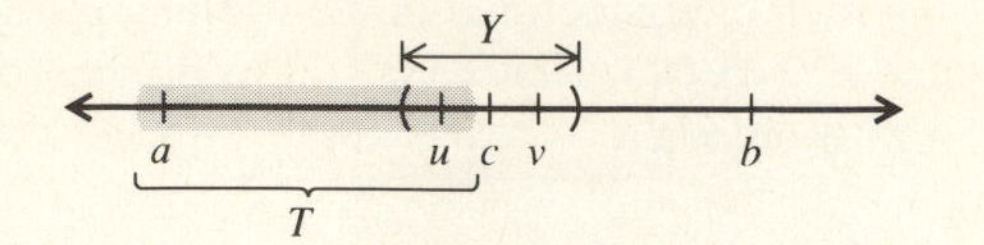

Figure 9.12. The shading represents a subcovering T for $[a, u]$, which can be combined with the interval Y to give a subcovering for $[a, c]$. In fact, if $v \in (c, b] \cap Y$, then $T \cup \{Y\}$ is a subcovering for $[a, v]$.

Since $c \in D$, and S is an open covering of D, there is some open interval $Y \in S$ such that $c \in Y$ (see Figure 9.12). Then $Y \cap [a, c) \neq \emptyset$, so there is some element $u \in Y \cap D^*$. Since $u \in D^*$, there is a subcovering T for $[a, u]$. But as Figure 9.12 shows, $T \cup \{Y\}$ is then a finite subcovering of S for the interval $[a, c]$, because any element of $[a, u]$ belongs to some interval in T, and any element of $[u, c]$ belongs to Y. In other words, $c \in D^*$.

We now claim $c = b$. For if $c < b$, then there is some number $v \in (c, b] \cap Y$. But then, as Figure 9.12 shows, $T \cup \{Y\}$ is a subcovering for $[a, v]$, since elements of $[a, u]$ belong to intervals in T, and elements of $[u, v]$ belong to Y. Therefore $v \in D^*$, contradicting the fact that c is an upper bound for D^*.

Therefore, $c = b$, and so $b \in D^*$, proving the theorem. ■

SUMMARY Once we defined D^* and its least upper bound c, the proof went as follows: since c is the lub of D^*, there are elements of D^* arbitrarily close to c; i.e., $[a, c) \subseteq D^*$. Therefore, any interval Y which contains c must also contain some element u of D^*. We then combine a subcovering T for $[a, u]$ with $\{Y\}$ to get a subcovering for $[a, c]$, showing $c \in D^*$.

We observed, moreover, that, if $c < b$, then $T \cup \{Y\}$ would actually be a subcovering for some interval $[a, v]$, with $v > c$. This would mean $v \in D^*$, contradicting the fact that c is an upper bound for D^*. Therefore, $c = b$, and so $b \in D^*$, as desired.

Topic 2: Cauchy Sequences

There is a concept for sequences which is somewhat analogous to the concept of uniform continuity for functions. Uniform continuity replaces the hypothesis "x is close to t", for a particular t, by the hypothesis "x and y are close to each other", and replaces the conclusion "$f(x)$ is close to $f(t)$" by the conclusion "$f(x)$ and $f(y)$ are close to each other".

The definition of limit for sequences uses the hypothesis "$m > N$" and has the conclusion "a_m is close to L". The following definition instead uses the hypothesis "$m, n > N$" and has the conclusion "a_m is close to a_n".

DEFINITION 9.11: For a sequence $\langle a_n \rangle$,

"$\langle a_n \rangle$ is a *Cauchy sequence*" means:

$$(\forall\, \varepsilon > 0)\,(\exists\, N \in \mathbf{N})\,(\forall\, m, n \in \mathbf{N})\,(m, n > N \Rightarrow |a_m - a_n| < \varepsilon)\,.$$

Historical Note: The French mathematician Augustin Cauchy (1789–1857) was one of the most prolific mathematicians of all time. He made contributions to virtually every branch of mathematics, as well as to the sciences of physics and astronomy. In addition to the term "Cauchy sequence", his name is associated with a variety of his famous contributions, including the Cauchy integral formula and the Cauchy–Riemann equations, in complex analysis; Cauchy's Theorem, concerning finite groups; and the Cauchy–Schwarz inequality, in the study of inner products.

The following theorem is fairly easy to prove, using the Triangle Inequality (recall that a *convergent sequence* is simply a sequence which has a limit):

THEOREM 9.24. Any convergent sequence is a Cauchy sequence.

For the proof, simply make both a_m and a_n within $\varepsilon/2$ of the limit, and they will be within ε of each other. Details are left to the reader as Exercise 5.

The converse of Theorem 9.24 is more interesting, and its proof, like that of Theorem 9.21, uses the completeness of **R**. It says the following:

THEOREM 9.25. Every Cauchy sequence is convergent.

We will adopt a different philosophy of presentation for this theorem from that used in discussing uniform continuity. In that discussion, we introduced the sophisticated concept of an open covering, and proved a powerful general theorem, the Heine–Borel Theorem. Here, we will take a more concrete approach, in which we use the concepts of least upper bound and greatest lower bound to construct the limit whose existence is asserted in Theorem 9.25.

Before proving Theorem 9.25, we prove the following lemma, which gets us part of the way toward our proof:

LEMMA 9.26. If $\langle a_n \rangle$ is a Cauchy sequence, then $\{a_n : n \in \mathbf{N}\}$ is bounded above and bounded below.

Proof. Let $\varepsilon = 1$. By Definition 9.11, there is some integer N such that, for all $m, n > N$, we have $|a_m - a_n| < 1$. In particular, using $n = N + 1$, we have that $|a_m - a_{N+1}| < 1$ for all $m > N$, so that $a_{N+1} - 1 < a_m < a_{N+1} + 1$.

In other words, $a_{N+1} - 1$ and $a_{N+1} + 1$ are, respectively, a lower bound and an upper bound for $\{a_t : t > N\}$, which is the "tail" of the sequence. We only need to take care of the earlier elements of the sequence.

Let u and v be, respectively, a lower bound and an upper bound for the finite set $\{a_t : t \leq N\}$. Then $\text{Min}(u, a_{N+1} - 1)$ and $\text{Max}(v, a_{N+1} + 1)$ are, respectively, a lower bound and an upper bound for $\{a_n : n \in \mathbf{N}\}$. ■

The upper bound and lower bound given by Lemma 9.26 suggest how we will construct the limit needed by Theorem 9.25. If b is an upper bound for $\{a_n\}$, then certainly the limit L we are looking for should be less than or equal to b. In fact, we should have $L \leq \text{lub}\{a_n : n \in \mathbf{N}\}$.

We then gradually discard initial elements of the sequence. As we do so, the least upper bound for the remaining terms of the sequence will become smaller (since there are "fewer" terms to bound).

The number that these successive least upper bounds decrease toward turns out to be our desired limit. (The least upper bounds are themselves bounded *below* by the lower bound of the original sequence.)

With that outline in mind, we turn to the proof of Theorem 9.25:

Proof of Theorem 9.25: Suppose $\langle a_n \rangle$ is a Cauchy seuqence. Let a and b be, respectively, lower and upper bounds for $\{a_n : n \in \mathbf{N}\}$. Let $S_n = \{a_t : t > n\}$, so that we have $S_1 \supseteq S_2 \supseteq S_3 \supseteq \cdots$.

Since each S_t is a subset of $[a, b]$, these sets are all bounded above. They are also nonempty, so we can let $b_n = \text{lub}(S_n)$. Since $S_1 \supseteq S_2 \supseteq S_3 \supseteq \cdots$, we have $b_1 \geq b_2 \geq b_3 \geq \cdots$.

Comment: The numbers b_t here are the "successive least upper bounds" described in our introduction to the proof. They are bounded below by a, and it seems like a reasonable guess that their greatest lower bound should be the desired limit. That's what we will prove.

Since $b_n \geq a$, for all n, we know that $\{b_n : n \in \mathbf{N}\}$ is bounded below, so, by completeness, we can let $L = \text{glb}(\{b_n : n \in \mathbf{N}\})$. We now show that $\lim_{n\to\infty} a_n = L$.

To do so, suppose $\varepsilon > 0$. We need to find an integer N such that, for all $m > N$, $|a_m - L| < \varepsilon$; i.e., such that $L - \varepsilon < a_m < L + \varepsilon$.

$L + \varepsilon$ is not a lower bound for $\{b_n : n \in \mathbf{N}\}$ (since L is the greatest lower bound), so there is some integer N_1 such that $b_{N_1} < L + \varepsilon$. But by the way the numbers b_t were defined, we know that, if $m > N_1$, we have $a_m \leq b_{N_1}$. Therefore, for all $m > N_1$, we have $a_m < L + \varepsilon$.

To get the other inequality, we use the fact that $\langle a_n \rangle$ is a Cauchy sequence. By Definition 9.11, there is thus an integer N_2 such that, if $m, n > N_2$, then $|a_m - a_n| < \varepsilon/2$.

We claim that, if $m > N_2$, then $a_m > L - \varepsilon$. We prove this by contradiction: suppose $m > N_2$, with $a_m \leq L - \varepsilon$.

Combining this inequality with the fact that $|a_m - a_n| < \varepsilon/2$ for $n > N_2$ gives (using the Triangle Inequality), $a_n \leq L - \varepsilon/2$ for $n > N_2$. In other words, $L - \varepsilon/2$ is an upper bound for $\{a_n : n > N_2\}$. But this set is precisely S_{N_2}, and b_{N_2} is the *least* upper bound for this set, so $b_{N_2} \leq L - \varepsilon/2$, which is impossible, since L is a lower bound for $\{b_n : n \in \mathbf{N}\}$.

Thus, we have shown that, if $m > N_2$, then $a_m > L - \varepsilon$. We also showed that, if $m > N_1$, we have $a_m < L + \varepsilon$. We now simply let $N = \text{Max}(N_1, N_2)$.

Thus, if $m > N$, then $|a_m - L| < \varepsilon$, completing the proof that $\lim_{n\to\infty} a_n = L$. Thus $\langle a_n \rangle$ has a limit, as desired. ■

EXERCISES

Topic 1: Uniform Continuity

1. Prove the following:

THEOREM 9.19. For a function f defined on an interval D, if f is uniformly continuous on D, then f is continuous on D.

2. a) (GYHD 2) Prove that the function defined by the equation $f(x) = 1/x$ is uniformly continuous on each of the following intervals:

i) $[\frac{1}{2}, 2]$,

ii) $(\frac{1}{8}, 2)$,

iii) $[\frac{1}{10}, 100]$.

b) Prove that the function defined by the equation $f(x) = 1/x$ is uniformly continuous on any interval of **R** that does not contain 0.

3. Prove the following: For functions f and g defined on an interval D, if f and g are uniformly continuous on D, then so is $f + g$.

4. Prove the following "Principle of Continuous Induction": Suppose $S \subseteq \mathbf{R}$, and $a \in \mathbf{R}$, and suppose S has the following properties:

i) $[a, t) \subseteq S$ for some $t > a$;

ii) $(\forall\, x \in (a, \infty))[[a, x) \subseteq S \Rightarrow (\exists\, y \in \mathbf{R})(y > x \text{ and } [a, y) \subseteq S)]$

Then $[a, \infty) \subseteq S$. (*Hint:* If not, let $c = \text{lub}\{x \in \mathbf{R} : [a, x) \subseteq S\}$, and get a contradiction.)

Topic 2: Cauchy Sequences

5. Prove: **THEOREM 9.24.** Any convergent sequence is a Cauchy sequence.

6. For a sequence $\langle a_n \rangle$, define b_n by $b_n = \text{lub}\{a_t : t > n\}$ (as in the proof of Theorem 9.25). Find the numbers b_n for each of the following sequences:

a) $a_n = \frac{1}{n}$,

b) $a_n = \frac{-1}{n}$,

c) $a_n = (-1)^n \cdot \frac{1}{n}$.

10

Groups

Introduction

We have talked often about the importance of *generalization*. For example, we saw that the set of multiples of 3 was closed under subtraction, and we generalized that by showing that the set of multiples of any particular integer is closed under subtraction (Theorem 2.11). We later proved the converse by showing that any nonempty subset of $\mathbf{Z}$ that is closed under subtraction has the form $n\mathbf{Z}$ for some $n \in \mathbf{Z}$ (Theorem 5.30).

We turn in this chapter to another level of generalization—a level in which we generalize, not from a particular object to a general object in that category, but from a mathematical *system* to a whole collection of similar systems. We will be talking simultaneously about various systems each of which fits some abstract structure involving a set, an operation, and assumptions about the set and operation. In a single theorem, we will make assertions about the set of integers, about sets of bijective functions, about power sets, about equivalence classes of integers, and about a thousand and one other seemingly disparate situations.

The study of such abstract structures began toward the end of the nineteenth century. Mathematicians recognized that various standard mathematical systems in which they were interested had many properties in common. They began to realize that these properties could be better understood if the essential common features of these systems were formalized. Thus was born *abstract algebra* (also known as "modern algebra").

One such abstract structure is a *group*. Specifically, a group is a set—*any* set: a set of numbers, a set of functions, a set of points, a set of sets, a set of any type of object whatsoever—together with a binary operation on that set which satisfies some simple requirements. The study of such systems is called *group theory*.

Mathematicians developed the particular requirements, or axioms, by looking at examples. These axioms are sufficiently restrictive that substantial results can be achieved, yet sufficiently general that the results can be applied to a wide range of examples. We will see that, within the theory of groups, there are subcategories of groups for which additional theorems can be proved.

The concept of a group is just one of several abstract structures examined in abstract algebra. Other branches of abstract algebra study sets with more than one operation, in which some assumptions are made both about each operation and about the relationship between the operations. For example, the theory of *rings* studies systems with two operations, usually called addition and multiplication, in which one of the assumptions is the distributive law, "$a \cdot (b + c) = (a \cdot b) + (a \cdot c)$".

Many mathematicians consider group theory the most fundamental of the abstract algebraic structures, and it is often the first topic in a course in abstract algebra.

Note: This chapter will make extensive use of several concepts from Section 7.1 concerning binary operations: specifically, closure (Definition 7.2), associativity (Definition 7.4), identity element (Definition 7.5), and inverse (Definition 7.6). You may wish to review this material before continuing.

10.1 Examples and Definitions

Developing a Definition

As we stated above, a group is a set of objects which are combined by means of some binary operation. Not every pair consisting of a set and an operation is acceptable—there are some restrictions. Before stating those restrictions, and the definition of a group, we list some examples of combinations of sets with operations that we want the definition to include:

Example 1. Some Groups Each of the following combinations of a set A and an operation $*$ on A is a group:

i) $A = \mathbf{Z}$, $* =$ addition.
ii) $A = \mathbf{Q}^+$ (the positive rational numbers), $* =$ multiplication.
iii) $A = \mathbf{Z}_n$, $* = +_n$ (addition mod n).
iv) $A =$ the set of bijective functions from $\{1, 2, \ldots, n\}$ to itself, $* =$ composition.
v) $A =$ the power set of $\{1, 2, \ldots, n\}$, $* =$ symmetric difference: this operation is often represented by the symbol "$\triangle$" and is defined for $X, Y \in A$ as follows:

$$X \triangle Y = (X - Y) \cup (Y - X).$$

Thus, $X \triangle Y$ consists of those elements of A which are in X or Y, but not both. (This operation was introduced in Exercise 14 of Section 2.1.)

vi) $A = \mathbf{R} \times \mathbf{R}$ (points in the plane); $* =$ *vector addition:* this operation is represented by the symbol "+", and is defined, for (x, y) and (u, v) in A, as follows:

$$(x, y) + (u, v) = (x + u, y + v).$$

(This equation uses the symbol "+" in two ways: on the left for the sum of ordered pairs, and on the right for sums of real numbers.)

vii) A = the set of 2×2 real matrices with nonzero determinant, $*$ = matrix multiplication. This example is mentioned primarily for the benefit of those already familiar with matrix algebra, but we give a brief explanation here of the terminology: a 2×2 real matrix is an array $\left(\begin{smallmatrix} a & b \\ c & d \end{smallmatrix}\right)$ with $a, b, c, d \in \mathbf{R}$; the determinant of this matrix is the real number $ad - bc$; and the product of two such matrices is written by placing them side by side, and is defined as follows:

$$\begin{pmatrix} a & b \\ c & d \end{pmatrix} \begin{pmatrix} e & f \\ g & h \end{pmatrix} = \begin{pmatrix} ae + bg & af + bh \\ ce + dg & cf + dh \end{pmatrix}.$$ □

We will look at example iv) above and some other groups in more detail later in this section. There are many, many more examples of groups with which you are already familiar.

For comparison with Example 1, we next list some sets with operations which we do not want to include under the classification of "group":

Example 2. **Some "Non-groups"** Each of the following combinations of a set A and an operation $*$ on A is not a group:

i) $A = \mathbf{Z}$, $*$ = subtraction.
ii) $A = 2\mathbf{Z}$ (the even integers), $*$ = multiplication.
iii) $A = \mathbf{Z}_n$, $* = \cdot_n$ (multiplication mod n).
iv) A = the set of all functions from $\{1, 2, \ldots, n\}$ to itself, $*$ = composition.
v) A = the power set of $\{1, 2, \ldots, n\}$, $*$ = union.
vi) $A = \mathbf{R} \times \mathbf{R} \times \mathbf{R}$ (points in space), $*$ = "cross product", represented by the symbol "$\times$", and defined for (x, y, z) and (u, v, w) in A, as follows:

$$(x, y, z) \times (u, v, w) = (yw - zv,\ zu - xw,\ xv - yu).$$

(This is a standard operation on 3-dimensional vectors. It is mentioned primarily for the benefit of those already familiar with vector algebra.)

vii) A = the set of all 2×2 real matrices (regardless of determinant), $*$ = matrix multiplication. (Again, this example is primarily for the benefit of those familiar with matrix algebra.) □

Each of the "nongroups" in Example 2 is based on the same set as the corresponding group in Example 1 or a closely related set. When the sets are the same, the operation has been changed.

QUESTION 1 Compare these two collections—the combinations of sets with operations in Example 1 and those in Example 2. We want to find a set of general properties of operations that are shared by every example in Example 1. At the same time, we want every example in Example 2 to fail to satisfy at least one of the properties. What collection of properties might we use?

You may wish to review the concepts from Section 7.1 mentioned in the introduction to this chapter. There is no "right answer" to this question—investigate and see what you come up with.

? ? ?

There are four properties which algebraists decided should be singled out as the essential ones for the concept of "group". (Other combinations of properties are used for such things as "monoids" and "semigroups", as well as for specialized types of groups.) We will state the formal definition, with some comments, and then ask you why none of the combinations of a set and an operation in Example 2 qualify as groups.

> **DEFINITION 10.1:** A *group* is an ordered pair $(A, *)$, where A is a set, called the *underlying set,* and $*$ is a function with domain $A \times A$, such that the following conditions are satisfied:
>
> **i)** *Closure.* A is closed under $*$ [i.e., $\text{Im}(*) \subseteq A$].
> **ii)** *Associativity.* $*$ is associative.
> **iii)** *Identity.* There is an identity element (in A) for $*$.
> **iv)** *Inverse.* Every element of A has an inverse (in A) for $*$.
>
> The *order* of a group is the cardinality of its underlying set.

COMMENTS

1. Up until this definition, we had described a group as a "set with an operation". In the formal definition above, we talk instead about an ordered pair consisting of a set A and a function on $A \times A$. We have chosen to make a separate requirement of the condition of closure, which is what makes that function into an operation on A. Although many books use a different formulation, we have chosen to state the definition with closure as a separate requirement because this better reflects the way the definition is usually implemented. Generally speaking, in proving that something is a group, we begin with a function on the set, and have to explicitly verify each of the four conditions of Definition 10.1, including the closure condition. We will sometimes still refer to a group as "a set with an operation", which it is, and sometimes, as in Examples 1 and 2, we will know from previous study that we have an operation—i.e., that the set is closed under the given function. However, as we shall see, if we change the set under consideration (e.g., by looking at a subset), verifying closure for the new set becomes an important task, and we have formulated the definition to emphasize the importance of this step.

2. Notice also that, formally, the set A is not, in itself, a group. A group is a *combination* of a set with a function or operation: thus our definition refers to the *ordered pair* $(A, *)$. However, when the context makes clear what the function is, or if the example is a standard one, we often relax our language and refer to the set itself as the group—e.g., saying "the group of integers" as shorthand for "the group consisting

of the pair ($\mathbf{Z}$, +)". We also will say "element of the group" when we mean "element of the underlying set". It is important to recognize that, if there are two different operations $*$ and $**$ on a set A which each satisfy the conditions i)–iv) of this definition, then we have to treat the ordered pairs $(A, *)$ and $(A, **)$ as two distinct groups. Also, as Examples 1 and 2 suggest, a set may be a group using one operation and not be a group using another.

3. We proved in Section 7.1 (Theorem 7.3) that an operation cannot have more than one identity element. Therefore, the identity element referred to in property iii) is unique.

4. We also proved in Section 7.1 (Theorem 7.7) that, for an associative operation with an identity element, an object cannot have more than one inverse. [This is one of the reasons for including associativity—property ii)—in Definition 10.1. Therefore, the inverses referred to in property iv) are also unique.

Get Your Hands Dirty 1

Write each of the conditions in Definition 10.1 using quantifier notation and the definitions of the terms involved. □

In each of the situations in Example 2, at least one of the four conditions in Definition 10.1 is not satisfied. For example, the operation of subtraction on $\mathbf{Z}$ is not associative—e.g., $(9 - 5) - 2 \neq 9 - (5 - 2)$—and so i) is not a group. In case vii) of Example 2, the matrix $\begin{pmatrix} 1 & 0 \\ 0 & 1 \end{pmatrix}$ is the identity element, but many matrices, such as $\begin{pmatrix} 0 & 0 \\ 0 & 0 \end{pmatrix}$, do not have an inverse (why not?).

We leave it to you in the following GYHD to complete the examination of Example 2:

Get Your Hands Dirty 2

For each set and operation in Example 2, determine which of conditions i–iv) of Definition 10.1 it fails to satisfy. There may be more than one such condition for a given example. [If an example fails to satisfy condition iii), you should not even consider condition iv).]

Justify your answers.

Warning: If you wish to show that a particular operation is not associative, you should give an explicit example of elements a, b, c in the set for which $(a * b) * c \neq a * (b * c)$. Do not just show that the general argument seems to fail. Similarly, to show that condition iv) is not satisfied, give a specific element which has no inverse, and indicate how you know that it has none.] □

Standard Notation for Groups

In discussing groups in general, with operation $*$, we will adopt the following conventions, which are standard in many textbooks:

i) The symbol **e** will represent the identity element.
ii) We will use language and notation appropriate for multiplication; specifically:

- The "result" $x * y$ will often be referred to as the "product" of x and y. (Many books also drop the "$*$" and just write "xy".)
- The notation t^{-1} will represent the inverse of the element t for $*$, and we will usually just say "inverse" to mean "inverse for $*$".
- We will use exponential notation to indicate "repeated starring": e.g., $x * x$ will be abbreviated as "x^2", $x * x * x$ as "x^3", etc. Similarly, x^0 will mean **e** (just as, for a real number x, $x^0 = 1$); and, for $n \in \mathbf{N}$, "x^{-n}" will mean $(x^n)^{-1}$. (We are somewhat casual about this definition; a more formal approach would be to define "x^n" using recursion.)

Get Your Hands Dirty 3

Verify *intuitively* that the following "laws of exponents" hold using the above exponential notation (for $m, n \in \mathbf{Z}$):

i) $x^m * x^n = x^{m+n}$,
ii) $(x^m)^n = x^{m \cdot n}$,
iii) $(x^m)^{-1} = (x^{-1})^m$.

(*Note:* For a formal approach, we would use a definition by recursion, and then prove these equations by induction.) □

The Symmetric Group $\mathbf{S}_n$

Some groups are of particular importance, and have been studied extensively by mathematicians. One of these is case iv) of Example 1, the set of all bijective functions from $\{1, 2, \ldots, n\}$ to itself, under the operation of composition. This group is generally represented by the notation $\mathbf{S}_n$, and is called the *symmetric group on n objects*. In this context, a bijective function from a set A to itself is sometimes referred to as a *permutation on A*, and $\mathbf{S}_n$ is called a *permutation group*.

We have already proved a collection of theorems that combine to demonstrate that $\mathbf{S}_n$ actually satisfies each of the four conditions of Definition 10.1 of a group. We will review those theorems here, condition by condition:

i) Corollary 4.7 states that the composition of bijective functions is bijective, so $\mathbf{S}_n$ satisfies condition i) of Definition 10.1.
ii) Theorem 7.1 states that composition is an associative operation, so $\mathbf{S}_n$ satisfies condition ii of Definition 10.1.
iii) Theorem 7.2 states that the identity function [i.e., the function defined by the equation $f(x) = x$, for $x = 1, 2, \ldots, n$] is an identity element for composition, so $\mathbf{S}_n$ satisfies condition iii) of Definition 10.1.
iv) And finally, Corollary 6.5 states that the inverse of a bijective function is bijective, so $\mathbf{S}_n$ satisfies condition iv) of Definition 10.1.

The above review shows that we have in fact proved the following theorem:

THEOREM 10.1. $\mathbf{S}_n$ is a group under composition.

Theorem 5.13 shows that the order of $\mathbf{S}_n$ is $n!$.

COMMENT In many cases, proving that a particular pair $(A, *)$ is a group will be very easy, and will follow from results already known. It is no accident that the theorems cited above were ready for our use—we've been preparing for Theorem 10.1 for some time. Other cases from Example 1 will be similar; for example, the results needed to prove that the integers are a group under addition are all included among our "Assumptions about Arithmetic" listed in Appendix B.

In other situations, you may need to do some new work to verify one or more of the four conditions in Definition 10.1. A proof that a set A and a function $*$ on $A \times A$ together form a group should include an explicit statement of what the identity element is, and an explicit formula or other appropriate demonstration of how to obtain the inverse of a general element of the set. Closure and associativity usually follow either from a previous result or from a straightforward general argument.

For example, in vii) of Example 1, we can prove that matrix multiplication is associative by a direct computation of both $(AB)C$ and $A(BC)$ for three general 2×2 real matrices A, B, and C. The key step in verifying the closure property is showing that, if A and B both have nonzero determinant, then their product AB also has nonzero determinant. (For 2×2 matrices, a straightforward computation shows that the determinant of the product of two matrices is the product of their determinants.)

Get Your Hands Dirty 4

Verify that each of the combinations of a set and operation in Example 1 is a group. [Case iv) has already been done.]

In particular, specify the identity element, and give an expression for the inverse of an arbitrary element. Indicate how you know that the set is closed under the operation and how you know that the operation is associative. [Venn diagrams may be helpful for case v).] □

You may be surprised to see that commutativity of $*$ is not included among the conditions for a group. Commutativity is intentionally excluded so that examples such as $\mathbf{S}_n$ will fit the definition. Groups with the added condition of commutativity are a specialized area of study within group theory. We make the following formal definition:

> **DEFINITION 10.2:** For a group A under an operation $*$, "A is an *Abelian* group" means: the operation $*$ is commutative.

A group for which $*$ is not commutative is called *non-Abelian*. Abelian groups are also sometimes simply called commutative groups.

Get Your Hands Dirty 5

Which of the groups in Example 1 are Abelian? Justify your answer. □

Historical Note: The term "Abelian group" comes from the name of the Norwegian mathematician Niels Henrik Abel (1802–1829). Abel made many enduring contributions to mathematics during his short lifetime. Perhaps his best-known achievement was proving that there is no formula using ordinary arithmetic operations and extraction of roots for the solutions of the general 5th-degree polynomial equation.

The quadratic formula, apparently known in some form as early as ancient Greek mathematics, gives the solution to the general 2nd-degree equation, and in the sixteenth century, general formulas were found for 3rd- and 4th-degree equations. Mathematicians continued to struggle to advance this work, searching for a general solution for equations of degree 5, until Abel demonstrated conclusively that such a formula could not exist.

Invertible Elements Form a Group

Compare each of the following pairs from Examples 1 and 2:

the set $\mathbf{S}_n$ of *bijective* functions from $\{1, 2, \ldots, n\}$ to itself, under composition,	vs.	the set of *all* functions from $\{1, 2, \ldots, n\}$ to itself, under composition,

and

the set of 2×2 real matrices *with nonzero determinant,* under matrix multiplication,	vs.	the set of *all* 2×2 real matrices, under matrix multiplication.

In each pair, the first (from Example 1) is a group and the second (from Example 2) is not. The second examples of each pair are "enlargements" of the first ones. Both second examples satisfy conditions i)–iii) of Definition 10.1. The reason they are not groups is that they contain objects which have no inverse.

There are two other examples among the "nongroups" listed in Example 2 in which conditions i)–iii) are satisfied, but condition iv) is not:

a) the set $\mathbf{Z}_n$ under multiplication mod n, and
b) the power set $\mathcal{P}(\{1, 2, \ldots, n\})$ under union.

QUESTION 2 Is there some subset for each of $\mathbf{Z}_n$ and $\mathcal{P}(\{1, 2, \ldots, n\})$ which continues to satisfy conditions i)–iii) (using the given function, $\cdot_n$ or $\cup$), but which also satisfies condition iv)? Is there a systematic way to create more groups in this way?

? **?** **?**

The answer to both questions is yes. In order to get a subset which satisfies condition iv), we need every element of the subset to have an inverse. The natural thing to do is to try the subset consisting precisely of all those elements which have an inverse, i.e., the invertible elements.

How can we formulate this idea as a carefully stated theorem?

What is the hypothesis? What is the conclusion? Write it down in detail.

We are suggesting the following result:

PROPOSED THEOREM. Suppose that A is a set and $*$ is a function on $A \times A$ such that conditions i)–iii) of Definition 10.1 are satisfied. Define the set B as follows:

$$B = \{t \in A : t \text{ has an inverse in } A \text{ (for } *)\}.$$

Then B is a group under $*$.

Is this proposed theorem true? What is involved in proving it?

The answer is not quite so obvious as it might appear to be. At first glance, it may seem that we have guaranteed condition iv) for B by its definition, and that we already have conditions i)–iii). The actual situation is somewhat more complicated.

The only one of the four conditions that is automatically satisfied for B and $*$ is condition ii)—associativity. Since the equation "$(x * y) * z = x * (y * z)$" holds for all elements of A, and B is a subset of A, the equation must hold, in particular, for the elements of B.

(The proof that associativity holds in B is essentially a use of specialization: our hypothesis includes a statement with the quantifier phrase "$(\forall x, y, z \in A)$", and we are applying this to specific objects, namely elements of B. This type of reasoning is sometimes described by the Latin phrase *a fortiori*.)

Each of the other three conditions requires some extra care (though none of them are difficult to verify). Let's look at them one at a time:

i) *Closure:* We need to show that B is closed under $*$. (We already know that $*$ is *defined* on $B \times B$—see comment 1 following Definition 10.1. The fact that A is closed does not guarantee that B is closed. So suppose that $x, y \in B$. Is $x * y \in B$?

What would that mean?

We need to show that $x * y$ has an inverse (in A). Fortunately, we have already proved a theorem which tells us precisely what we need. Theorem 7.5 says that, if $*$ is an associative operation with an identity element, and if x and y are two elements with inverses x^{-1} and y^{-1} respectively, then the element $x * y$ has an inverse, namely $y^{-1} * x^{-1}$.

Thus, this theorem tells us that the set B in our proposed theorem is closed under $*$.

iii) *Identity:* Condition iii) requires that there be an identity element for $*$. You might say: "But we know that. It's part of the hypothesis of the proposed theorem." The difficulty is in the location of the identity element. Yes, $*$ has an identity element, which we will call **e**. But is **e** an element of B? We need to verify this. In other words, we need to be sure that **e** is an element that has an inverse (in A).

Does *e* have an inverse? If so, what is it?

The answer is yes: the inverse of **e** is simply **e** itself. (Verify by referring to the definitions of inverse and identity.)

iv) *Inverse:* This may seem like the most foolish of all. Why do we have to check condition iv) for B when, by definition of B, every one of its elements has an inverse?

Read condition iv) and the definition of *B* carefully.

Once again, the issue is "location". Yes, every element of B has an inverse, but we only know that it has an inverse *in* A. For B to satisfy condition iv), each of its elements must have an inverse *in* B. So we have the following question to resolve: If $t \in A$, and t has an inverse t^{-1} (in A), does that element t^{-1} belong to B? In other words, does t^{-1} have an inverse (in A)?

If so, what is its inverse?

The inverse of t^{-1} is simply t itself. This is precisely what Theorem 7.4 says.

We have thus verified our proposed theorem, and therefore restate it as a theorem:

THEOREM 10.2. Suppose that A is a set and $*$ is an operation on A such that conditions i)–iii) of Definition 10.1 are satisfied. Define the set B as follows:

$$B = \{t \in A : t \text{ has an inverse in } A \text{ (for } *)\}.$$

Then B is a group under $*$.

COMMENT All this discussion leading to Theorem 10.2 may seem like much ado about nothing. But details such as these are very important in proving theorems. Simply recognizing that "having an inverse" and "having an inverse *in B*" are not necessarily the same requirement is the kind of sensitivity often required in writing a complete proof. Paying attention to the set from which objects come is a sign of mathematical sophistication, and the awareness of this type of distinction generally comes from experience working with quantifiers and formal mathematical statements.

In Question 2, which led to Theorem 10.2, we mentioned two set-and-operation combinations, from Example 2, that satisfied i)–iii) but not iv). Let's see what happens when we apply this theorem to these examples. We'll start with the power-set example:

Power Sets and Union

Let $Y = \{1, 2, \ldots, n\}$ and let $A = \mathcal{P}(Y)$. A is closed under the operation of union, and this operation is associative (verify). The identity element is $\varnothing$.

We are left with the question of inverses:

Which elements of $\mathcal{P}(Y)$ have inverses under union?

If $S \subseteq Y$, we want to know if there is a set $T \subseteq Y$ such that $S \cup T = \varnothing$. The only way for this to happen is if S and T are themselves both equal to $\varnothing$.

In other words, the only invertible element of A is $\varnothing$ itself.

COMMENTS

1. We thus have a group with only one element. You should verify that, for any group, the set consisting of just the identity element is a group under the original operation.
2. There is nothing special about using $\{1, 2, \ldots, n\}$ for Y. The above discussion applies to the power set of any set.

Z_n and Multiplication mod n

Our next example is more challenging:

Let n be an integer greater than 1, and $\mathbf{Z}_n$ the corresponding partition of $\mathbf{Z}$, based on the equivalence relation of congruence mod n. Thus,

$$\mathbf{Z}_n = \{[0], [1], \ldots, [n-1]\}.$$

This set is closed under the operation of multiplication mod n, and the operation is associative (see Definition 7.7 and Theorem 7.9 in Section 7.1). Further, the equivalence class [1] is the identity element for this operation (verify).

The problem of identifying the invertible elements in the case of $\mathbf{Z}_n$ under $\cdot_n$ is not as easy as with $\mathcal{P}(Y)$ and union. Identifying the invertible elements of $\mathbf{Z}_n$ was precisely the question posed in Exploration 1 at the end of Section 7.1.

Review your work on that Exploration. Look at specific values of n, and make an operation table for multiplication mod n to find the invertible elements.

Let's take the specific case $n = 12$. Figure 10.1 shows the operation of multiplication mod 12, with all products that are equal to the identity element boxed (brackets—"[ ]"—have been omitted). The boxed 1's show that the only elements with inverses are [1] itself, [5], [7], and [11]. These four elements of $\mathbf{Z}_{12}$ thus form a group under multiplication mod 12, whose table is given in Figure 10.2 (again, brackets are omitted).

In general, the following theorem describes the set of invertible elements of $\mathbf{Z}_n$:

THEOREM 10.3. Suppose $n \in \mathbf{N}$ and $[t] \in \mathbf{Z}_n$. Then $[t]$ has an inverse for $\cdot_n$ iff $\mathrm{GCD}(t, n) = 1$.

$\cdot_{12}$	0	1	2	3	4	5	6	7	8	9	10	11
0	0	0	0	0	0	0	0	0	0	0	0	0
1	0	1	2	3	4	5	6	7	8	9	10	11
2	0	2	4	6	8	10	0	2	4	6	8	10
3	0	3	6	9	0	3	6	9	0	3	6	9
4	0	4	8	0	4	8	0	4	8	0	4	8
5	0	5	10	3	8	1	6	11	4	9	2	7
6	0	6	0	6	0	6	0	6	0	6	0	6
7	0	7	2	9	4	11	6	1	8	3	10	5
8	0	8	4	0	8	4	0	8	4	0	8	4
9	0	9	6	3	0	9	6	3	0	9	6	3
10	0	10	8	6	4	2	0	10	8	6	4	2
11	0	11	10	9	8	7	6	5	4	3	2	1

Figure 10.1. Operation table for multiplication mod 12.

We leave the proof of this theorem as Exercise 8. The proof is based on the GCD–Linear–Combinations Theorem (Theorem 5.25), or more specifically, on Corollary 5.28 to that theorem. That corollary states that, for integers a, b, not both zero,

$$\text{GCD}(a, b) = 1 \quad \text{iff} \quad \text{there exist } x, y \in \mathbf{Z} \text{ such that } ax + by = 1 .$$

Motivated by Theorem 10.3, we define

$$\mathbf{U}_n = \{[t] \in \mathbf{Z}_n : \text{GCD}(t, n) = 1\} .$$

This group is sometimes called the *group of units* of $\mathbf{Z}_n$. This terminology comes from the traditional use of the word "unit" in mathematics as a synonym for "invertible element".

Thus, combining Theorem 10.3 with Theorem 10.2, we have:

COROLLARY 10.4. $\mathbf{U}_n$ is a group under multiplication mod n.

COMMENT Each of the elements of $\mathbf{U}_{12}$ happens to be its own inverse. This phenomenon does not occur with all of the $\mathbf{U}_n$ groups.

$\cdot_{12}$	1	5	7	11
1	1	5	7	11
5	5	1	11	7
7	7	11	1	5
11	11	7	5	1

Figure 10.2. Multiplication table for the invertible elements of $\mathbf{Z}_{12}$.

Get Your Hands Dirty 6

Identify the elements of $\mathbf{U}_n$ for $n = 4$, $n = 7$, $n = 10$, and $n = 20$ and make an operation table for each of these groups. □

More Examples of Groups

As we have stated before, the ability to explore an idea successfully often depends on having the experience of working with a wide enough variety of examples. With that goal in mind, we summarize and expand our list of groups. Here's our list so far:

1. $\mathbf{Z}$, addition;
2. $\mathbf{Q}^+$, multiplication;
3. $\mathbf{Z}_n$, addition mod n;
4. $\mathbf{S}_n$, composition;
5. $\mathcal{P}(Y)$, symmetric difference (for any set Y);
6. $\mathbf{R} \times \mathbf{R}$, vector addition;
7. 2×2 real matrices with nonzero determinant, matrix multiplication;
8. $\{\varnothing\}$, union;
9. $\mathbf{U}_n$, multiplication mod n (for any $n \in \mathbf{N}$).

We will now look at some ways to enlarge this list.

Cartesian Product of Groups

The "component addition" idea of example 6 in this list can be generalized considerably, by the following method. We begin with any two groups, say $(A, *)$ and $(B, **)$, and make a group out of the set $A \times B$, by defining an operation $\bullet$ (read "bullet") as follows: for (x, y) and $(u, v) \in A \times B$, define

$$(x, y) \bullet (u, v) = (x * u, y ** v) .$$

We leave it to you to verify the following statements:

a) $A \times B$ is closed under $\bullet$; i.e., $(x * u, y ** v)$ is an element of $A \times B$, for all $x, u \in A$, $y, v \in B$.
b) $\bullet$ is associative.
c) if $\mathbf{e}_A$ and $\mathbf{e}_B$ are the respective identity elements of $(A, *)$ and $(B, **)$, then $(\mathbf{e}_A, \mathbf{e}_B)$ is an identity element for $(A \times B, \bullet)$.
d) if x and y are elements, respectively, of A and B, with inverses x^{-1} and y^{-1} (under $*$ and $**$), then (x^{-1}, y^{-1}) is the inverse of (x, y) under $\bullet$.

Together, these four assertions state that $(A \times B, \bullet)$ is a group.

We can generalize this method further by taking a Cartesian product of more than two groups, and defining new operations in a similar component-wise manner.

More Groups under Addition

There are many sets, besides those listed above, that form groups using an operation known as addition. The word "addition" actually has a variety of meanings, depending on the domain, but we will include any of its different forms. Here are some such sets:

10. $\mathbf{R}$;
11. $\mathbf{Q}$;
12. $\mathbf{C}$;
13. $n\mathbf{Z}$, for any $n \in \mathbf{N}$;
14. $\{0\}$;
15. the set of polynomials with coefficients in $\mathbf{R}$ (or $\mathbf{Z}$, or $\mathbf{Q}$, etc.—in fact, the coefficient set can be any set which is itself a group under addition);
16. the polynomials (with an appropriate coefficient set) of degree n or less, for any $n \in \mathbf{N}$;
17. the set of functions from X to $(A, +)$, where X is any set, and $(A,+)$ is any group under addition. (Arithmetic operations on functions were defined by Definition 9.6 in Section 9.3, using $A = \mathbf{R}$).

NOTATION For all groups in which the operation is some form of addition, we use notational conventions that follow ordinary addition. In particular, for $x \in G$ and $n \in \mathbf{N}$, "nx" means the "repeated sum" $x + x + \cdots + x$, and "$-x$" means the inverse of x under the group operation.

More Groups under Multiplication

There are also many important groups that use the operation of multiplication. (As with addition, there are many "types" of multiplication.) The following sets each form a group under the appropriate interpretation of multiplication:

18. $\mathbf{R}^+$;
19. $\mathbf{R} - \{0\}$;
20. $\{1, -1\}$;
21. $\mathbf{C} - \{0\}$;
22. $\{x \in \mathbf{C} : |x| = 1\}$ (the absolute value of a complex number $a + b\sqrt{-1}$ is $\sqrt{a^2 + b^2}$);
23. the set of functions from X to $\mathbf{R} - \{0\}$, where X is any set (see Definition 9.6 in Section 9.3);
24. $\{2^n : n \in \mathbf{Z}\}$ $(= \{\ldots, \frac{1}{8}, \frac{1}{4}, \frac{1}{2}, 1, 2, 4, 8, \ldots\})$;
25. the set of 2×2 real matrices with determinant 1.

Dihedral Groups under Composition

We conclude our list of examples for now with a family of groups known as the *dihedral groups*. (We'll have some other examples in the exercises.) The dihedral groups, like the groups $\mathbf{S}_n$, use the operation of composition.

Consider a regular polygon with n sides. Picture the polygon like a stiff piece of cardboard, and imagine picking it up, putting the polygon through some kind of "rigid motion", and then putting it down in a way that superimposes it on its original position.

This overall movement of the polygon is known more formally as a *symmetry*. A regular n-sided polygon has two types of symmetries—*rotations* and *reflections*. A *rotation* consists of turning the polygon around its center by some multiple of $360/n$ degrees. A *reflection* consists of flipping the polygon around one of its lines of symmetry.

Let's look at the case $n = 4$. Figure 10.3 shows a square, with its vertices labelled A, B, C, and D, together with the eight symmetries, showing the possible "final positions" in which the square can land after a symmetry. Four of these are the result of rotations, and four are the result of reflections. We have labeled each symmetry with a name and description.

As noted earlier, there are two types of symmetries: rotations and reflections. In general, the rotations are turns by multiples of $360/n$ degrees. For 360° and beyond, we begin repeating the "final positions", and so there are exactly n rotations.

The description of the reflections is a little more complicated, since it depends on whether n is odd or even. When n is odd, each line of symmetry connects a vertex to the midpoint of the opposite side. When n is even, a line of symmetry either connects two opposite vertices or connects the midpoints of two opposite sides. Figure 10.4 shows the lines of symmetry for the cases $n = 5$ and $n = 6$.

However, we see that, whether n is odd or even, there are exactly n reflections, and so altogether there are exactly $2n$ symmetries.

Get Your Hands Dirty 7

Using the model of Figure 10.3, describe and label each of the symmetries for an equilateral triangle, showing the result of applying each symmetry to the original position. Use Figure 10.5 for the original position. □

Symmetries as Functions

A symmetry can be thought of more formally as a function on the set of points that make up the polygon, or, more simply, as a function on the set of vertices of the polygon. In expressing the symmetry as such a function, we think of the original position of the polygon as providing labels for the locations of the vertices. If a symmetry moves the vertex in location X to location Y, its function has $f(X) = Y$.

For example, consider the 90° clockwise rotation of the square in Figure 10.3: the vertex in original position A gets moved to original position B; the vertex in position B gets moved to position C; and so on. Thus, the symmetry $\mathbf{R}_{90}$ can be represented as a function by the following values:

$$\mathbf{R}_{90}(A) = B, \quad \mathbf{R}_{90}(B) = C, \quad \mathbf{R}_{90}(C) = D, \quad \text{and} \quad \mathbf{R}_{90}(D) = A\,.$$

Similarly, the flip around a horizontal line, called H, is given by

$$\mathbf{H}(A) = D, \quad \mathbf{H}(B) = C, \quad \mathbf{H}(C) = B, \quad \text{and} \quad \mathbf{H}(D) = A\,.$$

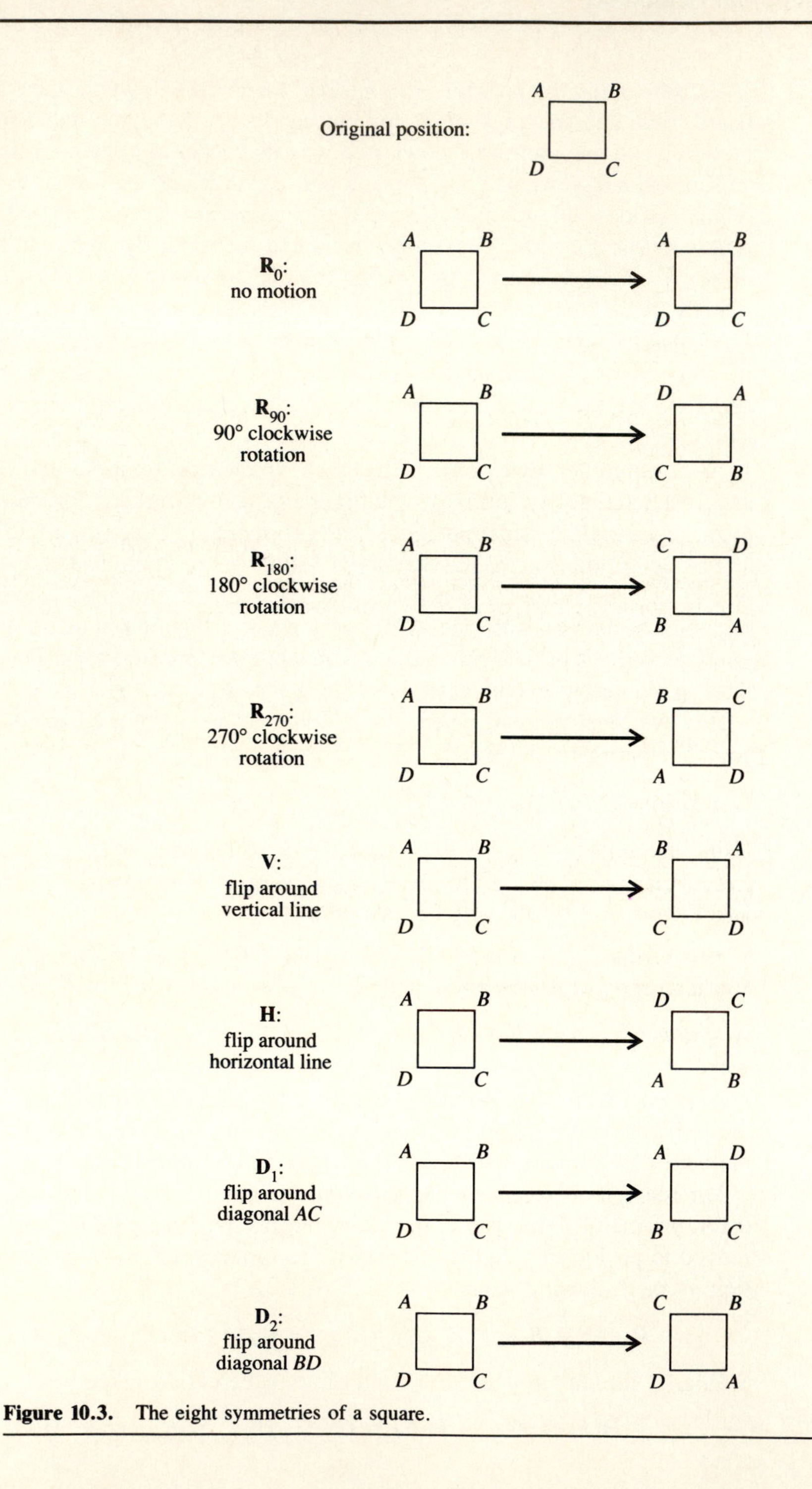

Figure 10.3. The eight symmetries of a square.

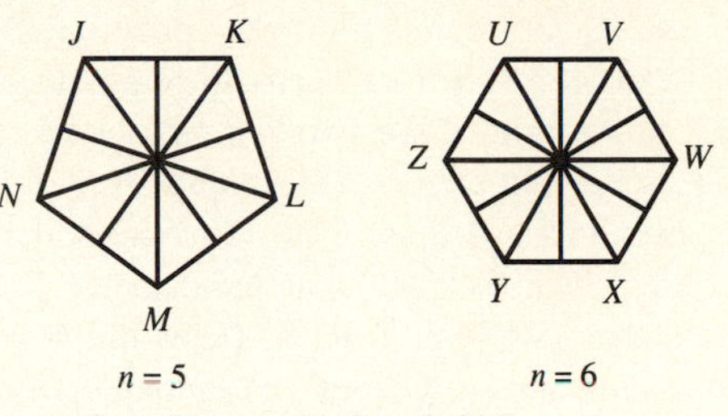

Figure 10.4. Lines of symmetry for polygons with 5 or 6 sides.

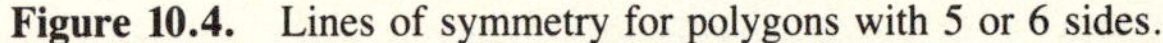

With this interpretation, and relabeling the vertices as 1 through n, the symmetries become functions from $\{1, 2, \ldots, n\}$ to $\{1, 2, \ldots, n\}$, and therefore can be thought of as elements of $\mathbf{S}_n$.

The composition of two such functions, say $\mathbf{R}_{90}$ and $\mathbf{H}$, is written $\mathbf{R}_{90} \circ \mathbf{H}$, and is the function describing the result of performing first the motion corresponding to $\mathbf{H}$ and then that of $\mathbf{R}_{90}$. Direct computation shows that $\mathbf{R}_{90} \circ \mathbf{H} = \mathbf{D}_1$. This can be analyzed by thinking of these as functions on the set of vertices, e.g.,

$$(\mathbf{R}_{90} \circ \mathbf{H})(A) = \mathbf{R}_{90}(\mathbf{H}(A)) = \mathbf{R}_{90}(D) = A = \mathbf{D}_1(A)$$

$$(\mathbf{R}_{90} \circ \mathbf{H})(B) = \mathbf{R}_{90}(\mathbf{H}(B)) = \mathbf{R}_{90}(C) = D = \mathbf{D}_1(B)$$

$$(\mathbf{R}_{90} \circ \mathbf{H})(C) = \mathbf{R}_{90}(\mathbf{H}(C)) = \mathbf{R}_{90}(B) = C = \mathbf{D}_1(C)$$

$$(\mathbf{R}_{90} \circ \mathbf{H})(D) = \mathbf{R}_{90}(\mathbf{H}(D)) = \mathbf{R}_{90}(A) = B = \mathbf{D}_1(D)$$

so that $\mathbf{R}_{90} \circ \mathbf{H}$ and $\mathbf{D}_1$ are seen to be the same function.

An alternative, more intuitive, approach is to look at the symmetries geometrically. To see how this is done, we give some examples:

Example 3 **Composition of Symmetries** Suppose we want to find the composition $\mathbf{R}_{90} \circ \mathbf{H}$. (Remember that this means "first do $\mathbf{H}$, and then do $\mathbf{R}_{90}$".) We can visualize this as follows:

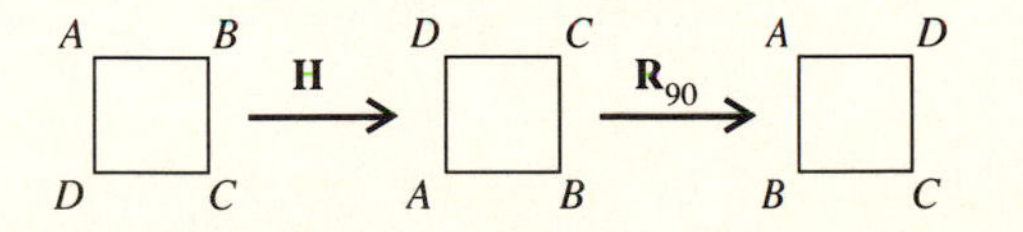

The first pair of squares are taken directly from the row for $\mathbf{H}$ in Figure 10.3. We then rotate the second square 90° clockwise to get the third square. (Notice that, although

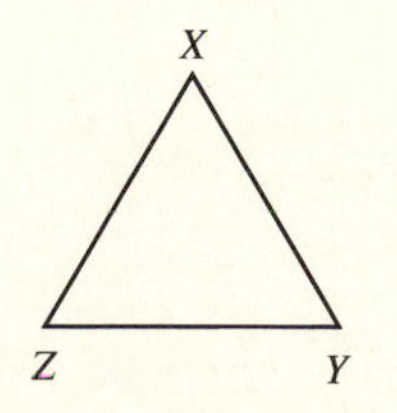

Figure 10.5.

the second square is not in original position, $\mathbf{R}_{90}$ still moves the vertex in original position D — the lower left corner — to original position A — the upper left corner, etc.)

The composition of the two symmetries has the same net result on the original square as the single symmetry $\mathbf{D}_1$ (see Figure 10.3). Therefore, we have $\mathbf{R}_{90} \circ \mathbf{H} = \mathbf{D}_1$.

The diagrams in Figure 10.6 illustrate some other examples of composition of these symmetries, with the results as indicated.

Notice that $\mathbf{H} \circ \mathbf{R}_{180} = \mathbf{R}_{180} \circ \mathbf{H}$, while $\mathbf{V} \circ \mathbf{D}_1 \neq \mathbf{D}_1 \circ \mathbf{V}$. As we have seen before, composition is not, in general, a commutative operation. (This does not mean that a particular pair of elements cannot commute; it just means that not every pair commutes.)

□

Since we can think of the symmetries of the square as elements of $\mathbf{S}_n$, we know that their composition is associative. The following GYHD asks you to show that these symmetries form a group.

Get Your Hands Dirty 8

Make a table for the operation of composition for the set of symmetries of a square. Show, using this table, that this set of symmetries is closed under composition, that there is an identity element, and that every symmetry has an inverse. (You can then conclude that these symmetries form a group.) □

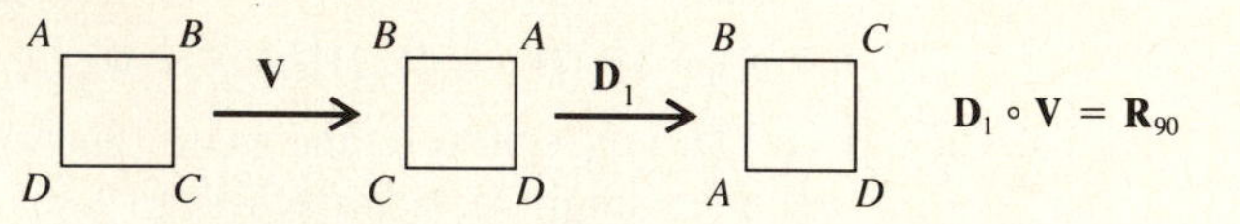

(Note that $\mathbf{D}_1$ flips the square around the diagonal corresponding to *original position AC*.)

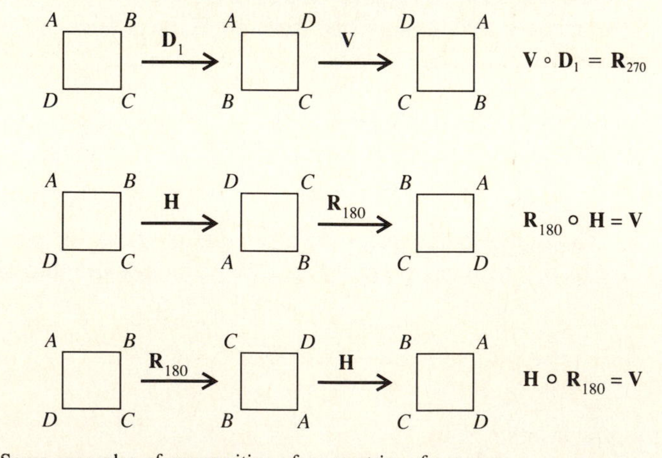

Figure 10.6. Some examples of composition of symmetries of a square.

More generally, we have the following result, which we state without formal proof:

THEOREM 10.5. For $n \in \mathbf{N}$, $n \geq 3$, the set of symmetries of a regular n-sided polygon forms a group of order $2n$ under the operation of composition.

COMMENT The only difficult part of this proof is showing that the set of symmetries is closed under composition, which requires a more formal discussion of the geometry than is appropriate here. Since the symmetries are essentially a subset of $\mathbf{S}_n$, we know that composition is associative. The identity function, which can be thought of as a 0° rotation, is the identity element for every n; the inverse of a rotation is the rotation in the opposite direction; and every reflection is its own inverse.

The group of symmetries of the regular n-sided polygon is usually represented by the symbol "$\mathbf{D}_n$", and the family of groups described in Theorem 10.5 are called *dihedral groups*. As already noted, we can think of $\mathbf{D}_n$ as a subset of $\mathbf{S}_n$.

Caution: Because of the number of its elements, $\mathbf{D}_n$ is sometimes referred to as the dihedral group of order $2n$. Be careful about the potential confusion between the number of sides of the polygon and the number of elements in the group.

Looking for Theorems about Groups

We conclude this section with a very open-ended exploration. By now you have seen quite a few different groups using different operations. In the last two sections, we will state and prove some theorems that apply to all of these groups, and introduce some general concepts to help understand them.

Meanwhile, we would like to give you a chance to play with the information we have given you here. Now that you have seen the definition and numerous examples of groups, it's time to look for some theorems. Feel free to add your own questions to the ones we suggest.

EXPLORATION 1: Exploring Finite Groups

Examine the finite groups discussed so far. Consider the following families of groups, among others:

$$\mathbf{D}_n, \mathbf{S}_n, \mathbf{U}_n, \mathbf{Z}_n.$$

What patterns do you find? What observations can you make?

Here are some specific questions to look at:

- Examine any given row or column of an operation table: what do you notice?
- Can you make a group using just some of the elements from a group?
- How many elements are their own inverses?

What other questions occur to you?

• E X P L O R E •

EXERCISES

1. (GYHD 2) For each set and operation in Example 2, determine which of conditions i)–iv) of Definition 10.1 it fails to satisfy. (There may be more than one such condition for a given example.) Justify your answers.
2. (GYHD 3) Prove the following "laws of exponents" for groups by induction (for $m, n \in \mathbf{Z}$):
   i) $x^m * x^n = x^{m+n}$,
   ii) $(x^m)^n = x^{m \cdot n}$,
   iii) $(x^m)^{-1} = (x^{-1})^m$.
3. (GYHD 4) For each of the following groups (from Example 1), state the identity element and give a formula or other means of finding the inverse of a general element:
   a) $A = \mathbf{Z}$, $* =$ addition.
   b) $A = \mathbf{Q}^+$ (the positive rational numbers), $* =$ multiplication.
   c) $A = \mathbf{Z}_n$, $* = +_n$ (addition mod n).
   d) $A =$ the power set of $\{1, 2, \ldots, n\}$, $* =$ symmetric difference.
   e) $A = \mathbf{R} \times \mathbf{R}$ ("points in the plane"), $* =$ vector addition.
4. Do the following for each of the values $n = 4$, $n = 7$, $n = 10$, and $n = 20$:
   a) (GYHD 10.6) Identify the elements of $\mathbf{U}_n$, and make an operation table for each of these groups.
   b) Identify the elements of $\mathbf{U}_n$ which are their own inverses.
   c) Show that, for each element $x \in \mathbf{U}_n$, there is a positive integer $t \leq n$ such that $x^t = [1]$.
5. (GYHD 7) Consider the triangle in Figure 10.7 as the "original position" for an equilateral triangle. Label the elements of $\mathbf{D}_3$ as follows:

$$\mathbf{R}_0 = \text{no motion;}$$
$$\mathbf{R}_{120} = 120^\circ \text{ clockwise rotation;}$$
$$\mathbf{R}_{240} = 240^\circ \text{ clockwise rotation;}$$
$$\mathbf{D}_X = \text{flip around the altitude from } X;$$
$$\mathbf{D}_Y = \text{flip around the altitude from } Y;$$
$$\mathbf{D}_Z = \text{flip around the altitude from } Z.$$

   a) Write each symmetry as a function on the set of vertices $\{X, Y, Z\}$.
   b) Make a table for the group $\mathbf{D}_3$.
6. Let G be the group of 2×2 real matrices with nonzero determinant. [See Example 1, vii), for the definition of "determinant".]
   a) What is the identity element for G?

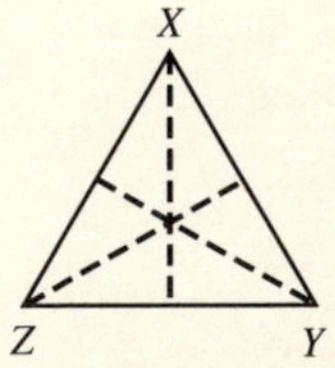

X

Z Y

Figure 10.7.

b) Find the determinant of the element $\begin{pmatrix} 2 & 1 \\ 4 & 3 \end{pmatrix}$.

c) Find the inverse in G for the element in b).

7. Suppose $G = \{\mathbf{e}, a, b\}$ is a group under an operation $*$, with $\mathbf{e}$ as the identity element. Show that there is only one possible way to complete the table in Figure 10.8.

8. Corollary 5.28 (in Section 5.3) asserts the following: For $a, b \in \mathbf{Z}$, not both zero,

$$\mathrm{GCD}(a, b) = 1 \text{ iff there exist } x, y \in \mathbf{Z} \text{ such that } ax + by = 1 .$$

Use this result to prove:

THEOREM 10.3. For $n \in \mathbf{N}$, $[t] \in \mathbf{Z}_n$,

$$[t] \text{ has an inverse for } \cdot_n \quad \text{iff} \quad \mathrm{GCD}(t, n) = 1 .$$

(*Hint:* Show and then use the following assertion, for $s, t \in \mathbf{Z}$:

$$[t] \cdot_n [s] = [1] \quad \text{iff} \quad ts + ny = 1 \quad \text{for some integer } y .)$$

9. Find the number of elements of $\mathbf{U}_n$ for the following values of n (s and t are positive integers); express your answers in terms of s, t, p, and q, as appropriate):
   - a) $n = 2^t$.
   - b) $n = 3^t$.
   - c) $n = p^t$, where p is a prime.
   - d) $n = 2 \cdot p^t$, where p is an odd prime.
   - e) $n = 2^s \cdot p^t$, where p is an odd prime.
   - f) $n = q^s \cdot p^t$, where q and p are distinct primes.

10. Show that, in any group, the condition $x = x^{-1}$ is the same as the condition $x^2 = \mathbf{e}$ (where $\mathbf{e}$ is the identity element).

11. The group $\mathbf{S}_3$ contains $3! = 6$ elements. Label these elements $f_1, \ldots, f_6$ as follows:

$$f_1: \begin{matrix} 1 \to 1 \\ 2 \to 2 \\ 3 \to 3 \end{matrix} \qquad f_2: \begin{matrix} 1 \to 1 \\ 2 \to 3 \\ 3 \to 2 \end{matrix} \qquad f_3: \begin{matrix} 1 \to 2 \\ 2 \to 1 \\ 3 \to 3 \end{matrix}$$

$$f_4: \begin{matrix} 1 \to 2 \\ 2 \to 3 \\ 3 \to 1 \end{matrix} \qquad f_5: \begin{matrix} 1 \to 3 \\ 2 \to 1 \\ 3 \to 2 \end{matrix} \qquad f_6: \begin{matrix} 1 \to 3 \\ 2 \to 2 \\ 3 \to 1 \end{matrix}$$

Using this notation, make a table for $\mathbf{S}_3$, and state the inverse of each element.

12. Let S be the set of six functions from $\mathbf{R} - \{0, 1\}$ to itself defined by the following equations:

*	**e**	*a*	*b*
e			
a			
b			

Figure 10.8. How can this table be completed to make a group?

$$g_1(x) = x \qquad g_2(x) = \frac{1}{x} \qquad g_3(x) = 1 - x$$

$$g_4(x) = \frac{x-1}{x} \qquad g_5(x) = \frac{1}{1-x} \qquad g_6(x) = \frac{x}{x-1}$$

a) Show that these equations do, in fact, define functions from $\mathbf{R} - \{0, 1\}$ to itself.

b) Show that these six functions form a group under composition. (*Suggestion:* Make a table for the operation.)

10.2 Basic Theorems

So far most of our discussion has been about individual groups. It is time to see what the abstraction process can accomplish for us—that is, to prove some theorems that apply to all groups.

Operation Tables and Cancellation Laws

We will begin by looking at operation tables. By now, you have seen or constructed tables for several different groups, of various sizes. Our first general theorem of this section states a property of these tables which you may have already noticed as part of Exploration 1:

THEOREM 10.6. Let G be any group. Then every row and every column of the operation table contains each element of G exactly once.

Discussion

Before proving this theorem, we will translate it into a form that is easier to work with. We'll look at rows first. What does it mean that every row contains each element of G exactly once?

What is this saying in terms of the group operation?

We can identify a row by its "label" at the left. This label represents an element of G which we will call s. The elements in the row for s have the form "$s * x$", where x is the label at the top of a given column.

We want to say that the row for s contains each element of G exactly once. The phrase "each element" represents yet another variable, say t. How do we say "t appears exactly once in the row for s"?

? ? ?

We want to say that there is one and only one value for x which gives $s * x = t$. In other words, we can rephrase the "row" portion of Theorem 10.6 as:

For every $s, t \in G$, the equation $s * x = t$ has a unique solution for x (with $x \in G$).

To state that every *column* contains each element of G exactly once, we just switch the roles of row and column, which replaces the equation "$s * x = t$" by "$x * s = t$". In other words, for each $s, t \in G$, we want the equation "$x * s = t$" to have a unique solution for x.

We can thus restate Theorem 10.6 as follows:

THEOREM 10.6′. Suppose G is a group, and $s, t \in G$. Then each of the equations

$$s * x = t \quad \text{and} \quad x * s = t.$$

has a unique solution for x (with $x \in G$).

Note: These two equations may have different solutions.

DISCUSSION (continued) Now we plan a proof of Theorem 10.6′. We begin by finding *one solution* to each equation.

?   ?   ?

You may be tempted to solve the equation "$s * x = t$" by "dividing by s". This is intuitively sound, but technically incorrect. Instead, you should think of the solving process as "multiplying both sides of the equation, on the left, by s^{-1}". This gives $x = s^{-1} * t$ [since $s^{-1} * (s * x) = x$]. The other equation is similar.

The following lemma is a useful result in its own right, and is a helpful tool for proving the uniqueness aspect of Theorem 10.6′.

LEMMA 10.7. Suppose G is a group and $s, x, y \in G$. Then:

i) if $s * x = s * y$, then $x = y$.
ii) if $x * s = y * s$, then $x = y$.

Proof:

i) Suppose $s * x = s * y$. Then $s^{-1} * (s * x) = s^{-1} * (s * y)$. But $s^{-1} * (s * x) = (s^{-1} * s) * x = \mathbf{e} * x = x$, and similarly, $s^{-1} * (s * y) = y$. Therefore $x = y$.

The proof of ii) is similar. ■

Lemma 10.7 is often called the *Cancellation Law,* with i) referred to as the *Left Cancellation Law* and ii) as the *Right Cancellation Law.*

We now prove Theorem 10.6′.

Proof of Theorem 10.6′: Let s and t be elements of G. Let $x = s^{-1} * t$. Then

$$s * x = s * (s^{-1} * t) = (s * s^{-1}) * t = \mathbf{e} * t = t.$$

Thus, $s^{-1} * t$ is a solution to the equation $s * x = t$.

To prove uniqueness, suppose that u and v are both solutions to the equation $s * x = t$. Then $s * u = s * v$, and therefore $u = v$, by the (Left) Cancellation Law of Lemma 10.7.

The equation $x * s = t$ is similar. ■

COMMENT Though Theorem 10.6′ is particularly useful for developing tables for small, finite groups, it is valid for every group, finite or infinite.

We can illustrate the use of Theorem 10.6′ with the following:

Example 1. **Groups of Size 3** Exercise 7 in the last section stated the following:

> Suppose $G = \{\mathbf{e}, a, b\}$ is a group under an operation $*$, with **e** as the identity element. Show that there is only one possible way to complete the table in Figure 10.8.

(The incomplete table appears with the exercise.) We can do this exercise very easily using Theorem 10.6′, as follows:

Let G be as given in the exercise. Since **e** is the identity element, the first row and column of the table must look like Figure 10.9. Next consider the entry in row b, column a, i.e., the value of $b * a$. The row containing $b * a$ already has an entry "b", so $b * a \neq b$. Similarly, the column containing $b * a$ already has an entry "a", so $b * a \neq a$. The only possibility remaining is $b * a = \mathbf{e}$. So now the table looks like Figure 10.10.

By Theorem 10.6, the remaining entry in the last row must be a. The missing entries in the middle row are found similarly: the entry for $a * a$ must be b, and so the entry for $a * b$ must be **e**.

Thus, the complete table must be as in Figure 10.11. □

The following corollary is a special case of Theorem 10.6′, and will be useful when we discuss subgroups:

COROLLARY 10.8. Suppose G is a group and $a \in G$. If $a^2 = a$, then $a = \mathbf{e}$.

Proof: If $a^2 = a$, then both **e** and a are solutions for x to the equation "$a * x = a$". By Theorem 10.6′, this equation has a unique solution, so it follows that $a = \mathbf{e}$. ■

Subgroups

In the last section, we proved a theorem about forming a group using some of the elements of a set on which an operation was defined. In that result (Theorem 10.2), the original set and operation did not necessarily form a group. We now look at the situation

$*$	**e**	a	b
e	**e**	a	b
a	a		
b	b		

Figure 10.9. The row and column for **e** have been filled in.

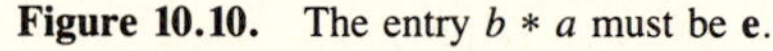

*	e	*a*	*b*
e	e	*a*	*b*
a	*a*		
b	*b*	e	

Figure 10.10. The entry $b * a$ must be e.

*	e	*a*	*b*
e	e	*a*	*b*
a	*a*	*b*	e
b	*b*	e	*a*

Figure 10.11. The complete table for the three-element group.

where we already have a group, say G, with operation $*$, and we ask whether this operation can be used with some of the elements of G to form a group.

So suppose H is a subset of a group G. We can think of $*$ as a function defined on $H \times H$, and ask whether the pair $(H, *)$ satisfies the four conditions that define a group (Definition 10.1). If so, then we will call H a *subgroup* of G.

Here is the formal definition:

DEFINITION 10.3: For a group G with operation $*$, and a subset $H \subseteq G$, "H is a *subgroup* of G" means: If we interpret $*$ as a function on $H \times H$, then $(H, *)$ is a group.

Here are some examples:

Example 2. Groups and Subgroups In each of the following situations, H is a subgroup of G:

a) $G = \mathbf{R}$, $* =$ addition, $H = \mathbf{Z}$.
b) $G = \mathbf{R} - \{0\}$, $* =$ multiplication, $H = \mathbf{Q} - \{0\}$.
c) $G = \mathbf{Z}_{12}$, $* =$ addition mod 12, $H = \{[0], [4], [8]\}$.
d) $G = \mathbf{S}_n$, $* =$ composition, $H = \{f \in \mathbf{S}_n : f(1) = 1\}$.
e) $G = \mathbf{R} \times \mathbf{R}$, $* =$ vector addition, $H = \{(x, y) \in \mathbf{R} \times \mathbf{R} : y = x\}$.
f) $G =$ the set of 2×2 real matrices with nonzero determinant, $* =$ matrix multiplication, $H =$ the set of those 2×2 real matrices whose determinant is 1.
g) $G = \mathcal{P}(\{1, 2, \ldots, n\})$, $* =$ symmetric difference, $H = \{A \in G : 1 \notin A\}$.
h) $G = \mathbf{D}_n$, $* =$ composition, $H =$ the set of rotations in $\mathbf{D}_n$.

In some of these cases, $(H, *)$ is already known to be a group. But in others, there is some work to be done. Let's consider example c). We need to verify the four conditions of the definition: closure, associativity, identity, and inverse. The discussion here is similar to that for Theorem 10.2 ("Invertible elements form a group"). Let's look at these conditions one at a time, for example c):

i) *Closure:* We need to establish that, if x and y are elements of H (i.e., if each is one of the elements $[0], [4], [8]$), then the sum $x +_n y$ is one of these three elements. Since H is so small, we can verify this explicitly for all nine combinations. We get the table in Figure 10.12 for the operation.

ii) *Associativity:* This requires no verification. As with Theorem 10.2, since the associativity equation holds for all elements of G, and H is a subset of G, the equation must hold for all elements of H.

iii) *Identity:* We need to show that H contains an identity element for $+_n$. Since [0] is the identity element in G, and $[0] \in H$, the element [0] continues to serve as identity element in H.

iv) *Inverse:* We need to show that, for $x \in H$, the inverse of H is also in H. Since [0] is its own inverse, and [4] and [8] are inverses of each other, this condition is satisfied. □

The procedure for showing that a subset of a group forms a subgroup is different from that for showing from scratch that some set and operation form a group. As we have seen, the associativity condition is free. In verifying the identity condition, we have the advantage that there already is an identity element; we just have to verify that this element belongs to H. Notice that H cannot have a different identity element from G, because such an element would have to satisfy the equation $x^2 = x$, and by Corollary 10.8, **e** is the only element of G which satisfies this equation.

Finally, in verifying the inverse condition, we have the advantage of knowing that every element has an inverse in G—we just have to verify that this inverse is in H. (By Theorem 7.7, an element can't have a different inverse in H from the one it had in G.)

Thus we have the following theorem for determining if a subset of a group is a subgroup:

THEOREM 10.9. Suppose G is a group with operation $*$, and H is a subset of G. Then H is a subgroup of G iff the following conditions hold:

i) H is closed under $*$ (closure);

ii) $\mathbf{e} \in H$ (identity);

iii) for all t in H, t^{-1} is also in H (inverse).

COMMENT It can be shown that, if H is nonempty and satisfies conditions i) and iii), then it must also satisfy condition ii). Many books prove this theorem for use as a shortcut in establishing that certain subsets of groups are subgroups. In our view, the effort saved by such a shortcut is not worth the loss in intuitive understanding that results. The three conditions in Theorem 10.9 are the natural result of interpreting the definition of "subgroup".

$+_{12}$	[0]	[4]	[8]
[0]	[0]	[4]	[8]
[4]	[4]	[8]	[0]
[8]	[8]	[0]	[4]

Figure 10.12. The table for $\{[0], [4], [8]\}$ under $+_{12}$.

Get Your Hands Dirty 1

Verify, using Theorem 10.9, that H is a subgroup of G in each of cases a)–h) of Example 2 above. □

The following theorem illustrates the use of Theorem 10.9:

THEOREM 10.10. Suppose G is a group, with subgroups H and K. Then $H \cap K$ is a subgroup of G.

Proof: We need to verify the three conditions of Theorem 10.9.

i) *Closure:* Suppose $x, y \in H \cap K$. Then $x, y \in H$, and H is a subgroup, so $x * y \in H$. Similarly, $x * y \in K$, so $x * y \in H \cap K$.
ii) *Identity:* We know that $\mathbf{e} \in H$ and $\mathbf{e} \in K$, since H and K are subgroups. Therefore $\mathbf{e} \in H \cap K$.
iii) *Inverse:* Suppose $x \in H \cap K$. Since $x \in H$, we have $x^{-1} \in H$, since H is a subgroup. Similarly, $x^{-1} \in K$, so we have $x^{-1} \in H \cap K$. ■

Get Your Hands Dirty 2

Give an example to show that "intersection" cannot be replaced by "union" in Theorem 10.10. That is, find a group G and subgroups H and K such that $H \cup K$ is not a subgroup of G. □

The question of when the union of subgroups *is* a subgroup is asked in Exercise 6.

Cyclic Groups

One of the questions that arises is how to *find* subgroups of a given group (if there are any). The following question leads to a general procedure for identifying a certain family of subgroups for any group.

QUESTION 1 If G is a group, and $x \in G$, what is the smallest subgroup of G which contains x?

Try each of the following cases, and then generalize:

a) $G = \mathbf{D}_4$, $x = \mathbf{R}_{90}$.
b) $G = \mathbf{U}_{20}$, $x = [3]$.
c) $G = \mathbf{S}_3$, $x =$ the function given by

$$x: \begin{array}{l} 1 \to 1 \\ 2 \to 3 \\ 3 \to 2 \end{array}$$

d) $G = \mathbf{Q} - \{0\}$ (under multiplication), $x = 2$.
e) $G = \mathbf{Z}$ (under addition), $x = 6$.

? ? ?

Theorem 10.9 is the key to this question. Let us suppose that there is an answer, say H, and investigate what we can learn about H.

First of all, by ii) and iii) of Theorem 10.9, we must have $\mathbf{e} \in H$ and we must have $x^{-1} \in H$. So H contains at least the set $\{\mathbf{e}, x, x^{-1}\}$. (We don't necessarily know that these elements are distinct — more on this later.)

Also, H must also be closed — i.e., $(\forall\, a, b \in H)(a * b \in H)$. Since the elements $\mathbf{e}$, x, and x^{-1} are all in H, the closure condition tells us that each of the possible products of these elements is also in H. Most of these nine products are repeats of the elements we already have (e.g., $x * x^{-1} = \mathbf{e}$), but we also get $x * x$ $(= x^2)$ in H and $x^{-1} * x^{-1}$ $(= x^{-2})$ in H.

So now we know that H must contain $\{x^{-2}, x^{-1}, \mathbf{e}, x, x^2\}$. Applying closure to this enlarged set gives four more elements: x^3, x^4, x^{-3}, and x^{-4}. We can keep on going like this, each time getting new powers of x. We can prove by induction (see Exercise 7) that all elements of the form x^n, with $n \in \mathbf{Z}$, must belong to H.

So our exploration suggests that the set $\{x^n : n \in \mathbf{Z}\}$ might be the answer to Question 1. So let's try it: let us *define* H to be this set, i.e.,

$$H = \{x^n : n \in \mathbf{Z}\} = \{\ldots, x^{-3}, x^{-2}, x^{-1}, x^0 (= \mathbf{e}), x, x^2, x^3, \ldots\}.$$

This set certainly contains x, and we have seen that no smaller set containing x can be a subgroup of G. We need only ask: is this set H a subgroup of G?

? ? ?

We check the three conditions of Theorem 10.9:

i) *Closure:* If $a, b \in H$, then we have $a = x^m$ and $b = x^n$, for some $m, n \in \mathbf{Z}$. Therefore, $a * b = x^m * x^n = x^{m+n} \in H$, as desired.
ii) *Identity:* We have $\mathbf{e} \in H$, by definition of H.
iii) *Inverse:* If $a \in H$, then $a = x^m$ for some $m \in \mathbf{Z}$. Therefore, $a^{-1} = x^{-m}$, and so we have $a^{-1} \in H$, as desired.

Thus we have proved the following:

THEOREM 10.11. Suppose G is a group and $x \in G$. Define the subset H of G by

$$H = \{x^n : n \in \mathbf{Z}\} = \{\ldots, x^{-2}, x^{-1}, x^0 (= \mathbf{e}), x, x^2, \ldots\}.$$

Then H is a subgroup of G containing x, and is the smallest such subgroup.

The subgroup H described in this theorem is called the *cyclic subgroup of G generated by* x, and is denoted "$\langle x \rangle$". If $H = \langle x \rangle$, we refer to x as a *generator* of H (see comment 1 below).

Get Your Hands Dirty 3

a) For each group G and element x given below, list explicitly each of the elements of $\langle x \rangle$:
  i) $G = \mathbf{Z}_{12}$, $x = [3]$.
  ii) $G = \mathbf{D}_4$, $x = \mathbf{R}_{90}$.

iii) $G = \mathbf{U}_{11}$, $x = [4]$.
iv) $G = \mathbf{S}_4$, $x =$ the function with $x(1) = 2, x(2) = 4, x(3) = 1, x(4) = 3$.
v) $G = \mathbf{C} - \{0\}$ (nonzero complex numbers) under multiplication, $x = (1 + \sqrt{-3})/2$. (It's possible—just be careful.)

b) For each group G and element x given below, describe the set $\langle x \rangle$:
i) $G = \mathbf{Z}$ under addition, $x = 5$.
ii) $G = \mathbf{Q}^+$ (positive rational numbers) under multiplication, $x = \frac{2}{3}$.
iii) $G = \mathbf{R}$ under addition, $x = \pi$. □

The terminology "cyclic subgroup" is consistent with the following definition:

> **DEFINITION 10.4:** For a group G, "G is *cyclic*" means: there is an element x in G such that $G = \langle x \rangle$.

Thus, the cyclic subgroup generated by an element is, in fact, a cyclic group.

COMMENTS

1. A cyclic group may have more than one element which qualifies as its generator. For example, if $H = \langle \mathbf{R}_{90} \rangle$ ($\subseteq \mathbf{D}_4$), then H can also be written as $\langle \mathbf{R}_{270} \rangle$ (verify). Exercises 9 and 10 will explore the question of different generators for the same group.
2. *Caution:* If G is a group with $x \in G$, and $G \neq \langle x \rangle$, that does not mean that G is not cyclic. There may be some other element y of G for which $G = \langle y \rangle$.
3. Other notations, such as $[x]$ and (x), are sometimes used instead of $\langle x \rangle$ for the subgroup generated by an element.

Get Your Hands Dirty 4

Prove that every cyclic group is Abelian. □

Order of an Element

Theorem 10.11 has an "optical illusion" in it. The set H appears to have infinitely many elements, one for each $n \in \mathbf{Z}$. But if you worked out the specific examples in GYHD 3, you saw that the answers can be finite sets.

For example, in item a)ii), with $G = \mathbf{D}_4$ and $x = \mathbf{R}_{90}$, we get $H = \{\mathbf{R}_0, \mathbf{R}_{90}, \mathbf{R}_{180}, \mathbf{R}_{270}\}$. Since the whole group G is finite, the subgroup H we are seeking must certainly be finite.

On the other hand, in item b)ii), with $G = \mathbf{Q}^+$ and $x = \frac{2}{3}$, the resulting set is

$$H = \left\{\ldots, \frac{27}{8}, \frac{9}{4}, \frac{3}{2}, 1, \frac{2}{3}, \frac{4}{9}, \frac{8}{27}, \ldots\right\} = \left\{\left(\frac{2}{3}\right)^n : n \in \mathbf{Z}\right\}.$$

which is infinite.

What's going on here?

? ? ?

The answer is that the apparent infiniteness of H is misleading. Although there is an element x^n for each $n \in \mathbf{Z}$, these elements are not necessarily all different. Returning to a)ii), with $H = \langle \mathbf{R}_{90} \rangle$, we have

$$\cdots = x^{-8} = x^{-4} = x^{0} = x^{4} = x^{8} = x^{12} = \cdots$$

(All of these elements are equal to $\mathbf{R}_0$.) Similarly,

$$\cdots = x^{-7} = x^{-3} = x = x^{5} = x^{9} = x^{13} = \cdots$$

(All of these elements are equal to $\mathbf{R}_{90}$.)

And so on. H only has four elements, namely the four rotations, $\mathbf{R}_0$, $\mathbf{R}_{90}$, $\mathbf{R}_{180}$, and $\mathbf{R}_{270}$. The infinitely many symbols of the form x^n fall into four families of elements, the elements in each family all equal to a particular one of the four rotations.

The way the various powers of $\mathbf{R}_{90}$ break up into families is typical of cyclic groups. In general, we can partition the set of symbols x^n by looking at which exponents give the same element of G.

In other words, we define an equivalence relation on $\mathbf{Z}$ by

$$u \sim v \quad \text{iff} \quad x^u = x^v.$$

Study this equivalence relation in the light of your work on GYHD 3. Does it look familiar?

In the example of $\langle \mathbf{R}_{90} \rangle$, the relation $\sim$ is the same as congruence mod 4; i.e., $(\mathbf{R}_{90})^u = (\mathbf{R}_{90})^v$ iff $u \equiv v \pmod 4$ (Verify). We will see more generally that, if the elements of the form x^t are not all distinct, then $\sim$ is in fact equal to the relation congruence mod n, for some n.

Why? How does this relation work? What is n?

If the elements of the form x^t are not all distinct, then there are integers $r > s$ with $x^r = x^s$, so $x^{r-s} = \mathbf{e}$. If we define $S = \{t \in \mathbf{N} : x^t = \mathbf{e}\}$, then we have $S \neq \varnothing$, and so S has a smallest element (by the Well-Ordering Principle).

This smallest integer in S plays an important role in group theory. We make the following definition:

DEFINITION 10.5: For a group G and an element $x \in G$ the *order* of x [written "$\mathbf{o}(x)$"] is defined as follows: Let $S = \{t \in \mathbf{N} : x^t = \mathbf{e}\}$. Then

i) if $S \neq \varnothing$, then $\mathbf{o}(x)$ is the smallest element of S;
ii) if $S = \varnothing$, we say x has *infinite order*, and write $\mathbf{o}(x) = \infty$.

Thus, if the powers of x are not all distinct, we can find the order of an element x by calculating $x^2, x^3, \ldots$, continuing until we get to $\mathbf{e}$ for the first time. The exponent needed to get back to $\mathbf{e}$ is the order.

We now look at the case where $\mathbf{o}(x)$ is finite, say $\mathbf{o}(x) = n$. We begin with the fact that the elements x^0 $(= \mathbf{e})$, x^1 $(= x)$, $x^2, \ldots, x^{n-1}$ are distinct. But then we start to repeat:

$$x^n = \mathbf{e}$$
$$x^{n+1} = x^n * x = \mathbf{e} * x = x$$
$$x^{n+2} = x^n * x^2 = \mathbf{e} * x^2 = x^2$$
$$\vdots$$

In general, we can use the Division Algorithm to write an arbitrary integer m in the form $nq + r$, with $q, r \in \mathbf{Z}$ and $0 \leq r < n$. Then

$$x^m = x^{nq+r} = (x^n)^q * x^r = \mathbf{e}^q * x^r = x^r$$

Thus, for $m \in \mathbf{Z}$, there is a (unique) integer r in $\{0, 1, \ldots, n-1\}$ such that $x^m = x^r$, and this is the same r such that $m \equiv r \pmod{n}$.

The discussion above leads to the following theorem, with some filling in of details needed for case 2:

THEOREM 10.12. Let G be a group and x an element of G. Then one of the following situations holds:

Case 1: $\mathbf{o}(x)$ is infinite, and the elements of the form x^t, $t \in \mathbf{Z}$, are all distinct.
Case 2: $\mathbf{o}(x) = n$ for some $n \in \mathbf{N}$, and $x^u = x^v$ iff $u \equiv v \pmod{n}$. In particular, $\langle x \rangle$ consists of the n distinct elements $\mathbf{e}, x, x^2, \ldots, x^{n-1}$.

Details for case 2 are left as Exercise 3.

COMMENT The description of the group $\langle x \rangle$ provided by this theorem explains the use of the term "cyclic", at least when $\mathbf{o}(x)$ is finite. The powers of x "cycle back" to $\mathbf{e}$ every nth term, repeating, over and over, the n distinct values $\mathbf{e}, x, x^2, \ldots, x^{n-1}$. This strongly resembles the operation of addition in the group $\mathbf{Z}_n$; we will discuss this resemblance with more precision in Section 10.3.

We defined the order of an element x as the smallest positive integer t such that $x^t = \mathbf{e}$. The following corollary to Theorem 10.12 gives another description of the order of an element:

COROLLARY 10.13. Let G be a group, and x an element of G with $\langle x \rangle$ finite. Then $\mathbf{o}(x) = |\langle x \rangle|$.

In other words, the order of an element is equal to the order of the cyclic subgroup it generates.

Lagrange's Theorem

We have used the term "order" in two different ways: the order of an element, as defined by Definition 10.5, and the order of a group, which is its cardinality. Corollary 10.13 suggests a basic consistency between these two meanings. The following gives you a chance to work with both of these concepts:

Get Your Hands Dirty 5

Find the order of each of these groups and the order of each element in each group:

a) $\mathbf{S}_n$, for $n = 2$ and 3;
b) $\mathbf{Z}_n$, for $n = 6, 9, 10$, and 12;
c) $\mathbf{U}_n$, for $n = 4, 7, 10$, and 20;
d) $\mathbf{D}_n$, for $n = 3$ and 4. □

What do you notice about the results from GYHD 5?

? ? ?

One of the most important basic observations in the study of finite groups is that the order of an individual element must be a divisor of the order of the group. This result is a special case of an even broader theorem, known as *Lagrange's Theorem*.

Lagrange's Theorem says the following:

THEOREM 10.14. Let G be any finite group and H any subgroup of G. Then the order of H is a divisor of the order of G. In symbols, $|H| \,|\, |G|$.

If $H = \langle x \rangle$, then we have $|H| = \mathbf{o}(x)$ (Corollary 10.13). This important special case of Lagrange's Theorem gives us the following:

COROLLARY 10.15. Let G be any finite group and x any element of G. Then $\mathbf{o}(x) \,|\, |G|$.

Before undertaking the proof of Lagrange's Theorem, we will look at some examples of groups and subgroups, in order to get some insight into why this divisibility relationship holds. The proof itself has one central idea: that we can create a partition of G,

Historical Note: Joseph Louis de Lagrange (1736–1813) was a French mathematician who made many contributions to the theory of equations, and first stated this theorem in 1770. He was also the first to prove that any positive integer can be written as the sum of the squares of four integers. (Some integers can be written as a sum of squares in many ways, some of which may use more than four squares, and others of which may use fewer. Lagrange proved that one never needs to use more than four squares.)

based on H, in which every part has size $|H|$. Our examples will be directed toward building an intuitive understanding of this partition.

Example 3. **The Symmetric Group $\mathbf{S}_n$** Recall that $\mathbf{S}_n$ is the group of all bijective functions from $\{1, 2, \ldots, n\}$ to itself, and that $|\mathbf{S}_n| = n!$.

Let L be the subgroup of $\mathbf{S}_n$ defined as follows:

$$L = \{f \in \mathbf{S}_n : f(n) = n\}.$$

(Verify that this is, in fact, a subgroup.)

The size of L is exactly $(n-1)!$, since L is essentially identical with the set $\mathbf{S}_{n-1}$. Thus $|L|$ is a divisor of $|\mathbf{S}_n|$—more precisely, $|\mathbf{S}_n| = n \cdot |L|$.

We can think of $\mathbf{S}_n$ partitioned according to the value a particular element has on n. That is, for $f \in \mathbf{S}_n$, $f(n)$ must be one of the numbers in the set $\{1, 2, \ldots, n\}$. We can form a partition of $\mathbf{S}_n$ by putting elements f and g in the same equivalence class iff $f(n) = g(n)$.

Thus, define $L_i = \{f \in \mathbf{S}_n : f(n) = i\}$, so that $L = L_n$. What is the size of each set L_i?

? ? ?

In fact, all of these n sets are the same size; namely, each has size $(n - 1)!$. We leave it to you to verify this, by finding one-to-one, onto functions from each L_i to L. □

Example 4. **The Dihedral Group $\mathbf{D}_n$** Recall that $\mathbf{D}_n$ is the group of all symmetries of an n-sided regular polygon, and $|\mathbf{D}_n| = 2n$.

Let T be the set of rotations in $\mathbf{D}_n$, so that $|T| = n$. Notice that $|T| \,\big|\, |\mathbf{D}_n|$: specifically, $|\mathbf{D}_n| = 2 \cdot |T|$.

We can partition $\mathbf{D}_n$ into the two sets T and $\mathbf{D}_n - T$, each of which has exactly n elements. $\mathbf{D}_n - T$ is the set of flips, or *reflections*. □

Example 5. **The Modular Group $\mathbf{Z}_n$** Recall that $\mathbf{Z}_n = \{[0], [1], [2], \ldots, [n - 1]\}$, so that $\mathbf{Z}_n$ is a group of order n.

Now, consider the specific case $n = 20$, and define the subgroup K as follows:

$$K = \{[0], [5], [10], [15]\},$$

so that K has order 4. (Verify that K is a subgroup.) Again, we have the order of the subgroup dividing the order of the group. Here, $|\mathbf{Z}_{20}| = 5 \cdot |K|$.

We will partition $\mathbf{Z}_{20}$ into five sets of size 4 each, as follows:

$$K_0 = \{[0], [5], [10], [15]\} \ (= K);$$

$$K_1 = \{[1], [6], [11], [16]\};$$

$$K_2 = \{[2], [7], [12], [17]\};$$

$$K_3 = \{[3], [8], [13], [18]\};$$

$$K_4 = \{[4], [9], [14], [19]\}.$$

We can describe K_t as the equivalence classes of those integers which, when divided by 5, give remainder t.

This example generalizes easily, replacing 20 by any integer n, and replacing K by the multiples of some divisor of n. □

In each of the above examples, we have a group G, a subgroup H, and a partition of G into subsets—equivalence classes—which each have the same size as H. In each case, the equivalence classes have the following interesting characteristic:

> If an element x of one of the equivalence classes is combined, using the group operation, with an element y of the subgroup, the result $x * y$ is in the same equivalence class as x.

Thus, for example, in Example 5, let $x = [2]$ and consider the sum $x +_{20} y$ for each element y in K:

$$[2] +_{20} [0] = [2]$$

$$[2] +_{20} [5] = [7]$$

$$[2] +_{20} [10] = [12]$$

$$[2] +_{20} [15] = [17]$$

The four sums, [2], [7], [12], and [17], are precisely the four elements of K_2, the equivalence class containing [2].

Get Your Hands Dirty 6

a) (From Example 3, with $n = 4$.) Let L be the subgroup of $\mathbf{S}_4$ consisting of all elements f such that $f(4) = 4$, and let x be the function given by $x(1) = 3$, $x(2) = 4$, $x(3) = 2$, $x(4) = 1$. In the notation of Example 3, we have $x \in L_1$, since $x(4) = 1$. Find the composition $x \circ y$ for each element $y \in L$, and verify that all such elements belong to L_1.

b) (From Example 4, with $n = 4$.) Let T be the group of rotations in $\mathbf{D}_4$, i.e., $T = \{\mathbf{R}_0, \mathbf{R}_{90}, \mathbf{R}_{180}, \mathbf{R}_{270}\}$, and let $x = \mathbf{H}$. Find the composition $x \circ y$ for each element $y \in T$, and verify that all such elements belong to $\mathbf{D}_4 - T$. □

These examples suggest the following definition:

DEFINITION 10.6: For a group G, a subgroup H, and an element $x \in G$, the *left coset xH* of H is

$$\{x * h : h \in H\}.$$

Right cosets can be defined by replacing "$x * h$" with "$h * x$". The two are not necessarily identical, but have similar properties. Either one will suffice for our needs.

We will prove the following two statements about cosets:

LEMMA 10.16. Suppose G is a group, H is a finite subgroup of G, and x an element of G. Then $|xH| = |H|$.

LEMMA 10.17. Suppose G is a group and H is a finite subgroup of G. Then $\{xH : x \in G\}$ forms a partition of G.

Assuming these two statements for the moment, we can easily prove Lagrange's Theorem (Theorem 10.14) as follows:

Proof of Lagrange's Theorem: Let G be a finite group, H a subgroup, and $\mathscr{C}$ the partition described by Lemma 10.17. Since each equivalence class has the same size $|H|$, we have

$$|G| = |\mathscr{C}| \cdot |H|,$$

where $|\mathscr{C}|$ is the number of parts in the partition, i.e., the number of equivalence classes.

This shows that $|H| \big| |G|$, as desired. ■

We next prove the two lemmas, beginning with Lemma 10.16:

Proof of Lemma 10.16: Define $f: H \to xH$ by $f(y) = x * y$. This function f is onto, by the definition of xH, and it is one-to-one, by the Cancellation Law (Lemma 10.7), i.e., if $f(y) = f(z)$, then $x * y = x * z$, and so $y = z$.

Since there is a bijective function from H to xH, these two sets have the same cardinality. ■

We will now show that the left coset xH is the equivalence class of x for a certain equivalence relation. Since we showed in Chapter 6 that the equivalence classes of an equivalence relation form a partition (Theorem 6.9), this will prove Lemma 10.17.

Proof of Lemma 10.17: Define a relation $\sim$ on G as follows:

$$y \sim z \quad \text{iff} \quad z^{-1} * y \in H.$$

First of all, we claim that $\sim$ is an equivalence relation. We need to prove three things:

a) $\sim$ is reflexive: If $y \in G$, then "$y \sim y$" means "$y^{-1} * y \in H$". But $y^{-1} * y = \mathbf{e}$, and $\mathbf{e} \in H$ (since H is a subgroup), so the statement "$y^{-1} * y \in H$" is true.

b) $\sim$ is symmetric: If $y, z \in G$, with $y \sim z$, then $z^{-1} * y \in H$. But $y^{-1} * z = (z^{-1} * y)^{-1}$. Since H is a subgroup, the inverse of its element $z^{-1} * y$ must also be in H. Thus $y^{-1} * z \in H$, and so $z \sim y$.

c) $\sim$ is transitive: If y, z, and $w \in H$, with $y \sim z$ and $z \sim w$, then $z^{-1} * y \in H$ and $w^{-1} * z \in H$. By the closure property of subgroups, we have $(w^{-1} * z) * (z^{-1} * y) \in H$. But this product simplifies to $w^{-1} * y$ (verify), so $y \sim w$, as desired.

Since $\sim$ is an equivalence relation, its equivalence classes form a partition. We claim that the equivalence class $[x]$ (with respect to $\sim$) is precisely xH. The proof is as follows:

$$\begin{aligned} y \in [x] \quad &\text{iff} \quad y \sim x \text{ (by definition of } [x]); \\ &\text{iff} \quad x^{-1} * y \in H \text{ (by definition of } \sim); \\ &\text{iff} \quad x^{-1} * y = h \text{ for some } h \in H; \\ &\text{iff} \quad y = x * h \text{ for some } h \in H \\ &\qquad \text{(multiplying the previous equation on the left by } x); \\ &\text{iff} \quad y \in xH \text{ (by definition of } xH). \end{aligned}$$

■

COMMENTS

1. Each of the three requirements of an equivalence relation followed directly from a specific one of the three requirements that make H a subgroup, as described in Theorem 10.9. The properties of an equivalence relation match up with the properties of a subgroup as follows:

   reflexivity ↔ identity element

   symmetry ↔ inverse elements

   transitivity ↔ closure

2. The fact that the cosets form a partition can be proved directly, without use of equivalence relations. You are asked for such a proof in Exercise 11.

3. You may find it helpful to compare the abstract description of the partition as described in this proof with the specific examples which preceded the proof (Examples 3, 4, and 5). Just as the specific examples are supposed to help you understand the abstraction, so also the general principles should help you understand and appreciate the specific examples.

We state here one more general consequence of Lagrange's Theorem. More precisely, this is a consequence of Corollary 10.15.

COROLLARY 10.18. Let G be a finite group and x an element of G. Then $x^{|G|} = \mathbf{e}$.

Proof: By Definition 10.5, we have that $x^{\mathbf{o}(x)} = \mathbf{e}$. Since $\mathbf{o}(x) \mid |G|$, we have $|G| = \mathbf{o}(x) \cdot t$ for some $t \in \mathbf{Z}$.

Therefore $x^{|G|} = (x^{\mathbf{o}(x)})^t = \mathbf{e}^t = \mathbf{e}$. ■

We conclude this section with a fascinating special case of this corollary, obtained by applying it to the group $\mathbf{U}_p$, where p is a prime.

We defined $\mathbf{U}_n$ as $\{[t] \in \mathbf{Z}_n : \text{GCD}(t, n) = 1\}$. Corollary 10.4 (Section 10.1) stated that this set is a group under multiplication mod n.

Now consider the case where $n = p$, a prime: we have $\text{GCD}(t, p) = 1$ except when $p \mid t$. Thus $\mathbf{U}_p = \{[1], [2], \ldots, [p-1]\}$, and so $|\mathbf{U}_p| = p - 1$. Thus we have the following:

COROLLARY 10.19. Suppose p is a prime, $t \in \mathbf{Z}$, and $[t]$ is the equivalence class of t in $\mathbf{Z}_p$. If $p \nmid t$, then $[t]^{p-1} = [1]$.

This result becomes more interesting with a little clarification. By the definition of $\cdot_n$ (Definition 7.7), we have $[t]^{p-1} = [t^{p-1}]$. Thus, the conclusion of Corollary 10.19 can be rewritten as "$[t^{p-1}] = [1]$". But for two integers to be in the same equivalence class of $\mathbf{Z}_n$ means that their difference is divisible by n. We can thus rewrite this conclusion one more way: $p \mid (t^{p-1} - 1)$.

Stating the result in this form, we get the following version of Corollary 10.19, which is known as *Fermat's Little Theorem:*

COROLLARY 10.19′. Suppose p is a prime and $t \in \mathbf{Z}$. If $p \nmid t$, then $p \mid (t^{p-1} - 1)$.

COMMENT The proof we have given of Fermat's Little Theorem is an excellent example of the power of abstraction applied to a concrete situation. We used such concepts as group and subgroup, equivalence class and partition, coset and order of an element, and applied them to the group $\mathbf{U}_n$. Our description of this group was itself based on the GCD–Linear Combinations Theorem. All these ideas have been combined to tell us, for example, that $24^{36} - 1$ is divisible by 37, without any computation besides the knowledge that 37 is a prime.

EXERCISES

1. (GYHD 2) Give an example of a group G and subgroups H and K such that $H \cup K$ is not a subgroup of G.
2. (GYHD 3)
   - **a)** For each group G and element x given below, list explicitly each of the elements of $\langle x \rangle$:
     - **i)** $G = \mathbf{Z}_{12}$, $x = [3]$.
     - **ii)** $G = \mathbf{D}_4$, $x = \mathbf{R}_{90}$.
     - **iii)** $G = \mathbf{U}_{11}$, $x = [4]$.
     - **iv)** $G = \mathbf{S}_4$, $x =$ the function with $x(1) = 2$, $x(2) = 4$, $x(3) = 1$, $x(4) = 3$.
     - **v)** $G = \mathbf{C} - \{0\}$ (nonzero complex numbers) under multiplication, $x = (1 + \sqrt{-3})/2$. (It's possible—just be careful.)
   - **b)** For each group G and element x given below, describe the set $\langle x \rangle$:
     - **i)** $G = \mathbf{Z}$ (under addition), $x = 5$.
     - **ii)** $G = \mathbf{Q}^+$ (under multiplication), $x = \frac{2}{3}$.
     - **iii)** $G = \mathbf{R}$ (under addition), $x = \pi$.

3. Prove the following:

 THEOREM 10.12. Case 2. Let G be a group and x an element of G, with $\mathbf{o}(x) = n$ for some $n \in \mathbf{N}$. Then $x^u = x^v$ iff $u \equiv v \pmod{n}$. In particular, $\langle x \rangle$ consists of the n distinct elements $\mathbf{e}, x, x^2, \ldots, x^{n-1}$.

4. (GYHD 5) Find the order of each of these groups and the order of each element in each group:
 a) $\mathbf{S}_n$, for $n = 2$ and 3.
 b) $\mathbf{Z}_n$, for $n = 6$, 9, 10, and 12.
 c) $\mathbf{U}_n$, for $n = 4$, 7, 10, and 20.
 d) $\mathbf{D}_n$, for $n = 3$ and 4.
5. (GYHD 4) Prove that every cyclic group is Abelian.
6. Exercise 1 asks you to show that the union of subgroups is not necessarily a subgroup. Suppose G is a group with subgroups H and K, such that $H \cup K$ *is* a subgroup of G. What can you conclude about the relationship of H and K? (*Hint:* Try some arbitrary pairs of subgroups, and see why $H \cup K$ is not generally a subgroup of G.)
7. Suppose G is a group, H is a subgroup of G, and $x \in H$. Prove that $x^n \in H$ for every $n \in \mathbf{Z}$. (*Hint:* Use induction to show $x^n \in H$ for every $n \in \mathbf{N}$, and then use the identity and inverse properties for subgroups for $n \leq 0$.)
8. Let G be a group, and x an element of finite order. Suppose that $x^t = \mathbf{e}$. Prove that t is a multiple of $\mathbf{o}(x)$. (*Hint:* Use the Division Algorithm.)
9. Let G be a cyclic group, with $G = \langle x \rangle$. Prove that $G = \langle x^{-1} \rangle$.
10. Recall the following result:

 COROLLARY 5.29. If a and b are integers that are relatively prime, then there are integers x and y such that $ax + by = 1$.

 Now, let $G = \langle x \rangle$ be a cyclic group of order n, and $r \in \mathbf{Z}$. Prove the following:

$$G = \langle x^r \rangle \quad \text{iff} \quad \mathrm{GCD}(r, n) = 1.$$

 (Exercise 9 looked at the case $r = -1$. These two exercises show that there may be several individual elements that each generate a particular cyclic group.) [*Hint:* The harder direction of this exercise is showing that if $\mathrm{GCD}(r, n) = 1$, then $G = \langle x^r \rangle$. To do this, use Corollary 5.29 to express an arbitrary integer t as $ur + vn$, and show $x^t = (x^r)^u$. For the other direction, suppose $d = \mathrm{GCD}(r, n)$, with $d > 1$, and show $\mathbf{o}(x^r) < n$.]
11. Let G be a group, and H a subgroup (not necessarily finite). Prove directly that the left cosets of H form a partition of G; that is, prove each of the following statements:
 a) For all $x \in G$, there exists $a \in G$ such that $x \in aH$.
 b) For all $a \in G$, $aH \neq \emptyset$.
 c) For all $a, b \in G$, if $aH \cap bH \neq \emptyset$, then $aH = bH$.

10.3 Isomorphism

We conclude this chapter, and the text, with a discussion of what might be described as "the ultimate equivalence relation for group theory".

Comparing Groups

We have discussed many different groups, whose elements may be numbers, functions, motions, vectors, sets, etc., and with operations including addition and its variants, multiplication and its variants, composition, symmetric difference, etc. Yet, from the mathematician's perspective, some of these seemingly disparate groups are actually identical. The concept of *isomorphism*—meaning "same form"—is designed to describe what mathematicians mean when they say that two mathematical systems are "essentially the same". We will first look at two pairs of groups which "have the same form", and then examine a pair which do not.

Two Similar Infinite Groups

Our first example concerns a pair of infinite groups:

Example 1. **Comparing Z and $\{2^n : n \in \mathbf{Z}\}$** Consider the following two groups:

$$\mathbf{Z} = \{\ldots, -3, -2, -1, 0, 1, 2, 3, \ldots\}$$

and

$$G = \{2^n : n \in \mathbf{Z}\} = \{\ldots, \tfrac{1}{8}, \tfrac{1}{4}, \tfrac{1}{2}, 1, 2, 4, 8, \ldots\}.$$

From the way G is defined, there is a natural bijective function from $\mathbf{Z}$ to G, namely, $f(n) = 2^n$. But this function is more than just a function: it also correlates with the operations in the two groups, because we multiply elements of G by adding their exponents. We can describe the relationship between f and the two operations by the diagram in Figure 10.13.

What's happening in Figure 10.13 is this: We begin with a pair of integers, (m, n), and follow one of two paths. Either we go across and then down, or we go down and then across. In the first case, we add the integers, in $\mathbf{Z}$, to get $m + n$, and then apply the function f to $m + n$, giving 2^{m+n}. In the second case, we apply f to the individual integers of the pair (m, n) to get the pair $(2^m, 2^n)$, and then multiply the two components, as elements of G, to get $2^m \cdot 2^n$.

The key observation is that we get the same answer either way; i.e., $2^m \cdot 2^n$ and 2^{m+n} are equal. The following equation expresses this succinctly:

$$f(m + n) = f(m) \cdot f(n). \qquad (*)$$

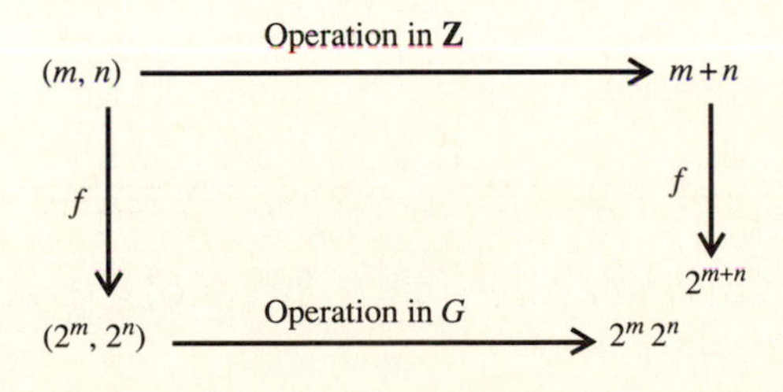

Figure 10.13. The two paths from (m, n) lead to equal results.

The operation on the left, addition, is the group operation in $\mathbf{Z}$; the operation on the right, multiplication, is the group operation in G.

We will see another version of the equation $(*)$ in our next example, and will generalize it in the definition of "isomorphism".

Two Similar Finite Groups

Our next example involves a pair of groups, each of which has exactly four elements:

Example 2. **Comparing $\mathbf{Z}_4$ and $\langle R_{90} \rangle$** Consider the following two groups:

$$\mathbf{Z}_4 = \{[0], [1], [2], [3]\}$$

and

$$G = \{\mathbf{R}_0, \mathbf{R}_{90}, \mathbf{R}_{180}, \mathbf{R}_{270}\} \subseteq \mathbf{D}_4 .$$

Once again, there is a natural function f from one to the other, with $f([t]) = (\mathbf{R}_{90})^t$. It may help to list f explicitly:

$$f: \begin{array}{l} [0] \to \mathbf{R}_0 \\ [1] \to \mathbf{R}_{90} \\ [2] \to \mathbf{R}_{180} \\ [3] \to \mathbf{R}_{270} \end{array}$$

Once again, the function behaves nicely with regard to the two operations. Combining rotations by composition matches up with combining elements of $\mathbf{Z}_4$ by addition mod 4. Specifically, for $[m]$ and $[n]$ in $\mathbf{Z}_4$, we have

$$f([m] +_4 [n]) = f([m]) \circ f([n]). \qquad (**)$$

This is the analog of the equation $(*)$ in Example 1. In $(**)$, we have the operation in $\mathbf{Z}_4$ on the left and the operation in $\mathbf{D}_4$ on the right. On the left, we add (mod 4) and then apply f; on the right, we apply f, and then do the composition.

Figure 10.14 is the analog of Figure 10.13. The left side of $(**)$ represents the path "across and down"; the right side, the path "down and across". Equation $(**)$ says that the results of the two paths are the same.

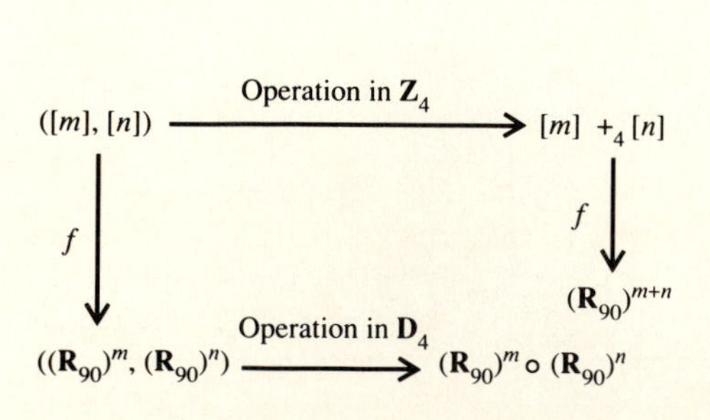

Figure 10.14. As in Figure 10.13, the "across and down" path gives the same result as the "down and across" path.

Two Dissimilar Finite Groups

Our third example, like the second, concerns two groups of order 4:

Example 3. **Comparing $\mathbf{Z}_4$ and $\{\mathbf{R}_0, \mathbf{R}_{180}, \mathbf{H}, \mathbf{V}\}$** It may help to look at an example which does not work out so nicely.

Let us look at a comparison of $\mathbf{Z}_4$ with a different subgroup of $\mathbf{D}_4$, namely, $G = \{\mathbf{R}_0, \mathbf{R}_{180}, \mathbf{H}, \mathbf{V}\}$. (Verify, using the operation table for $\mathbf{D}_4$, that this is actually a subgroup of $\mathbf{D}_4$.)

Once again, G has four elements, and we can find a bijective function between the two groups. Unfortunately, it is not so obvious what the right function is. Let us try the following:

$$f: \begin{array}{ll} [0] \to \mathbf{R}_0 & \text{(at least we'll match up the identity elements)} \\ [1] \to \mathbf{R}_{180} & \\ [2] \to \mathbf{H} & \\ [3] \to \mathbf{V} & \end{array}$$

If we use $m = 1$ and $n = 2$, we have the following:

$$[m] +_4 [n] = [3] \quad \text{and} \quad f([m] +_4 [n]) = f[3] = \mathbf{V}$$

while

$$f([m]) = \mathbf{R}_{180}, \qquad f([n]) = \mathbf{H}, \quad \text{and} \quad \mathbf{R}_{180} \circ \mathbf{H} = \mathbf{V}.$$

Thus, in this case, we have

$$f([m] +_4 [n]) = f([m]) \circ f([n]), \tag{**}$$

as in Example 2.

On the other hand, suppose we take $m = 1$ and $n = 3$. Then

$$[m] +_4 [n] = [0] \quad \text{and} \quad f([m] +_4 [n]) = f([0]) = \mathbf{R}_0$$

while

$$f([m]) = \mathbf{R}_{180}, \qquad f([n]) = \mathbf{V}, \quad \text{and} \quad \mathbf{R}_{180} \circ \mathbf{V} = \mathbf{H}.$$

We thus have

$$f([m] +_4 [n]) \neq f([m]) \circ f([n]).$$

You may suggest that perhaps we used the wrong function from $\mathbf{Z}_4$ to G. This is a healthy sort of skeptical response. There are actually 24 (= 4!) distinct bijective functions between these two sets. It is certainly possible that one of these, say g, actually does satisfy the equation "$g([m] +_4 [n]) = g([m]) \circ g([n])$", for all m and n, even if f does not.

But in fact no bijective function from $\mathbf{Z}_4$ to G has this property.

That is, if g is *any* bijective function from $\mathbf{Z}_4$ to G, there will be *at least one pair* of elements $[m], [n] \in \mathbf{Z}_4$ which does not satisfy the equation

$$g([m] +_4 [n]) = g([m]) \circ g([n]) . \qquad (***)$$

One method of proving that there is no bijective function satisfying (***) is "by brute force": that is, for each of the 24 bijective functions from $\mathbf{Z}_4$ to G, we explicitly find a pair of elements that make (***) false.

An alternative approach is to prove this by means of general theorems. One such theorem states that there cannot be a bijective function between two groups, satisfying the appropriate generalization of (*), if one of the groups is cyclic and the other is not (verify that $\mathbf{Z}_4$ is cyclic and G is not). The proof of this theorem is part of Exercise 4, which includes other theorems about the nonexistence of bijective functions satisfying the appropriate analog of (*).

Thus, there is something fundamentally different about the two groups $\mathbf{Z}_4$ and G in this case. They cannot be "matched up" as groups.

Comparison Using Operation Tables

From the mathematician's point of view—or here, more particularly, the group theorist's point of view—the names of the elements in a group are irrelevant, and the type of operation is also irrelevant. What matters is how the elements are interrelated by the operation—that is, what the *internal structure* of the group is like.

We can illustrate this idea graphically by looking at the operation tables for the two groups $\mathbf{Z}_4$ and $G = \langle \mathbf{R}_{90} \rangle$ from Example 2 (see Figure 10.15). These two tables are connected in the following way:

Recall that our function $f: \mathbf{Z}_4 \to \langle \mathbf{R}_{90} \rangle$ in Example 2 was defined as follows:

$$f: \begin{array}{l} [0] \to \mathbf{R}_0 \\ [1] \to \mathbf{R}_{90} \\ [2] \to \mathbf{R}_{180} \\ [3] \to \mathbf{R}_{270} \end{array}$$

$+_4$	[0]	[1]	[2]	[3]
[0]	[0]	[1]	[2]	[3]
[1]	[1]	[2]	[3]	[0]
[2]	[2]	[3]	[0]	[1]
[3]	[3]	[0]	[1]	[2]

Operation table for $\mathbf{Z}_4$

$\xrightarrow{f}$

$\circ$	$\mathbf{R}_0$	$\mathbf{R}_{90}$	$\mathbf{R}_{180}$	$\mathbf{R}_{270}$
$\mathbf{R}_0$	$\mathbf{R}_0$	$\mathbf{R}_{90}$	$\mathbf{R}_{180}$	$\mathbf{R}_{270}$
$\mathbf{R}_{90}$	$\mathbf{R}_{90}$	$\mathbf{R}_{180}$	$\mathbf{R}_{270}$	$\mathbf{R}_0$
$\mathbf{R}_{180}$	$\mathbf{R}_{180}$	$\mathbf{R}_{270}$	$\mathbf{R}_0$	$\mathbf{R}_{90}$
$\mathbf{R}_{270}$	$\mathbf{R}_{270}$	$\mathbf{R}_0$	$\mathbf{R}_{90}$	$\mathbf{R}_{180}$

Operation table for $G = \langle \mathbf{R}_{90} \rangle$

Figure 10.15. If every element in the table for $\mathbf{Z}_4$ is replaced by its image under f, we get precisely the table for G.

If we replace every entry in the operation table for $\mathbf{Z}_4$ by the corresponding element of G, we get precisely the second table, the operation table for G.

In other words: *the only difference between the two groups is the names of the elements*. The *structure,* or *form,* of the groups is identical.

Now suppose we apply this replacement process to the table for $\mathbf{Z}_4$, but using a different function, namely our function in Example 3 from $\mathbf{Z}_4$ to $\{\mathbf{R}_0, \mathbf{R}_{180}, \mathbf{H}, \mathbf{V}\}$ given by

$$f: \begin{array}{l} [0] \to \mathbf{R}_0 \\ [1] \to \mathbf{R}_{180} \\ [2] \to \mathbf{H} \\ [3] \to \mathbf{V} \end{array}$$

If we replace each element of $\mathbf{Z}_4$ by its image in $\mathbf{D}_4$ under this function, we get the correspondence shown in Figure 10.16.

The table on the right of Figure 10.16 is not the table for the operation of composition on the set $\{\mathbf{R}_0, \mathbf{R}_{180}, \mathbf{H}, \mathbf{V}\}$. (Although that table defines a perfectly good operation—in fact, the operation for a group—that operation is not the composition operation of $\mathbf{D}_4$). The correct table for G is shown in Figure 10.17.

Isomorphism—Definition and Theorems

The question we have been discussing here—whether certain groups have the same form—can be expressed very succinctly in terms of the operations on the groups and a function from one to the other. We have the following:

> **DEFINITION 10.7:** For groups G and H, with operations $*$ and $**$, respectively, an *isomorphism* from G to H is a function $f: G \to H$ such that:
>
> **i)** f is bijective,
> **ii)** for all $x, y \in G$, $f(x * y) = f(x) ** f(y)$.
>
> Any function satisfying condition ii) is said to *preserve the operation,* and is called *operation-preserving*.
>
> "G is *isomorphic* to H" (written $G \cong H$) means: there is an isomorphism from G to H.

We have discussed condition ii)—that f is "operation-preserving"—in terms of the specific operations and functions in Examples 1 and 2. In Example 1, we had the function $f(n) = 2^n$, which we saw satisfied the equation

$$f(m + n) = f(m) \cdot f(n),$$

$+_4$	[0]	[1]	[2]	[3]
[0]	[0]	[1]	[2]	[3]
[1]	[1]	[2]	[3]	[0]
[2]	[2]	[3]	[0]	[1]
[3]	[3]	[0]	[1]	[2]

Operation table for $\mathbf{Z}_4$

$\xrightarrow{f}$

?	$\mathbf{R}_0$	$\mathbf{R}_{180}$	**H**	**V**
$\mathbf{R}_0$	$\mathbf{R}_0$	$\mathbf{R}_{180}$	**H**	**V**
$\mathbf{R}_{180}$	$\mathbf{R}_{180}$	**H**	**V**	$\mathbf{R}_0$
H	**H**	**V**	$\mathbf{R}_0$	$\mathbf{R}_{180}$
V	**V**	$\mathbf{R}_0$	$\mathbf{R}_{180}$	**H**

This is *not* the operation table for $G = \{\mathbf{R}_0, \mathbf{R}_{180}, \mathbf{H}, \mathbf{V}\}$

Figure 10.16. Replacing elements of $\mathbf{Z}_4$ by their images under f gives something which is *not* the operation table for $\{\mathbf{R}_0, \mathbf{R}_{180}, \mathbf{H}, \mathbf{V}\}$.

and in Example 2, we had the function $f([t]) = (\mathbf{R}_{90})^t$, which we saw satisfied the equation

$$f([m] +_4 [n]) = f([m]) \circ f([n]) .$$

Thus, in the course of those examples, we proved that the group $\mathbf{Z}$ (under addition) is isomorphic to the group $\{2^n : n \in \mathbf{Z}\}$ (under the operation of multiplication), and that the group $\mathbf{Z}_4$ (under addition mod 4) is isomorphic to the subgroup of $\mathbf{D}_4$ consisting of the rotations $\{\mathbf{R}_0, \mathbf{R}_{90}, \mathbf{R}_{180}, \mathbf{R}_{270}\}$ (under composition).

On the other hand, the function of Example 3 was not operation-preserving. In fact, as we asserted there, there is no bijective, operation-preserving function between the groups $\mathbf{Z}_4$ and $\{\mathbf{R}_0, \mathbf{R}_{180}, \mathbf{H}, \mathbf{V}\}$, and so those groups are not isomorphic.

Get Your Hands Dirty 1

In each of the following examples, G_1 and G_2 are groups and f is a function from G_1 to G_2. Determine if f is operation-preserving:

a) $G_1 = \mathbf{Z}_4, G_2 = \mathbf{U}_5, f([0]) = [1], f([1]) = [2], f([2]) = [4], f([3]) = [3]$.
b) $G_1 = \mathbf{Z}_4, G_2 = \mathbf{U}_5, f([0]) = [1], f([1]) = [3], f([2]) = [2], f([3]) = [4]$.
c) $G_1 = \mathbf{Z}_6, G_2 = \mathbf{Z}_6, f([0]) = [0], f([1]) = [2], f([2]) = [4], f([3]) = [0]$, $f([4]) = [2], f([5]) = [4]$. □

The groups discussed in Examples 1 and 2 were all cyclic, and we will be proving a powerful theorem about isomorphism of cyclic groups. First we want to make some general observations about the concept of isomorphism.

∘	$\mathbf{R}_0$	$\mathbf{R}_{180}$	**H**	**V**
$\mathbf{R}_0$	$\mathbf{R}_0$	$\mathbf{R}_{180}$	**H**	**V**
$\mathbf{R}_{180}$	$\mathbf{R}_{180}$	$\mathbf{R}_0$	**V**	**H**
H	**H**	**V**	$\mathbf{R}_0$	$\mathbf{R}_{180}$
V	**V**	**H**	$\mathbf{R}_{180}$	$\mathbf{R}_0$

Figure 10.17. The correct operation table for the group $\{\mathbf{R}_0, \mathbf{R}_{180}, \mathbf{H}, \mathbf{V}\}$

COMMENTS

1. We want to emphasize that the statement "G is isomorphic to H" is an *existence* statement. To prove this for a particular pair of groups, we need to identify a specific bijective function, either explicitly or through other theorems, which has the operation-preserving property. This is precisely what we did in Examples 1 and 2. The fact that "G is isomorphic to H" is an existence statement means that its negation is a universal statement. In other words, as we pointed out in our discussion of Example 3, to prove that two groups are not isomorphic, we must show that every function from one to the other fails to satisfy at least one of the conditions of Definition 10.7. Exercise 4 suggest some ways of doing this indirectly.

2. The term "isomorphism" is used in many other mathematical contexts besides group theory. For groups, the defining features are a set and an operation, and so the concept of isomorphism of groups is defined in terms of these aspects. The condition that f be bijective says that the underlying *sets* are "essentially the same"—that is, they have the same number of elements; and the condition that f be operation-preserving says that the *operations* on the two underlying sets are "essentially the same"—that is, if the elements of the sets are matched up properly, then the operations will be identical.

3. An operation-preserving function (not necessarily bijective) from one group to another is called a *homomorphism*. Many results about isomorphisms can be proved more generally about homomorphisms. Exercises 1, 2, 10, and 11 give some of the basic results.

Isomorphism as an Equivalence Relation

We described isomorphism at the beginning of this section as "the ultimate equivalence relation for group theory". We now justify this (at least partially—the word "ultimate" is perhaps hyperbole).

We have a technical difficulty in that we cannot talk about "the set of all groups", just as we couldn't refer to the "set of all sets" when we wanted to show that equivalence of sets was an equivalence relation. For the purposes of stating this theorem correctly, we let **U** be some unspecified universe of discourse and **G** be the set of all groups whose underlying sets are subsets of **U**.

THEOREM 10.20. Isomorphism is an equivalence relation on **G**.

Proof: As usual, we must verify three conditions.

i) $\sim$ is reflexive: Suppose $G \in \mathbf{G}$, with operation $(*)$. We need to find an isomorphism from G to G. The identity function I_G on G is bijective, and the requirement that I_G be operation-preserving simply says "$I_G(x * y) = I_G(x) * I_G(y)$". Since $I_G(t) = t$ for all t, this condition becomes just "$x * y = x * y$", which is true (by the definition of equality).

ii) $\sim$ is symmetric: Suppose $G, H \in \mathbf{G}$, with operations $*$ and $**$ respectively, and suppose that $G \cong H$, so that there is an isomorphism $f: G \to H$. We must show that $H \cong G$, which we will do by showing that $f^{-1}: H \to G$ is an isomorphism. Since f is bijective, its inverse $f^{-1}: H \to G$ is also bijective, by Corollary 6.5.

To show that f^{-1} preserves the operation, suppose $u, v \in H$. We need to show $f^{-1}(u ** v) = f^{-1}(u) * f^{-1}(v)$. Let $x = f^{-1}(u ** v)$ and $y = f^{-1}(u) * f^{-1}(v)$, so we want $x = y$. We will prove this by showing that $f(x) = f(y)$. Since f is injective, the equation $x = y$ will follow. So we compute $f(x)$ and $f(y)$:

We have $f(x) = f(f^{-1}(u ** v)) = u ** v$, since the composition of f and f^{-1} is the identity function (Theorem 6.7).

On the other hand, $f(y) = f(f^{-1}(u) * f^{-1}(v))$. We now apply the fact that f is operation-preserving to the elements $f^{-1}(u)$ and $f^{-1}(v)$ of G. This gives

$$\begin{aligned} f(f^{-1}(u) * f^{-1}(v)) &= f(f^{-1}(u)) ** f(f^{-1}(v)) \\ &= u ** v \end{aligned}$$

(again using the fact that $f \circ f^{-1}$ is the identity function).

Thus $f(x) = f(y)$, and so $x = y$.

iii) $\sim$ is transitive: Suppose $G, H, K \in \mathbf{G}$, with operations $*$, $**$, and $***$ respectively. Suppose further that $G \cong H$ and $H \cong K$, so that there are isomorphisms $f: G \to H$ and $g: H \to K$. We claim that the composition function $g \circ f: G \to K$ is an isomorphism from G to K.

Since g and f are bijective, it follows that $g \circ f$ is bijective (Corollary 4.7).

To show that $g \circ f$ is operation-preserving, suppose that $x, y \in G$. Then

$$\begin{aligned} (g \circ f)(x * y) &= g(f(x * y)) && \text{(by the definition of composition)} \\ &= g(f(x) ** f(y)) && \text{(since } f \text{ is operation-preserving)} \\ &= g(f(x)) *** g(f(y)) && \text{(since } g \text{ is operation-preserving)} \\ &= (g \circ f)(x) *** (g \circ f)(y) && \text{(by the definition of composition)}. \end{aligned}$$

Comparing the first and last terms in this sequence, we have precisely the statement that $g \circ f$ is operation-preserving. ∎

Isomorphism of Cyclic Groups

Now we return to cyclic groups for our final results. The following theorem generalizes Examples 1 and 2.

THEOREM 10.21. Suppose $G = \langle x \rangle$ is a cyclic group. Then

i) if G is infinite, then $\mathbf{Z} \cong G$;
ii) if G is finite, with $|G| = n$, then $\mathbf{Z}_n \cong G$.

Proof: We first consider case i), where G is infinite. Define a function $f: \mathbf{Z} \to G$ by

$$f(t) = x^t.$$

Since G is cyclic, f is surjective. By case 1 of Theorem 10.12, f is one-to-one. We need to show that f is operation-preserving; i.e., we must prove

$$f(s + t) = f(s) * f(t).$$

But this simply says $x^{s+t} = x^s * x^t$, which is the rule for exponents. Thus f is operation-preserving, and so f is an isomorphism.

Now suppose G is finite, with $|G| = n$. Let $\overline{f}$ be the *relation* from $\mathbf{Z}_n$ to G defined by

$$\overline{f} = \{([u], x^u) : u \in \mathbf{Z}\}.$$

Since $[u]$ can be any element of $\mathbf{Z}_n$, $\overline{f}$ is domain-complete. Also, case 2 of Theorem 10.12 states that $x^u = x^v$ iff $u \equiv v \pmod{n}$, i.e., iff $[u] = [v]$. Therefore, both $\overline{f}$ and $[\overline{f}]^{-1}$ are functionlike. Since $\overline{f}$ is both domain-complete and functionlike, it is a function (by Theorem 6.1). Since $[\overline{f}]^{-1}$ is functionlike, $\overline{f}$ is one-to-one (by Theorem 6.3).

As in case i), $\overline{f}$ is surjective, because G is cyclic. We need to show that $\overline{f}$ is operation-preserving; i.e., we must prove

$$\overline{f}([s] + [t]) = \overline{f}([s]) * \overline{f}([t]).$$

But $[s] + [t] = [s + t]$, and $\overline{f}([u]) = x^u$, so

$$\overline{f}([s] + [t]) = \overline{f}([s + t]) = x^{s+t} = x^s * x^t = \overline{f}([s]) * \overline{f}([t]),$$

which completes the proof. ■

Finally, Theorem 10.21 combines with Theorem 10.20 to prove the following:

COROLLARY 10.22. Suppose G and H are cyclic groups with $|G| = |H|$ (including the case where both groups are infinite). Then $G \cong H$.

Proof: By Theorem 10.21, G and H are either both isomorphic to $\mathbf{Z}$ (if they are infinite) or both isomorphic to $\mathbf{Z}_n$ for the same n (if they are finite). Since isomorphism is an equivalence relation on any set of groups, it follows that $G \cong H$. ■

EXERCISES

Throughout these exercises, G and H are groups, with identity elements $\mathbf{e}_G$ and $\mathbf{e}_H$ respectively.

1. Suppose that $f: G \to H$ is an operation-preserving function. Prove that $f(\mathbf{e}_G) = \mathbf{e}_H$. (*Hint:* Show that $[f(\mathbf{e}_G)]^2 = f(\mathbf{e}_G)$, and apply Corollary 10.8.)
2. Suppose that $f: G \to H$ is an operation-preserving function, and that $x \in G$. Prove each of the following:
   a) $f(x^{-1}) = [f(x)]^{-1}$. (*Hint:* Use Exercise 1.)
   b) $f(x^n) = [f(x)]^n$.
   c) $\mathbf{o}(f(x)) \mid \mathbf{o}(x)$. [*Hint:* Use Exercise 1, Exercise 8 of Section 10.2, and part b) of this exercise.]
3. Suppose that $f: G \to H$ is an isomorphism, and that $x \in G$. Prove that $\mathbf{o}(f(x)) = \mathbf{o}(x)$.
4. **a)** Suppose G and H are two isomorphic groups. Prove each of the following statements:
   i) If G is Abelian, then H is Abelian.
   ii) If G is cyclic, then H is cyclic.
   iii) For any $n \in \mathbf{N}$, G and H have the same number of elements of order n.

b) Using part a), state three theorems each of which has the conclusion "G and H are not isomorphic".

5. Let L be the group consisting of the rotations of $\mathbf{D}_n$; i.e., $L = \langle \mathbf{R}_{360/n} \rangle$. Prove that $L \cong \mathbf{Z}_n$.
6. Prove that $\mathbf{D}_3 \cong \mathbf{S}_3$.
7. Let G be the group of six functions described in Exercise 12 of Section 10.1. Prove that $G \cong \mathbf{S}_3$.
8. Prove that $\mathbf{Z}_4 \cong \mathbf{U}_5$.
9. Prove that $\mathbf{Z}_4 \not\cong \mathbf{U}_{12}$.
10. Suppose that $f:G \to H$ is an operation-preserving function. Prove each of the following statements:
    a) If L is a subgroup of G, then $\{f(x): x \in L\}$ is a subgroup of H. [In particular, Im(f) is a subgroup of H.]
    b) If K is a subgroup of H, then $\{x \in G: f(x) \in K\}$ is a subgroup of G. [In particular, $\{x \in G: f(x) = \mathbf{e}_H\}$ is a subgroup of G. This is known as the *kernel* of f, written Ker(f).]
11. Suppose that $f: G \to H$ is an operation-preserving function, that $K = \text{Ker}(f)$ (see Exercise 10), and that $u \in \text{Im}(f)$. Let $T = \{x \in G: f(x) = u\}$. Prove that $T = yK$ for any $y \in T$.
12. We conclude with another context for the concept of isomorphism. Consider the following definition:

DEFINITION. For sets A and B, with relations $\sim_A$ and $\sim_B$ defined on A and B respectively, "$\sim_A$ is *isomorphic* to $\sim_B$" means: there is a bijective function $f: A \to B$ such that

$$(\forall\, x, y \in A)(x \sim_A y \Leftrightarrow f(x) \sim_B f(y)) .$$

Thus, for any sets A and B, isomorphism is a relation from $\mathcal{P}(A \times A)$ (the set of relations on A) to $\mathcal{P}(B \times B)$ (the set of relations on B).

a) If $A = B$, isomorphism becomes a relation on the set of relations on A. We will call this relation E. Specifically, if $\sim_1$ and $\sim_2$ are two relations on A, we define "$(\sim_1, \sim_2) \in E$" (or "$\sim_1 E \sim_2$") to mean

$$(\exists\, f: A \to A)\,[f \text{ is bijective and } (\forall\, x, y \in A)(x \sim_1 y \Leftrightarrow f(x) \sim_2 f(y))] .$$

Prove that, for any set A, this relation E is an equivalence relation on the set of all relations on A.

b) Prove that, if $\sim_A$ is isomorphic to $\sim_B$, and $\sim_A$ has any of the properties "reflexive", "irreflexive", "symmetric", "antisymmetric", or "transitive", then $\sim_B$ has the given property also.

c) Use part b) to prove that, if $\sim_A$ is isomorphic to $\sim_B$, and $\sim_A$ is a partial ordering, a total ordering, a strict ordering, or an equivalence relation, then $\sim_B$ is the same type of relation.

d) For $A = \{1, 2\}$, there are 16 relations on A. Explicitly list the equivalence classes of this set of relations under the equivalence relation of isomorphism.

e) In Exercise 5 of Section 7.2, we asked you to list explicitly all the partial orderings on $A = \{1, 2, 3\}$, and form a partition of them by putting together those partial orderings whose lattices were "the same shape".

   i) Review your work on that problem, and list the parts of that partition of the set of partial orderings.

ii) Next, use the equivalence relation described in part a) to form a partition of the set of partial orderings, listing the parts explicitly. [By part c), partial orderings can only be isomorphic to other partial orderings.]

iii) Compare the partitions in i) and ii).

Appendix A

Formal Logic

Introduction

In this appendix, we present a different approach to logic from that given in the main body of this book. The discussion here is more formal, and is based on the concept of a *proposition*. In Section A.1, we will examine the algebra for combining simple propositions into compound propositions, and introduce the idea of a *truth table*—a simple device for analyzing such combinations. In Section A.2, our main goal is to define what it means to say that two statements are *logically equivalent;* intuitively, this means that they "say the same thing" by virtue of their formal structure. We will discuss some important examples of logical equivalence, and then see how these facts about propositions can be used to justify the methods of proof used in the main body of the text.

This appendix and the presentation of logic in the main body of the text can be read independently of each other. However, the intuitive meaning of conditional sentences, as discussed in Section 1.3, is the motivating force for the formal definition of implication. This definition is an element of formal logic which students often find difficult, and even counter to their intuition. Therefore, you may want to read the discussion in Section 1.3 in connection with this appendix.

A.1 Propositions, Logical Operations, and Truth Tables

Propositions

We begin with the concept of a *proposition*. A proposition is a statement that is either true or false.

Example 1. Some Propositions The following are examples of propositions:

i) $2 + 3 = 5$.
ii) Emmy Noether was a twentieth-century algebraist.

iii) Every real number is positive.
iv) Euclid's *Elements* are only about geometry.

The first two of these statements are true; the other two are false. □

Since a proposition is, by definition, either true or false, we can assign a *truth value* to it. A true proposition is given the truth value "T" and a false proposition is given the truth value "F".

There are several types of sentences which are not propositions. One important example is an *open sentence,* which is a statement involving *variables*—symbols such as "x" or "y" which represented unspecified objects.

Example 2. Some Open Sentences The following are examples of open sentences:

i) $x + 3 > 9$,
ii) $y^2 - 4y - 5 = 0$.

These statements are neither true nor false, because we do not know what numbers are represented by "x" or "y". □

Other word combinations are not considered propositions because they do not make an assertion. Questions, commands, and incomplete sentences are examples of this.

Example 3. Other Types of Nonpropositions The following are neither propositions nor open sentences:

i) Why is every differentiable function continuous?
ii) Find the cardinality of the set of integers.
iii) $(11 - 3)^2$
iv) Once upon a time □

Logical Operations and Truth Tables

Just as there are various *arithmetic* operations, by which numbers are combined to yield other numbers, so there are also *logical* operations, by which existing propositions are combined to create other propositions, called *compound propositions*. The propositions being combined are called *component propositions*.

Formal logic is primarily concerned with the truth value of propositions, rather than their intrinsic meaning, and operations on propositions are defined in terms of truth values. The formal definitions of the operations just state how the truth value of each compound proposition is determined by the truth values of its components. Fortunately, each operation has an intuitive meaning as well, which we will give following the formal definition.

Conjunction and Disjunction

We begin with the following:

DEFINITION A.1: For propositions p and q, the *conjunction* of p and q (written "$p \wedge q$") is a proposition which is true if both p and q are true, and false otherwise.

Intuitively, the conjunction "$p \wedge q$" is the proposition "p and q"—i.e., the assertion that both p and q are true. For example, if p is the proposition "$3 < 7$" and q is the proposition "$4 + 2 = 9$", then "$p \wedge q$" is the proposition "$3 < 7$ and $4 + 2 = 9$". This proposition is false, since its second component is false.

DEFINITION A.2: For propositions p and q, the *disjunction* of p and q (written "$p \vee q$") is a proposition which is true if either p or q is true (or both are true), and false otherwise.

Intuitively, the conjunction "$p \vee q$" is the proposition "p or q"—i.e., the assertion that either p or q is true (or both are true). For example, if p is the proposition "$3 < 7$" and q is the proposition "$4 + 2 = 9$", then "$p \vee q$" is the proposition "$3 < 7$ or $4 + 2 = 9$". This proposition is true, since its first component is true.

Note: We want to emphasize that, in mathematics, "or" has the meaning of "inclusive or"—that is, it includes the situation when both components are true. If p is, as above, the proposition "$3 < 7$", and r is the proposition "$4 + 2 = 6$", then $p \vee r$ is defined to be true. The operation of "exclusive or" is discussed in Exercise 11.

Truth Tables

The truth value of the conjunction and disjunction of two propositions depends only on the truth values of the components, and there are only two possibilities for the truth value of each component. Therefore, Definitions A.1 and A.2 could be made by simply listing each possible combination of truth values for the components, and indicating the truth value of the compound proposition. A table giving this type of information is called a *truth table*.

Figures A.1 and A.2 show truth tables which can be used to define conjunction and disjunction. For example, the boxed entry in Figure A.1 states that, if p is false and q is true, then $p \wedge q$ is false.

As with arithmetic operations, the use of parentheses can have a significant effect on logical operations. The following GYHD asks you, among other things, to compare $p \wedge (q \vee r)$ with $(p \wedge q) \vee r$.

p	q	$p \wedge q$
T	T	T
T	F	F
F	T	$\boxed{\text{F}}$
F	F	F

Figure A.1. Truth table for $p \wedge q$. The boxed entry states that, if p is false and q is true, then $p \wedge q$ is false.

p	q	$p \vee q$
T	T	T
T	F	T
F	T	T
F	F	F

Figure A.2. Truth table for $p \vee q$.

Get Your Hands Dirty 1

a) Find truth values for p, q, and r such that $p \wedge (q \vee r)$ and $(p \wedge q) \vee r$ have different truth values.

b) Find two compound propositions s and t (with components in $\{p, q, r\}$) such that $s \vee t$ always has the same truth value as $p \wedge (q \vee r)$. □

Negation

Our next logical operation only has one component. (Technically, we would call it a *unary* operation, whereas conjunction and disjunction are *binary* operations, since they each involve two components.)

DEFINITION A.3: For a proposition p, the *negation* of p (written "$\sim p$") is a proposition which is true if p is false, and false if p is true.

Intuitively, the negation "$\sim p$" is the assertion "it is not the case that p". For example, if p is the proposition "$3 < 7$" (which is true), then $\sim p$ is the proposition "it is not the case that $3 < 7$" (which is false, since it is the case that $3 < 7$). For this choice of p, we can express $\sim p$ more simply as "$3 \geq 7$". If the original proposition is complicated, the task of simplifying its negation can be a difficult one. This process is discussed in Section 3.1 and subsequent sections.

Figure A.3 gives a truth table defining $\sim p$.

p	$\sim p$
T	F
F	T

Figure A.3. Truth table for $\sim p$.

De Morgan's Laws

The three operations defined so far—conjunction, disjunction, and negation—can be combined to express some important principles, which give a foretaste of Section A.2.

Consider the following (in this discussion, n is supposed to represent some fixed integer): Let p be the statement "n is even" and let q be the statement "n is greater than or equal to 10". Then the compound proposition "$p \wedge q$" says:

"n is even and n is greater than or equal to 10"

What is the negation of this statement? What is $\sim(p \wedge q)$?

Under what circumstance(s) would $p \wedge q$ be false?

The compound proposition $p \wedge q$ is false if n is either odd (i.e., not even) or less than 10 (i.e., not "greater than or equal to 10"). In other words, $p \wedge q$ is false if either p is false or q is false.

There is nothing special about the specific propositions p and q used above. No matter what the propositions p and q are, $p \wedge q$ will be false precisely if either p is false or q is false. We can state this using the language of truth values:

THEOREM A.1. For any propositions p and q, the propositions $\sim(p \wedge q)$ and $\sim p \vee \sim q$ have the same truth value.

Comment: Since this is a theorem about truth values, we can prove it simply by looking at all the possible cases; i.e., by means of a truth table.

Proof: Figure A.4 gives a truth table for both $\sim(p \wedge q)$ and $\sim p \vee \sim q$. (The boldfaced columns give truth values for these two compound propositions; the other columns show intermediate steps.)

We see that these two compound propositions have the same truth value in every case.

p	q	$p \wedge q$	$\sim(p \wedge q)$	$\sim p$	$\sim q$	$\sim p \vee \sim q$
T	T	T	**F**	F	F	**F**
T	F	F	**T**	F	T	**T**
F	T	F	**T**	T	F	**T**
F	F	F	**T**	T	T	**T**

Figure A.4. ■

In Section A.2, we will express the idea of Theorem A.1 in terms of the concept of logical equivalence. Theorem A.1 involves one of two formulas known as *De Morgan's Laws*. The following GYHD asks you to find the other one:

Get Your Hands Dirty 2

Find a compound proposition using $\sim$ and $\wedge$ which always has the same truth as $\sim(p \vee q)$. □

Implication

We turn now to a more complex logical operation, the operation of implication:

DEFINITION A.4: For propositions p and q, the *implication* of p and q (written "$p \Rightarrow q$") is a proposition which is false if p is true and q is false, and true otherwise.
The proposition p is called the *hypothesis* of the implication, and the proposition q is called the *conclusion* of the implication.

The implication of P and Q is also known as the *conditional* of p and q.
Figure A.5 gives a truth table for implication.

DISCUSSION Intuitively, the implication "$p \Rightarrow q$" expresses roughly the same idea as the statement "if p, then q". However, in ordinary English, and in most mathematical contexts, we use the "if . . . , then . . ." form for situations where the hypothesis and conclusion are not propositions, but are more like open sentences. That is, we are usually interested in situations in which the truth values of p and q depend on some other information.

For example, in ordinary speech, we might say "if it is sunny tomorrow, I will go to the beach". The hypothesis, "it is sunny tomorrow", is not exactly true or false when it is spoken. If the next day is rainy, the "if . . . , then . . ." statement seems more "irrelevant" than either true or false. Nevertheless, according to Definition A.4 "it is sunny tomorrow $\Rightarrow$ I will go to the beach" will be considered a true implication if the next day is rainy, regardless of whether the speaker goes to the beach or not.

A more mathematical example may give some additional insight. Consider the statement "if $x > 2$, then $x^2 > 4$". The hypothesis and conclusion here, "$x > 2$" and "$x^2 > 4$", are open sentences, neither true nor false; their truth values depends on the number used to replace the variable x. There are values of x which make the hypothesis and conclusion both true (e.g., $x = 7$), there are values that make both false (e.g., $x = 1$), and there are values that make the hypothesis false and the conclusion true (e.g., $x = -4$).

p	q	$p \Rightarrow q$
T	T	T
T	F	F
F	T	T
F	F	T

Figure A.5. Truth table for $p \Rightarrow q$.

Historical Note: The English mathematician Augustus De Morgan (1806–1871) was one of the pioneers of symbolic logic. His work, together with that of his contemporary, George Boole (1815–1864), forms the foundation on which modern propositional logic rests. De Morgan was the founding president of the London Mathematical Society, and is credited with introducing and defining the term "mathematical induction" (see Section 5.1), as well as work which suggested the idea of quaternions.

We have defined the implication "$p \Rightarrow q$" so that all three of these combinations make the implication true. That is, each of the following implications has the truth value "T":

$$7 > 2 \Rightarrow 7^2 > 4$$
$$1 > 2 \Rightarrow 1^2 > 4$$
$$-4 > 2 \Rightarrow (-4)^2 > 4$$

According to Definition A.4 and Figure A.5, the only case in which "$p \Rightarrow q$" is considered false is the case where p is true and q is false. In the statement "if $x > 2$, then $x^2 > 4$", there are no values of x which make the hypothesis true and the conclusion false, and therefore this "if . . . , then . . ." statement is considered true.

We get somewhat strange results if we express the individual implications above in "if . . . , then . . ." form. For example, why would we say "if $7 > 2$, then $7^2 > 4$", when we know that 7 is greater than 2 and 7^2 is greater than 4? Why not use conjunction, and say that both hypothesis and conclusion are true?

The answer lies in the use of "if . . . , then . . ." for open sentences. The definition of implication works because it leads to the results we want for statements like "if $x > 2$, then $x^2 > 4$". We repeat our earlier suggestion that you read the intuitive discussion of conditional sentences in Section 1.3 as motivation for Definition A.4.

FURTHER COMMENTS

1. Another difference between implication and "everyday" use of "if . . . , then . . ." is that our everyday use normally involves a causal connection between the hypothesis and the conclusion. For example, when we say "if it is sunny tomorrow, I will go to the beach", it is implicit that we will go to the beach because it is sunny. The formal definition of "$p \Rightarrow q$" makes no such assumption of causality. For example, the implication "$2 + 4 = 6 \Rightarrow 3 \times 7 = 21$" is given the truth value "T" even though the conclusion has no apparent connection with the hypothesis.

2. The phrase "if . . . , then . . ." is used in many computer programming languages, with a significantly different meaning from that described here. The construction is "if p, then X" where X is some program segment. The meaning of such a statement is that, if the statement p is true, then the computer is to execute the program segment X, while if p is false, then the computer is to proceed to its next instruction, without executing X.

There are some related implication statements that can be formed from a given implication statement "$p \Rightarrow q$". We have the following definitions:

DEFINITION A.5: For propositions p and q, the *converse* of the implication "$p \Rightarrow q$" is the implication "$q \Rightarrow p$"; the *contrapositive* of the implication "$p \Rightarrow q$" is the implication "$\sim q \Rightarrow \sim p$".

As with implication itself, converse and contrapositive make more intuitive sense in terms of open sentences than in terms of propositions. For example, consider the true statement "if $x > 2$, then $x^2 > 4$". Its converse is the statement "if $x^2 > 4$, then $x > 2$". This converse is false, since the value $x = -3$ makes its hypothesis true and its conclusion false. The contrapositive of the original implication is: "if $x^2 \ngtr 4$, then $x \ngtr 2$". This can be simplified to say "if $x^2 \leq 4$, then $x \leq 2$", which is true.

We shall see in the next section that the contrapositive of an implication is *logically equivalent* to the implication, i.e., an implication and its contrapositive always have the same truth value, regardless of the specific propositions p and q involved. On the other hand, an implication and its converse can have different truth values. For example, "$2 + 4 = 9 \Rightarrow 1 + 1 = 2$" has the truth value "T", while "$1 + 1 = 2 \Rightarrow 2 + 4 = 9$" has the truth value "F".

Get Your Hands Dirty 3

State the converse and the contrapositive of each of the following:

a) If x is even, then $2x$ is even.
b) If A is a subset of B, then $A \cup B = B$.
c) If f is continuous on $[a, b]$, then f is uniformly continuous on $[a, b]$. □

There is one other logical operation which we introduce here:

DEFINITION A.6: For propositions p and q, the *double implication* of p and q (written "$p \Leftrightarrow q$") is a proposition which is true if p and q are both true and if p and q are both false, and false otherwise.

The double implication of p and q is also known as the *biconditional* of p and q.

Intuitively, the double implication "$p \Leftrightarrow q$" expresses the statement "p if and only if q"; that is, p and q have the same truth values. Figure A.6 gives a truth table defining double implication.

p	q	$p \Leftrightarrow q$
T	T	T
T	F	F
F	T	F
F	F	T

Figure A.6. Truth table for $p \Leftrightarrow q$.

Bit Operations

In some situations, the symbols "1" and "0" are used to represent truth values, instead of the symbols "T" and "F" respectively. This idea is particularly suitable for computer use, in which all information is represented by sequences of 0's and 1's. Each 0 or 1 is known as a *bit*. (The word "bit" is short for "binary digit". The system of binary arithmetic involves representing numbers using only the digits zero and one, instead of the usual $0, 1, 2, \ldots, 9$.)

The operations of conjunction and disjunction can be interpreted as *bit operations;* that is, we apply them to the digits "1" and "0" according to the same principle that we use for "T" and "F" respectively. These operations are defined by "operation tables" which are written in a similar way to ordinary addition and multiplication tables. (For example, the boxed entry in Figure A.7 states that $0 \wedge 1 = 0$.) We will continue to represent these operations by the symbols "$\wedge$" and "$\vee$"; they are defined by the tables in Figures A.7 and A.8.

A *bit string* is simply a sequence of 0's and 1's; the *length* of a bit string is the number of bits it contains. (For example, "1 0 1 1" is a bit string of length 4.) The bit operations defined above can be extended to apply to pairs of bit strings that have the same length. The operations are defined on bit strings by applying the operation to corresponding bits of the two strings, and making a string from the results.

For example, consider the two bit strings 1 1 0 1 0 and 0 1 1 0 0: the first bits (reading from the left) for these strings are 1 and 0 respectively. Since $1 \wedge 0 = 0$, the conjunction $11010 \wedge 01100$ is a string whose first digit is 0. Similarly, both strings have

$\wedge$	0	1
0	0	[0]
1	0	1

Figure A.7. Bit operation table for $\wedge$. The boxed entry states that $0 \wedge 1 = 0$.

$\vee$	0	1
0	0	1
1	1	1

Figure A.8. Bit operation table for $\vee$.

their second bit equal to 1, and $1 \wedge 1 = 1$, so the second bit of the conjunction is 1. Continuing, we get $11010 \wedge 01100 = 01000$.

Similarly, we get $11010 \vee 01100 = 11110$, since, in every place but the last, at least one of the original strings has a bit of 1.

EXERCISES

1. Which of these are propositions? Give the truth values of those which are propositions.
   - **a)** $3 + 7 \neq 11$.
   - **b)** $2^5 + 1$.
   - **c)** $x > 17$.
   - **d)** All integers are positive.

2. Let p be the proposition "today is Tuesday", and let q be the proposition "today is June 4". Express each of the following concisely as an English sentence:
   - **a)** $p \wedge q$.
   - **b)** $\sim p$.
   - **c)** $p \vee \sim q$.
   - **d)** $\sim(p \wedge q)$.

3. Let p be the proposition "today is Tuesday", and let q be the proposition "today is June 4". Express each of these sentences symbolically in terms of p, q, and the logical operations:
   - **a)** Today is Tuesday, June 4.
   - **b)** If today is Tuesday, then today is not June 4.
   - **c)** If today is June 4, then today is not Tuesday.

4. (GYHD 1)
   - **a)** Find truth values for p, q, and r such that $p \wedge (q \vee r)$ and $(p \wedge q) \vee r$ have different truth values.
   - **b)** Find compound propositions s and t (with components in $\{p, q, r\}$) such that $s \vee t$ always has the same truth value as $p \wedge (q \vee r)$.

5.
   - **a)** How many rows are needed to make a complete truth table for the compound proposition "$p \wedge (q \vee r)$"?
   - **b)** More generally, how many rows are needed to make a complete truth table for a compound proposition that has n components?

6. In this exercise, p and q represent propositions with truth value "T", and r and s represent propositions with truth value "F". Find the truth value of each of the following compound propositions:
   - **a)** $(p \vee r) \vee q$.
   - **b)** $(\sim p \wedge q) \vee (\sim q \wedge \sim s)$.
   - **c)** $p \Rightarrow (q \wedge \sim r)$.
   - **d)** $(p \wedge s) \Leftrightarrow (q \wedge r)$.

7. Construct truth tables for each of the following compound propositions:
   - **a)** $p \wedge (\sim p \vee q)$.
   - **b)** $(q \Rightarrow p) \wedge p$.
   - **c)** $p \vee (q \Rightarrow p)$.
   - **d)** $(q \wedge \sim r) \vee p$.

8. State the converse and the contrapositive of each of the following "if..., then..." statements:

a) If x is even, then x^2 is even.

b) If p is a prime, then p is either odd or 2.

9. Show that, if p and q are any propositions, then either "$p \Rightarrow q$" or its converse, "$q \Rightarrow p$", must have the truth value "T". (*Note:* If p and q are open sentences rather than propositions, then it is possible for both "if p, then q" and "if q, then p" to be false.)

10. **a)** Find a compound proposition of p and q, using only $\sim$ and $\wedge$, which will always give the same truth value as $p \vee q$.

b) Find a compound proposition of p and q, using only $\sim$ and $\vee$, which will always give the same truth value as $p \wedge q$.

c) Find a compound proposition of p and q, using only $\sim$ and $\wedge$, which will always give the same truth value as $p \Rightarrow q$.

d) Find a compound proposition of p and q, using only $\sim$ and $\vee$, which will always give the same truth value as $p \Rightarrow q$.

11. The logical operation "exclusive or" is defined as follows:

DEFINITION: For propositions p and q, the *exclusive or* of p and q (written "$p \oplus q$") is a proposition which is true when exactly one of p and q is true, and false otherwise.

Figure A.9 gives a truth table for $p \oplus q$.

a) Make a truth table for the following compound propositions:

i) $p \oplus p$.

ii) $p \oplus (q \vee r)$.

iii) $(p \oplus q) \oplus r$.

b) Find a compound proposition of p and q, using only $\sim$, $\wedge$, and $\vee$, which will always give the same truth value as $p \oplus q$.

c) Make an operation table for the bit operation that corresponds to $\oplus$.

12. Find the result of each of these bit-string expressions:

a) $101 \wedge 110$.

b) $110 \vee 010$.

c) $011 \wedge 101$.

d) $110 \oplus 010$ (see Exercise 11).

13. (This exercise is for readers who are familiar with mod 2 arithmetic, which is discussed in Section 7.1.) Express the bit operations $\wedge$, $\vee$, and $\oplus$ in terms of the operations of addition mod 2 (written $+_2$) and multiplication mod 2 (written $\cdot_2$).

14. The notation "$x := y$" is used in some programming languages to mean "replace the value of the variable x by the value of the variable y". For example, if x initially has the value 3, then the instruction "$x := x + 4$" would give x the new value $3 + 4$. For each of the following

p	q	$p \oplus q$
T	T	F
T	F	T
F	T	T
F	F	F

Figure A.9. Truth table for $p \oplus q$.

"if . . . , then . . ." statements (see further comment 2 following Definition A.4), give the resulting value for x if the initial value is as shown:

a) If $x > 10$, then $x := x + 3$; initial value $x = 12$.
b) If $x < 3$ and x is even, then $x := x^2$; initial value $x = 4$.
c) If $x < 3$ or x is even, then $x := 2x + 1$; initial value $x = 4$.
d) If $x^2 < 3$ or x is not even, then $x := x + 5$; initial value $x = 1$.

A.2 Logical Equivalence and Predicate Logic

In mathematics, we often prove one statement by proving something equivalent instead. For example, the method of proof by contrapositive allows us to prove the following statement (for integers x):

"if x^2 is even, then x is even"

by instead proving the statement

"if x is not even, then x^2 is not even".

Without using any knowledge or understanding about integers, mathematical logic can guarantee that these two statements are "equivalent". It does this by focusing on the form of the statements, rather than on their content. Our primary goal here is to explain this idea of "logical equivalence".

In order to focus on form, we want to think of "p" and "q" as variables—as generic propositions. When we want to emphasize this generic character, we will refer to such propositions as *propositional variables*. Rather than substitute specific propositions for "p" and "q", we will simply substitute truth values. Since a proposition can only take one of two possible truth values, this gives us a simple way of analyzing all possibilities. (A variable taking on the two values T and F—or 1 and 0—is also known as a *Boolean variable* [see Historical Note in Section A.1 on George Boole and Augustus De Morgan].)

We will refer to expressions using these propositional variables as *propositional forms*. (For example, "$p \Rightarrow (q \vee p)$" is a propositional form.) We will use capital Greek letters Φ (phi) and Ψ (psi) to represent propositional forms.

Logical Equivalence

We saw in Theorem A.1 that the propositional forms "$\sim(p \wedge q)$" and "$\sim p \vee \sim q$" always have the same truth value, regardless of the truth values of p and q. Intuitively, this means that these two propositional forms are "saying the same thing".

This suggests the following formal definition for the intuitive idea of "logical equivalence":

DEFINITION A.7: For propositional forms Φ and Ψ, "Φ and Ψ are *logically equivalent*" means: Φ and Ψ have the same truth value regardless of the truth values of their components.

Thus, we can restate Theorem A.1 as follows:

THEOREM A.1′. The propositional forms "$\sim(p \wedge q)$" and "$\sim p \vee \sim q$" are logically equivalent.

Later in this section, we will see how statements about logical equivalence of propositional forms lead to principles that are useful for proving ordinary mathematical theorems (as opposed to theorems about logic).

Tautologies and Contradictions

The logical operation of double implication ("$\Leftrightarrow$") can be combined with the concept of logical equivalence in a useful way. For if two propositional forms always have the same truth value, then their double implication will always have the truth value "T", regardless of the truth values of their components. This leads us to make the following definition:

DEFINITION A.8: A *tautology* is a propositional form which is true regardless of the truth values assigned to its propositional variables.

Using this terminology, we can restate Theorem A.1 in yet another way:

THEOREM A.1″. The propositional form "$[\sim(p \wedge q)] \Leftrightarrow [\sim p \vee \sim q]$" is a tautology.

Though double implications are an important way to form tautologies, not all tautologies are of this type. The following gives a simple alternative example:

Example 1. ***"$p \vee \sim p$" Is a Tautology*** We can see that the propositional form "$p \vee \sim p$" is a tautology, as follows: if p is assigned the truth value T, then the first component of "$p \vee \sim p$" is true and the second component is false (by Definition A.3). Therefore, the disjunction is true (by Definition A.2). Similarly, if p is assigned the value F, then the first component of "$p \vee \sim p$" is false and the second component is true, and again the disjunction is true. This is summarized in the truth table of Figure A.10. □

p	$\sim p$	$p \vee \sim p$
T	F	T
F	T	T

Figure A.10. Truth table for the propositional form "$p \vee \sim p$".

The following result summarizes the relationship between tautologies, logical equivalence, and double implication. It is an immediate consequence of the last three definitions.

THEOREM A.2. Propositional forms Φ and Ψ are logically equivalent if and only if "$\Phi \Leftrightarrow \Psi$" is a tautology.

The following definition expresses the opposite extreme from a tautology:

DEFINITION A.9: A *contradiction* is a propositional form which is false regardless of the truth values assigned to its propositional variables.

Example 2. **"$q \wedge \sim q$" Is a Contradiction** The prototypical contradiction is "$q \wedge \sim q$". The truth table in Figure A.11 shows that "$q \wedge \sim q$" is a contradiction, since either choice of truth value for q gives a truth value of "F" for "$q \wedge \sim q$". □

The following theorem lists many of the basic tautologies, some of which are analogous to various theorems about open sentences stated in Chapter 3. Many of these tautologies are commonly referred to by names as indicated.

THEOREM A.3. The following are tautologies:

a) $p \Leftrightarrow \sim(\sim p)$ (Law of Double Negation)
b) $(p \wedge q) \Leftrightarrow (q \wedge p)$ (Commutativity of Conjunction)
c) $(p \vee q) \Leftrightarrow (q \vee p)$ (Commutativity of Disjunction)
d) $[p \wedge (p \Rightarrow q)] \Rightarrow q$ (Modus Ponens)
e) $[(p \Rightarrow q) \wedge \sim q] \Rightarrow \sim p$ (Modus Tollens)
f) $[\sim p \wedge (p \vee q)] \Rightarrow q$ (Disjunctive Syllogism)

q	$\sim q$	$q \wedge \sim q$
T	F	F
F	T	F

Figure A.11. Truth table for the propositional form "$q \wedge \sim q$".

g) $[(p \Rightarrow q) \wedge (q \Rightarrow r)] \Rightarrow (p \Rightarrow r)$ (Hypothetical syllogism)
h) $p \Leftrightarrow [\sim p \Rightarrow (q \wedge \sim q)]$ (Law of Contradiction)
i) $[p \Rightarrow q] \Leftrightarrow [\sim q \Rightarrow \sim p]$ (Law of Contraposition)
j) $[\sim(p \wedge q)] \Leftrightarrow [\sim p \vee \sim q]$ (De Morgan's Laws)
k) $[\sim(p \vee q)] \Leftrightarrow [\sim p \wedge \sim q]$ (De Morgan's Laws)
l) $[p \Leftrightarrow q] \Leftrightarrow [(p \Rightarrow q) \wedge (q \Rightarrow p)]$ (Law of Double Implication)
m) $[p \Rightarrow (q \vee r)] \Leftrightarrow [(p \wedge \sim q) \Rightarrow r]$
n) $[p \Rightarrow (q \Rightarrow r)] \Leftrightarrow [(p \wedge q) \Rightarrow r]$
o) $[(p \vee q) \Rightarrow r] \Leftrightarrow [(p \Rightarrow r) \wedge (q \Rightarrow r)]$

The proof of this theorem requires nothing more than a series of truth tables. For example, the truth table in our proof of Theorem A.1 showed that $\sim(p \wedge q)$ and $\sim p \vee \sim q$ always have the same truth value. One more column would show that $[\sim(p \wedge q)] \Leftrightarrow [\sim p \vee \sim q]$ always has the truth value T. This proves item j) of Theorem A.3. The other tautologies are similar.

COMMENTS

1. Tautology d) of Theorem A.3 intuitively says that if p and $p \Rightarrow q$ are both true, then we are guaranteed that q must be true. This basic syllogism rule is the principle we use when applying a previously proved "if . . . , then . . ." theorem or employing the method of specialization. Namely, we show that the hypothesis of some proved or assumed "if . . . , then . . ." statement is true, and we are guaranteed that the conclusion of that "if . . . , then . . ." statement is true.

2. Tautology e) says that, if $p \Rightarrow q$ is true, but q is false, then p must also be false.

3. Tautology f) says that, if we know that one of p and q is true, and p isn't true, then q must be true.

4. Tautology g) is the formal basis for "linking" theorems. For example, suppose we know that if x is a real number, then $x^2 \geq 0$ and also know that if $x^2 \geq 0$, then $x^2 + 1 \geq 1$. Tautology g) would allow us to conclude that if x is a real number, then $x^2 + 1 \geq 1$.

All of the tautologies in Theorem A.3 except d)–g) (just discussed) use the double implication "$\Leftrightarrow$", and so we get pairs of propositional forms that are logically equivalent. Items a), b), and c) follow immediately from the definitions of the operations $\sim$, $\wedge$, and $\vee$.

Items h) through o) are more complex. Each of these tautologies is the basis of a general principle for proof, discussed in Section 4.2. For example, item i) is the basis for the method of proof by contrapositive. It tells us that any implication is logically equivalent to its contrapositive. In other words, an implication and its contrapositive are either both true or both false. In order to use this fact to prove theorems, we need to extend its scope from propositional forms to open sentences. We turn now to see how this is formally justified.

From Propositional Forms to Propositions to Open Sentences to Conditional Sentences

Theorem A.3 becomes useful to mathematicians when they can use it to prove theorems. Most theorems are expressed, not in terms of propositions, but in terms of open sentences, and, more specifically, as conditional sentences. We will trace a series of steps by which Theorem A.3, which is a theorem about propositional forms, leads to results which help us prove theorems, such as the method of proof by contrapositive.

From Propositional Forms to Propositions

We begin by applying the term "logically equivalent" to the situation of specific propositions, rather than just to propositional forms. We make the following definition:

DEFINITION A.10: Two propositions are called *logically equivalent* if they can be obtained from logically equivalent propositional forms by replacing the propositional variables of those forms with specific propositions.

The following result is an immediate consequence of the definitions:

THEOREM A.4. Two propositions that are logically equivalent must have the same truth value.

COMMENT It is certainly possible for propositions to have the same truth value without being logically equivalent. For example, "$5 < 7$" and "$\mathbf{N} \subseteq \mathbf{Z}$" have the same truth value, since both are true, but they are not logically equivalent. The importance of logical equivalence is that it allows us to deduce that certain propositions have the same truth value without thinking about their content. For example, there is a tautology which tells us that "$5 < 7$ and $5 < 7$" is logically equivalent to "$5 < 7$".

From Propositions to Open Sentences

The next step in using the collection of logical equivalences that result from Theorem A.3 is to replace the component propositions by open sentences, that is, by statements involving variables, which become propositions when the variables are replaced by specific objects. We will use notation such as $p(x)$ or $q(x)$ to represent generic open sentences.

We begin with formal definitions for operations on open sentences that correspond to those in Definitions A.1, A.2, A.3, A.4, and A.6 for propositions. In each part of the definition below, the truth value of a compound open sentence for a given value of the variable is obtained by substituting that value for the variable in each of the components, and finding the truth value of the corresponding compound proposition.

DEFINITION A.11: For open sentences $p(x)$ and $q(x)$, and an object c:

a) the *conjunction* of $p(x)$ and $q(x)$ [written "$p(x) \wedge q(x)$"] is the open sentence whose truth value for $x = c$ is the truth value of $p(c) \wedge q(c)$.
b) the *disjunction* of $p(x)$ and $q(x)$ [written "$p(x) \vee q(x)$"] is the open sentence whose truth value for $x = c$ is the truth value of $p(c) \vee q(c)$.
c) the *negation* of $p(x)$ [written "$\sim p(x)$"] is the open sentence whose truth value for $x = c$ is the truth value of $\sim p(c)$.
d) the *implication* of $p(x)$ and $q(x)$ [written "$p(x) \Rightarrow q(x)$"] is the open sentence whose truth value for $x = c$ is the truth value of $p(c) \Rightarrow q(c)$.
e) the *double implication* of $p(x)$ and $q(x)$ [written "$p(x) \Leftrightarrow q(x)$"] is the open sentence whose truth value for $x = c$ is the truth value of $p(c) \Leftrightarrow q(c)$.

For example, let $p(x)$ be the open sentence "$x > 3$", and let $q(x)$ be the open sentence "$x \in \mathbf{Z}$". The truth value of the open sentence "$p(x) \wedge q(x)$" for a substitution $x = c$ is found simply by computing the truth value of $p(c) \wedge q(c)$. For instance, suppose $x = 2$: $p(2)$ is false and $q(2)$ is true. Therefore "$p(x) \wedge q(x)$" is false for $x = 2$. On the other hand, "$p(x) \wedge q(x)$" is true for $x = 7$.

Using Definition A.11, every propositional form has a corresponding *open-sentence form,* obtained by simply replacing proposition variables such as p and q with generic open sentences, such as $p(x)$ and $q(x)$.

The following definition extends the concept of logical equivalence to open-sentence forms:

DEFINITION A.12: For open-sentence forms $\Phi(x)$ and $\Psi(x)$, "$\Phi(x)$ and $\Psi(x)$ are *logically equivalent*" means: the propositional forms corresponding to $\Phi(x)$ and $\Psi(x)$ are logically equivalent.

As with propositions and propositional forms, we are really interested in replacing the generic components of logically equivalent open-sentence forms by specific open sentences. We therefore make the following definition:

DEFINITION A.13: Two open sentences are called *logically equivalent* if they can be obtained from logically equivalent open-sentence forms by replacing the generic open sentences of those forms with specific open sentences.

If $p(x)$ and $q(x)$ are two logically equivalent open sentences, they must come from logically equivalent forms $\Phi(x)$ and $\Psi(x)$. Therefore, if we substitute some value $x = c$

into $p(x)$ and $q(x)$, the resulting propositions $p(c)$ and $q(c)$ are also logically equivalent. Thus $p(c)$ and $q(c)$ have the same truth value.

The essential feature of an open sentence is that it only achieves a truth value when its variables are replaced by specific objects. We are therefore interested in knowing which objects make the open sentence true. In the following definition, **U** represents some *universe of discourse*, that is, a general set from which the replacement objects may be taken.

DEFINITION A.14: For an open sentence $p(x)$, the *truth set* of $p(x)$ is the set of all objects $c \in \mathbf{U}$ such that $p(c)$ is true.

We can summarize the above discussion as follows:

THEOREM A.5. For open sentences $p(x)$ and $q(x)$, if $p(x)$ and $q(x)$ are logically equivalent, then they have the same truth set.

COMMENT We pointed out earlier that propositions can have the same truth value without being logically equivalent. Similarly, open sentences can have the same truth set without being logically equivalent. For example, the open sentences "$x + 3 = 7$" and "$36x = 144$" both have $\{4\}$ as their truth set, but they are not logically equivalent. We have to deal with the mathematical content of these open sentences to find out that the truth sets are the same.

On the other hand, the two open sentences "x is not both odd and greater than 5" and "x is either not odd or not greater than 5" are logically equivalent [by De Morgan's Law—item j) of Theorem A.3]. We can be sure that the two open sentences have the same truth set, without actually finding the truth set of either open sentence.

From Open Sentences to Conditional Sentences

We now move on to the final step in demonstrating the practical importance of this material for mathematicians—its use in proving theorems. Theorems are generally stated in "if . . . , then . . ." form: "if $p(x)$, then $q(x)$". (This is known as a *conditional sentence*.) As discussed in Section 3.1, such a statement is defined to be true precisely when the implication "$p(x) \Rightarrow q(x)$" is true for all x in the "universe of discourse" **U**.

We illustrate the significance of Theorem A.3 by giving a formal proof of Theorem 4.8, concerning contrapositives:

THEOREM A.6: (Theorem 4.8). For any open sentences $p(x)$ and $q(x)$, if either "if $p(x)$, then $q(x)$" or its contrapositive "if $\sim q(x)$, then $\sim p(x)$" is true, then so is the other.

Proof: The two open sentences "$p(x) \Rightarrow q(x)$" and "$\sim q(x) \Rightarrow \sim p(x)$" are logically equivalent [by item i) of Theorem A.3, and the definitions of "implication" for propositions and for open sentences (Definitions A.4 and A.11d))], and so they have the same truth set (by Theorem A.5). Therefore, if one of them has **U** as its truth set, so does the other. Thus, if either conditional sentence is true, so is the other. ∎

It is important to recognize that Theorem A.6 is true no matter what the open sentences are. That's the essential idea of logical equivalence. When trying to determine if a mathematical statement is true or false (or prove that it is true or false), we can replace the statement by something which is logically equivalent to it, before we get involved in the details of the actual meaning of the statement. Any results we get about the truth of the logically equivalent statement will also apply to the original.

Several other results of Section 4.2 are proved using other tautologies from Theorem A.3. Specifically, we have the following pairing of tautologies from Theorem A.3 and theorems from Section 4.2:

Tautology	*Theorem*
$[\sim(p \wedge q)] \Leftrightarrow [\sim p \vee \sim q]$ [item j)]	4.13
$[\sim(p \vee q)] \Leftrightarrow [\sim p \wedge \sim q]$ [item k)]	4.14
$[p \Rightarrow (q \vee r)] \Leftrightarrow [(p \wedge \sim q) \Rightarrow r]$ [item m)]	4.15
$[p \Rightarrow (q \Rightarrow r)] \Leftrightarrow [(p \wedge q) \Rightarrow r]$ [item n)]	4.16
$[(p \vee q) \Rightarrow r] \Leftrightarrow [(p \Rightarrow r) \wedge (q \Rightarrow r)]$ [item o)]	4.17

The proofs of these theorems can be obtained by appropriately modifying the proof of Theorem 4.8, replacing the tautology "$[p(x) \Rightarrow q(x)] \Leftrightarrow [\sim q(x) \Rightarrow \sim p(x)]$" by the correct item from Theorem A.3.

There are two logical equivalences from Theorem A.3 which do not correspond to specific theorems in the text. These do give rise to important principles of proof, however, as indicated in the following observations:

i) The logical equivalence of "p" and "$\sim p \Rightarrow (q \wedge \sim q)$" [from item h)] describes the idea behind proof by contradiction. This logical equivalence essentially tells us: showing that the assumption that p is false leads to a contradiction is the same as showing that p itself is true.

ii) The logical equivalence of "$p \Leftrightarrow q$" with "$[p \Rightarrow q] \wedge [q \Rightarrow p]$" (from item l)] corresponds to the method of proving an "if and only if" statement by proving two "if . . . , then . . ." statements.

EXERCISES

1. Use truth tables to prove that each of the propositional forms of Theorem A.3 is a tautology.

2. **a)** Use truth tables to prove that each of the following propositional forms is a tautology:

i) $p \Leftrightarrow [p \vee (p \wedge q)]$.

ii) $p \Leftrightarrow [p \wedge (p \vee q)]$.

(These tautologies are known as the *absorption laws*.)

b) Give a verbal explanation for each of the logical equivalences in part a).

3. **a)** Use truth tables to prove that each of the following pairs of propositional forms are logically equivalent:

i) "$p \wedge (q \vee r)$" and "$(p \wedge q) \vee (p \wedge r)$".

ii) "$p \vee (q \wedge r)$" and "$(p \vee q) \wedge (p \vee r)$".

[These logical equivalences are known as the *distributive laws:* i) is the distributive law of conjunction over disjunction; ii) is the distributive law of disjunction over conjunction. Notice the resemblance in form to the ordinary arithmetical distributive law $a \cdot (b + c) = (a \cdot b) + (a \cdot c)$.]

b) Give a verbal explanation for each of the logical equivalences in part a).

4. **a)** Use truth tables to prove that each of the following propositional forms is a tautology:

i) $(p \Rightarrow q) \Leftrightarrow (\sim p \vee q)$.

ii) $(p \Rightarrow q) \Leftrightarrow \sim(p \wedge \sim q)$.

b) Give a verbal explanation for each of the logical equivalences in part a).

5. Classify each of the following compound propositions as i) a tautology, ii) a contradiction, or iii) neither of these:

a) $p \Rightarrow (p \vee \sim q)$.

b) $(p \Rightarrow q) \vee (q \Rightarrow p)$.

c) $(p \wedge q) \Leftrightarrow (p \vee \sim q)$.

d) $p \wedge (q \Rightarrow \sim p) \wedge q$.

6. Assume that the following statement is true (which it is):

If a matrix is invertible, then its determinant must be nonzero.

Which of the following statements follow *by logic* from this? (No knowledge of matrices or determinants is needed for this question.)

i) Any matrix with nonzero determinant is invertible.

ii) Every matrix is either invertible or has a nonzero determinant.

iii) If the determinant of a matrix is zero, then the matrix is not invertible.

iv) There is no invertible matrix whose determinant is zero.

v) If a matrix is not invertible, then its determinant is zero.

7. Assume that the following statement is true (which it is):

Every differentiable function is continuous.

Which of the following statements follow *by logic* from this? (No knowledge of functions, differentiability, or continuity is needed for this question.)

i) There are functions which are both continuous and differentiable.

ii) If there is a differentiable function, then there is a continuous function.

iii) If a function is not continuous, then it must be differentiable.

iv) If a function is not continuous, then it must not be differentiable.

Appendix B

Basic Ideas about Sets and Arithmetic

Introduction

The language of sets is used throughout mathematics. Although there is an axiomatic, formal way of talking about sets, our use of sets in this book is primarily intuitive. There are several simple ideas and notations that you should be familiar and comfortable with, which are presented in this appendix. Some of these ideas—especially *set-builder notation, subset,* and the *empty set*—are discussed more fully in Section 2.1. This appendix also lists some basic assumptions about arithmetic which will be used in this text.

How to Describe a Set

A set is simply a collection of objects. The objects in a set are usually called its *elements* or *members*. There are several common ways to describe a set or specify what its elements are.

a) *Listing:* A small set can often best be specified by actually listing the elements that belong to it. Braces—{ }—are used to enclose the elements. For example:

$\{1, 2, 3, 4\}$ is the set containing the first four positive integers;
$\{1, 2, 3, 4, 6, 12\}$ is the set of positive integers that divide "evenly" into 12.

The order in which objects are listed does not matter; the set $\{1, 2, 3\}$ is the same as the set $\{2, 3, 1\}$. Also, repeating an element in the listing does not change the set: $\{1, 2, 3\}$ is the same as the set $\{1, 2, 1, 3\}$.

b) *Listing with ellipsis:* "Ellipsis" is the name for the mathematical "etcetera" symbolized by three dots (. . .). It means that a list should continue in the pattern already indicated. The set goes on "forever" if nothing comes after the ellipsis, or the three dots can be used to represent the "middle" of the set, if some object is given at the end. For example:

$\{2, 4, 6, 8, \ldots\}$ is the set of positive even integers, which goes on indefinitely.

Because it is infinite, this set could not be described by a list without ellipsis.

$\{a, b, c, d, \ldots, m\}$ is the set of letters in the first half of the alphabet. Here ellipsis is simply a convenience, to avoid having to list all the letters individually.

Ellipsis must be used with caution, to be sure it is not ambiguous. $\{2, 4, \ldots\}$ is not a good use of ellipsis, because this could represent the set of powers of 2—$\{2, 4, 8, 16, 32, \ldots\}$—as well as the set of positive even integers—$\{2, 4, 6, 8, 10, \ldots\}$.

c) *Verbal description:* Often this is the simplest way to describe a set. It helps give a sense of the relationship between the elements, i.e., explain why they are being grouped together. For example: "the set of points in the plane with integer coordinates" describes a set that includes such points as $(4, -7)$, $(0, 2)$, and $(-3, 0)$, but excludes $(3.2, 6)$ or $(2, \pi)$.

d) *Set-builder notation:* This is a special type of verbal description, in which a symbol such as x or y is used to represent a possible element, and then a condition about that symbol is given to indicate whether an element is in the set or not. For example:

$$\{x : x \text{ is a real number and } 4 \leq x < 7\}.$$

The colon (:) is read as "such that". Thus, this set is "the set of all x such that x is a real number and 4 is less than or equal to x which is less than 7." More colloquially, we would describe this as "the set of all real numbers which are greater than or equal to 4 and less than 7." The real numbers 4, 5.3, and 6.19 belong to this set; 7, -2.6, and 8.3 do not. Set-builder notation is a powerful tool, and is discussed more fully in the text (see Section 2.1).

No matter what method is used to describe a set, a set is defined by what its elements are. Therefore, two sets are considered *equal* precisely when they have the same elements, even if the descriptions are different.

For example, "the set of odd integers between 2 and 8" and "the set of the first three odd prime numbers" both describe the set whose elements are 3, 5, and 7. Therefore these two sets are equal.

Elements and Subsets

No matter what method is used to indicate which objects belong to a set, it is often convenient to designate the set itself by a symbol; capital letters are usually used. For example, we might use the letter A to represent the set described in d) above:

$$A = \{x : x \text{ is a real number and } 4 \leq x < 7\}$$

The symbol "$\in$" is used to indicate that a particular object is an element of a particular set. For example, the real number 5.3 is an element of the set A just described. We can represent this fact symbolically by the notation "$5.3 \in A$". (The symbol "$\in$" is variously read as "is an element of", "is a member of", "is in", or "belongs to".) The symbol "$\notin$" means "is not a member of".

If every element of one set is also an element of a second set, we say that the first set is a *subset* of the second. For example, let B represent the set $\{4.5, 5, 5.5, 6, 6.5\}$. Since each one of these numbers belongs to A, we say that B is a subset of A; this is written symbolically "$B \subseteq A$" (read "B is a subset of A", or "B is contained in A"). Any set is considered to be a subset of itself. The notation "$A \not\subseteq B$" means "A is not a subset of B".

Caution: Both the "element of" relationship ($\in$) and the "subset of" relationship ($\subseteq$) are sometimes expressed verbally by such phrases as "is in", "is part of", or "is contained in". Keep in mind that the two relationships are quite distinct and are not interchangeable. To avoid confusion, it is sometimes important to insist on the more formal language. Similarly, an individual object, such as the number 6, is different from the set with that object as its only member: $\{6\}$.

A set X is called a *proper subset* of some set Y if X is a subset of Y other than Y itself. This is represented by the notation "$X \subset Y$" (read "X is a proper subset of Y" or "X is properly contained in Y"). (Notice how the "subset" and "proper subset" symbols resemble the symbols for "less than or equal" and "less than".)

Note: some books use the symbol "$\subset$" to mean subset, and write "$\subsetneq$" to mean proper subset.

If X is a subset of Y, we can turn the notation around and write $Y \supseteq X$, and we say that Y is a *superset* of X.

Special Sets

Certain commonly discussed sets have standard symbols associated with them. The following symbols are used throughout this book, and in most mathematics texts, to mean the specific sets described here:

$\varnothing$: this represents the *empty set* (also known as the *null set*), that is, the set with no elements in it. The empty set is a subset of every set.

N: this represents the *natural numbers,* that is, the set of positive integers $\{1, 2, 3, 4, \ldots\}$.

W: this represents the *whole numbers,* that is, the positive integers together with zero: $\{0, 1, 2, 3, \ldots\}$.

Z: this represents the *integers,* that is, positive, negative, and zero: $\{\ldots, -3, -2, -1, 0, 1, 2, 3, \ldots\}$.

Q: this represents the *rational numbers,* that is, all numbers which can be represented as fractions, whether positive or negative (this includes all the integers as well).

R: this represents the *real numbers,* that is, all the numbers that correspond to points on the real number line.

C: this represents the *complex numbers,* that is, all numbers which are combinations of real and imaginary numbers.

We occasionally use notation such as $\mathbf{Q}^{+}$ for positive rational numbers or $\mathbf{R}^{\geq 0}$ for nonnegative real numbers.

In addition, we use the following standard notation for intervals on the real number line:

$$[a, b] = \{x : a \leq x \leq b\}$$
$$[a, b) = \{x : a \leq x < b\}$$
$$(a, b] = \{x : a < x \leq b\}$$
$$(a, b) = \{x : a < x < b\}$$
$$[a, \infty) = \{x : a \leq x\}$$
$$(a, \infty) = \{x : a < x\}$$
$$(-\infty, b] = \{x : x \leq b\}$$
$$(-\infty, b) = \{x : x < b\}$$
$$(-\infty, \infty) = \mathbf{R}$$

Making New Sets from Old Ones

Sets can be combined to create other sets, just as numbers are combined arithmetically to give other numbers. A general process by which objects are combined to make a new object is called an *operation*. (Addition, subtraction, etc., are operations on numbers.) If an operation works with two objects to create a third, it is called a *binary operation*. (We define this term formally in Section 7.1.) There are several standard binary operations for combining sets:

The *union* of two sets X and Y is the set which contains precisely those elements which belongs to either X or Y (or both). The combined set is written $X \cup Y$ (read "X union Y"). In set-builder notation, we can write

$$X \cup Y = \{w : w \in X \text{ or } w \in Y\}.$$

For example, if $X = \{1, 4, 6\}$ and $Y = \{2, 4, 9\}$, then

$$X \cup Y = \{1, 2, 4, 6, 9\}.$$

Notice that the element 4 belongs to both X and Y, and it is included (once) in the union.

The *intersection* of two sets X and Y is the set which contains precisely those elements which belong to both X and Y. This set is written $X \cap Y$ (read "X intersect Y"). In set-builder notation, we can write

$$X \cap Y = \{w : w \in X \text{ and } w \in Y\}.$$

If $X \cap Y = \varnothing$, that is, if X and Y have no elements in common, then we say that X and Y are *disjoint*. Notice that $X \cap Y$ is always a subset of X, Y, and $X \cup Y$. For example, if $X = \{1, 4, 6\}$ and $Y = \{2, 4, 9\}$, then

$$X \cap Y = \{4\}.$$

The two binary operations just described—union and intersection—can be extended naturally to work with more than two sets at a time. Thus, the union of several sets consists of those elements which belong to any of the given sets, and the intersection consists of those elements which belong to all of the given sets.

The next operation can only be applied to two sets at a time:

The *difference* of two sets X and Y is the set which contains precisely those elements which belong to X but not to Y. The combined set is written $X - Y$ (read "X minus Y"). (There may be objects in Y which are not in X; we can't "remove" these from X.) In set-builder notation, we can write

$$X - Y = \{w : w \in X \text{ and } w \notin Y\}.$$

For example, if $X = \{1, 4, 6\}$ and $Y = \{2, 4, 9\}$, then

$$X - Y = \{1, 6\}.$$

Often a mathematical discussion will have a *universe of discourse,* a set $\mathbf{U}$ which represents the overall framework of the given situation. Usually this universe is some standard set such as $\mathbf{R}$ or $\mathbf{Z}$. In this context, we define the *complement* of an individual set A to be the set difference $\mathbf{U} - A$, i.e., the set consisting of those elements of the given universe which do not belong to A. This difference is abbreviated symbolically as A' (read "A prime" or "A complement"). For example, if the universe is $\mathbf{R}$, and A is the interval $[4, 7)$, then A' is the set $(-\infty, 4) \cup [7, \infty)$.

Keep in mind that the idea of the complement of a set is only meaningful if some universe has been specified, and that the meaning of A' will change if the universe is changed.

While union, intersection, and difference are binary operations, complementation is an example of a *unary operation*—a way of defining a new set which starts with just one set.

We define two other important set operations—*Cartesian product* and *power set*—in Section 2.1.

Example 1. Operations on Sets Suppose $\mathbf{U} = \{1, 2, 3, \ldots, 12\}$. Define the sets $A, B, C, D,$ and E as follows:

$$A = \{1, 2, 3, 4, 5\}, \quad B = \{2, 4, 6, 8\}, \quad C = \{9, 10, 11\},$$
$$D = \{2, 6\}, \quad \text{and} \quad E = \varnothing.$$

Find each of the following:

a) $A \cup B$.
b) $A \cup C$.
c) $B \cup E$.
d) $A \cap B$.
e) $B \cap C$.
f) $A \cap E$.

g) $A - B$.
h) $A - C$.
i) A'.

SOLUTIONS

a) $\{1, 2, 3, 4, 5, 6, 8\}$.
b) $(1, 2, 3, 4, 5, 9, 10, 11\}$.
c) $\{2, 4, 6, 8\}$.
d) $\{2, 4\}$.
e) $\varnothing$.
f) $\varnothing$.
g) $\{1, 3, 5\}$.
h) $\{1, 2, 3, 4, 5\}$.
i) $\{6, 7, 8, 9, 10, 11, 12\}$. □

Cardinality

The number of elements in a finite set X is called its *cardinality*. We will use the notation $|X|$ for the cardinality of the set X, although there are other commonly used notations, such as $n(X)$ and $\#(X)$. For example, if T is the set $\{2, 4, 7, 11, 16\}$, then the cardinality of T is 5, and we can write $|T| = 5$.

In Section 8.1, we give a more formal and abstract definition of cardinality, which can be applied to infinite sets as well. However, the definition given here is used everywhere except Chapter 8.

Assumptions about Arithmetic

Throughout this text, we will be assuming certain basic facts about arithmetic. With certain exceptions that will be clearly noted, we will prove all the other properties that we develop. If a proof done in this text seems to leave out a step, or to make an assumption that seems unjustified, the missing reasoning should be found in the list here, or be a simple variation of, or deduction from, these assumptions. You should refer to this list if you are in doubt, and seek justification of any reasoning that seems unclear.

Since the set **C** of complex numbers includes all the other sets of numbers we are concerned with, we will state our most general properties in terms of that set. In other words, all objects in the following conditions are assumed to be elements of **C**. If a particular property applies only to some smaller set, we will indicate that. (The labels for the properties listed here are defined and discussed primarily in Section 7.1. Many may be familiar to you.)

a) General properties of operations:

i) $a + b = b + a$ (commutativity of addition).
ii) $a \cdot b = b \cdot a$ (commutativity of multiplication).

iii) $(a + b) + c = a + (b + c)$ (associativity of addition).
iv) $(a \cdot b) \cdot c = a \cdot (b \cdot c)$ (associativity of multiplication).
v) $a(b + c) = ab + ac$ (distributivity of multiplication over addition).
vi) $a(-b) = -(ab)$ (rule of signs).
vii) If $a + b = a + c$, then $b = c$ (cancellation property for addition).
viii) If $ab = ac$ and $a \neq 0$, then $b = c$ (cancellation property for multiplication).

b) Properties of 0, 1, and inverses:
i) $a + 0 = a$ (identity for addition).
ii) $a \cdot 0 = 0$.
iii) $a \cdot 1 = a$ (identity for multiplication).
iv) $a + (-a) = 0$ (inverse for addition).
v) $-(-a) = a$.
vi) $a - b = a + (-b)$ (definition of subtraction).

For properties vii) and viii), $a \neq 0$:

vii) $a \cdot a^{-1} = 1$ (inverse for multiplication).
viii) $(a^{-1})^{-1} = a$.

For property ix), $b \neq 0$:

ix) $a \div b = a \cdot b^{-1}$ (definition of division).

c) Closure properties:

i) If a and b are both real numbers, then $a + b$, $a - b$, and $a \cdot b$ are real numbers; if, further, $b \neq 0$, then $a \div b$ is also a real number.
ii) If a and b are both rational numbers, then $a + b$, $a - b$, and $a \cdot b$ are rational numbers; if, further, $b \neq 0$, then $a \div b$ is also a rational number.
iii) If a and b are both integers, then $a + b$, $a - b$, and $a \cdot b$ are integers.
iv) If a and b are both natural numbers, then $a + b$ and $a \cdot b$ are natural numbers.

d) Properties of real numbers—inequalities (for $a, b, c \in \mathbf{R}$):
i) If $a > 0$, then $1/a > 0$ and $-a < 0$; if $a < 0$, then $1/a < 0$ and $-a > 0$.
ii) If $a > b$, then $a + c > b + c$.
iii) If $a > b$ and $c > 0$, then $ac > bc$ and $a \div c > b \div c$.
iv) If $a > b$ and $c < 0$, then $ac < bc$.
v) For any two real numbers, exactly one of the following is true: $a < b$, $a = b$, $a > b$ (trichotomy property).
vi) If $a > b$ and $b > c$, then $a > c$ (transitivity property of inequality).

e) Properties of real numbers—exponents (for $a, b, c \in \mathbf{R}$, $a > 0$):
i) $a^{b+c} = a^b a^c$.
ii) $(a^b)^c = a^{bc}$.

iii) $(ab)^c = a^c b^c$.
iv) $a^{-b} = 1/a^b$.

f) Properties of natural numbers (for $a, b \in \mathbf{N}$):
i) $0 < 1 \le a$ (i.e., 1 is positive, and is the smallest natural number).
ii) If $a > b$, then $a \ge b + 1$ (i.e., there are no natural numbers between b and $b + 1$).

Symbols and Abbreviations Used in This Text

The following list shows the place in the text where each symbol or abbreviation is defined. (Set notations that are introduced in Appendix B are listed according to the definitions there.) For symbols used in more than one way (e.g., logic symbols), the location of the definition for each meaning is given.

$\mathbf{U}$ universe, **p. 28**

$\mathcal{P}(A)$ power set of A, **p. 40**

$A \times B$ Cartesian product of A and B, **p. 43**

$A \triangle B$ symmetric difference of A and B, **p. 54**

$n\mathbf{Z}$ set of multiples of n, **p. 67**

$x \mid y$ "x is a divisor of y", **p. 67**

$x \nmid y$ "x is not a divisor of y", **p. 67**

GCD(x, y) greatest common divisor of x and y, **p. 76**

LCM(x, y) least common multiple of x and y, **pp. 80–81**

$\forall$ "for all"—universal quantifier, **p. 90**

$\exists$ "there exists"—existential quantifier, **p. 90**

$\Rightarrow$ implication for open sentences, **p. 92**

$\sim$ negation for open sentences, **p. 101**

lub(S) least upper bound of S, **p. 111**

glb(S) greatest lower bound of S, **p. 111**

$f: A \to B$ "f is a function from A to B", **p. 114**

$f(u) = v$ "v is the image of u under f", **p. 114**

Im(f) image set of f, **p. 115**

$f(A)$ image of the set A under f, **p. 115**

$g \circ f$ composition of functions f and g, **pp. 120, 122**

$\exists!$ unique existential quantifier, **p. 124**

$\overline{\overline{S}} \leq \overline{\overline{T}}$ used in comparing cardinalities, **p. 147**

iff "if and only if", **p. 157**

$\Leftrightarrow$ double implication for open sentences, **p. 158**

$\mathbf{N}_{[k, \infty)}$ set of integers greater than or equal to k, **p. 182**

Σ summation symbol, **p. 184**

$n!$ n factorial, **p. 193**

F_n nth term in sequence of Fibonacci numbers, **p. 195**

$\mathbf{N}_{[1, t]}$ set of integers from 1 to t, inclusive, **p. 200**

$L_{a,b}$ set of linear combinations of a and b, **p. 215**

$x \sim y$ "x is related to y under the relation $\sim$", **p. 228**

Dom(R) domain of a relation R, **p. 229**

Im(R) image of a relation R, **p. 229**

R^{-1} inverse of the relation R, **p. 231**

I_A identity function on set A, **p. 234**

$f^{-1}(T)$ inverse image of a set T under f, **p. 245**

$f|_C$ restriction of the function f to the set C, **p. 245**

$\sim_{\mathscr{C}}$ relation associated with a collection $\mathscr{C}$ of subsets of a given set, **p. 249**

$\mathscr{C}^{\sim}$ collection of sets associated with the relation $\sim$, **p. 251**

$[t]$ or $[t]_{\sim}$ equivalence class of t (under the relation $\sim$), **p. 253**

$x \equiv y \pmod{n}$ "x is congruent to y, modulo n", **p. 255**

$\mathbf{Z}_n$ the integers modulo n, **p. 257**

$\mathscr{F}(S)$ set of functions from S to S, **p. 262**

max(x, y) maximum of x and y, **p. 263**

P perimeter operation, **p. 267**

e identity element, **p. 270**

x^{-1} inverse of an element x under an operation, **p. 271**

$+_n$ addition modulo n, **p. 275**

$\cdot_n$ multiplication modulo n, **p. 275**

$\lesssim, \gtrsim$ symbols for a partial ordering, **p. 282**

$<, >$ symbols for a strict ordering, **p. 282**

$S \approx T$ set S is equivalent to set T, **p. 294**

$S \not\approx T$ set S is not equivalent to set T, **p. 294**

$\overline{\overline{S}}$ cardinality of the set S, **p. 295**

$S \lesssim T$ as a relation on sets, **p. 297**

$\leq$ as a relation on cardinalities, **p. 298**

$\aleph_0$ cardinality of the natural numbers, **p. 304**

$x|_n$ nth digit in the decimal expansion of x, **p. 309**

$\langle x_n \rangle$ sequence whose nth term is x_n, **p. 321**

$\lim_{n\to\infty} x_n$ limit of the sequence $\langle x_n \rangle$ as n approaches infinity, **p. 330**

$\lim_{x\to c} f(x)$ limit of $f(x)$ as x approaches c, **p. 347**

$\lim_{x\to\infty} f(x)$ limit of $f(x)$ as x approaches ∞, **p. 360**

$\mathbf{S}_n$ symmetric group on n objects, **p. 378**

$\mathbf{U}_n$ group of units of $\mathbf{Z}_n$, **p. 384**

$\mathbf{R}_0, \mathbf{R}_{90}, \mathbf{R}_{180}, \mathbf{R}_{270}, \mathbf{H}, \mathbf{V}, \mathbf{D}_1, \mathbf{D}_2$ elements of the dihedral group of order 8, **p. 388**

$\mathbf{D}_n$ dihedral group of order $2n$, **p. 391**

$\mathbf{R}_0, \mathbf{R}_{120}, \mathbf{R}_{240}, \mathbf{D}_X, \mathbf{D}_Y, \mathbf{D}_Z$ elements of the dihedral group of order 6, **p. 392**

$\langle x \rangle$ group generated by x, **p. 400**

$\mathbf{o}(x)$ order of the element x, **p. 402**

xH left coset of the subgroup H containing the element x, **p. 406**

$G \cong H$ G is isomorphic to H (for groups), **p. 415**

$\mathrm{Ker}(f)$ kernel of f, **p. 420**

$\wedge$ conjunction, **p. 425**

$\vee$ disjunction, **p. 425**

$\sim$ negation for propositions, **p. 426**

$\Rightarrow$ implication for propositions, **p. 428**

$\Leftrightarrow$ double implication for propositions, **p. 430**

$\oplus$ exclusive or, **p. 433**

$\{\ldots\}$ set notation, **p. 443**

$\{x \in S : p(x)\}$ set-builder notation, **p. 444**

$=$ equality for sets, **p. 444**

$\in$ "is an element of", **p. 444**

$\notin$ "is not an element of", **p. 444**

$\subseteq$ "is a subset of", **p. 445**

$\not\subseteq$ "is not a subset of", **p. 445**

$\subset$ "is a proper subset of", **p. 445**

$\supseteq$ "is a superset of", **p. 445**

$\varnothing$ empty set, **p. 445**

$\mathbf{N}$ set of natural numbers: $\{1, 2, 3, \ldots\}$, **p. 445**

$\mathbf{W}$ set of whole numbers: $\{0, 1, 2, \ldots\}$, **p. 445**

$\mathbf{Z}$ set of integers: $\{\ldots, -2, -1, 0, 1, 2, \ldots\}$, **p. 445**

$\mathbf{Q}$ set of rational numbers, **p. 445**

$\mathbf{R}$ set of real numbers, **p. 445**

$\mathbf{C}$ set of complex numbers, **p. 445**

$\mathbf{Q}^+$ set of positive rational numbers, **p. 445**

$\mathbf{R}^{\geq 0}$ set of nonnegative real numbers, **p. 445**

∞ infinity—used in naming intervals, **p. 446**

$[a, b], [a, b), (a, b], (a, b), [a, \infty), (a, \infty), (-\infty, b), (-\infty, b], (-\infty, \infty)$ intervals of real numbers, **p. 446**

$A \cup B$ union of A and B, **p. 446**

$A \cap B$ intersection of A and B, **p. 446**

$A - B$ difference of A and B, **p. 447**

A' complement of A, **p. 447**

$|A|$ cardinality of A, **p. 448**

Index

Q

R

S